Lecture Notes in Computer Science   13513

Qiufen Ni · Weili Wu (Eds.)

Algorithmic Aspects in Information and Management

16th International Conference, AAIM 2022
Guangzhou, China, August 13–14, 2022
Proceedings

 Springer

Editors
Qiufen Ni 🆔
Guangdong University of Technology
Guangzhou, China

Weili Wu 🆔
University of Texas at Dallas
Richardson, TX, USA

ISSN 0302-9743   ISSN 1611-3349 (electronic)
Lecture Notes in Computer Science
ISBN 978-3-031-16080-6   ISBN 978-3-031-16081-3 (eBook)
https://doi.org/10.1007/978-3-031-16081-3

This Springer imprint is published by the registered company Springer Nature Switzerland AG
The registered company address is: Gewerbestrasse 11, 6330 Cham, Switzerland

Preface

The 16th International Conference on Algorithmic Aspects in Information and Management (AAIM 2022), took place at Guangzhou, China, August 13–14, 2022. The conference was held virtually due to the COVID-19 pandemic.

The AAIM conference series, which started in 2005 in Xi'an, China, aims to stimulate various fields for which algorithmics has become a crucial enabler, and to strengthen the ties of various research communities of algorithmics and applications. AAIM 2022 seeks to address emerging and important algorithmic problems by focusing on the fundamental background, theoretical technological development, and real-world applications associated with information and management analysis, modeling and data mining. Special considerations are given to algorithmic research that was motivated by real-world applications. We received 59 submissions, out of which 41 papers were accepted for publication. Each submission was reviewed by at least three reviewers.

We would like to thank the four keynote speakers, Hui Xiong (IEEE fellow, a Distinguished Professor at Rutgers University, USA and a Distinguished Guest Professor at the University of Science and Technology of China), Kui Ren (ACM Fellow, IEEE Fellow, a Professor at Zhejiang University), Cong Tian (a Professor at Xidian University), and Xiaoming Sun (Professor at the Institute of Computing Technology, Chinese Academy of Sciences) for their contributions to the conference.

We would like to express our appreciation to all members of the Program Committee and the external referees whose efforts enabled us to achieve a high scientific standard for the proceedings. We would also like to thank all members of the Organizing Committee for their assistance and contribution which attributed to the success of the conference. Particularly, we would like to thank Anna Kramer and her colleagues at Springer for meticulously supporting us in the timely production of this volume. Last but not least, our special thanks go to all the authors and participants for their contributions to the success of this event.

July 2022   Qiufen Ni
                                                                                                        Weili Wu

Organization

Program Committee Chairs

Qiufen Ni   Guangdong University of Technology, China
Weili Wu   University of Texas at Dallas, USA

Program Committee Members

Wolfgang Bein   University of Nevada, USA
Gruia Calinescu   Illinois Institute of Technology, USA
Xujin Chen   Academy of Mathematics and Systems Science,
                         Chinese Academy of Sciences, China
Zhizhong Chen   Tokyo Denki University, Japan
Bhaskar DasGupta   University of Illinois at Chicago, USA
Xingjian Ding   Beijing University of Technology, Faculty of
                         Information Technology, China
Hongwei Du   Harbin Institute of Technology (Shenzhen), China
Rudolf Fleischer   Heinrich Heine University Dusseldorf, Germany
Shuyang Gu   Texas A&M University - Central Texas, USA
Jianxiong Guo   Beijing Normal University, Advanced Institute of
                         Natural Sciences, China
Sun-Yuan Hsieh   National Cheng Kung University, China
Liying Kang   Shanghai University, China
Michael Khachay   Krasovsky Institute of Mathematics and
                         Mechanics, Russia
Chia-Wei Lee   National Taitung University, Computer Science
                         and Information Engineering, Taitung, China
Xianyue Li   Lanzhou University, Lanzhou, China
Xiao Li   University of Texas at Dallas, USA
Shengxin Liu   Harbin Institute of Technology (Shenzhen), China
Chuanwen Luo   Beijing Forestry University, Beijing, China
Viet Hung Nguyen   University of Clermont-Auvergne, France
Ghosh Smita   Santa Clara University, USA
Zhiyi Tan   Zhejiang University, China
Weitian Tong   Georgia Southern University, USA
Weili Wu   University of Texas at Dallas, USA
Yicheng Xu   Shenzhen Institute of Advanced Technology,
                         Chinese Academy of Sciences, China
Boting Yang   University of Regina, Canada

Ruiqi Yang	Beijing University of Technology, Beijing Institute for Scientific and Engineering Computing, China
Nan Zhang	Xidian University, China
Yapu Zhang	Beijing University of Technology, China
Yong Zhang	Shenzhen Institute of Advanced Technology, Chinese Academy of Sciences, China
Fay Zhong	California State University, China
Yuqing Zhu	California State University, Los Angeles, USA

Reviewers

Wolfgang Bein	University of Nevada, USA
Gruia Calinescu	Illinois Institute of Technology, USA
Xujin Chen	Academy of Mathematics and Systems Science, Chinese Academy of Sciences, China
Sijia Dai	Shenzhen Institute of Advanced Technology, Chinese Academy of Sciences, China
Bhaskar DasGupta	University of Illinois at Chicago, USA
Xingjian Ding	Beijing University of Technology, Faculty of Information Technology, China
Dingzhu Du	University of Texas at Dallas, USA
Hongwei Du	Harbin Institute of Technology (Shenzhen), China
Liman Du	University of Chinese Academy of Sciences, China
Guichen Gao	Shenzhen Institute of Advanced Technology, Chinese Academy of Sciences, China
Yuping Gao	Lanzhou University, China
Shuyang Gu	Texas A&M University - Central Texas, USA
Jianxiong Guo	Beijing Normal University, Advanced Institute of Natural Sciences, China
Xinxin Han	Shenzhen Institute of Advanced Technology, Chinese Academy of Sciences, China
Yi Hong	Beijing Forestry University, China
Sun-Yuan Hsieh	National Cheng Kung University, China
Liying Kang	Shanghai University, China
Michael Khachay	Krasovsky Institute of Mathematics and Mechanics, Russia
Chia-Wei Lee	National Taitung University, Computer Science and Information Engineering, China
Xianyue Li	Lanzhou University, China
Xiao Li	University of Texas at Dallas, USA
Shengxin Liu	Harbin Institute of Technology (Shenzhen), China
Chuanwen Luo	Beijing Forestry University, China

Contents

Graph Theory

Logic and Machine Learning

Approximation Algorithms

A Binary Search Double Greedy Algorithm for Non-monotone DR-submodular Maximization

Shuyang Gu[1]($\boxtimes$) (iD), Chuangen Gao[2], and Weili Wu[3]

[1] Department of Computer Information Systems,
Texas A&M University - Central Texas, Killeen, TX 76549, USA
s.gu@tamuct.edu
[2] School of Computer Science and Technology,
Qilu Technology University, Jinan, China
[3] Department of Computer Science,
The University of Texas at Dallas, Dallas, TX, USA
weiliwu@utdallas.edu

Abstract. In this paper, we study the non-monotone DR-submodular function maximization over integer lattice. Functions over integer lattice have been defined submodular property that is similar to submodularity of set functions. DR-submodular is a further extended submodular concept for functions over the integer lattice, which captures the diminishing return property. Such functions finds many applications in machine learning, social networks, wireless networks, etc. The techniques for submodular set function maximization can be applied to DR-submodular function maximization, e.g., the double greedy algorithm has a 1/2-approximation ratio, whose running time is $O(nB)$, where n is the size of the ground set, B is the integer bound of a coordinate. In our study, we design a 1/2-approximate binary search double greedy algorithm, and we prove that its time complexity is $O(n \log B)$, which significantly improves the running time.

Keywords: Non-monotone · DR-submodular · Binary search double greedy · Approximation algorithm

1 Introduction

A lot of real-world problems have objective functions with a so-called submodular property, which reflects the diminish return nature for the problems. Since such property exists in a vast amount of applications, submodular optimization has caught a lot of attention during the past two decades. A set function $f : 2^E \to \mathbb{R}$ is submodular if $f(X) + f(Y) \geq f(X \cup Y) - f(X \cap Y)$ holds for any two sets $X, Y \subseteq E$ where E is a ground set. Submodular function has an equivalent definition in terms of the diminishing return property: $f(X \cup \{e\}) - f(X) \geq f(Y \cup \{e\}) - f(Y)$ $X \subseteq Y$, the element $e \in E \setminus Y$.

Q. Ni and W. Wu (Eds.): AAIM 2022, LNCS 13513, pp. 3–14, 2022.
https://doi.org/10.1007/978-3-031-16081-3_1

Submodular set function optimization includes maximizing or minimizing a submodular function with or without some constraints. One of the directions is non-monotone submodular maximization without constraint. Given a non-negative submodular function f, the goal is to find a subset S that maximizes $f(S)$. Since this problem captures many applications in machine learning, viral marketing, etc., it has been studied extensively. A deterministic local search gives a $1/2$ -approximation and a randomized smoothed local search algorithm gives $2/5$-approximation [3]. Buchbinder et al. further improve that result, they show that a deterministic double greedy algorithm provides $1/3$-approximation, and the randomized version of it gives a $1/2$-approximation, both in linear time [5].

Recently, submodular optimization has been extended to functions over integer lattice, which considers the situation that each element in the ground set can be selected as multiple copies. The functions over integer lattice may also have submodular property, which is defined similarly to set functions' submodular. A function defined over the integer lattice $\mathbb{Z}_+^E$ is lattice submodular if the following inequality holds:

$$f(\boldsymbol{x}) + f(\boldsymbol{y}) \geq f(\boldsymbol{x} \vee \boldsymbol{y}) + f(\boldsymbol{x} \wedge \boldsymbol{y}), \boldsymbol{x}, \boldsymbol{y} \in \mathbb{Z}_+^E.$$

The techniques for submodular set function optimization can be applied to lattice submodular optimization. Based on the double greedy algorithm in [5], an algorithm for submodular functions over the bounded integer lattice is designed with $1/3$ approximation ratio [2].

Although the definition of lattice submodular is similar to set function submodular, the lattice submodular does not imply diminish return property. Soma et al. [4] give a stronger generalization of submodularity on integer lattice, which is called diminishing return submodular (DR-submodular) functions, such functions capture various applications with diminishing return property. DR submodular function on a bounded integer lattice satisfies $f(\boldsymbol{x} + \boldsymbol{\chi}_e) - f(\boldsymbol{x}) \geq f(\boldsymbol{y} + \boldsymbol{\chi}_e) - f(\boldsymbol{y})$ for any $\boldsymbol{x} \leq \boldsymbol{y}$ and $e \in E$, where $\boldsymbol{\chi}_e$ denotes a unit vector, i.e. $\chi_{i_e} \in \mathbb{Z}^E$ is the vector with $\boldsymbol{\chi}_e(e) = 1$ and $\boldsymbol{\chi}_e(a) = 0$ for every $a \neq e$. In this paper, we specifically study the profit maximization problem in social networks as an application of non-monoton DR-submodular maximization.

The contributions of this paper are summarized as follows.

- To solve the non-monotone DR-submodular maximization problem, we propose the binary search double greedy algorithm.
- We prove the algorithm has a $1/2$- approximate ratio and the time complexity is polynomial($n \log B$). To the best of our knowledge, this is the fastest algorithm with the least queries to the objective function.

2   Related Work

Non-monotone DR-submodular Maximization is closely related to non-monotone submodular set function maximization because the algorithm for the latter problem can be applied to the former problem directly. The non-monotone submodular maximization is also called Unconstrained Submodular Maximization (USM).

USM has various applications, such as marketing strategies over social networks [18], Max-Cut [19], and maximum facility location [20]. USM problem has been studied extensively [21–23]. Buchbinder gives a tight linear time randomized $(1/2)$-approximation for the problem [5].

The topic of functions over integer lattice optimization has attracted much attention recently. Monotone submodular functions over integer lattice with cardinality constraint are addressed in [10,12,14]. Sahin et al. study lattice submodular functions subject to a discrete (integer) polymatroid constraint [13]. Zhang et al. study the problem of maximizing the sum of a monotone nonnegative DR-submodular function and a supermodular function on the integer lattice subject to a cardinality constraint [11]. The non-submodular functions on the integer lattice are addressed in [16]. Nong et al. focus on maximizing a nonmonotone weak-submodular function on a bounded integer lattice [17]. For the problem addressed in this paper, non-monotone DR-submodular function maximization, Soma et al. design a $\frac{1}{2+\epsilon}$-approximation algorithm with a running time of $O(\frac{n}{\epsilon}\log^2 B)$ [6].

In the meantime, the discrete domains of submodular functions over integer lattice are further extended to continuous domains, Hassani et al. study stochastic projected gradient methods for maximizing continuous submodular functions with convex constraints [15]. In [7,8], the authors consider maximizing a continuous and nonnegative submodular function over a hypercube.

3 Preliminaries

We say that a set function $g : f : 2^E \rightarrow \mathbb{R}_+$ is submodular if it satisfies a natural "diminishing returns" property: the marginal gain from adding an element to a set X is at least as high as the marginal gain from adding the same element to a superset of X. Formally, for every set X, Y such that $X \subseteq Y \subseteq E$ and every $e \in E \setminus Y$, it follows that

$$g(X \cup \{e\}) - g(X) \geq g(Y \cup \{e\}) - g(Y)$$

An equivalent definition of the submodularity is

$$g(X) + g(Y) \geq g(X \cup Y) + f(X \cap Y), \forall X, Y \subseteq E$$

A set function is monotone if $g(X) \leq g(Y)$ for all $X \subseteq Y$.

Functions over integer lattice has similar property. A function $h : \mathbb{Z}_+^E \rightarrow \mathbb{R}_+$ that is defined over the integer lattice is submodular if the following holds [2]:

$$h(\boldsymbol{x}) + h(\boldsymbol{y}) \geq h(\boldsymbol{x} \vee \boldsymbol{y}) + h(\boldsymbol{x} \wedge \boldsymbol{y}), \boldsymbol{x}, \boldsymbol{y} \in \mathbb{Z}_+^E.$$

where $(\boldsymbol{x} \vee \boldsymbol{y})(i) = \max\{\boldsymbol{x}(i), \boldsymbol{y}(i)\}$ and $(\boldsymbol{x} \wedge \boldsymbol{y})(i) = \min\{\boldsymbol{x}(i), \boldsymbol{y}(i)\}$. Hence $\boldsymbol{x} \vee \boldsymbol{y}$ represents coordinate-wise maximum, and $\boldsymbol{x} \wedge \boldsymbol{y}$ denote the coordinate-wise minimum. We can see this form of submodularity is a more generalized definition of submodularity that covers set functions submodular, because vectors with all entries equal to either 0 or 1 can be seen as a subset including the elements that are equal to 1 while excluding the elements that are equal to 0, in that case, $\boldsymbol{x} \wedge \boldsymbol{y}$ and $\boldsymbol{x} \wedge \boldsymbol{y}$ transform to set intersection and set union of the subsets that $\boldsymbol{x}$ and $\boldsymbol{y}$ represent respectively.

The submodular function over integer lattice does not have the diminishing return property. To capture such property in real-world problems, a stronger version of submodularity has been introduced, which is called DR-submodular [4]. DR submodular function on a bounded integer lattice satisfies the following diminish return property:

$$h(\boldsymbol{x} + \chi_e) - h(\boldsymbol{x}) \geq h(\boldsymbol{y} + \chi_e) - h(\boldsymbol{y}), \forall \boldsymbol{x} \leq \boldsymbol{y}, \forall e \in E$$

where χ_e denotes a unit vector, i.e. $\chi_e \in \mathbb{Z}^E$ is the vector with $\chi_e(e) = 1$ and $\chi_e(a) = 0$ for every $a \neq e$.

The problem we consider is maximizing (non-monotone) DR-submodular functions. Formally, we study the optimization problem

$$\begin{aligned} \max \quad & f(x) \\ \text{subject to} \quad & \boldsymbol{0} \leq x \leq \boldsymbol{B}, \end{aligned} \tag{1}$$

where $f : \mathbb{Z}_+^E \to \mathbb{R}_+$ is a non-negative DR-submodular function and not necessarily monotone. $\boldsymbol{0}$ is the all zero vector, and $\boldsymbol{B} \in \mathbb{Z}_+^E$ is a vector representing the maximum value for each coordinate. When $\boldsymbol{B}$ is the all-ones vector, the problem is equivalent to the original unconstrained submodular set function maximization. We assume that f is given as an evaluation oracle; when we specify $\boldsymbol{x} \in \mathbb{Z}_+^E$, the oracle returns the value of $f(\boldsymbol{x})$. We define $f(\boldsymbol{x}|\boldsymbol{y}) = f(\boldsymbol{y} + \boldsymbol{x}) - f(\boldsymbol{y})$.

4   Algorithm

In this section, we present the algorithm for non-monotone DR-submodular function maximization. The main idea is inspired by the double greedy algorithm for the unconstrained submodular maximization (USM) [5] on set functions. The algorithm can be extended to accommodate DR-submodular function over integer lattice [6], because DR-submodular function can be treated as submodular set function on each coordinate. We investigate some interesting properties for DR-submodular functions to further speed up the algorithm. In the rest of the paper some notations will be used. We define two new functions $\phi(b) := f(\chi_e|\boldsymbol{x} + b\chi_e)$, $\psi(b) := f(-\chi_e|\boldsymbol{y} - b\chi_e)$, where $b \in \mathbb{Z}^+$. Both functions are non-increasing functions of b because the function f is DR-submodular.

The algorithm starts with two vectors, $x = \boldsymbol{0}$ and $y = \boldsymbol{c}$. For each coordinate $e \in E$ it iteratively either increase $\mathbf{x}(e)$ or decrease $\mathbf{y}(e)$ by σ, which depends on the marginal gain by adding σ to $\mathbf{x}(e)$ and the marginal gain by removing σ to $\mathbf{y}(e)$. This procedure continues until $\mathbf{x}(e) = \mathbf{y}(e)$. Then it moves on to work on

---

Algorithm 1. Binary Search Greedy Algorithm

Input: $f : \mathbb{Z}_+^E \to \mathbb{R}^+$, $c \in \mathbb{Z}_+^E$
Assumption: f is DR-submodular

1: $\boldsymbol{x} \leftarrow \boldsymbol{0}$, $\boldsymbol{y} \leftarrow \boldsymbol{c}$;
2: **for** $e \in E$ **do**
3:    Find $\arg\min_b \phi(b)$ such that $\phi(b) < 0$ by binary search.
4:    $u \leftarrow \boldsymbol{x}(e) + \arg\min_b \phi(b) - 1$.
5:    Find $\arg\min_b \psi(b)$ such that $\psi(b) < 0$ by binary search.
6:    $v \leftarrow \boldsymbol{y}(e) - \arg\min_b \psi(b) + 1$.
7:    **while** $\boldsymbol{x}(e) < \boldsymbol{y}(e)$ **do**
8:       $\sigma \leftarrow \max(\lfloor \frac{\boldsymbol{y}(e) - \boldsymbol{x}(e)}{2} \rfloor, 1)$
9:       $\alpha \leftarrow f(\sigma\boldsymbol{\chi}_e | \boldsymbol{x})$ and $\beta \leftarrow f(-\sigma\boldsymbol{\chi}_e | \boldsymbol{y})$
10:      **if** $\beta \leq 0$ **then**
11:         $\boldsymbol{x}(e) \leftarrow \boldsymbol{x}(e) + \sigma$
12:      **else if** $\alpha \leq 0$ **then**
13:         $\boldsymbol{y}(e) \leftarrow \boldsymbol{y}(e) - \sigma$
14:      **else**
15:         Randomly update $\boldsymbol{x}(e) \leftarrow \boldsymbol{x}(e) + \sigma$ or $\boldsymbol{y}(e) \leftarrow \boldsymbol{y}(e) - \sigma$; the former case occurs with probability $\frac{\alpha}{\alpha+\beta}$, the later case with the probability $\frac{\beta}{\alpha+\beta}$.
16:      **end if**
17:   **end while**
18:   **if** $\boldsymbol{x}(e) \geq u$ **then**
19:      $\boldsymbol{x}(e) \leftarrow u$, $\boldsymbol{y}(e) \leftarrow u$
20:   **end if**
21:   **if** $\boldsymbol{y}(e) \leq v$ **then**
22:      $\boldsymbol{y}(e) \leftarrow v$, $\boldsymbol{x}(e) \leftarrow v$
23:   **end if**
24: **end for**
25: **return** $\boldsymbol{x}$

---

the next coordinate, after $\mathbf{x}$ and $\mathbf{y}$ agrees on all coordinates $e \in E$, $\mathbf{x} = \mathbf{y}$, and the vector is the output of the algorithm. Different from applying the double greedy algorithm directly, which tightens the gap one unit per step, Algorithm 1 tightens it by half in each iteration. The binary search nature of this algorithm guarantees that the number of iterations needed is a logarithm of B. Next let us firstly give a few results based on the diminish return property, which will be used in proving the theoretical guarantee of the algorithm later.

Lemma 1. *For* $\forall \boldsymbol{x} \leq \boldsymbol{y}, k \geq 1, k \in \mathbb{Z}^+$, *and the value of* k *does not violate the integer bound. We have*

$$f(\boldsymbol{x} + k\boldsymbol{\chi}_e) - f(\boldsymbol{x}) \geq f(\boldsymbol{y} + k\boldsymbol{\chi}_e) - f(\boldsymbol{y})$$

Proof.

$$f(x + \chi_e) - f(x) \geq f(y + \chi_e) - f(y)$$
$$f(x + 2\chi_e) - f(x + \chi_e) \geq f(y + 2\chi_e) - f(y + \chi_e)$$

$$\vdots$$

$$f(x + k\chi_e) - f(x + (k-1)\chi_e) \geq f(y + k\chi_e) - f(y + (k-1)\chi_e)$$

Sum up the above inequalities, the lemma holds. □

Note that from Lemma 1, we have

$$\alpha + \beta = f(x + \sigma\chi_e) - f(x) - (f(y + \sigma\chi_e) - f(y)) \geq 0 \qquad (2)$$

Lemma 2. *For* $\forall x \leq y, k \geq 1, k \in Z^+, k \leq y(e)$, *we have*

$$f(x - k\chi_e) - f(x) \leq f(y - k\chi_e) - f(y)$$

Lemma 2 can be easily obtained from Lemma 1.

Lemma 3. *Given* $p \leq q, p, q \in Z^+$, *if* $f(\chi_e|x + (q-1)\chi_e) \geq 0$, *then*

$$0 \leq f(p\chi_e|x) \leq f(q\chi_e|x)$$

Proof.

$$f(p\chi_e|x) = f(\chi_e|x) + f(\chi_e|x + \chi_e) + \cdots + f(\chi_e|x + (p-1)\chi_e)$$
$$f(q\chi_e|x) = f(\chi_e|x) + f(\chi_e|x + \chi_e) + \cdots + f(\chi_e|x + (q-1)\chi_e)$$

Since the values of the terms in the above equations are non-increasing from left to right, $f(\chi_e|x + (q-1)\chi_e) \geq 0$, so all terms are greater than or equal to 0. And $f(q\chi_e|x)$ has more terms, so the lemma holds. □

We can obtain a similar property in terms of the vector y.

Lemma 4. *Given* $p \leq q, p, q \in Z^+$, *if* $f(-\chi_e|y - (q-1)\chi_e) \geq 0$, *then*

$$0 \leq f(-p\chi_e|y) \leq f(-q\chi_e|y)$$

The rest of this section is devoted to proving that Algorithm 1 provides an approximation ratio of $1/2$ for DR-submodular maximization. Let us begin the analysis of Algorithm 1 with the introduction of some notation. Let x_i^e and y_i^e be random variables denoting the vectors generated by the algorithm at the end of the i-th iteration for coordinate e, let the number of iterations for coordinate e is θ_e, note that $1 \leq i \leq \theta_e \leq \log B$. Denote by $\boldsymbol{opt}$ the optimal solution. Let us define the following random variable: $\boldsymbol{opt}_i^e \triangleq (\boldsymbol{opt} \vee x_i^e) \wedge y_i^e$. Note that $x_0^e(e) = 0$, $y_0^e(e) = B$, and $\boldsymbol{opt}_0^e(e) = \boldsymbol{opt}(e)$. Additionally, the following always holds: $\boldsymbol{opt}_{\theta_e}^e(e) = x_{\theta_e}^e(e) = y_{\theta_e}^e(e), \forall e \in E$.

Let us analyze the approximation ratio of the randomized algorithm. We consider the subsequence $\mathbb{E}[f(opt_0^e)], \ldots, \mathbb{E}[f(opt_{\theta_e}^e)]$ for any dimension $e \in E$, and a whole sequence which is a combination of every such subsequence for each element $e \in E$. This sequence starts with $f(opt)$ and ends with the expected value of the algorithm's output. The following lemma upper bounds the loss between every two consecutive elements in the sequence. Formally, $\mathbb{E}[f(opt_{i-1}^e) - f(opt_i^e)]$ is upper bounded by the average expected change in the value of the two solutions maintained by the algorithm, i.e., $\frac{1}{2}\mathbb{E}[f(x_i^e) - f(x_{i-1}^e) + f(y_i^e) - f(y_{i-1}^e)]$.

Lemma 5. *For every $1 \leq i \leq \theta_e$,*

$$\mathbb{E}[f(opt_{i-1}^e) - f(opt_i^e)] \leq \frac{1}{2}\mathbb{E}[f(x_i^e) - f(x_{i-1}^e) + f(y_i^e) - f(y_{i-1}^e)] \qquad (3)$$

where expectations are taken over the random choices of the algorithm.

Proof. Notice that it suffices to prove the inequality conditioned on any event of the form $x_{i-1}^e = s_{i-1}^e$, where $s_{i-1}^e \in \mathbb{Z}_+^E$, $(s_{i-1}^e(e) \leq \sigma_1 + \cdots + \sigma_{i-1})$, for which the probability that $x_{i-1}^e = s_{i-1}^e$ is nonzero. Hence, fix such an event corresponding to an integer vector s_{i-1}^e. The rest of the proof implicitly assumes everything is conditioned on this event. Since the analysis is same for every coordinate, we omit the superscript e in x_i^e, y_i^e, s_i^e and opt_i^e in the following proof. After an iteration i on coordinate e, denote by δ_i the distance between $x_i(e)$ and $y_i(e)$, which can be calculate as $\delta_i = y_i(e) - x_i(e) = B - \sum_{k=1}^{i} \sigma_k$. The parameter σ_i can be obtained iteratively as $\sigma_1 = \lfloor \frac{B}{2} \rfloor$, $\sigma_i = \lfloor \frac{B - \sum_{k=1}^{i-1} \sigma_k}{2} \rfloor = \lfloor \frac{\delta_i}{2} \rfloor$. Due to the conditioning, the following random variables become constants:

1. y_{i-1}, where $y_{i-1}(e) = s_{i-1}(e) + \delta_{i-1}$
2. $opt_{i-1} \triangleq (opt \vee x_{i-1}) \wedge y_{i-1}$, where $opt_{i-1}(e) = s_{i-1}(e) + \min(opt(e) - s_{i-1}(e), \delta_{i-1})$
3. α_i and β_i, which refer to α, β at the iteration i .

By Lemma 1, $\alpha_i + \beta_i \geq 0$. Thus at most one of α_i and β_i is strictly less than zero. We need to consider the following three cases for the value of α_i and β_i:

Case 1: $(\alpha_i \geq 0$ and $\beta_i \leq 0)$. In this case the vector y does not change: $y_i = y_{i-1}$. The vector x changes. $x_i \leftarrow x_{i-1} + \sigma_i \chi_e$. Hence, $f(y_i) - f(y_{i-1}) = 0$. Also, by our definition $opt_i \triangleq (opt \vee x_i) \wedge y_i = (opt \vee (x_{i-1} + \sigma_i \chi_e)) \wedge y_i$. Thus, we are left to prove that

$$f((opt \vee x_{i-1}) \wedge y_{i-1}) - f((opt \vee (x_{i-1} + \sigma_i \chi_e)) \wedge y_i) \leq \frac{1}{2}[f(x_i) - f(x_{i-1})] = \frac{\alpha_i}{2}$$

We prove it considering the relationship among $x_{i-1}(e)$, $x_i(e)$ and $opt_i(e)$.

Case 1.1: $x_i(e) = s_{i-1}(e) + \sigma_i \leq opt(e)$.
   This condition implies $x_{i-1}(e) \leq opt(e)$. Since $y_i = y_{i-1}$, the left-hand side of the last inequality is 0, which is definitely not greater than the nonnegative $\frac{\alpha_i}{2}$.

Case 1.2: $s_{i-1}(e) \geq opt(e)$.

This condition implies that $x_i(e) = s_{i-1}(e) + \sigma_i > opt(e)$.

We can see that $(opt \vee (x_{i-1} + \sigma_i \chi_e)) \wedge y_i = (opt \vee x_{i-1}) \wedge y_{i-1} + \sigma_i \chi_e = opt_{i-1} + \sigma_i \chi_e \leq y_{i-1}$, by diminish return submodularity we have

$$f((opt \vee x_{i-1}) \wedge y_{i-1}) - f((opt \vee (x_{i-1} + \sigma_i \chi_e)) \wedge y_i)$$
$$= f(-\sigma_i \chi_e | opt_{i-1} + \sigma_i \chi_e) \leq f(-\sigma_i \chi_e | y_{i-1}) = \beta \leq 0 \leq \frac{\alpha_i}{2}$$

Case 1.3: $s_{i-1}(e) \leq opt(e)$ and $x_i(e) = s_{i-1}(e) + \sigma_i > opt(e)$.

Then we have

$$((opt \vee x_{i-1}) \wedge y_{i-1}) - (opt \vee (x_{i-1} + \sigma_i \chi_e) \wedge y_i) = -(s_{i-1}(e) + \sigma_i - opt(e))\chi_e$$

Let $s_{i-1}(e) + \sigma_i - opt(e) = \delta$, then

$$((opt \vee x_{i-1}) \wedge y_{i-1}) - (opt \vee (x_{i-1} + \sigma_i \chi_e) \wedge y_i) = -\delta \chi_e.$$

And

$$f((opt \vee x_{i-1}) \wedge y_{i-1}) - f((opt \vee (x_{i-1} + \sigma_i \chi_e)) \wedge y_i)$$
$$= f(-\delta \chi_e | (opt \vee (x_{i-1} + \sigma_i \chi_e)) \wedge y_i) \qquad (4)$$
$$= f(-\delta \chi_e | opt_i).$$

By the definition of the random variable opt_i, we have

$$x_i \leq opt_i \leq y_i$$

Note that $0 \leq \delta < \sigma_i$. Since $\psi(b) := f(-\chi_e | y - b\chi_e)$ is a non-increasing function on b. $f(-\sigma_i \chi_e | y_i) \leq 0$ implies $f(-\delta \chi_e | y_i - \sigma_i \chi_e) \leq 0$. We note that in the $i$the iteration, $y_i = y_{i-1}$ and due to the condition of case 1.3, we have $y_i - \sigma_i \chi_e \geq opt_i$, thus $f(-\delta \chi_e | opt_i) \leq f(-\delta \chi_e | y_i - \sigma_i \chi_e) \leq 0 \leq \frac{\alpha_i}{2}$.

Case 2: ($\alpha_i < 0$ and $\beta_i > 0$). This case is analogous to the previous one, and therefore we omit its proof.

Case 3: ($\alpha_i \geq 0$ and $\beta_i > 0$). With probability $\frac{\alpha_i}{\alpha_i + \beta_i}$ the following events happen: $x_i \leftarrow x_{i-1} + \sigma_i \chi_e$ and $y_i \leftarrow y_{i-1}$; while with probability $\frac{\beta_i}{\alpha_i + \beta_i}$ the following events happen: $x_i \leftarrow x_{i-1}$ and $y_i \leftarrow y_{i-1} - \sigma_i \chi_e$ Thus,

$$\mathbb{E}[f(x_i) - f(x_{i-1}) + f(y_i) - f(y_{i-1})] = \frac{\alpha_i}{\alpha_i + \beta_i}[f(x_{i-1} + \sigma_i \chi_e) - f(x_{i-1})]$$
$$+ \frac{\beta_i}{\alpha_i + \beta_i}[f(y_{i-1} - \sigma_i \chi_e) - f(y_{i-1})] \quad (5)$$
$$= \frac{\alpha_i^2 + \beta_i^2}{\alpha_i + \beta_i}$$

Next, we upper bound $\mathbb{E}[f(\boldsymbol{opt}_{i-1}) - f(\boldsymbol{opt}_i)]$.

$$
\begin{aligned}
&\mathbb{E}[f(\boldsymbol{opt}_{i-1}) - f(\boldsymbol{opt}_i)] \\
&= \frac{\alpha_i}{\alpha_i + \beta_i}[f((\boldsymbol{opt} \vee \boldsymbol{x}_{i-1}) \wedge \boldsymbol{y}_{i-1}) - f((\boldsymbol{opt} \vee (\boldsymbol{x}_{i-1} + \sigma_i \chi_e)) \wedge \boldsymbol{y}_i)] \\
&\quad + \frac{\beta_i}{\alpha_i + \beta_i}[f((\boldsymbol{opt} \vee \boldsymbol{x}_{i-1}) \wedge \boldsymbol{y}_{i-1}) - f((\boldsymbol{opt} \vee \boldsymbol{x}_i) \wedge (\boldsymbol{y}_{i-1} - \sigma_i \chi_e))] \\
&\leq \frac{\alpha_i \beta_i}{\alpha_i + \beta_i}
\end{aligned} \tag{6}
$$

The final inequality follows by considering two cases. The first case is: $\boldsymbol{y}_i(e) = \boldsymbol{y}_{i-1}(e) - \sigma_i$, $\boldsymbol{x}_i(e) = \boldsymbol{x}_{i-1}(e)$, the first term of the left-hand side of the last inequality equals zero. There are three subcases,

Case 3.1: $(\boldsymbol{x}_i(e) \leq \boldsymbol{y}_i(e) = \boldsymbol{y}_{i-1}(e) - \sigma_i < \boldsymbol{y}_{i-1}(e) \leq \boldsymbol{opt}(e))$.
   Thus $(\boldsymbol{opt} \vee \boldsymbol{x}_{i-1}) \wedge \boldsymbol{y}_{i-1} = (\boldsymbol{opt} \vee \boldsymbol{x}_i) \wedge (\boldsymbol{y}_{i-1} - \sigma_i \chi_e) + \sigma_i \chi_e$, and $(\boldsymbol{opt} \vee \boldsymbol{x}_i) \wedge (\boldsymbol{y}_{i-1} - \sigma_i \chi_e) \geq \boldsymbol{x}_{i-1}$, hence

$$
\begin{aligned}
&f((\boldsymbol{opt} \vee \boldsymbol{x}_{i-1}) \wedge \boldsymbol{y}_{i-1}) - f((\boldsymbol{opt} \vee \boldsymbol{x}_i) \wedge (\boldsymbol{y}_{i-1} - \sigma_i \chi_e)) \\
&= f(\sigma_i \chi_e | (\boldsymbol{opt} \vee \boldsymbol{x}_i) \wedge (\boldsymbol{y}_{i-1} - \sigma_i \chi_e)) \leq f(\sigma_i \chi_e | \boldsymbol{x}_{i-1}) = \alpha_i
\end{aligned}
$$

Case 3.2: $(\boldsymbol{opt}(e) \leq \boldsymbol{y}_i(e) = \boldsymbol{y}_{i-1}(e) - \sigma_i < \boldsymbol{y}_{i-1}(e))$.
   In this case, the second term of the left-hand side of inequality (6) also equals zero, thus inequality (6) follows.

Case 3.3: $(\boldsymbol{x}_i(e) \leq \boldsymbol{y}_i(e) = \boldsymbol{y}_{i-1}(e) - \sigma_i \leq \boldsymbol{opt}(e) < \boldsymbol{y}_{i-1}(e))$.
   We have $(\boldsymbol{opt} \vee \boldsymbol{x}_{i-1}) \wedge \boldsymbol{y}_{i-1} = (\boldsymbol{opt} \vee \boldsymbol{x}_i) \wedge (\boldsymbol{y}_{i-1} - \sigma_i \chi_e) + (\boldsymbol{opt}(e) - \boldsymbol{y}_{i-1}(e) + \sigma_i) \chi_e$. Let $\mu = \boldsymbol{opt}(e) - \boldsymbol{y}_{i-1}(e) + \sigma_i$, then $0 < \mu \leq \sigma_i$, and

$$
\begin{aligned}
&f((\boldsymbol{opt} \vee \boldsymbol{x}_{i-1}) \wedge \boldsymbol{y}_{i-1}) - f((\boldsymbol{opt} \vee \boldsymbol{x}_i) \wedge (\boldsymbol{y}_{i-1} - \sigma_i \chi_e)) \\
&= f(\mu \chi_e | (\boldsymbol{opt} \vee \boldsymbol{x}_i) \wedge (\boldsymbol{y}_{i-1} - \sigma_i \chi_e)) \\
&\leq f(\mu \chi_e | \boldsymbol{y}_{i-1} - \sigma_i \chi_e) \leq 0 \leq \alpha_i
\end{aligned}
$$

The line 18–23 in the algorithm guarantees that $f(-\chi_e | \boldsymbol{y}_{i-1} - (\sigma - 1)\chi_e)) \geq 0$. By Lemma 3, we have $f(-\sigma \chi_e | \boldsymbol{y}_{i-1}) \geq 0$. Also $f(-\mu \chi_e | \boldsymbol{y}_{i-1} - (\sigma - \mu)\chi_e) \geq 0$. Thus,

$$
f(-\mu \chi_e | \boldsymbol{y}_{i-1} - (\sigma - \mu)\chi_e) = -f(\mu \chi_e | \boldsymbol{y}_{i-1} - \sigma_i \chi_e) \geq 0
$$

By now, we show that the inequality (6) holds for the case $\boldsymbol{y}_i(e) = \boldsymbol{y}_{i-1}(e) - \sigma_i$, $\boldsymbol{x}_i(e) = \boldsymbol{x}_{i-1}(e)$. The other case is that $\boldsymbol{y}_i(e) = \boldsymbol{y}_{i-1}(e)$, $\boldsymbol{x}_i(e) = \boldsymbol{x}_{i-1}(e) + \sigma_i$, which is analogous to the previous case, we omit the proof here.

   Now we show inequality (6) follows for all situations. By (5) and (6) inequality (3) holds if

$$
\frac{\alpha_i \beta_i}{\alpha_i + \beta_i} \leq \frac{1}{2} \cdot \frac{\alpha_i^2 + \beta_i^2}{\alpha_i + \beta_i}
$$

which can easily be verified.

Theorem 1. *Algorithm 1 is a randomized $O(n \log B)$ time (1/2)-approximation algorithm for the DR-Submodular Maximization problem.*

Proof. Summing up Lemma 4 for every $1 \le i \le \theta_e$ for each $e \in E$ gives

$$
\sum_{e \in E} \sum_{i=1}^{\theta_e} \mathbb{E}[f(\boldsymbol{opt}_{i-1}^e) - f(\boldsymbol{opt}_i^e)]
$$
$$
\le \frac{1}{2} \sum_{e \in E} \sum_{i=1}^{\theta_e} \mathbb{E}[f(\boldsymbol{x}_i^e) - f(\boldsymbol{x}_{i-1}^e) + f(\boldsymbol{y}_i^e) - f(\boldsymbol{y}_{i-1}^e)] \tag{7}
$$

The above sum is telescopic. We define that the algorithm executes on the vector coordinates ordered by $e_1, \ldots, e_n$. Collapse the inequality, we get

$$
f(\boldsymbol{opt}_0^{e_1}) - f(\boldsymbol{opt}_{\theta_e}^{e_n}) \le \frac{1}{2} \mathbb{E}[f(\boldsymbol{x}_{\theta_{e_n}}^{e_n}) - f(\boldsymbol{x}_0^{e_1}) + f(\boldsymbol{y}_{\theta_{e_n}}^{e_n}) - f(\boldsymbol{y}_0^{e_1})]
$$
$$
\le \frac{1}{2} \mathbb{E}[f(\boldsymbol{x}_{\theta_{e_n}}^{e_n}) + f(\boldsymbol{y}_{\theta_{e_n}}^{e_n})] \tag{8}
$$

Recalling the definitions of $\boldsymbol{opt}_i^e$, $\boldsymbol{opt}_0^{e_1} = \boldsymbol{opt}$, $\boldsymbol{opt}_{\theta_e}^{e_n} = \boldsymbol{x}_{\theta_{e_n}}^{e_n} = \boldsymbol{y}_{\theta_{e_n}}^{e_n}$ is the output solution, thus $\mathbb{E}[f(\boldsymbol{x}_{\theta_{e_n}}^{e_n})] = \mathbb{E}[f(\boldsymbol{y}_{\theta_{e_n}}^{e_n})] \ge f(\boldsymbol{opt})/2$. It is clear that the algorithm makes $O(n \log B)$ oracle calls since for each coordinate $e \in E$ the number of oracle calls is at most $\log B$ and there are $n = |E|$ coordinates.

5   Conclusions

In this paper, we propose a binary search double greedy algorithm for non-monotone DR-submodular function maximization over bounded integer lattice. Our algorithm improves the approximation ratio and significantly reduces the time complexity.

One interesting direction for our future work is to explore the problems in social networks that fall into non-monotone DR-submodular maximization, exploit our proposed algorithm and test with real datasets. Another direction we are interested in is maximizing DR-submodular function with some types of constraints such as cardinality constraint, matroid constraint and knapsack constraint.

References

1. Alon, N., Gamzu, I., Tennenholtz, M.: Optimizing budget allocation among channels and influencers. In: Proceedings of the 21st International Conference on World Wide Web, pp. 381–388 (2012)
2. Gottschalk, C., Peis, B.: Submodular function maximization on the bounded integer lattice. In: Sanità, L., Skutella, M. (eds.) WAOA 2015. LNCS, vol. 9499, pp. 133–144. Springer, Cham (2015). https://doi.org/10.1007/978-3-319-28684-6_12

3. Feige, U., Mirrokni, V.S., Vondrák, J.: Maximizing non-monotone submodular functions. SIAM J. Comput. **40**(4), 1133–1153 (2011)
4. Soma, T., Yoshida, Y.: A generalization of submodular cover via the diminishing return property on the integer lattice. In: Advances in Neural Information Processing Systems, vol. 28 (2015)
5. Buchbinder, N., Feldman, M., Seffi, J., Schwartz, R.: A tight linear time (1/2)-approximation for unconstrained submodular maximization. SIAM J. Comput. **44**(5), 1384–1402 (2015)
6. Soma, T., Yoshida, Y.: Non-monotone DR-submodular function maximization. In: Proceedings of the AAAI Conference on Artificial Intelligence, vol. 31, no. 1 (2017)
7. Niazadeh, R., Roughgarden, T., Wang, J.: Optimal algorithms for continuous non-monotone submodular and DR-submodular maximization. In: Advances in Neural Information Processing Systems, vol. 31 (2018)
8. Bian, A., Levy, K., Krause, A., Buhmann, J.M.: Continuous DR-submodular maximization: structure and algorithms. In: Advances in Neural Information Processing Systems, vol. 30 (2017)
9. Soma, T., Kakimura, N., Inaba, K., Kawarabayashi, K.-I.: Optimal budget allocation: theoretical guarantee and efficient algorithm. In: International Conference on Machine Learning, pp. 351–359. PMLR (2014)
10. Soma, T., Yoshida, Y.: Maximizing monotone submodular functions over the integer lattice. Math. Program. **172**(1), 539–563 (2018)
11. Zhang, Z., Du, D., Jiang, Y., Wu, C.: Maximizing DR-submodular+ supermodular functions on the integer lattice subject to a cardinality constraint. J. Glob. Optim. **80**(3), 595–616 (2021)
12. Lai, L., Ni, Q., Lu, C., Huang, C., Wu, W.: Monotone submodular maximization over the bounded integer lattice with cardinality constraints. Discrete Math. Algorithms Appl. **11**(06), 1950075 (2019)
13. Sahin, A., Buhmann, J., Krause, A.: Constrained maximization of lattice submodular functions. In: ICML 2020 Workshop on Negative Dependence and Submodularity for ML, Vienna, Austria, PMLR, vol. 119 (2020)
14. Zhang, Z., Guo, L., Wang, Y., Xu, D., Zhang, D.: Streaming algorithms for maximizing monotone DR-submodular functions with a cardinality constraint on the integer lattice. Asia-Pacific J. Oper. Res. **38**(05), 2140004 (2021)
15. Hassani, H., Soltanolkotabi, M., Karbasi, A.: Gradient methods for submodular maximization. In: Advances in Neural Information Processing Systems, vol. 30 (2017)
16. Kuhnle, A., Smith, J.D., Crawford, V., Thai, M.: Fast maximization of non-submodular, monotonic functions on the integer lattice. In: International Conference on Machine Learning, pp. 2786–2795. PMLR (2018)
17. Nong, Q., Fang, J., Gong, S., Du, D., Feng, Y., Qu, X.: A 1/2-approximation algorithm for maximizing a non-monotone weak-submodular function on a bounded integer lattice. J. Comb. Optim. **39**(4), 1208–1220 (2020)
18. Hartline, J., Mirrokni, V., Sundararajan, M.: Optimal marketing strategies over social networks. In: Proceedings of the 17th International Conference on World Wide Web, pp. 189–198 (2008)
19. Goemans, M.X., Williamson, D.P.: Improved approximation algorithms for maximum cut and satisfiability problems using semidefinite programming. J. ACM (JACM) **42**(6), 1115–1145 (1995)
20. Ageev, A.A., Sviridenko, M.I.: An 0.828-approximation algorithm for the uncapacitated facility location problem. Discret. Appl. Math. **93**(2–3), 149–156 (1999)

21. Gharan, S.O., Vondrák, J.: Submodular maximization by simulated annealing. In: Proceedings of the 22nd Annual ACM-SIAM Symposium on Discrete Algorithms, pp. 1098–1116. SIAM (2011)
22. Buchbinder, N., Feldman, M.: Deterministic algorithms for submodular maximization problems. ACM Trans. Algorithms (TALG) **14**(3), 1–20 (2018)
23. Pan, X., Jegelka, S., Gonzalez, J.E., Bradley, J.K., Jordan, M.I.: Parallel double greedy submodular maximization. In: Advances in Neural Information Processing Systems, vol. 27 (2014)

An Approximation Algorithm for the Clustered Path Travelling Salesman Problem

Jiaxuan Zhang, Suogang Gao, Bo Hou, and Wen Liu[✉]

Hebei Key Laboratory of Computational Mathematics and Applications,
School of Mathematical Sciences, Hebei Normal University,
Shijiazhuang 050024, People's Republic of China
liuwen1975@126.com

Abstract. In this paper, we consider the clustered path travelling salesman problem. In this problem, we are given a complete graph $G = (V, E)$ with edge weight satisfying the triangle inequality. In addition, the vertex set V is partitioned into clusters $V_1, \cdots, V_k$. The objective of the problem is to find a minimum Hamiltonian path in G, and in the path all vertices of each cluster are visited consecutively. We provide a polynomial-time approximation algorithm for the problem.

Keywords: Travelling salesman problem · Stacker crane problem · Path · Cluster

1 Introduction

The travelling salesman problem (TSP) is a best-known combinatorial optimization problem. In this problem, we are given a complete graph $G = (V, E)$ with vertex set V and edge set E, and there is an edge weight function ω satisfying the triangle inequality. The task of the TSP is to find a minimum Hamiltonian cycle. This problem is NP-hard and has multitudes of applications [4,6].

Meanwhile, TSP has quite a lot variants [11,18]. Among these variants there is an important one called the path travelling salesman problem (PTSP), whose task is to find a minimum Hamiltonian path. For the PTSP, Hoogeveen [15] presented a $\frac{5}{3}$-approximation algorithm. An et al. [1] improved this result and gave a $\frac{1+\sqrt{5}}{2}$-approximation algorithm. Recently, Zenklusen [22] developed the best known $\frac{3}{2}$-approximation algorithm. For more work on this problem, one can see [12,19,20].

Supported by the NSF of China (No. 11971146), the NSF of Hebei Province of China (No. A2019205089, No. A2019205092), Overseas Expertise Introduction Program of Hebei Auspices (25305008) and the Graduate Innovation Grant Program of Hebei Normal University (No. CXZZSS2022052).

Q. Ni and W. Wu (Eds.): AAIM 2022, LNCS 13513, pp. 15–27, 2022.
https://doi.org/10.1007/978-3-031-16081-3_2

Another best-studied variant of the TSP is the clustered travelling salesman problem (CTSP). In this problem, the vertex set V is partitioned into clusters $V_1, \cdots, V_k$. The goal of the CTSP is to find a minimum Hamiltonian cycle in G, and in the cycle all vertices of each cluster are visited consecutively. Note that if $k = 1$, the CTSP is exactly the TSP. So in the following we assume $k \geq 2$. Chisman [5] first introduced the CTSP and gave some applications about it. Arkin et al. [3] developed the first approximation algorithm for the CTSP with a performance guarantee of $\frac{7}{2}$. Guttmann-Beck et al. [13] designed approximation algorithms for several cases of the CTSP by decomposing them into the PTSP together with the stacker crane problem, or the PTSP together with the rural postman problem. Then, Kawasaki and Takazawa [17] improved approximation ratios by applying an improved approximation algorithm for the PTSP given by Zenklusen [22]. Applications and other related work for the CTSP may be found in [9, 16].

Motivated by the work of Kawasaki and Takazawa [17] and Anily et al. [2], we study the clustered path travelling salesman problem (CPTSP). In the CPTSP, we are given a complete graph $G = (V, E)$ with edge weight satisfying the triangle inequality, and the vertex set V is partitioned into clusters $V_1, \cdots, V_k$. The goal is to find a minimum Hamiltonian path in G, and in the path all vertices of each cluster are visited consecutively. For the CPTSP, we get three corresponding problems when we specify neither, one or both endpoints of the Hamiltonian path. But we only deal with the case that both endpoints of the path are specified. The reason is by guessing an endpoint, one can use an algorithm for the case with two specified endpoints to solve the case with only one specified endpoint, and by the algorithm for the case with one specified endpoint to solve that with no specified endpoint. Specifically, let s (t) be the start (end) vertex of the Hamiltonian path. The goal for the CPTSP is to find a minimum Hamiltonian path in G from s to t, and in the path all vertices of each cluster are visited consecutively. Note that s must be the start vertex for one cluster and t must be the end vertex for another. We might as well assume $s \in V_1$ and $t \in V_k$. Then V_1 and V_k are the first and the last clusters to be visited respectively. For other clusters, we visit them in any order. Note that there are k clusters in the CPTSP and a Hamiltonian path of this problem induces a Hamiltonian path in each cluster. For simplicity, we assume that in each cluster the start vertex and the end vertex for the Hamiltonian path are both specified. In this paper, we design an approximation algorithm with an approximation ratio $\frac{8}{3}$ for the CPTSP by decomposing it into the path travelling salesman problem and the path version of the stacker crane problem.

The paper is organized as follows. In Sect. 2, we give some definitions and results. In Sect. 3, we design two approximation algorithms for the path stacker crane problem. Based on these two algorithms, we design an algorithm for the clustered path travelling salesman problem and analyze its approximation ratio in Sect. 4. In Sect. 5, we provide several future research problems.

2   Preliminaries

In this section, we introduce some terminology, concepts and related results.

A graph G is a pair (V, E) where V is a set of objects called vertices and E is a set of edges. Each edge is a pair $\{v, w\}$ of vertices $v, w \in V$. If an edge $\{v, w\}$ is associated with a direction, a directed edge (arc) is obtained and we denote it by (v, w) if the direction is from v to w or (w, v) otherwise. Vertices v and w are called the tail and the head of the arc (v, w), respectively. A directed graph has a set of vertices and a set of directed edges. In a directed graph, the outdegree (indegree) of a vertex is the number of edges directed out of (into) the vertex. A mixed graph in which both directed and undirected edges may exist. A mixed multigraph is a graph, possibly with parallel edges, each of which is either undirected or directed.

A walk in a graph G from vertex v_1 to v_l is a sequence $(v_1, \{v_1, v_2\}, v_2, \{v_2, v_3\}, v_3, \cdots, v_{l-1}, \{v_{l-1}, v_l\}, v_l)$, in which all $v_i's$ are vertices and $\{v_{i-1}, v_i\} \in E(G)$ for $i = 2, 3, \cdots, l$. We also call v_1, v_l the start vertex and the end vertex of the walk. A path is a walk with no repeated vertices. A trail is a walk with no repeated edges. An Eulerian trail is a walk passing through every edge of G exactly once. If this walk is closed (starts and ends at the same vertex), it is called an Eulerian tour. A Hamiltonian path/cycle in G is a path/cycle visiting every vertex of G exactly once. A directed walk in a directed graph is a sequence $(v_1, (v_1, v_2), v_2, (v_2, v_3), v_3, \cdots, v_{l-1}, (v_{l-1}, v_l), v_l)$. A directed path is a directed walk with no repeated vertices. A directed trail is a directed walk with no repeated arcs. A directed Eulerian trail is a directed walk passing through every arc of G exactly once. Similarly, we can define that on the mixed multigraph.

In the stacker crane problem (SCP), we are given a mixed multigraph $G' = (V', E', D)$, where $V' = \{s_i, t_i \mid i \in [k]\}$, (V', E') is an undirected complete graph with an edge weight function satisfying the triangle inequality, and $D = \{(s_i, t_i) \mid i \in [k]\}$. Here $[k]$ is the notation for the set $\{1, 2, \cdots, k\}$. The objective is to find a minimum Hamiltonian cycle that traverses each arc (s_i, t_i) in the specified direction from s_i to t_i, $i \in [k]$, where we identify the weight of the arc (s_i, t_i) with that of the corresponding edge $\{s_i, t_i\}$.

If the objective "Hamiltonian cycle" substitutes for "Hamiltonian path" in the definition of the SCP, we get the path version of the stacker crane problem, and we call it the path stacker crane problem (PSCP).

The following results are important for the discussion in Sects. 3 and 4.

Theorem 2.1 *[7]. A connected multigraph has an Eulerian trail if and only if it has either 0 or 2 vertices of odd degree.*

By a simple deduction, we get the following result similar to that in [14].

Theorem 2.2. *A connected directed multigraph has an Eulerian trail if and only if every vertex has the same indegree and outdegree or the indegree of one vertex is equal to the outdegree of this vertex plus one, the outdegree of another vertex is equal to the indegree of this vertex plus one, and the outdegree of other vertices is equal to the indegree of them.*

Theorem 2.3 *[17].* *For the PTSP with u_1 and u_2 being the two given endpoints, there exists a polynomial time approximation algorithm that finds Hamiltonian paths S_1 and S_2 from u_1 to u_2 such that $w(S_1) \leq 2OPT' - w\{u_1, u_2\}, w(S_2) \leq \frac{3}{2}OPT'$, where OPT' denotes the weight of an optimal solution of the problem.*

3   Approximation Algorithms for the PSCP

In this section, we give two polynomial-time algorithms for the PSCP which will be used when we design the algorithm for the CPTSP in next section.

Recall that in the PSCP, we are given a mixed multigraph $G' = (V', E', D)$, where $V' = \{s_i, t_i \mid i \in [k]\}$, (V', E') is an undirected complete graph with an edge weight function ω satisfying the triangle inequality, and $D = \{(s_i, t_i) \mid i \in [k]\}$. The objective is to find a minimum Hamiltonian path from s_1 to t_k that traverses each arc (s_i, t_i) in the specified direction from s_i to t_i, $i \in [k]$. The first algorithm for the PSCP is as follows.

Algorithm 1.

Step 1: Find a minimum bipartite matching between the head set $T = \{t_1, t_2, \cdots, t_{k-1}\}$ and the tail set $S = \{s_2, s_3, \cdots, s_k\}$.

Step 2: Initialize E_1 to be empty. For each edge included in the above matching, associate a direction with it, going from T to S, and insert it into E_1. (This results in $m \geq 1$ disjoint connected components, each of which consists of edges with the associated directions in the matching and arcs in D, and we denote these m disjoint connected components by R_i, $i \in [m]$.)

Step 3: Condense each R_i into a single node n_i. Define

$$d\{n_i, n_j\} = \min\{\omega\{u, v\} \mid u \in R_i{}', v \in R_j{}'\}$$

where $R_i{}'$ represents the set of all vertices in R_i except for s_1, t_k.

Step 4: Find a minimum spanning tree for the nodes $\{n_i \mid i \in [m]\}$. Here the minimum is with respect to the distance function d defined in Step 3.

Step 5: Firstly, make two copies of each edge in the spanning tree. Secondly, associate one direction with one copy, and the opposite direction with the other. Thirdly, insert these edges with the associated directions into E_1. (This results in a directed graph $G'_1 = (V', E_1 \bigcup D)$.)

Step 6: Find an Eulerian trail from s_1 to t_k in G'_1.

Step 7: Using the triangle inequality, we get a Hamiltonian path P_{PSCP1} from s_1 to t_k traversing each arc (s_i, t_i) in the specified direction from s_i to t_i, $i \in [k]$.

Example 3.1. Assume $G' = (V', E', D)$ with (V', E') being a complete graph, $V' = \{s_i, t_i \mid i \in [4]\}$, and $D = \{(s_i, t_i) \mid i \in [4]\}$. The weight of edges is as follows: $\omega\{s_1, t_1\} = 1.6, \omega\{s_1, t_2\} = 1.5, \omega\{s_1, t_3\} = 1.4, \omega\{s_1, t_4\} = 1.3, \omega\{s_1, s_2\} = 1.2, \omega\{s_1, s_3\} = 1.5, \omega\{s_1, s_4\} = 1.4, \omega\{s_2, t_1\} = 1.5, \omega\{s_2, t_2\} = 1.9, \omega\{s_2, t_3\} = 1.8, \omega\{s_2, t_4\} = 1.6, \omega\{s_2, s_3\} = 1.1, \omega\{s_2, s_4\} = 1.2, \omega\{s_3, t_1\} = 1.6, \omega\{s_3, t_2\} = 1.9, \omega\{s_3, t_3\} = 1.8, \omega\{s_3, t_4\} = 1.7, \omega\{s_3, s_4\} = 2, \omega\{s_4, t_1\} = 1.5, \omega\{s_4, t_2\} = 1.7, \omega\{s_4, t_3\} = 2, \omega\{s_4, t_4\} = 1.7, \omega\{t_1, t_2\} = 1.2, \omega\{t_1, t_3\} = 1.7, \omega\{t_1, t_4\} = 1.2, \omega\{t_2, t_3\} = 1.8, \omega\{t_2, t_4\} = 1.3, \omega\{t_3, t_4\} = 1.4.$

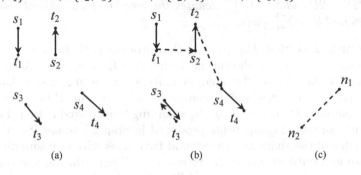

(a)   (b)   (c)

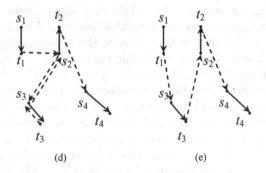

(d)   (e)

At the beginning of our algorithm, there are four arcs. $T = \{t_1, t_2, t_3\}$, and $S = \{s_2, s_3, s_4\}$ (a). Find a minimum bipartite matching between T and S. By Step 2 in Algorithm 1, we get two connected components (b). Condense each connected component into a single node and by using the distance function d, we find a minimum spanning tree for nodes n_1, n_2, i.e., the edge $\{s_2, s_3\}$ (c). Make two copies of the edge $\{s_2, s_3\}$ and the two copies are in opposite directions: one is (s_2, s_3) and the other is (s_3, s_2). Then we find an Eulerian trail

$(s_1, (s_1, t_1), t_1, (t_1, s_2), s_2, (s_2, s_3), s_3, (s_3, t_3), t_3, (t_3, s_3), s_3, (s_3, s_2), s_2, (s_2, t_2), t_2,$
$(t_2, s_4), s_4, (s_4, t_4), t_4)$ in the graph G'_1 with vertex set $V' = \{s_i, t_i \mid i \in$
$[4]\}$ and directed edge set $\{(s_1, t_1), (t_1, s_2), (s_2, s_3), (s_3, t_3), (t_3, s_3), (s_3, s_2),$
$(s_2, t_2), (t_2, s_4), (s_4, t_4)\}$ (d). Using the triangle inequality, we get a path
$(s_1, (s_1, t_1), t_1, (t_1, s_3), s_3, (s_3, t_3), t_3, (t_3, s_2), s_2, (s_2, t_2), t_2, (t_2, s_4), s_4, (s_4, t_4), t_4)$
(e).

Lemma 3.2. *Algorithm 1 outputs a Hamiltonian path P_{PSCP1} with weight at most $3OPT'' - 2U$, where OPT'' denotes the weight of an optimal solution of the problem and $U = \sum_{i=1}^{k} \omega(s_i, t_i)$.*

Proof. We first show that P_{PSCP1} is a Hamiltonian path from s_1 to t_k that traverses each arc (s_i, t_i) in the specified direction from s_i to t_i. According to the directions of the edges of the bipartite matching in Step 2, we can know that except for s_1, t_k, the indegree and outdegree of each vertex are thus equal in these connected components. The edges of the spanning tree created in Step 4 connect these disjoint connected components produced in Step 2 into one. Since in Step 5 for each edge of the spanning tree we add two edges with opposite directions, the indegree and outdegree of each vertex are still equal in the one connected component, except for s_1, t_k. Step 1 and Step 3 of the algorithm guarantee that s_1 and t_k are the only two vertices with odd degree. For specific, the indegree of $s_1(t_k)$ is 0(1) and the outdegree of $s_1(t_k)$ is 1(0). In Step 6, by Theorem 2.2, we get the Eulerian trail from s_1 to t_k in the directed graph $G'_1 = (V', E_1 \bigcup D)$. In Step 7, using the triangle inequality, we get a desired Hamiltonian path.

In the following, we consider the weight of P_{PSCP1}. Since an optimal path of this problem contains a bipartite matching between T and S and all arcs in D, its weight can not be smaller than that of the minimum bipartite matching obtained in Step 1 and the arcs in D. For convenience, we denote the weight of the minimum bipartite matching by M. Then $M + U \leq OPT''$. Since the edges with the associated directions in the optimal path except for the arcs in D can connect these disjoint connected components into one and the spanning tree created in Step 4 also connects these disjoint connected components into one, the weight of the minimum spanning tree in Step 4 must be no greater than $OPT'' - U$. Therefore, the weight of the Eulerian trail in Step 6 is at most $M + U + 2(OPT'' - U)$, and then is at most $3OPT'' - 2U$. Using the triangle inequality, we get the weight of the path P_{PSCP1} is at most the weight of the Eulerian trail. The results follows. $\qquad\square$

The following is the second algorithm for the PSCP.

Algorithm 2.

Step 1: Condense each arc (s_i, t_i) into a node n_i for each $i \in [k]$. For each pair i, j with $i, j \in [k]$, define

$$
d\{n_i, n_j\} = \begin{cases}
\min\{\omega\{s_i, s_j\}, \omega\{s_i, t_j\}, \omega\{t_i, s_j\}, \omega\{t_i, t_j\}\}, & \text{if } s_1, t_k \notin \{s_i, s_j, t_i, t_j\}, \\
\min\{\omega\{t_i, s_j\}, \omega\{t_i, t_j\}\}, & \text{if } s_1 = s_i, \\
\min\{\omega\{s_i, t_j\}, \omega\{t_i, t_j\}\}, & \text{if } s_1 = s_j, \\
\min\{\omega\{s_i, t_j\}, \omega\{s_i, s_j\}\}, & \text{if } t_k = t_i, \\
\min\{\omega\{t_i, s_j\}, \omega\{s_i, s_j\}\}, & \text{if } t_k = t_j, \\
\omega\{t_i, s_j\}, & \text{if } s_1 = s_i, t_k = t_j, \\
\omega\{s_i, t_j\}, & \text{if } s_1 = s_j, t_k = t_i.
\end{cases}
$$

Step 2: Find a minimum spanning tree for the nodes $\{n_i \mid i \in [k]\}$. Here the minimum is with respect to the distance function d defined in Step 1.

Step 3: Initialize E_1 to be empty. Insert all edges in the above spanning tree into E_1. Replace the node n_i with the arc (s_i, t_i) for each $i \in [k]$. (This results in a graph with vertex set V' and edge set $E_1 \bigcup D$.)

Step 4: Identify vertices with odd degree in the graph obtained in Step 3. Then find a minimum perfect matching for all these vertices with odd degree, except for s_1, t_k.

Step 5: Insert all edges in the above matching into E_1. (This results in a graph with vertex set V' and edge set $E_1 \bigcup D$.)

Step 6: Find an Eulerian trail from s_1 to t_k in the graph obtained in Step 5, ignoring the directions of the arcs in D.

Step 7: Associate each edge in E_1 with the direction of the Eulerian trail. For each arc in D that is incorrectly traversed, add two directed edges to E_1, both with direction opposite to that of the arc. (This results in a directed graph $G_2' = (V', E_1 \bigcup D)$.)

Step 8: Find an Eulerian trail from s_1 to t_k in G_2'.

Step 9: Using the triangle inequality, we get a Hamiltonian path P_{PSCP2} from s_1 to t_k traversing each arc (s_i, t_i) in the specified direction from s_i to t_i, $i \in [k]$.

The following is an example to show the process of Algorithm 2.

Example 3.3. Assume $G'' = (V'', E'', D)$ with (V'', E'') being a complete graph, $V'' = \{s_i, t_i \mid i \in [4]\}$, and $D = \{(s_i, t_i) \mid i \in [4]\}$. The weight of edges is as follows: $\omega\{s_1, t_1\} = 2.2, \omega\{s_1, t_2\} = 1.2, \omega\{s_1, t_3\} = 1.3, \omega\{s_1, t_4\} = 1.2, \omega\{s_1, s_2\} = 1.4, \omega\{s_1, s_3\} = 1.6, \omega\{s_1, s_4\} = 1.7, \omega\{s_2, t_1\} = 1.5, \omega\{s_2, t_2\} = 1.5, \omega\{s_2, t_3\} = 1.4, \omega\{s_2, t_4\} = 1.4, \omega\{s_2, s_3\} = 1.3, \omega\{s_2, s_4\} = 1.2, \omega\{s_3, t_1\} = 2, \omega\{s_3, t_2\} = 1, \omega\{s_3, t_3\} = 1.8, \omega\{s_3, t_4\} = 2, \omega\{s_3, s_4\} = 2.4, \omega\{s_4, t_1\} = 1.8, \omega\{s_4, t_2\} = 1.5, \omega\{s_4, t_3\} = 1.8, \omega\{s_4, t_4\} = 2.5, \omega\{t_1, t_2\} = 1.4, \omega\{t_1, t_3\} = 2.6, \omega\{t_1, t_4\} = 1.7, \omega\{t_2, t_3\} = 2.2, \omega\{t_2, t_4\} = 2, \omega\{t_3, t_4\} = 2.1$.

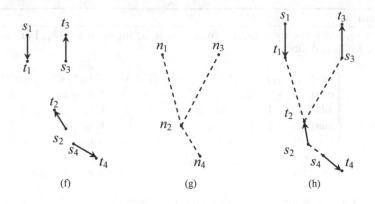

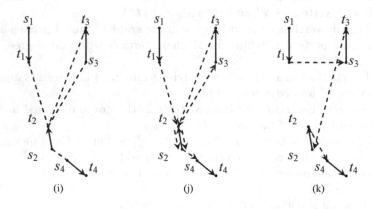

With a similar discussion with Example 3.1, we get a desired Hamiltonian path $(s_1, (s_1, t_1), t_1, (t_1, s_3), s_3, (s_3, t_3), t_3, (t_3, s_2), s_2, (s_2, t_2), t_2, (t_2, s_4), s_4, (s_4, t_4), t_4)$.

Lemma 3.4. *Algorithm 2 outputs a Hamiltonian path P_{PSCP2} with weight at most $2OPT'' + 2U$, where OPT'' denotes the weight of an optimal solution of the problem and $U = \sum_{i=1}^{k} \omega(s_i, t_i)$.*

Proof. With a similar discussion with Lemma 3.2, after completion of Step 5 all vertices are of even degree except for s_1, t_k. The degree of s_1, t_k is 1. This is obtained by the definition of d in Step 1 and by ignoring these two odd degree vertices s_1, t_k in Step 4. According to Theorem 2.1, we find an Eulerian trail from s_1 to t_k in the graph obtained in Step 5 by ignoring the directions of arcs in D. Step 7 shows how to augment the graph to allow the Eulerian trail to traverse arcs in D in the proper direction. Similarly, the definition of d in Step 1 and omission of these two odd degree vertices s_1, t_k in Step 4 ensure that arcs $(s_1, t_1), (s_k, t_k)$ are correctly traversed. When there exists the arc in D that is incorrectly traversed, add two directed edges to E_1, both with direction opposite to that of the arc, the indegree and outdegree of each vertex are equal, except for s_1, t_k. The indegree of $s_1(t_k)$ is 0(1) and the outdegree of $s_1(t_k)$ is 1(0). Then we get the Eulerian trail from s_1 to t_k traversing each arc (s_i, t_i) in the specified direction from s_i to t_i in the directed graph $G_2' = (V', E_1 \bigcup D)$. In Step 9, using the triangle inequality, we get a desired Hamiltonian path.

In the following, we consider the weight of P_{PSCP2}. Since the edges with the associated directions in the optimal path except for the arcs in D can connect the nodes, the weight of the spanning tree in Step 2 is at most $OPT'' - U$. Then the weight of the graph obtained in Step 3 is at most OPT''. So the minimum perfect matching weight on these vertices of odd degree except for s_1, t_k is at most OPT''. Therefore the weight of the graph obtained in Step 5 is at most $2OPT''$, and so the weight of the Eulerian trail in Step 6 is at most $2OPT''$. Note that the weight of all directed edges added in Step 7 is at most $2U$. Then the weight of the Eulerian trail in Step 8 is at most $2OPT'' + 2U$. Also using the triangle inequality, we get the weight of the Hamiltonian path P_{PSCP2} is at most $2OPT'' + 2U$, as desired.   □

Remark 3.5. The outputs of the above two algorithms both have weights related to the weight U of the arcs in D. Each of Algorithm 1 and Algorithm 2 is run, and the value U relative to OPT'' will determine which algorithm we will choose. For specific, if $U \geq \frac{1}{4} OPT''$, we choose Algorithm 1. Choose the other one instead.

Lemma 3.6. *Both Algorithm 1 and Algorithm 2 can be implemented in polynomial time.*

Proof. The running time of Algorithm 1 depends on the running time of finding the minimum bipartite matching and the minimum spanning tree. According to Grinman [10], the running time of the Hungarian algorithm for finding the minimum bipartite matching is $O(|V^3|)$. According to Yao [21], the minimum spanning tree algorithm runs in polynomial time. So Algorithm 1 can be implemented in polynomial time.

The running time of Algorithm 2 depends on the running time of finding the minimum perfect matching and the minimum spanning tree. There is a polynomial-time blossom algorithm for computing minimum perfect matching by Edmonds [8]. So Algorithm 2 can be implemented in polynomial time   □

4 Approximation Algorithm for the CPTSP

In this section, we first design an approximation algorithm for the CPTSP, then we analyze its approximation ratio.

Recall that in the CPTSP, the start vertex and end vertex are both specified for each cluster V_i, $i \in [k]$. In order to apply the PSCP problem in the algorithm, we assume that the start (end) vertex is s_i (t_i) for each V_i, $i \in [k]$. Obviously, s_1, t_k are s, t. Our algorithm mainly consists of four parts. In the first part, for each fixed $i \in [k]$, we find the $path_i$, a path from s_i to t_i that goes through all vertices in V_i. This is exactly the PTSP with given start and end vertices. In the second part, we replace the $path_i$ by the arc (s_i, t_i). In the third part, we apply Algorithm 1 or Algorithm 2 to find a Hamiltonian path from s_1 to t_k that traverses each arc (s_i, t_i) for $i \in [k]$. In the fourth part, we replace the arc (s_i, t_i) by the path $path_i$ obtained in the first part.

Algorithm 3.

Input: A complete graph $G = (V, E)$ with weight function $\omega : E \to \mathbb{R}_+$, clusters $V_1, \cdots, V_k$, and the start (end) vertex s_i (t_i) for each V_i, $i \in [k]$.

begin

Step 1: For each V_i, $i \in [k]$, compute $path_i$, a Hamiltonian path with start vertex s_i and end vertex t_i.

Step 2: Replace the $path_i$ with the arc (s_i, t_i). (This results in a mixed multigraph with vertex set $\{s_i, t_i \mid i \in [k]\}$ and $D = \{(s_i, t_i) \mid i \in [k]\}$.)

Step 3: Find a Hamiltonian path P_{PSCP} in the mixed multigraph obtained above.

Step 4: In P_{PSCP}, replace the arc (s_i, t_i) by the $path_i$, $i \in [k]$.

Output: Return the resulting path T

In this algorithm, we are involved in the PTSP, and the PSCP that are polynomial-time solvable, so our algorithm runs in polynomial time.

Example 4.1. See this example for a sample execution of the algorithm.

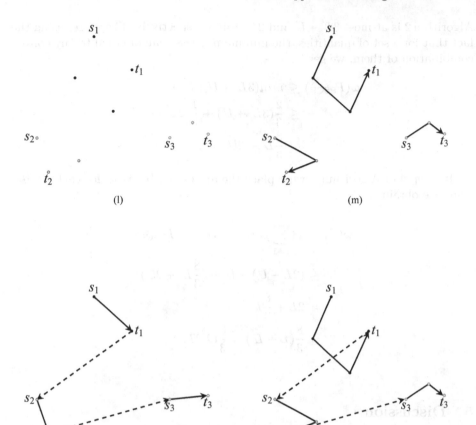

In this example, we give three clusters (l). In (m), we compute paths in each cluster. In (n), we solve the PSCP instance. In (o), we replace the arcs by the paths in each cluster.

Since there are two solutions for the PSCP, we either take one of them or take a combination of them, in order to get a better approximation ratio. Let OPT denote both an optimal solution of the CPTSP and its total weight.

Theorem 4.2. *Let T be the path output by Algorithm 3. Then $w(T) \leq \frac{8}{3}OPT$.*

Proof. The algorithm consists of two subproblems: the PTSP with given start and end vertices, and the PSCP. Let P_i denote the induced path of OPT on V_i, $i \in [k]$. Let $L = \sum_{i\in[k]} \sum_{e\in P_i \cap OPT} w(e)$, $L' = OPT - L$. Recall that $U = \sum_{i=1}^{k} w(s_i, t_i)$. By Theorem 2.3, we get $w(\sum_{i=1}^{k} path_i) \leq min(2L - U, \frac{3}{2}L)$. Therefore, $w(\sum_{i=1}^{k} path_i) \leq min(2L - U, \frac{3}{2}L) \leq 2L - U$.

Note that there exists a solution to the PSCP with weight $L' + U$. By Lemmas 3.2 and 3.4, the weight of the two solutions returned by the Algorithm 1 and

Algorithm 2 is at most $3L' + U$ and $2L' + 4U$, respectively. Therefore, using the fact that for a set of quantities, the minimum is less than or equal to any convex combination of them, we get

$$\omega(P_{\text{PSCP}}) \leq min(3L' + U, 2L' + 4U)$$
$$\leq \frac{2}{3}(3L' + U) + \frac{1}{3}(2L' + 4U)$$
$$= \frac{8}{3}L' + 2U.$$

In Step 4 of Algorithm 3, we replace the arc (s_i, t_i) by $path_i$ for each $i \in [k]$. Then we obtain

$$\omega(T) = \omega(\sum_{i=1}^{k} path_i) - U + \omega(P_{\text{PSCP}})$$
$$\leq (2L - U) - U + (\frac{8}{3}L' + 2U)$$
$$= 2L + \frac{8}{3}L'$$
$$\leq \frac{8}{3}(L + L') = \frac{8}{3}OPT.$$

$\square$

5   Discussion

In this paper, we only consider the case that the start vertex and the end vertex are both given in each cluster. Other cases including two endpoints in each cluster are given but we are free to choose the start vertex, only the start vertex is given in each cluster, and neither of the endpoints is given in each cluster, are also interesting problems to consider.

References

1. An, H.-C., Kleinberg, R., Shmoys, D.B.: Improving Christofide' algorithm for the s-t path TSP. J. ACM **62**(5), 1–28 (2015)
2. Anily, S., Bramel, J., Hertz, A.: A 5/3-approximation algorithm for the clustered traveling salesman tour and path problems. Oper. Res. Lett. **24**, 29–35 (1999)
3. Arkin, E.M., Hassin, R., Klein, L.: Restricted delivery problems on a network. Networks **29**, 205–216 (1994)
4. Bland, R.G., Shallcross, D.F.: Large travelling salesman problems arising from experiments in X-ray crystallography: a preliminary report on computation. Oper. Res. Lett. **8**(3), 125–128 (1989)
5. Chisman, J.A.: The clustered traveling salesman problem. Comput. Oper. Res. **2**, 115–119 (1975)

6. Christofides, N.: Worst-case analysis of a new heuristic for the Travelling Sales-man Problem. Technical report 388, Graduate School of Industrial Administration, Carnegie Mellon University (1976)
7. Diestel, R.: Graph Theory. Springer, New York (2017). https://doi.org/10.1007/978-3-662-53622-3
8. Edmonds, J.: Paths, trees and flowers. Can. J. Math. **17**, 449–467 (1965)
9. Gendreau, M., Hertz, A., Laporte, G.: The traveling salesman problem with back-hauls. Comput. Oper. Res. **23**, 501–508 (1996)
10. Grinman, A.: The Hungarian algorithm for weighted bipartite graphs. Seminar in Theoretical Computer Science (2015)
11. Grötschel, M., Holland, O.: Solution of large-scale symmetric traveling salesman problems. Math. Program. **51**, 141–202 (1991)
12. Gottschalk, C., Vygen, J.: Better s-t-tours by Gao trees. Math. Program. **172**, 191–207 (2018)
13. Guttmann-Beck, N., Hassin, R., Khuller, S., Raghavachari, B.: Approximation algorithms with bounded performance guarantees for the clustered traveling sales-man problem. Algorithmica **28**, 422–437 (2000)
14. Hong, Y.M., Lai, H.J., Liu, Q.H.: Supereulerian digraphs. Discrete Math. **330**, 87–95 (2014)
15. Hoogeveen, J.A.: Analysis of Christofides' heuristic: some paths are more difficult than cycles. Oper. Res. Lett. **10**, 291–295 (1991)
16. Jongens, K., Volgenant, T.: The symmetric clustered traveling salesman problem. Eur. J. Oper. Res. **19**, 68–75 (1985)
17. Kawasaki, M., Takazawa, T.: Improving approximation ratios for the clustered travelling salesman problem. J. Oper. Res. Soc. Jpn. **63**(2), 60–70 (2020)
18. Plante, R.D., Lowe, T.J., Chandrasekaran, R.: The product matrix travelling sales-man problem: an application and solution heuristics. Oper. Res. **35**, 772–783 (1987)
19. Sebő, A., van Zuylen, A.: The salesman's improved paths: a 3/2+1/34 approxima-tion. In: Proceedings of 57th Annual IEEE Symposium on Foundations of Com-puter Science (FOCS), pp. 118–127 (2016)
20. Traub, V., Vygen, J.: Approaching 3/2 for the s-t path TSP. J. ACM **66**(2), 1–17 (2019)
21. Yao, A.: An $O(|E| \log \log |V|)$ algorithm for finding minimum spanning trees. Inf. Process. Lett. **4**, 21–23 (1975)
22. Zenklusen, R.: A 1.5-approximation for path TSP. In: Proceedings of the 30th Annual ACM-SIAM Symposium on Discrete Algorithms, pp. 1539–1549 (2019)

Improved Approximation Algorithm for the Asymmetric Prize-Collecting TSP

Bo Hou$^{(\boxtimes)}$, Zhenzhen Pang, Suogang Gao, and Wen Liu

Hebei Key Laboratory of Computational Mathematics and Applications,
School of Mathematical Sciences, Hebei Normal University,
Shijiazhuang 050024, People's Republic of China
houbo1969@163.com

Abstract. We present a $\frac{4\lceil \log(n) \rceil}{0.698\lceil \log(n) \rceil + 1.302}$-approximation algorithm for the asymmetric prize-collecting traveling salesman problem. This is obtained by combining a randomized variant of a rounding algorithm of N.H. Nguyen and T.T. Nguyen [6] and a primal-dual algorithm of N.H. Nguyen [7].

Keywords: Asymmetric prize-collecting traveling salesman problem · Approximation algorithm

1 Introduction

In this paper, we mainly study the version of the asymmetric prize-collecting traveling salesman problem (APC-TSP). In the APC-TSP, we are given a complete directed graph $G = (V, A)$ with vertex set $V = \{1, 2, \cdots, n\}$ and arc set A. Let $c : A \to \mathbb{R}_{\geq 0}$ be an arc cost function and let $\pi : V \to \mathbb{R}_{\geq 0}$ be a penalty function. Assume that the arc costs satisfy the triangle inequality, that is, $c_{(i,j)} \leq c_{(i,k)} + c_{(k,j)}$ for all $i, j, k \in V$. Fix a vertex $j \in V$. The goal is to find a tour T with $j \in V(T)$ such that $c(T) + \pi(V \setminus V(T))$ is minimized, where $c(T) = \sum_{(i,k) \in T} c_{(i,k)}$, $\pi(S) = \sum_{i \in S} \pi_i$, and $V(T)$ denotes the vertices spanned by T.

If $c_{(i,k)} = c_{(k,i)}$ for all $i, k \in V$, then the APC-TSP is exactly the prize-collecting traveling salesman problem (PC-TSP). For the PC-TSP, in 1993, D. Bienstock et al. [2] presented the first constant approximation algorithm with an approximation ratio of $\frac{5}{2}$ based on the solution of a linear program problem. In 1995, D.P. Goemans and M.X. Willamson [4] presented a primal-dual approximation algorithm with an approximation ratio of $2 - \frac{1}{n-1}$. In 2009, A. Archer et al. [1] and D.P. Goemans [5] respectively improved the algorithm of D.P. Goemans and M.X. Williamson to 1.990283 and 1.91456.

Supported by the NSF of China (No. 11971146), the NSF of Hebei Province of China (No. A2019205089, No. A2019205092), Overseas Expertise Introduction Program of Hebei Auspices (25305008) and the Graduate Innovation Grant Program of Hebei Normal University (No. CXZZSS2022053).

Q. Ni and W. Wu (Eds.): AAIM 2022, LNCS 13513, pp. 28–32, 2022.
https://doi.org/10.1007/978-3-031-16081-3_3

For the APC-TSP, in 2012, V.H. Nguyen and T.T. Nguyen [6] presented an $(1 + \lceil \log(n) \rceil)$-approximation algorithm based on Frieze et al.'s heuristic for the asymmetric traveling salesman problem as well as a method to round fractional solutions of a linear programming relaxation to integers. In 2013, V.H. Nguyen [7] presented a combinatorial approximation algorithm with an approximation ratio of $\lceil \log(n) \rceil$ based on the primal-dual method.

Motivated by the above work, in this paper, we focus our attention on the APC-TSP problem and propose a $\frac{4\lceil \log(n) \rceil}{0.698\lceil \log(n) \rceil + 1.302}$-approximation algorithm by combining a randomized variant of the rounding algorithm of N.H. Nguyen and T.T. Nguyen [6] and the primal-dual algorithm of N.H. Nguyen [7].

2   Preliminaries

In this section, we review the results of V.H. Nguyen and T.T. Nguyen [6] and those of V.H. Nguyen [7]. We start by considering a classical LP relaxation of the APC-TSP.

Let $y_i = 1$ if i is in the tour and 0 otherwise. Let $x_e = 1$ if e is in the tour and 0 otherwise. The LP relaxation is as follows:

$$
\begin{aligned}
\min \ & \textstyle\sum_{e \in E} c_e x_e + \sum_{i \in V} \pi_i (1 - y_i) \\
\text{s.t.} \ & \textstyle\sum_{e \in \delta^+(j)} x_e \geq 1, \\
& \textstyle\sum_{e \in \delta^+(j)} x_e \geq 1, \\
& \textstyle\sum_{e \in \delta^+(S)} x_e \geq y_i, && \emptyset \neq S \subset V\backslash\{j\}, \forall i \in S, \\
& \textstyle\sum_{e \in \delta^-(S)} x_e \geq y_i, && \emptyset \neq S \subset V\backslash\{j\}, \forall i \in S, && (1) \\
& y_j = 1, \\
& y_i \leq 1, && \forall i \in V\backslash\{j\}, \\
& y_i \geq 0, && \forall i \in V\backslash\{j\}, \\
& x_e \geq 0, && \forall e \in A.
\end{aligned}
$$

For conciseness, we use $c(x) + \pi(1 - y)$ to denote the objective function of this LP relaxation. Let x^*, y^* be an optimum solution of this LP relaxation, and let $Z^* = c(x^*) + \pi(1 - y^*)$ denote the optimum value.

Next, we briefly review the rounding result of V.H. Nguyen and T.T. Nguyen [6]. V.H. Nguyen and T.T. Nguyen [6] showed the following result which is based on the analysis of Frieze et al.'s heuristic algorithm for the asymmetric traveling salesman problem [3].

Proposition 2.1. *Let* $0 \leq \gamma \leq 1$ *and let* $S(\gamma) = \{i : y_i^* \geq \gamma\}$. *Let* T_γ *denote the tour on* $S(\gamma)$ *output by Frieze et al.'s heuristic algorithm when given* $S(\gamma)$ *as vertex set. Then we have*

$$
c(T_\gamma) \leq \frac{\lceil \log(n) \rceil}{\gamma} c(x^*).
$$

Note that the $(1 + \lceil \log(n) \rceil)$-approximation algorithm for the APC-TSP proposed in [6] can be derived by setting $\gamma = \frac{\lceil \log(n) \rceil}{1 + \lceil \log(n) \rceil}$.

Finally, we review the primal-dual results of V.H. Nguyen [7]. Let T denote the tour returned by the primal-dual algorithm [7]. The following lemmas hold.

Lemma 2.2 *[7, Lemma 1]. There are at most $\lceil \log_2(n) \rceil$ iterations.*

Lemma 2.3 *[7, Lemma 2]. For every iteration except the last, the total cost of the arcs added to T at this iteration is at most Z^*.*

Lemma 2.4 *[7, Lemma 3]. For the last iteration, the cost of the arcs added to T at this iteration plus the penalties associated to the vertices eliminated from T (from the first iteration) is at most Z^*.*

Corollary 2.5. *With reference to the above notation, we have*

$$c(T) + \lceil \log(n) \rceil \pi(V \setminus V(T)) \leq 2\lceil \log(n) \rceil Z^*. \tag{2}$$

Proof. Immediate from Lemmas 2.2–2.4. □

Now we apply the primal-dual algorithm [7] to an instance in which we replace the penalties $\pi(\cdot)$ by $\pi'(\cdot)$ given by

$$\pi_i' = \frac{1}{\lceil \log(n) \rceil} \pi_i. \tag{3}$$

Let T_{pd} denote the output tour for the penalties $\pi'(\cdot)$. By (2) and (3), we obtain

$$c(T_{pd}) + \pi(V \setminus V(T_{pd})) = c(T_{pd}) + \lceil \log(n) \rceil \pi'(V \setminus V(T_{pd})) \leq 2\lceil \log(n) \rceil Z^{*\prime}, \tag{4}$$

where $Z^{*\prime}$ denotes the optimum value for the penalties $\pi'(\cdot)$.

As the optimum solution x^*, y^* for the penalties $\pi(\cdot)$ is feasible for the linear programming relaxation with penalties $\pi'(\cdot)$, we have

$$Z^{*\prime} \leq c(x^*) + \pi'(1 - y^*). \tag{5}$$

Combining (4) and (5), we have

$$c(T_{pd}) + \pi(V \setminus V(T_{pd})) \leq 2\lceil \log(n) \rceil c(x^*) + 2\pi(1 - y^*). \tag{6}$$

3   Approximation Algorithms

In this section, we present our approximation algorithm for the APC-TSP and analyze its approximate ratio.

In our algorithm, we will use as subroutines the rounding algorithm for the APC-TSP proposed by V.H. Nguyen and T.T. Nguyen [6] and the primal-dual algorithm for the APC-TSP proposed by V.H. Nguyen [7].

Algorithm 1

1: Given a complete directed graph $G = (V, A)$ with the vertex set $V = \{1, 2, \cdots, n\}$ and the arc set A, a special vertex $j \in V$, a cost function $c : A \to R_+$, and a penalty function $\pi : V \to R_+$.
2: Solve the LP relaxation (1) to obtain an optimal fractional solution (x^*, y^*).
3: Select a parameter γ uniformly at random from the interval $[e^{-\frac{1}{3}}, 1]$.
4: Let $S(\gamma) = \{i : y_i^* \geq \gamma\}$. Run Frieze et al.'s heuristic algorithm on the instance in which $S(\gamma)$ is the vertex set to obtain the tour T_γ.
5: Define the function $\pi' : V \to R_+$ by (3).
6: Run the primal-dual algorithm on the instance in which we replace the penalties $\pi(\cdot)$ by $\pi'(\cdot)$ to obtain the tour T_{pd}.
7: Let $p = \dfrac{(1-\frac{\lceil \log(n) \rceil}{3})\frac{1}{1-e^{-1/3}}}{2\lceil \log(n) \rceil - 2 + \frac{1}{1-e^{-1/3}}(1-\frac{\lceil \log(n) \rceil}{3})}$.
8: Select T_{pd} with probability p or select T_γ with probability $1 - p$ as the output tour.

Theorem 3.1. *The expected output value of Algorithm 1 is an α-approximation to the APC-TSP, where $\alpha \leq \frac{4\lceil \log(n) \rceil}{0.698\lceil \log(n) \rceil + 1.302}$.*

Proof. First, assume that we select γ randomly according to a certain distribution to be specified. Then, by Proposition 2.1, we have that

$$E[c(T_\gamma)] \leq E[\frac{\lceil \log(n) \rceil}{\gamma} c(x^*)] = \lceil \log(n) \rceil E[\frac{1}{\gamma}] c(x^*),$$

while the expected penalty we have to pay is

$$E[\pi(V \setminus V(T_\gamma))] = \sum_{i \in V} Pr[\gamma > y_i^*] \pi_i.$$

Thus, the overall expected cost is:

$$E[c(T_\gamma) + \pi(V \setminus V(T_\gamma))] \leq \lceil \log(n) \rceil E[\frac{1}{\gamma}] c(x^*) + \sum_{i \in V} Pr[\gamma > y_i^*] \pi_i. \quad (7)$$

Now, we assume that γ is chosen uniformly at random from the interval $[e^{-\frac{1}{3}}, 1]$. Then,

$$E[\frac{1}{\gamma}] = \int_a^1 \frac{1}{1-a} \frac{1}{x} dx = -\frac{\ln a}{1-a} = \frac{1}{3(1-a)} = \frac{1}{3(1-e^{-1/3})},$$

and $Pr[\gamma > y] = \frac{1-y}{1-a}$ when $a \leq y \leq 1$ and $Pr[\gamma > y] = 1 \leq \frac{1-y}{1-a}$ when $0 \leq y \leq a$. Therefore, (7) becomes:

$$\begin{aligned}
E[c(T_\gamma) + \pi(V \setminus V(T_\gamma))] &\leq \lceil \log(n) \rceil \frac{1}{3(1-e^{-1/3})} c(x^*) + \sum_{i \in V} \frac{1-y_i^*}{1-e^{-1/3}} \pi_i \\
&= \lceil \log(n) \rceil \frac{1}{3(1-e^{-1/3})} c(x^*) + \frac{1}{1-e^{-1/3}} \pi(1 - y^*).
\end{aligned} \quad (8)$$

Next, let

$$p = \frac{(1 - \frac{\lceil \log(n) \rceil}{3})\frac{1}{1 - e^{-1/3}}}{2\lceil \log(n) \rceil - 2 + \frac{1}{1 - e^{-1/3}}(1 - \frac{\lceil \log(n) \rceil}{3})}.$$

Observe that

$$2p\lceil \log(n) \rceil + (1 - p)\lceil \log(n) \rceil \frac{1}{3(1 - e^{-1/3})} = [2p + (1 - p)\frac{1}{1 - e^{-1/3}}]. \quad (9)$$

We select T_{pd} with probability p or select T_γ with probability $1 - p$ as the output tour. From (6), (8) and (9), we get that the expected cost E^* of the resulting algorithm satisfies:

$$E^* \leq [2\lceil \log(n) \rceil pc(x^*) + 2p\pi(1 - y^*)] + [(1 - p)\lceil \log(n) \rceil \frac{1}{3(1 - e^{-1/3})}c(x^*)$$

$$+ (1 - p)\frac{1}{1 - e^{-1/3}}\pi(1 - y^*)]$$

$$= [2p\lceil \log(n) \rceil + (1 - p)\lceil \log(n) \rceil \frac{1}{3(1 - e^{-1/3})}]c(x^*) + [2p + (1 - p)\frac{1}{1 - e^{-1/3}}]\pi(1 - y^*)$$

$$= [2p + (1 - p)\frac{1}{1 - e^{-1/3}}](c(x^*) + \pi(1 - y^*)).$$

Therefore, Algorithm 1 outputs a solution of cost at most αZ^* where

$$\alpha = 2p + (1 - p)\frac{1}{1 - e^{-1/3}} = \frac{\frac{4\lceil \log(n) \rceil}{3} \cdot \frac{1}{1 - e^{-1/3}}}{2\lceil \log(n) \rceil - 2 + \frac{1}{1 - e^{-1/3}}(1 - \frac{\lceil \log(n) \rceil}{3})} < \frac{4\lceil \log(n) \rceil}{0.698\lceil \log(n) \rceil + 1.302}.$$

□

We remark that the probability distribution given in the proof is optimal and the approximation ratio of Algorithm 1 is stronger than those of algorithms proposed in [6] and [7] when n is greater than 8.

References

1. Archer, A., Bateni, M., Hajiaghayi, M., Karloff, H.: Improved approximation algorithms for prize-collecting Steiner tree and TSP. In: Proceedings of the 50th Annual Symposium on Foundations of Computer Science (2009)
2. Bienstock, D., Goemans, M.X., Simchi-Levi, D., Williamon, D.P.: A note on the prize collecting traveling salesman problem. Math. Prog. **59**, 413–420 (1993)
3. Frieze, A.M., Galbiati, G., Maffioli, F.: On the worst case performance of some algorithms for the asymmetric traveling salesman problem. Networks **12**, 23–39 (1982)
4. Goemans, M.X., Williamson, D.P.: A general approximation technique for constrained forest problems. SIAM J. Comput. **24**(2), 296–317 (1995)
5. Goemans, M.X.: Combining approximation algorithms for the prize-collecting TSP. arXiv arXiv: 0910.0553v1 (2009)
6. Nguyen, V.H., Nguyen, T.T.: Approximating the asymmetric profitable tour. Int. J. Math. Oper. Res. **4**(3), 294–301 (2012)
7. Nguyen, V.H.: A primal-dual approximation algorithm for the asymmetric prize-collecting TSP. J. Comb. Optim. **25**, 265–278 (2013)

Scheduling Problem and Game Theory

Approximation Scheme for Single-Machine Rescheduling with Job Delay and Rejection

Ruiqing Sun[1] and Xiaofei Liu[2(✉)]

[1] School of Mathematics and Statistics, Yunnan University, Kunming, China
[2] School of Information Science and Engineering, Yunnan University,
Kunming, China
lxfjl2016@163.com

Abstract. In this paper, we consider the single-machine rescheduling with job delay and rejection. In this problem, we are given a set of jobs and a single machine, where each job has a processing time and weight. We can get an original scheduling with the minimal the total weighted completion time based on the shortest weighted processing time, if all jobs are available at time zero. For each job, the completion time in the original scheduling is defined as its due date. In the real world, some jobs may be delayed when the formal processing begins. In order to ensure a reasonable service level, it is allowed to reject some delayed jobs, but a rejection cost should be paid. This problem is to reschedule the jobs such that the maximum tardiness is bounded by a given threshold and the sum of the following three components: the total weighted tardiness time, the total rejection cost, and the maximum tardiness for the accepted jobs, is minimized. We present a pseudo-polynomial time dynamic programming algorithm, and a fully polynomial time approximation scheme.

Keywords: Rescheduling · Rejection · Dynamic programming · Fully polynomial time approximation scheme

1 Introduction

Given a set M of machines and a set J of jobs such that each job has to be processed on one of the machines, the classical scheduling problem is to find a scheduling such that the makespan is minimized [3,4]. Starting from classical list scheduling algorithm in [4], numerous algorithms have been proposed [1,6,7,20]. However, some jobs may be delayed when the formal processing begins. Bean et al. [2] proposed the rescheduling problem where the objective is to reschedule the jobs to reduce the negative impact of the delay. This problem has been found a lot of applications in many areas such as the automobile industry, the aviation industry, the medical industry and so on [16,19,22,24].

Due to different applications, the specific definition of the rescheduling problem has some differences. Hall and Potts [5] considered the rescheduling problem

Q. Ni and W. Wu (Eds.): AAIM 2022, LNCS 13513, pp. 35–45, 2022.
https://doi.org/10.1007/978-3-031-16081-3_4

on a single machine, where any job can be allowed disruption and the objective is to minimize the total weighted completion time. They proposed an approximation algorithm, and a fully polynomial time approximation scheme (FPTAS). Liu and Ro [10] considered the rescheduling problem on single machine with machine unavailability, where the objective is to minimize the makespan under a maximum allowable time deviation. They provided a pseudo-polynomial time algorithm, a constant factor approximation algorithm, and an FPTAS. Luo et al. [13] considered the rescheduling problem on single machine, where the objective is to find a rescheduling such that the total weighted completion time is minimized. They presented a pseudo-polynomial time algorithm and an FPTAS. More related work can be found in [11,21].

In the real life, the manufacturer can handle a part of delayed jobs by outsourcing, but the manufacturer needs to pay extra. Thus, the rescheduling with rejection is considered, where the penalty of the rejected jobs is regarded as the outsourcing costs. Wang et al. [17] considered the rescheduling on multiple identical parallel machines and rejection, where the objective is to find a rescheduling such that the total cost, which consists of the total completion time of the accepted jobs, the total rejection cost and the maximum time deviation, is minimized. They presented a pseudo-polynomial time algorithm for the number of machines is fixed. Li et al. [8] studied constrained penalty scheduling problem in which the total penalty of the rejected jobs is no more than a given bound.

From the perspective of fairness, the maximum tardiness is an objective that should not be ignored for the rescheduling problem. Yu et al. [23] considered the rescheduling on single machine, where the objective is to find a rescheduling such that the maximum tardiness is bounded by a given threshold and the total cost which consists of the total weighted completion time, the total rejection cost and the maximum tardiness, is minimized. They presented a pseudo-polynomial time algorithm, and an FPTAS. Luo et al. [12] considered a generalization of this problem, in which the rescheduling is required that the total rejection cost is also no more than a given threshold. They presented an FPTAS by using the sparse technique.

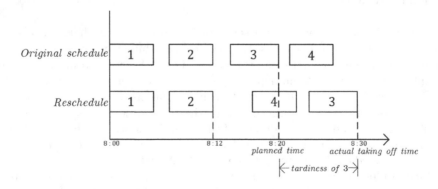

Fig. 1. Airport reschedule

Normal circumstances, the passenger flights take off from the airport according to the flight schedule timetable. However, due to weather and other reasons, some previous flight did not arrive on time, causing the subsequent flight to be delayed. The airport control center needs to reschedule the flight schedule timetable to reduce negative impact, where the negative impact is mainly caused by the tardiness which is difference between the actual taking off time and the original planned time, see Fig. 1. Motivated by the studies in [12,23], we generalize the first part of the objective in [12] to the total weighted tardiness time, and consider the single-machine rescheduling with job delay and rejection, where the objective is to find a rescheduling such that the maximum tardiness is no more than a given threshold and the total cost, which consists of the total weighted tardiness time, the total rejection cost and the maximum tardiness, is minimized.

The remainder of the paper is organized as follows. In Sect. 2, we describe the definition of the single-machine rescheduling with job delay and rejection and some important properties. Section 3, we present a dynamic programming algorithm for this problem. Section 4, we modify the dynamic programming algorithm in Sect. 3 and present an FPTAS. We conclude the paper and suggest some possible future research in the last Section.

2   Preliminaries

We are given a job set, $J = \{1, 2, \cdots, n\}$, and a single machine. Each job $j \in J$ has a processing time p_j and a weight w_j, where p_j and w_j are positive integers. For convenience, we sort the jobs in J such that

$$\frac{p_1}{w_1} \le \frac{p_2}{w_2} \le \cdots \le \frac{p_n}{w_n}.$$

Let π^* be a scheduling, denoted by original scheduling, based on the shortest weighted processing time (SWPT) [15], if all jobs are available at time zero, where

SWPT schedules the jobs in order of nondecreasing $\dfrac{p_j}{w_j}$.

Each job j has a due date

$$d_j = C_j(\pi^*),$$

where $C_j(\pi^*)$ is the completion time of job j in π^*. When the formal processing starts, a delayed job set $D \subseteq J$ and a delayed threshold k are given, where each job j in D has a rejected cost e_j and is available at time r. Only jobs in D can be rejected, however, we need to pay e_j if job j is rejected.

The single-machine rescheduling with job delay and rejection is to find a rescheduling $(\sigma; A, R)$ such that

$$\max_{j:j \in A} T_j \le k,$$

where

$$T_j = \max\{C_j - d_j, 0\}$$

is the tardiness of job j, C_j is the completion time of job j in σ, A is the subset of jobs accepted of $(\sigma; A, R)$, R is the subset of jobs rejected of $(\sigma; A, R)$, and σ is the rescheduling for J. The objective is to minimize the sum of the following three components: the total weighted tardiness time, the total rejection cost, and the penalty on the maximum tardiness for the accepted jobs, $i.e.$,

$$\min(\sum_{j:j\in A} w_j T_j + \sum_{j:j\in R} e_j + T_{\max}),$$

where

$$T_{\max} = \max_{j:j\in A} T_j$$

is the maximum tardiness in σ. By using the general notation for scheduling problems, the problem is denoted by

$$1|\text{rej}, r, T_{\max} \leq k| \sum_{j\in A} w_j T_j + \sum_{j\in R} e_j + T_{\max}. \tag{1}$$

Lemma 1. *When $\sum_{j=1}^{j_{\min}-1} p_j \geq r$, $(\pi^*, J, \emptyset)$ is an optimal solution of Problem (1), where r is the available time of each delay job, and $j_{\min} = \min\{j|j \in D\}$.*

Proof. When $\sum_{j=1}^{j_{\min}-1} p_j \geq r$, all jobs in J can be processed based on the original scheduling π^*, and the maximum tardiness is 0. The objective function value of $(\pi^*, J, \emptyset)$ is 0 and the lemma holds. □

Let (σ^*, A^*, R^*) be an optimal rescheduling for Problem (1), if all jobs in D^* are rejected, then $(\delta; J \setminus D, D)$ is an optimal rescheduling, where all jobs in $J \setminus D$ are processed by SWPT in δ. Thus, in the following paper, we assume that

$$\sum_{j=1}^{j_{\min}-1} p_j < r, \text{ and } D \cap A^* \neq \emptyset \tag{2}$$

in Problem (1).

Using the available time r and optimal rescheduling σ^*, we can partition the accepted set A^* of σ^* into two parts as follows: The *earlier* part

$$A^*_{<r} = \{j \in A^*|C_j^* - p_j < r\}$$

is a set of jobs which are started processing strictly before time point r, where C_j^* is the completion time of J by σ^*; The *later* part

$$A^*_{\geq r} = \{j \in A^*|C_j^* - p_j \geq r\} = A^* \setminus A^*_{<r}$$

is set of the remainder jobs, $i.e.$, $A^*_{\geq r} = A^* \setminus A^*_{<r}$. Similar to [23], we have the following properties:

Property 1. *If job j' is processed before job j by σ^* for $j', j \in A^*_{<r}$, then $j' < j$;*

Property 2. If job j' is processed before job j by σ^* for $j', j \in A^*_{\geq r}$, then $j' < j$. Let

$$j_* = \arg\min_j\{j | j \in A^*_{\geq r}\}$$

be the minimum job in set $A^*_{\geq r}$. Similar to [12], we have the following lemma.

Lemma 2. *The tardiness of (σ^*, A^*, R^*) is*

$$T_{\max} = \max\{C^*_{j_*} - d_{j_*}, 0\},$$

where C^*_j is the completion time of job j by σ^*.

Then, in the optimal rescheduling, job j_* may be two processed possibilities, continuous or discontinuous processed, see Fig. 2.

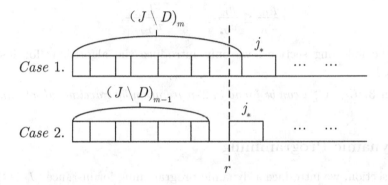

Fig. 2. Continuous or discontinuous processed

Case 1. Continuous processed, let m be the minimum integer satisfying

$$\sum_{j: j \in (J \setminus D)_m} p_j \geq r,$$

where $(J \setminus D)_m$ is the set of the first m jobs with minimum index in $J \setminus D$. Thus, $A^*_{<r} = (J \setminus D)_m$. Let

$$D_m = \{j \in D | j < j_*\},$$

and all jobs in D_m should be rejected by Property 2, and only jobs in $J_{j_*} = J \setminus ((J \setminus D)_m \cup D_m \cup \{j_*\})$ need to be considered. Furthermore, the completion processing time of job j_* is

$$C^*_{j_*} = \sum_{j: j \in (J \setminus D)_m} p_j + p_{j_*}.$$

Case 2: Discontinuous processed, let $m-1$ be the maximum integer satisfying

$$\sum_{j:j\in(J\backslash D)_{m-1}} p_j < r.$$

Only jobs in $J_{j_*} = J \backslash ((J \backslash D)_{m-1} \cup D_m \cup \{j_*\})$ need to be considered, where $(J \backslash D)_{m-1}$ is the set of the first $m-1$ jobs with minimum index in $J \backslash D$ and $D_m = \{j \in D | j < j_*\}$. Furthermore, the completion processing time of job j_* is

$$C_{j_*}^* = r + p_{j_*}.$$

In order to describe more clearly, we need a preprocessing step: guessing the first job j_* in $A_{\geq r}^*$ and its processing case. Let $C_{j_*}^*$ be the completion processing time of job j_*, and let J_{j_*} be the set of only jobs needed to be considered, where we resort the jobs in J_{j_*} such that

$$\frac{p_{j_1}}{w_{j_1}} \leq \frac{p_{j_2}}{w_{j_2}} \leq \cdots \leq \frac{p_{j_{|J_{j_*}|}}}{w_{j_{|J_{j_*}|}}}.$$

In the following sections, we only introduce the algorithms for instance $(J_{j_*}, C_{j_*}^*)$.

Lemma 3. $(J_{j_*}, C_{j_*}^*)$ can be found in $2 \cdot n$ iteration by traveling all jobs and two cases.

3   Dynamic Programming

In this section, we introduce a dynamic programming for instance $(J_{j_*}, C_{j_*}^*)$ by modifying the algorithm [23].

We define

$$\begin{cases} j_0 = j_*, \\ C_{j_0} = C_{j_*}^*, \\ E_{j_0} = \sum_{j:j\in D\backslash J_{j_*}} e_j, \end{cases}$$

and use $F(j_l; C, E)$ to denote the minimum total weighted tardiness time, when the completion time of job j_l is C and the total rejected cost is E among $\{j_1, j_2, \ldots, j_l\}$.

Dynamic programming algorithm (DP1) is described as follows:
Initially,

$$F(j_0; C, E) = \begin{cases} \max\{w_{j_0}(C_{j_0} - d_{j_0}), 0\}, \text{ if } C = C_{j_0}, \text{ and } E = E_{j_0} \\ \infty, \text{ otherwise.} \end{cases}$$

For $i = 1, 2, \ldots, |J_{j_*}|$;

$$F(j_{i+1}; C, E)$$
$$= \min\{F(j_i; C - p_{j_{i+1}}, E) + \max\{w_{j_{i+1}}(C - d_{j_{i+1}}), 0\}, F(j_i; C, E - e_{j_{i+1}})\}.$$

Let

$$OUT = \min\{F(j_{|J_{j_*}|}; C, E) + E + \max\{C_{j_*} - d_{j_*}, 0\}|$$
$$0 \le C \le n \cdot p_{\max}, \text{ and } 0 \le E \le n \cdot e_{\max}\},$$

where $p_{\max} = \max_{j:j\in J} p_j$ and $e_{\max} = \max_{j:j\in J} e_j$. Thus, the optimal value is OUT.

Lemma 4. *OUT can be found in time $O(p_{\max}\cdot e_{\max}\cdot n^3)$, where $p_{\max} = \max_{j:j\in J} p_j$ and $e_{\max} = \max_{j:j\in J} e_j$.*

Proof. Since both p_j and e_j are integers for any job $j \in J$, then C is an element of the set $\{0, 1, \ldots, n \cdot p_{\max}\}$ and E is an element of the set $\{0, 1, \ldots, n \cdot e_{\max}\}$ at every iteration. Thus, we can compute the value OUT in time $O(p_{\max} \cdot e_{\max} \cdot n^3)$. □

Combining Lemma 3 and 4, we have the following theorem.

Theorem 1. *Problem (1) can be solved in time $O(p_{\max} \cdot e_{\max} \cdot n^4)$.*

4 A Fully Polynomial Time Approximation Scheme

In this section, we present a fully polynomial time approximation scheme (FPTAS) for instance $(J_{j_*}, C_{j_*}^*)$, which is based on a modified DP1 mentioned in Sect. 3 and a rounding technique in [14,18].

We define

$$\begin{cases} j_0 = j_*, \\ C_{j_0} = C_{j_*}^*, \\ E_{j_0} = \sum_{j:j\in D\backslash J_{j_*}} e_j, \\ F_{j_0} = \max\{w_{j_*} C_{j_*}^*, 0\}. \end{cases}$$

We use a 4-dimensional vector (j_l, C, E, F) to denote a station to represent a state, which is true if and only if there is a feasible rescheduling on the job subset $\{j_1, j_2 \ldots, j_l\}$ such that

$$\begin{cases} \text{the completion time is exactly } C, \\ \text{the total rejected cost is exactly } E, \\ \text{the minimum total weighted tardiness time is exactly } F. \end{cases}$$

Dynamic programming algorithm (DP2) is described as follows:
Initially,

$$s(j_0, C, E, F) = \begin{cases} 1, \text{ if } C = C_{j_0}, E = E_{j_0} \text{ and } F = F_{j_0}, \\ 0, \text{ otherwise.} \end{cases}$$

For $i = 1, 2, \ldots, |J_{j_*}|$;

$s(j_{i+1}, C, E, F)$
$= \max\{s(j_i, C - p_{j_{j+1}}, E, F - w_{j_{j+1}}(\max\{C - d_{j_{j+1}}, 0\})), s(j_i, C, E - e_{j_{j+1}}, F)\}.$

At the end, we check all the true states of the form $(j_{|J_{j_*}|}, C, E, F)$, i.e., $s(j_{|J_{j_*}|}, C, E, F) = 1$, and calculate its objective value

$$OUT_{(j_{|J_{j_*}|}, C, E, F)} = F + E + T_{max},$$

where $T_{max} = \max\{C_{j_*}^* - d_{j_*}, 0\}$. Thus, the optimal value is equal to the minimum objective value among all true states, i.e.,

$$OUT = \min_{s(j_{|J_{j_*}|}, C, E, F)=1} \{OUT_{(j_{|J_{j_*}|}, C, E, F)}\}.$$

Lemma 5. *OUT can be found by DP2 in time* $O(p_{max}^2 \cdot e_{max} \cdot w_{max} \cdot n^4)$, *where* $p_{max} = \max_{j:j \in J} p_j$, $e_{max} = \max_{j:j \in J} e_j$ *and* $w_{max} = \max_{j:j \in J} w_j$.

Proof. Since both p_j and w_j are integers for any job $j \in J$, then F is an element of the set $\{0, 1, \ldots, n \cdot p_{max} \cdot w_{max}\}$ at every iteration. By Lemma 4, we can compute the value OUT in time $O(p_{max}^2 \cdot e_{max} \cdot w_{max} \cdot n^4)$.   □

Given any parameter $\varepsilon > 0$, let $\delta = (1 + \frac{\varepsilon}{2n})$, let $X = \lceil \log_\delta \frac{n \cdot p_{max}}{C_{j_0}} \rceil$, then we partition the interval $[C_{j_0}, n \cdot p_{max}]$ into X intervals as follows:

$$\begin{cases} \text{the 1 interval is } (C_{j_0}, \delta C_{j_0}], \\ \text{the 2 interval is } (\delta C_{j_0}, \delta^2 C_{j_0}], \\ \cdots \\ \text{the } X - 1 \text{ interval is } (\delta^{X-2} C_{j_0}, \delta^{X-1} C_{j_0}], \\ \text{the } X \text{ interval is } (\delta^{X-1} C_{j_0}, n \cdot p_{max}]. \end{cases}$$

Let $Y = \lceil \log_\delta \frac{n \cdot e_{max}}{E_{j_0}} \rceil$, then we partition the interval $[E_{j_0}, n \cdot e_{max}]$ into Y intervals as follows:

$$\begin{cases} \text{the 1 interval is } (E_{j_0}, \delta E_{j_0}], \\ \text{the 2 interval is } (\delta E_{j_0}, \delta^2 E_{j_0}], \\ \cdots \\ \text{the } Y - 1 \text{ interval is } (\delta^{Y-2} E_{j_0}, \delta^{Y-1} E_{j_0}], \\ \text{the } Y \text{ interval is } (\delta^{Y-1} E_{j_0}, n \cdot e_{max}]. \end{cases}$$

Let $Z = \lceil \log_\delta \frac{n \cdot w_{max} \cdot p_{max}}{F_{j_0}} \rceil$, then we partition the interval $[F_{j_0}, n \cdot w_{max} \cdot p_{max}]$ into Z intervals as follows:

$$\begin{cases} \text{the 1 interval is } (F_{j_0}, \delta F_{j_0}], \\ \text{the 2 interval is } (\delta F_{j_0}, \delta^2 F_{j_0}], \\ \cdots \\ \text{the } Z - 1 \text{ interval is } (\delta^{Z-2} F_{j_0}, \delta^{Z-1} F_{j_0}], \\ \text{the } Z \text{ interval is } (\delta^{Z-1} F_{j_0}, n \cdot w_{max} p_{max}]. \end{cases}$$

For any 4-dimensional vector (j_l, C, E, F) in DP2, we can find three positive integers x, y and z satisfying that $C \in (\delta^{x-1} C_{j_0}, \delta^x C_{j_0}]$, $E \in (\delta^{y-1} E_{j_0}, \delta^y E_{j_0}]$ and $F \in (\delta^{z-1} F_{j_0}, \delta^z F_{j_0}]$, and we define this 4-dimensional vector (j_l, C, E, F) fell in the state of (l, x, y, z). If there are several 4-dimensional vectors fell in same state (l, x, y, z), our FPTAS algorithm only stores the 4-dimensional vector with the minimum second component, and let $C(l, x, y, z)$ be the value of the second component of the 4-dimensional vector store in state (l, x, y, z). We propose the detailed FPTAS algorithm A_ε in Algorithm.

Algorithm 1: A_ε

Input: An instance $(J_{j_*}, C_{j_*}^*)$.
Output: A objective value OUT_ε.
1 **Initially**,

$$(l, x, y, z) = \begin{cases} (j_0, C_{j_0}, E_{j_0}, F_{j_0}), & \text{if } x = y = z = 0, \\ (j_0, \infty, \infty, \infty), & \text{otherwise}. \end{cases}$$

2 **for** $l = 1$ **to** $|J_{j_*}|$ **do**
3   **for** (l, x, y, z) *with integer* $x \in [1, X]$, $y \in [1, Y]$ *and* $z \in [1, Z]$ **do**
4   Let (j_l, C, E, F) be the 4-dimensional vector stored in this state, and construct two 4-dimensional vectors

$$\begin{cases} v_1 = (j_{l+1}, C + p_{j_{l+1}}, E, F + \max\{w_{j_{l+1}}(C + p_{j_{l+1}} - d_{j_{l+1}}), 0\}), \\ v_2 = (j_{l+1}; C, E + e_{j_{l+1}}, F). \end{cases}$$

5   Let (l, x_1, y_1, z_1) be the state contained the vector v_1, and let (l, x_2, y_2, z_2) be the state contained the vector v_2
6   **if** $C(l, x_1, y_1, z_1) > C(v_1)$ **then**
7   let $(l, x_1, y_1, z_1) = v_1$.
8   **if** $C(l, x_2, y_2, z_2) > C(v_2)$ **then**
9   let $(l, x_2, y_2, z_2) = v_2$.

10 Check each the state $(|J_{j_*}|, x, y, z)$ and calculate its objective value of the 4-dimensional vector $(j_{|J_{j_*}|}, C, E, F)$ as $F + E + \max\{C_{j_0} - d_{j_0}, 0\}$, and output OUT_ε which is the minimum objective value.

By mathematical induction, we have the following lemma.

Lemma 6. *For any* $l \in \{1, 2, \ldots, |J_{j_*}|\}$ *and any true state* (j_*, j_l, C, E, F) *in DP2, there exists a state* (l, x, y, z) *in Algorithm 1, and its stored vector* (j_l, C', E', F') *satisfying* $C' \leq C$, $E' \leq \delta^l E$ *and* $F' \leq \delta^l F$.

By Lemma 6, there exists a state $(|J_{j_*}|, x, y, z)$, and its stored vector (j_q, C', E', F') satisfying

$$\begin{cases} C' \leq C^*; \\ E' \leq \delta^l E^* \\ F' \leq \delta^l F^*. \end{cases} \tag{3}$$

Thus, the following lemma is obvious.

Lemma 7. *Algorithm 1 is an FPTAS for instance $(J_{j_*}, C^*_{j_*})$ and its time complexity is $O(\frac{n^4}{\varepsilon^3} \log(np_{\max}) \log(ne_{\max}) \log(nw_{\max}p_{\max}))$.*

Combining Lemma 3 and 7, we have the following theorem.

Theorem 2. *Problem (1) possesses an FPTAS with running time*

$$O(\frac{n^5}{\varepsilon^3} \log(np_{\max}) \log(ne_{\max}) \log(nw_{\max}p_{\max})),$$

where $p_{\max} = \max_{j:j \in J} p_j$, $e_{\max} = \max_{j:j \in J} e_j$ and $w_{\max} = \max_{j:j \in J} w_j$.

5   Conclusion

In this paper, we study the single-machine rescheduling with job delay and rejection, and design a pseudo-polynomial time dynamic programming algorithm. Furthermore, we present a fully polynomial time approximation scheme.

It would be interesting to design a better algorithm for this problem. Recently, problems with submodular penalties become a research hotpot, in which the penalty is determined by a submodular function. The generalization of this problem to submodular penalties is worth considering. In [9], each job is characterized by a d-dimension vector, and the generalization of this problem to vector is also worth considering.

References

1. Alon, N., Azar, Y., Woeginger, G.J., Yadid, T.: Approximation schemes for scheduling on parallel machines. J. Sched. **1**(1), 55–66 (1998)
2. Bean, J.C., Birge, J.R., Mittenthal, J., Noon, C.E.: Matchup scheduling with multiple resources, release dates and disruptions. Oper. Res. **39**(3), 470–483 (1991)
3. Graham, R.L.: Bounds for certain multiprocessing anomalies. Bell Syst. Tech. J. **45**(9), 1563–1581 (1966)
4. Graham, R.L.: Bounds on multiprocessing timing anomalies. SIAM J. Appl. Math. **17**(2), 416–429 (1969)
5. Hall, N.G., Potts, C.N.: Rescheduling for job unavailability. Oper. Res. **58**(3), 746–755 (2010)
6. Kones, I., Levin, A.: A unified framework for designing EPTAS for load balancing on parallel machines. Algorithmica **81**(7), 3025–3046 (2019)

7. Li, W., Li, J., Zhang, T.: Two approximation schemes for scheduling on parallel machines under a grade of service provision. Asia-Pacific J. Oper. Res. **29**(05), 1250029 (2012)
8. Li, W., Li, J., Zhang, X., Chen, Z.: Penalty cost constrained identical parallel machine scheduling problem. Theor. Comput. Sci. **607**, 181–192 (2015)
9. Li, W., Cui, Q.: Vector scheduling with rejection on a single machine. 4OR **16**(1), 95–104 (2018)
10. Liu, Z., Ro, Y.K.: Rescheduling for machine disruption to minimize makespan and maximum lateness. J. Sched. **17**(4), 339–352 (2014)
11. Liu, Z., Lu, L., Qi, X.: Cost allocation in rescheduling with machine unavailable period. Eur. J. Oper. Res. **266**(1), 16–28 (2018)
12. Luo, W., Jin, M., Su, B., Lin, G.: An approximation scheme for rejection-allowed single-machine rescheduling. Comput. Ind. Eng. **146**, 106574 (2020)
13. Luo, W., Luo, T., Goebel, R., Lin, G.: Rescheduling due to machine disruption to minimize the total weighted completion time. J. Sched. **21**(5), 565–578 (2018)
14. Schuurman, P., Woeginger, G.J.: Approximation schemes - a tutorial. In: Wolsey, L.A., et al. (eds.) Lectures on Scheduling (2009)
15. Smith, W.E.: Various optimizers for single-stage production. Nav. Res. Logist. Q. **3**, 59–66 (1956)
16. Thomson, S., Nunez, M., Garfinkel, R., Dean, M.D.: Efficient short term allocation and reallocation of patients to floor of a hospital during demand surges. Oper. Res. **57**(2), 261–273 (2009)
17. Wang, D., Yin, Y., Cheng, T.C.E.: Parallel-machine rescheduling with job unavailability and rejection. Omega **81**, 246–260 (2018)
18. Woeginger, G.J.: When does a dynamic programming formulation guarantee the existence of an FPTAS? INFORMS J. Comput. **12**(1), 57–74 (2000)
19. Wu, S.D., Storer, R.H., Chang, P.C.: A rescheduling procedure for manufacturing systems under random disruptions. In: Proceedings Joint USA/German Conference on New Directions for Operations Research in Manufacturing, pp. 292–306 (1991)
20. Xiao, M., Ding, L., Zhao, S., Li, W.: Semi-online algorithms for hierarchical scheduling on three parallel machines with a buffer size of 1. In: He, K., Zhong, C., Cai, Z., Yin, Y. (eds.) NCTCS 2020. CCIS, vol. 1352, pp. 47–56. Springer, Singapore (2021). https://doi.org/10.1007/978-981-16-1877-2_4
21. Yin, Y., Cheng, T.C.E., Wang, D.J.: Rescheduling on identical parallel machines with machine disruptions to minimize total completion time. Eur. J. Oper. Res. **252**(3), 737–749 (2016)
22. Yu, G., Argüello, M., Song, G., McCowan, S.M., White, A.: A new era for crew recovery at Continental Airlines. Interfaces **33**(1), 5–22 (2003)
23. Yu, S., Jin, M., Luo, W.: Approximation scheme for rescheduling on a single machine with job delay and rejection. OR Trans. **25**(2), 104–114 (2021)
24. Zweben, M., Davis, E., Daun, B., Deale, M.: Scheduling and rescheduling with iterative repair. IEEE Trans. Syst. Man Cybern. **23**(6), 1588–1596 (1993)

Online Early Work Maximization Problem on Two Hierarchical Machines with Buffer or Rearrangements

Man Xiao, Xihua Bai, and Weidong Li$^{(\boxtimes)}$

School of Mathematics and Statistics, Yunnan University, Kunming 650504,
People's Republic of China
weidongmath@126.com

Abstract. In this paper, we consider two semi-online models of online early work maximization problem on two hierarchical machines. When a buffer size of K is available, we propose an optimal online algorithm with a competitive ratio of $\frac{4}{3}$. If we are allowed to reassign at most K jobs after all the jobs have been scheduled, we propose an optimal online algorithm with a competitive ratio of $\frac{4}{3}$.

Keywords: Semi-online · Early work · Hierarchy · Competitive ratio · Two machines

1 Introduction

We are given two hierarchical machines M_1, M_2 and a job set $\mathcal{J} = \{J_1, J_2, \cdots, J_n\}$, where jobs arrive one by one in a list. A new job arrives only when the current job is scheduled. Machine M_1 can process all jobs, while machine M_2 can only process part of the jobs. Each job can only be processed by one machine. For an online maximization (minimization) problem, the competitive ratio of online algorithm A is defined as the minimum value of ρ satisfying $C^{OPT}(I) \leq \rho C^A(I)$ ($C^A(I) \leq \rho C^{OPT}(I)$) for any instance I, where $C^A(I)$ (C^A, for short) denotes the output value of online algorithm A, and $C^{OPT}(I)$ (C^{OPT}, for short) denotes the off-line optimal value. If there is no online algorithm with a competitive ratio strictly less than ρ, then ρ is called a lower bound of the problem. If there is an algorithm whose competitive ratio is equal to ρ, this algorithm is called the optimal online algorithm.

A typical objective is to minimize the maximum machine completion time, which is denoted by $P2|GoS|C_{max}$. Park et al. [11] and Jiang et al. [7] independently gave an optimal online algorithm with a competitive ratio of $\frac{5}{3}$. Park et al. [11] also considered a semi-online version with known total job processing time, and proposed an optimal algorithm with a competitive ratio of $\frac{3}{2}$. Zhang et al. [18] studied the semi-online model with bounded processing times and gave an optimal online algorithm. Wu et al. [14] considered two semi-online versions where the off-line optimal value or the largest processing time of all jobs is known, and gave two optimal online algorithms with competitive ratios of $\frac{3}{2}$ and $\frac{\sqrt{5}+1}{2}$

Q. Ni and W. Wu (Eds.): AAIM 2022, LNCS 13513, pp. 46–54, 2022.
https://doi.org/10.1007/978-3-031-16081-3_5

respectively. Chen et al. [2] studied several semi-online versions where the total processing time of low-hierarchy jobs is known, and gave several optimal online algorithms. Chen et al. [4] considered two semi-online problems with buffer or rearrangements, and designed two optimal algorithms with a competitive ratio of $\frac{3}{2}$. Dai et al. [5] studied semi-online hierarchical scheduling for bag-of-tasks on two machines. Xiao et al. [16] considered two semi-online models with a buffer of size 1 on three hierarchical machines. Li et al. [8] proposed an efficient polynomial time approximation schemes for scheduling on m hierarchical machines with two hierarchies.

Another typical objective is to maximize the minimum machine completion time, which is denoted by $P2|GoS|C_{min}$. Chassid and Epstein [1] proved that the competitive ratio of any online algorithm is infinite. Epstein et al. [6] designed an optimal online algorithm for the restricted assignment version with a buffer size of 1 on two machines, whose competitive ratio is 2. Wu et al. [13] considered several semi-online versions on two hierarchical machines. Luo and Xu [10] studied a semi-online version with bounded processing times on two hierarchical machines, and designed an optimal online algorithm. Wu and Li [12] considered a semi-online model where the processing times are in $\{1, 2, 2^2, ..., 2^k\}$ with $k \geq 2$, and gave an optimal online algorithm with a competitive ratio of 2^k. Xiao et al. [17] gave several optimal online algorithms for semi-online versions with known total processing time of low-hierarchy jobs.

For the objective of maximizing the total early work on two identical machines, denoted by $P2|GoS|X$, Xiao et al. [15] proposed three optimal online algorithms when the total size of low-hierarchy jobs, the total size of high-hierarchy jobs, the total size of low-hierarchy and high-hierarchy jobs are known in advance, respectively. In addition, Chen et al. [3] designed several offline and online algorithms and Li [9] designed several improved approximation schemes for early work scheduling on identical parallel machines with a common due date.

Motivated by [4,15], we consider the early work maximization problem on two hierarchical machines with buffer or rearrangements. The rest of the paper is organized as follows. In Sect. 2, we give problem statement and symbol description. In Sect. 3, we consider the buffer model and propose an optimal online algorithm. In Sect. 4, we consider the rearrangement model and propose an optimal online algorithm. Finally, we make a summary.

2 Preliminaries

We are given two hierarchical machines M_1 and M_2, and a series of jobs arriving online which are to be scheduled irrevocably at the time of their arrivals. The arrival of a new job occurs only after the current job is scheduled. Let $\mathcal{J} = \{J_1, J_2, ..., J_n\}$ be the set of all jobs arranged in the order of arrival time. We denote the j-th job as $J_j = (p_j, g_j)$, where p_j is processing time (also called size) of the job J_j, and $g_j \in \{1, 2\}$ is the hierarchy of the job J_j. If $g_j = k$, we call J_j as a job of hierarchy k, $k \in \{1, 2\}$. M_1 can process all jobs, and M_2 can only process the jobs of hierarchy 2.

As in [3], we assume that each job has a common due date $d > 0$, and

$$p_j \leq d, \text{ for } j = 1, 2, \ldots, n.$$

The early work of job J_j is denoted by $X_j \in [0, p_j]$. If job J_j is completed before the due date d, the job is called **totally early** and $X_j = p_j$. If job J_j starts at the time of $s_j < d$, but finishes after the due date d, the job is called **partially early** and $X_j = d - s_j$. If job J_j starts at the time of $s_j \geq d$, the job is called **totally late** and $X_j = 0$.

A feasible schedule is actually a partition (S_1, S_2) of the job set $\mathcal{J}$, such that $S_1 \cup S_2 = \mathcal{J}$ and $S_1 \cap S_2 = \emptyset$. Let $L_i = \sum_{J_j \in S_i} p_j$ be the load of M_i, $i \in \{1, 2\}$. The objective is to find a schedule such that total early work

$$X = \sum_{j=1}^{n} X_j = \sum_{i=1}^{2} \min \{L_i, d\}$$

is maximized. Let T be the total size of the jobs in $\mathcal{J}$, and L_i^j be the load of M_i after job J_j is assigned to a machine or stored in buffer.

From the above definitions, we have

Lemma 1. The optimal objective value C^{OPT} is at most $\min \{2d, T\}$.

3   Buffer Model

In this section, we have a buffer of size K, where the buffer can temporarily store at most K jobs. When the buffer is full, a new job can be stored in the buffer only when an earlier job is removed from the buffer and assigned to some machine. Denote this problem as $P2|GoS, online, d_j = d, buffer|max(X)$. We give a lower bound $\frac{4}{3}$ for any constant K, and propose an online algorithm with a competitive ratio of $\frac{4}{3}$ for $K = 1$.

Theorem 2. Any online algorithm A for $P2|GoS, online, d_j = d, buffer| max(X)$ has a competitive ratio at least $\frac{4}{3}$.

Proof. Let $d = 1$ and $\varepsilon = \frac{1}{2N}$, where N is a sufficiently large integer. The first N identical jobs with $(\varepsilon, 2)$ arrive. Since the buffer can only store at most K jobs, at least $N - K$ jobs are allocated to M_1 or M_2.

Case 1. $L_2^N \geq \frac{1}{2} - K\varepsilon$ or $L_1^N \geq \frac{1}{2} - K\varepsilon$.

If $L_2^N \geq \frac{1}{2} - K\varepsilon$, the last two jobs $J_{N+1} = (\frac{1}{2}, 1)$ and $J_{N+2} = (1, 2)$ arrive. Since there are at most K jobs with $(\varepsilon, 2)$ in the buffer, we have $C^A \leq \frac{3}{2} + K\varepsilon$ and $C^{OPT} = 2$. Thus,

$$\lim_{\varepsilon \to 0} \frac{C^{OPT}}{C^A} \geq \lim_{\varepsilon \to 0} \frac{2}{\frac{3}{2} + K\varepsilon} = \frac{4}{3}.$$

If $L_1^N \geq \frac{1}{2} - K\varepsilon$, the last job $J_{N+1} = (1, 1)$ arrives. Since there are at most K jobs $(\varepsilon, 2)$ in the buffer, we have $C^A \leq 1 + K\varepsilon$ and $C^{OPT} = \frac{3}{2}$, implying that

$$\lim_{\varepsilon \to 0} \frac{C^{OPT}}{C^A} \geq \lim_{\varepsilon \to 0} \frac{\frac{3}{2}}{1 + K\varepsilon} = \frac{3}{2}.$$

Case 2. $L_2^N < \frac{1}{2} - K\varepsilon$ and $L_1^N < \frac{1}{2} - K\varepsilon$.

The next $t - N$ identical jobs with $(\varepsilon, 2)$ arrive one by one until that $L_2^t \in [\frac{1}{2} - K\varepsilon, \frac{1}{2}]$ or $L_1^t \in [\frac{1}{2} - K\varepsilon, \frac{1}{2}]$, where $t \geq N$ is the minimal integer satisfying the condition.

If $L_2^t \in [\frac{1}{2} - K\varepsilon, \frac{1}{2}]$ and $L_1^t < \frac{1}{2} - K\varepsilon$, the last two jobs $J_{t+1} = (\alpha, 1)$ and $J_{t+2} = (1, 2)$ arrive, where $\alpha = \frac{1}{2} - L_1^t$. Since there are at most K jobs with $(\varepsilon, 2)$ in the buffer, we have $C^A \leq \frac{3}{2} + K\varepsilon$ and $C^{OPT} \geq 2 - K\varepsilon$. Thus,

$$\lim_{\varepsilon \to 0} \frac{C^{OPT}}{C^A} \geq \lim_{\varepsilon \to 0} \frac{2 - K\varepsilon}{\frac{3}{2} + K\varepsilon} = \frac{4}{3}.$$

If $L_2^t \in [\frac{1}{2} - K\varepsilon, \frac{1}{2}]$ and $L_1^t \in [\frac{1}{2} - K\varepsilon, \frac{1}{2}]$, the last job $J_{t+1} = (1, 1)$ arrives. Since there are at most K jobs with $(\varepsilon, 2)$ in the buffer, we have $C^A \leq \frac{3}{2} + K\varepsilon$ and $C^{OPT} \geq 2 - 2K\varepsilon$. Thus,

$$\lim_{\varepsilon \to 0} \frac{C^{OPT}}{C^A} \geq \lim_{\varepsilon \to 0} \frac{2 - 2K\varepsilon}{\frac{3}{2} + K\varepsilon} = \frac{4}{3}.$$

If $L_2^t < \frac{1}{2} - K\varepsilon$ and $L_1^t \in [\frac{1}{2} - K\varepsilon, \frac{1}{2}]$, the last job $J_{t+1} = (1, 1)$ arrives. Since there are at most K jobs with $(\varepsilon, 2)$ in the buffer, we have $C^A \leq 1 + L_2^t + K\varepsilon$ and $C^{OPT} \geq \frac{3}{2} - K\varepsilon + L_2^t$. Thus,

$$\lim_{\varepsilon \to 0} \frac{C^{OPT}}{C^A} \geq \lim_{\varepsilon \to 0} \frac{\frac{3}{2} - K\varepsilon + L_2^t}{1 + L_2^t + K\varepsilon} = \lim_{\varepsilon \to 0} 1 + \frac{\frac{1}{2} - 2K\varepsilon}{1 + L_2^t + K\varepsilon} \geq \lim_{\varepsilon \to 0} 1 + \frac{\frac{1}{2} - 2K\varepsilon}{\frac{3}{2}} = \frac{4}{3}.$$

∎

Our online algorithm for $P2|GoS, online, d_j = d, buffer|max(X)$ with a buffer of size 1 is described as follows, where we always put the current largest job of hierarchy 2 into the buffer.

Algorithm 1:

---

1   Initially, let $L_2^0 = 0$ and $b_0 = 0$.
2   When a new job J_j arrives,
3   **if** $g_j = 1$ **then**
4     $\lfloor$ Assign job J_j to machine M_1.
5   **else**
6   Compare job J_j with job $B_j = (b_j, 2)$ which is in the buffer. Put the bigger one into buffer, and the other one is also denoted by J_j for convenience.
7   **if** $L_2^{j-1} < \frac{d}{2}$ **then**
8       $\lfloor$ Assign job J_j to machine M_2
9   **else**
10      $\lfloor$ Assign J_j to M_1.
11   If there is no job, assign job B_j to the least loaded machine, and stop.

Theorem 3. The competitive ratio of **Algorithm** 1 is at most $\frac{4}{3}$.

Proof. Based on Lemma 1, if $L_1 \geq d$ and $L_2 \geq d$, we have $C^{A1} = 2d \geq C^{OPT}$. If $L_1 \leq d$ and $L_2 \leq d$, we have $C^{A1} = T \geq C^{OPT}$. It implies that we only need to consider the following two cases.

Case 1. $L_1 > d$ and $L_2 < d$. In this case, we have $C^{A1} = d + L_2$. If there is no job of hierarchy 2 assigned to M_1, then Algorithm 1 reaches the optimal. Else, let J_l be the last job of hierarchy 2 assigned to M_1. If J_l is assigned to M_1 at line 10, according to the choice of Algorithm 1, we have $L_2 \geq L_2^{l-1} \geq \frac{d}{2}$. Thus, by Lemma 1, we have

$$\frac{C^{OPT}}{C^{A1}} \leq \frac{2d}{d + L_2} \leq \frac{2d}{d + \frac{d}{2}} = \frac{4}{3}.$$

Else, J_l is assigned to M_1 at line 11. According to the choice of Algorithm 1, we have $L_2 \geq L_1 - p_l$. If $L_2 \geq \frac{d}{2}$, as above, we have $\frac{C^{OPT}}{C^{A1}} \leq \frac{4}{3}$. If $L_2 < \frac{d}{2}$, then $L_1 - p_l \leq L_2 < \frac{d}{2}$. By Lemma 1, we have

$$\frac{C^{OPT}}{C^{A1}} \leq \frac{T}{d + L_2} = \frac{L_1 + L_2}{d + L_2} = 1 + \frac{L_1 - d}{d + L_2} \leq 1 + \frac{L_1 - p_l}{d + L_2} \leq 1 + \frac{L_2}{d + L_2} \leq \frac{4}{3}.$$

Case 2. $L_1 < d$ and $L_2 > d$.

In this case, $C^{A1} = L_1 + d$. Let J_t be the last job of hierarchy 2 assigned to M_2. If J_t is assigned to M_2 at line 8, according to the choice of Algorithm 1, job B^n (the last job in the buffer) is assigned to M_1, and $L_2 = L_2^{t-1} + p_t < \frac{d}{2} + p_t$. Since $L_2 > d$, we have $p_t > \frac{d}{2}$ implying that $L_1 \geq b_n \geq p_t > \frac{d}{2}$. Thus, we have

$$\frac{C^{OPT}}{C^{A1}} \leq \frac{2d}{L_1 + d} \leq \frac{2d}{\frac{d}{2} + d} = \frac{4}{3}.$$

If J_t is assigned to M_2 at line 11, according to the choice of Algorithm 1, we have $L_2 - p_t \leq L_1$. If $L_1 \geq \frac{d}{2}$, as above, we have $\frac{C^{OPT}}{C^{A1}} \leq \frac{4}{3}$. Else, we have $L_1 < \frac{d}{2}$ and $L_2 - p_t \leq L_1 < \frac{d}{2}$. Therefore,

$$\frac{C^{OPT}}{C^{A1}} \leq \frac{T}{L_1 + d} = \frac{L_1 + L_2}{L_1 + d} = 1 + \frac{L_2 - d}{L_1 + d} \leq 1 + \frac{L_2 - p_t}{L_1 + d} \leq 1 + \frac{L_1}{L_1 + d} \leq \frac{4}{3}.$$

∎

4   Rearrangement Model

In this section, after all the jobs have been arrived and scheduled, we are allowed to rearrange at most K jobs from M_2 to M_1, or from M_1 to M_2. Denote this problem as $P2|GoS, online, d_j = d, rearrangement|max(X)$. We give a lower bound $\frac{4}{3}$ for any constant K, and propose an online algorithm with a competitive ratio of $\frac{4}{3}$ for $K = 1$.

Theorem 4. Any online algorithm A for $P2|GoS, online, d_j = d, rearrangement| max(X)$ has a competitive ratio at least $\frac{4}{3}$.

Proof. Let $d = 1$ and $\varepsilon = \frac{1}{2N}$, where N is a sufficiently large integer. The first N identical jobs with $(\varepsilon, 2)$ arrive.

Case 1. $L_2^N = \frac{1}{2}$ or $L_1^N = \frac{1}{2}$.

If $L_2^N = \frac{1}{2}$, the last two jobs $J_{N+1} = (\frac{1}{2}, 1)$ and $J_{N+2} = (1, 2)$ arrive. Since at most K jobs of hierarchy 2 are rearranged, we have $C^A \leq \frac{3}{2} + K\varepsilon$ and $C^{OPT} = 2$. Thus,

$$\lim_{\varepsilon \to 0} \frac{C^{OPT}}{C^A} \geq \lim_{\varepsilon \to 0} \frac{2}{\frac{3}{2} + K\varepsilon} = \frac{4}{3}.$$

If $L_1^N = \frac{1}{2}$, the last job $J_{N+1} = (1, 1)$ arrives. Since at most K jobs of hierarchy 2 are rearranged, we have $C^A \leq 1 + K\varepsilon$ and $C^{OPT} = \frac{3}{2}$. Hence,

$$\lim_{\varepsilon \to 0} \frac{C^{OPT}}{C^A} \geq \lim_{\varepsilon \to 0} \frac{\frac{3}{2}}{1 + K\varepsilon} = \frac{3}{2}.$$

Case 2. $L_2^N < \frac{1}{2}$ and $L_1^N < \frac{1}{2}$.

The next $t - N$ identical jobs $(\varepsilon, 2)$ arrive one by one until (1) $L_2^t = \frac{1}{2}$ and $L_1^t < \frac{1}{2}$ or (2) $L_1^t = \frac{1}{2}$ and $L_2^t < \frac{1}{2}$, where $t \geq N$ is the minimal integer satisfying the condition. If $L_2^t = \frac{1}{2}$ and $L_1^t < \frac{1}{2}$, the last two jobs $J_{t+1} = (\alpha, 1)$ and $J_{t+2} = (1, 2)$ arrive, where $\alpha = \frac{1}{2} - L_1^t$. Since at most K jobs of hierarchy 2 are rearranged, we have $C^{OPT} = 2$ and $C^A \leq \frac{3}{2} + K\varepsilon$. Therefore,

$$\lim_{\varepsilon \to 0} \frac{C^{OPT}}{C^A} \geq \lim_{\varepsilon \to 0} \frac{2}{\frac{3}{2} + K\varepsilon} = \frac{4}{3}.$$

If $L_1^t = \frac{1}{2}$ and $L_2^t < \frac{1}{2}$, the last job $J_{t+1} = (1, 1)$ arrives, implying $C^{OPT} = \frac{3}{2} + L_2^t$. Since at most K jobs of hierarchy 2 are rearranged, we have $C^A \leq 1 + L_2^t + K\varepsilon$. Therefore,

$$\lim_{\varepsilon \to 0} \frac{C^{OPT}}{C^A} \geq \lim_{\varepsilon \to 0} \frac{\frac{3}{2} + L_2^t}{1 + L_2^t + K\varepsilon} = 1 + \frac{\frac{1}{2}}{1 + L_2^t} \geq \frac{4}{3}.$$

∎

Our algorithm for $P2|GoS, online, d_j = d, rearrangement|max(X)$ is described as follows, where $p_{max,2}^j$ is the processing time of the largest job scheduled on M_2 after job J_j is scheduled.

Algorithm 2:

1 Initially, let $L_2^0 = 0$ and $p_{max,2}^0 = 0$.
2 When a new job J_j arrives,
3 **if** $g_j = 1$ **then**
4   Assign job J_j to machine M_1.

5 **else**
6   **if** $L_2^{j-1} - p_{max,2}^{j-1} < \frac{d}{2}$ **then**
7   Assign the job J_j to M_2

8   **else**
9   Assign job J_j to machine M_1.

10 If $L_2^j - p_{max,2}^j > L_1^j$, rearrange the largest job on M_2 to M_1. Else, do nothing.

Theorem 5. The competitive ratio of **Algorithm** 2 is at most $\frac{4}{3}$.

Proof. Based on Lemma 1, if $L_1 \geq d$ and $L_2 \geq d$, we have $C^{A2} = 2d \geq C^{OPT}$. If $L_1 \leq d$ and $L_2 \leq d$, we have $C^{A2} = T \geq C^{OPT}$. It implies that we only need to consider the following two cases.

Case 1. $L_1 > d$ and $L_2 < d$. In this case, we have $C^{A2} = d + L_2$. If there is no job of hierarchy 2 assigned to M_1, then Algorithm 2 reaches the optimal. Else, let J_l be the last job of hierarchy 2 assigned to M_1. If J_l is assigned to M_1 at line 9, according to the choice of Algorithm 2, we have $L_2 \geq L_2^{l-1} - p_{max,2}^{l-1} \geq \frac{d}{2}$. Thus, by Lemma 1, we have

$$\frac{C^{OPT}}{C^{A2}} \leq \frac{2d}{d + L_2} \leq \frac{2d}{d + \frac{d}{2}} = \frac{4}{3}.$$

If J_l is assigned to M_1 at line 10, according to the choice of Algorithm 2, we have $L_2 \geq L_2^n - p_{max,2}^n > L_1^n$ and $p_{max,2}^n = p_l \leq d$. If $L_2 \geq \frac{d}{2}$, as above, we have $\frac{C^{OPT}}{C^{A2}} \leq \frac{4}{3}$. If $L_2 < \frac{d}{2}$, then $L_1 = L_1^n + p_{max,2}^n \leq L_1^n + d$. Thus, by Lemma 1, we have

$$\frac{C^{OPT}}{C^{A2}} \leq \frac{T}{d + L_2} \leq \frac{L_1 + L_2}{d + L_2} = 1 + \frac{L_1 - d}{d + L_2} \leq 1 + \frac{L_1^n}{d + L_2} \leq 1 + \frac{L_2}{d + L_2} \leq \frac{4}{3}.$$

Case 2. $L_1 < d$ and $L_2 > d$. In this case, we have $C^{A2} = L_1 + d$. Let $J_t = (p_t, 2)$ be the last job of hierarchy 2 assigned to M_2. Since J_t is assigned to M_2 at line 7, according to the choice of Algorithm 2, we have $L_2^{t-1} - p_{max,2}^{t-1} < \frac{d}{2}$ and $L_2^n = L_2^t = L_2^{t-1} + p_t < \frac{d}{2} + p_{max,2}^{t-1} + p_t$. If the rearrangement happens, we have

$$d < L_2 = L_2^n - \max\{p_{max,2}^{t-1}, p_t\} < \frac{d}{2} + \min\{p_{max,2}^{t-1}, p_t\},$$

which implies that $\min\{p_{max,2}^{t-1}, p_t\} > \frac{d}{2}$ and $L_1 \geq \max\{p_{max,2}^{t-1}, p_t\} > \frac{d}{2}$. Thus, by Lemma 1, we have

$$\frac{C^{OPT}}{C^{A2}} \leq \frac{2d}{L_1 + d} \leq \frac{2d}{\frac{d}{2} + d} = \frac{4}{3}.$$

If the rearrangement does not happen, we have $L_2 = L_2^n$ and $L_2^n - \max\{p_{max,2}^{t-1}, p_t\} \leq L_1^n = L_1$. If $L_1 \geq \frac{d}{2}$, as above, we have $\frac{C^{OPT}}{C^{A2}} \leq \frac{4}{3}$. If $L_1 < \frac{d}{2}$, we have

$$\frac{C^{OPT}}{C^{A2}} \leq \frac{T}{d + L_1} \leq \frac{L_1 + L_2}{d + L_1} = 1 + \frac{L_2 - d}{d + L_1}$$
$$\leq 1 + \frac{L_2 - \max\{p_{max,2}^{t-1}, p_t\}}{d + L_1} \leq 1 + \frac{L_1}{d + L_1} \leq \frac{4}{3}.$$

∎

5 Discussion

In this paper, we considered two semi-online versions for early work maximization problem on two hierarchical machines. When a buffer is available, we proposed an optimal online algorithm with competitive ratio of $\frac{4}{3}$. When we are allowed to reassign at most K jobs at the end, we also proposed an optimal online algorithm with a competitive ratio of $\frac{4}{3}$. In the future, it is interesting to study the online early work maximization problem on m hierarchical machines with buffer or rearrangements.

Acknowledgment. The work is supported in part by the National Natural Science Foundation of China [No. 12071417].

References

1. Chassid, O., Epstein, L.: The hierarchical model for load balancing on two machines. J. Comb. Optim. **15**(4), 305–314 (2008)
2. Chen, X., Ding, N., Dosa, G., Han, X., Jiang, H.: Online hierarchical scheduling on two machines with known total size of low-hierarchy jobs. Int. J. Comput. Math. **92**(5–6), 873–881 (2015)
3. Chen, X., Sterna, M., Han, X., Blazewicz, J.: Scheduling on parallel identical machines with late work criterion: offline and online cases. J. Sched. **19**(6), 729–736 (2016)
4. Chen, X., Xu, Z., Dosa, G., Han, X., Jiang, H.: Semi-online hierarchical scheduling problems with buffer or rearrangements. Inf. Process. Lett. **113**, 127–131 (2013)
5. Dai, B., Li, J., Li, W.: Semi-online hierarchical scheduling for bag-of-tasks on two machines. In: Proceedings of the 2018 2nd International Conference on Computer Science and Artificial Intelligence, pp. 609–614 (2018)
6. Epstein, L., Levin, A., Stee, R.V.: Max-min online allocations with a reordering buffer. SIAM J. Discret. Math. **25**(3–4), 1230–1250 (2011)
7. Jiang, Y., He, Y., Tang, C.: Optimal online algorithms for scheduling on two identical machines under a grade of service. J. Zhejiang Univ. Sci. A **7**, 309–314 (2006)
8. Li, W., Li, J., Zhang, T.: Two approximation schemes for scheduling on parallel machines under a grade of service provision. Asia-Pacific J. Oper. Res. **29**(05), 1250029 (2012)

9. Li, W.: Improved approximation schemes for early work scheduling on identical parallel machines with a common due date. J. Oper. Res. Soc. Chin. (2022). https://doi.org/10.1007/s40305-022-00402-y
10. Luo, T., Xu, Y.: Semi-online hierarchical load balancing problem with bounded processing times. Theoret. Comput. Sci. **607**, 75–82 (2015)
11. Park, J., Chang, S.Y., Lee, K.: Online and semi-online scheduling of two machines under a grade of service provision. Oper. Res. Lett. **34**(6), 692–696 (2006)
12. Wu, G., Li, W.: Semi-online machine covering on two hierarchical machines with discrete processing times. In: Li, L., Lu, P., He, K. (eds.) NCTCS 2018. CCIS, vol. 882, pp. 1–7. Springer, Singapore (2018). https://doi.org/10.1007/978-981-13-2712-4_1
13. Wu, Y., Cheng, T.C.E., Ji, M.: Optimal algorithms for semi-online machine covering on two hierarchical machines. Theoret. Comput. Sci. **531**(6), 37–46 (2014)
14. Wu, Y., Ji, M., Yang, Q.F.: Optimal semi-online scheduling algorithms on two parallel identical machines under a grade of service provision. Int. J. Prod. Econ. **135**(1), 367–371 (2012)
15. Xiao, M., Liu, X., Li, W.: Semi-online early work maximization problem on two hierarchical machines with partial information of processing time. In: Wu, W., Du, H. (eds.) AAIM 2021. LNCS, vol. 13153, pp. 146–156. Springer, Cham (2021). https://doi.org/10.1007/978-3-030-93176-6_13
16. Xiao, M., Ding, L., Zhao, S., Li, W.: Semi-online algorithms for hierarchical scheduling on three parallel machines with a buffer size of 1. In: He, K., Zhong, C., Cai, Z., Yin, Y. (eds.) NCTCS 2020. CCIS, vol. 1352, pp. 47–56. Springer, Singapore (2021). https://doi.org/10.1007/978-981-16-1877-2_4
17. Xiao, M., Wu, G., Li, W.: Semi-online machine covering on two hierarchical machines with known total size of low-hierarchy jobs. In: Sun, X., He, K., Chen, X. (eds.) NCTCS 2019. CCIS, vol. 1069, pp. 95–108. Springer, Singapore (2019). https://doi.org/10.1007/978-981-15-0105-0_7
18. Zhang, A., Jiang, Y., Fan, L., Hu, J.: Optimal online algorithms on two hierarchical machines with tightly-grouped processing times. J. Comb. Optim. **29**(4), 781–795 (2013). https://doi.org/10.1007/s10878-013-9627-7

On-line Single Machine Scheduling with Release Dates and Submodular Rejection Penalties

Xiaofei Liu[1], Yaoyu Zhu[2], Weidong Li[3], and Lei Ma[4,5(✉)]

[1] School of Information Science and Engineering, Yunnan University,
Kunming, People's Republic of China
[2] School of Electronic Engineering and Computer Science, Peking University,
Beijing, People's Republic of China
[3] School of Mathematics and Statistics, Yunnan University,
Kunming, People's Republic of China
[4] Beijing Academy of Artificial Intelligence, Beijing, People's Republic of China
[5] National Biomedical Imaging Center, Peking University,
Beijing, People's Republic of China
lei.ma@pku.edu.cn

Abstract. In this paper, we consider the on-line single machine scheduling problem with release dates and submodular rejection penalties. We are given a single machine and a sequence of jobs that arrive on-line and must be immediately and irrevocably either assigned on the machine or rejected. The objective is to minimize the sum of the makespan of the accepted jobs and the penalty of the rejected jobs which is determined by a submodular function. We prove that there is no on-line algorithm with a constant competitive ratio if the penalty submodular function is nonmonotone. When the penalty submodular function is monotone, we present an on-line algorithm with a competitive ratio 3.

Keywords: On-line scheduling · Submodular penalties · On-line algorithm · Competitive ratio

1 Introduction

Multiprocessor scheduling with rejection (MSR), which is first proposed by Bartal et al. [1], is a classical and important problem in operations research and combinatorial optimization. On this prblem, we are given m identical parallel machines and a set of n jobs, where each job can be either accepted and processed on the machine or rejected and paid a penalty. The objective is to minimize the makespan of accepted jobs plus the rejected penalty. Bartal et al. [1], proposed a 2-approximation algorithm and a polynomial-time approximation scheme (PTAS). Ou et al. [12] proposed a $(3/2 + \epsilon)$-approximation algorithm, where ϵ is a small given positive constant. Li et al. [6] designed a PTAS for a variant of the MSR, where the objective is to minimize the makespan when the

Q. Ni and W. Wu (Eds.): AAIM 2022, LNCS 13513, pp. 55–65, 2022.
https://doi.org/10.1007/978-3-031-16081-3_6

rejection cost is bounded by a given constant. When the number of machine is 1, Zhang et al. [14] considered the scheduling with rejection and release dates, in which each job has a release date, and they designed a PTAS. Li and Cui [5] considered the vector MSR, where each job is characterized by a d-dimension vector, and designed a PTAS.

In all the above problems, the information of all jobs can be achieved before making a decision. Since the market becomes more competitive, a sequence of jobs that arrive on-line and must be immediately and irrevocably either assigned on the machine or rejected. Bartal et al. [1] considered the on-line MSR problem, and presented an on-line algorithm with a competitive ratio $(\sqrt{5}+3)/2 \approx 2.618$; in particular, for $m = 2$, they presented an on-line algorithm with a competitive ratio $(\sqrt{5}+1)/2 \approx 1.618$. If preemption is allowed, Seiden [13] presented an on-line algorithm with a competitive ratio $(4+\sqrt{10})/3 \approx 2.3874$. Epstein et al. [3] considered the on-line MSR problem with unit processing jobs and proved that there does not exist an on-line algorithm with a competitive ratio less than 1.63784, and presented an on-line algorithm with a competitive ratio $(2+\sqrt{3})/2 \approx 1.866$. Lu et al. [11] designed an optimal on-line algorithm with a competitive ratio 2 for the online single machine scheduling with rejection and release dates. Dai and Li [2] designed an on-line algorithm with a competitive ratio $1.62d$ for the vector MSR on two machines, where d is the number of dimensions.

Recently, problems with submodular penalties have gradually become a research hotspot in the field of theoretical computers and combinatorial optimization [8,10]. Zhang et al. [15] proposed a 3-approximation algorithm for precedence-constrained scheduling with submodular rejection on parallel machines. Liu and Li [7] proposed a 2-approximation algorithm for the single machine scheduling problem with release dates and submodular rejection. Liu et al. [9] proposed a 2-approximation algorithm for the off-line single machine vector scheduling problem with submodular rejection penalties.

In this paper, we generalize the problem in [7] to on-line setting, and consider the on-line single machine scheduling problem with release dates and submodular rejection penalties The remainder of this paper is structured as follows. In Sect. 2, we first provide basic definitions, and give a formal on-line problem statement. Then, we prove the lower bound of this problem. In Sect. 3, first, we recall the algorithm for the offline case. Second, we present the on-line algorithm. In Sect. 4, we give a brief conclusion.

2 Preliminaries

Let J be a given ground set and $\pi(\cdot) : 2^J \to \mathbb{R}$ be a real-valued function defined on all the subsets of J. Function $\pi(\cdot)$ is called *submodular* if $\pi(S) + \pi(T) \geq \pi(S \cup T) + \pi(S \cap T), \forall S, T \subseteq J$, i.e.,

$$\pi(S \cup X) - \pi(S) \geq \pi(T \cup X) - \pi(T), \ \forall X \subseteq J \text{ and } \forall \ S \subseteq T \subseteq J \setminus X. \quad (1)$$

In particular, it is called *monotone* if function $\pi(\cdot)$ satisfies $\pi(S) \leq \pi(T), \forall S \subseteq T \subseteq J$. Moreover, we assume that $\pi(\cdot)$ can be computed in polynomial time for any subset $S \subseteq J$, where the 'polynomial' we use is with regard to the size n.

The on-line single machine scheduling problem with release dates and sub-modular rejection penalties is defined as follows: we are given a single machine and a sequence of n jobs, $J = \{1, 2, \ldots, n\}$, arriving online, where each job j in J has a processing time p_j and a release date r_j, without loss of generality, we assume that

$$r_j \leq r_{j'} \text{ for any } 1 \leq j < j' \leq n.$$

The job is to be either scheduled or rejected irrevocably at the time of their arrivals. This problem is to find a feasible schedule (A, R), where A is the set of accepted jobs that are processed on machine and $R = J \setminus A$ is the set of the rejected jobs. The objective is to minimize the sum of the makespan of the accepted jobs and the total penalty of the rejected jobs, which is determined by a penalty submodular function $\pi(\cdot)$. Thus, we use $1|r_j, on\text{-}line, reject|C_{max} + \pi(R)$ to denote this problem based on the three field notation.

Theorem 1. *There is no on-line algorithm with a constant competitive ratio for the $1|r_j, on\text{-}line, reject|C_{max} + \pi(R)$ if the penalty submodular function $\pi(\cdot)$ is nonmonotone.*

Proof. Consider a penalty submodular function $\pi(\cdot)$ defined on all the subsets of job set $J = \{1, 2\}$, where

$$\begin{cases} \pi(\emptyset) = 0; \\ \pi(\{1\}) = P^2 + 1; \\ \pi(\{2\}) = 1; \\ \pi(\{1, 2\}) = 1. \end{cases}$$

For any algorithm $\mathcal{A}$, the first job 1 in the sequence arrives at 0 and its process time $p_1 = P$. If $\mathcal{A}$ rejects job 1, then no job arrives, and we have $Z_{\mathcal{A}} = \pi(\{1\}) = P^2 + 1$ and $Z^* = r_1 + p_1 = P$, where $Z_{\mathcal{A}}$ is the value of schedule generated by $\mathcal{A}$ and Z^* is the optimal value. This implies that

$$\frac{Z_{\mathcal{A}}}{Z^*} = \frac{P^2 + 1}{P} > P.$$

If job 1 is accepted by $\mathcal{A}$, then the next job 2 arrives at 1 and its process time $p_2 = P$. The optimal schedule is to reject all jobs and its objective value $Z^* = \pi(\{1, 2\}) = 1$. If $\mathcal{A}$ rejects job 2, we have $Z_{\mathcal{A}} = r_1 + p_1 + \pi(\{2\}) = P + 1$; otherwise $\mathcal{A}$ processes job 2 on the machine, we have $Z_{\mathcal{A}} = r_1 + p_1 + p_2 = 2P$. These statements imply that

$$\frac{Z_{\mathcal{A}}}{Z^*} \geq \frac{P + 1}{1} > P.$$

Since P is positive number, the lemma holds.   □

Therefore, in the following part of this paper, we assume that the penalty submodular function is monotone.

3 $1|r_j, on\text{-}line, reject|C_{max} + \pi(R)$

In this section, first, we recall the 2-approximation algorithm for the off-line case [7], and prove some key lemmas. Second, we present the online algorithm for the $1|r, on\text{-}line, reject|C_{max} + \pi(R)$ based on the off-line algorithm, and prove that the competitive ratio of this algorithm is 3.

3.1 The Offline Problem

In this subsection, we recall the definition of the off-line problem. We are given a single machine and n jobs, $J = \{1, 2, \ldots, n\}$, where each job j has a processing time p_j and a release date r_j. The off-line problem is to find a feasible schedule (A, R), where A is the set of accepted jobs that are processed on machine and $R = J \setminus A$ is the set of the rejected jobs. The objective is to minimize the sum of the makespan of the accepted jobs and the total penalty of the rejected jobs, which is determined by a monotone submodular penalty function $\pi(\cdot)$.

For convenience, for each $r \in \{1, 2, 3, \ldots, n+1\}$, let

$$B_r = \{j \in J | j \geq r\},$$

then $B_1 = J$ and $B_{n+1} = \emptyset$. We display the algorithm as follows. For each $r \in \{1, 2, 3, \ldots, n+1\}$, algorithm needs to find a feasible solution (A_r, R_r) based on an auxiliary function. Then, output the best feasible solution.

For each $r \in \{1, 2, 3, \ldots, n+1\}$, we construct the auxiliary function $p\pi_r(\cdot)$ defined on all the subset of $J \setminus B_r$, where

$$p\pi_r(S) = p((J \setminus B_r) \setminus S) + \pi(S), \ \forall \ S \subseteq J \setminus B_r.$$

Then, $p\pi_r(S)$ is a submodular function, which is proven in Lemma 1, and the set

$$S_r := \arg \min_{S:S \subseteq J \setminus B_r} p\pi_j(S)$$

can be found in polynomial-time by modifying the methods in [4] slight. Then, set $R_r = B_r \cup S_r$ and $A_r = J \setminus R_r$.

We propose the detailed algorithm in Algorithm 1.

Lemma 1. *For any* $r \in \{1, 2, 3, \ldots, n+1\}$, $p\pi_r(\cdot)$ *is a submodular function.*

Proof. For any $r \in \{1, 2, 3, \ldots, n+1\}$ and any two job set $S_1, S_2 \subseteq J \setminus B_r$, we have

$$
\begin{aligned}
&p\pi_r(S_1) + p\pi_r(S_2) \\
&= p((J \setminus B_r) \setminus S_1) + \pi(S_1) + p((J \setminus B_r) \setminus S_2) + \pi(S_2) \\
&\leq p((J \setminus B_r) \setminus (S_1 \cup S_2)) + p((J \setminus B_r) \setminus (S_1 \cap S_2)) + \pi(S_1 \cup S_2) + \pi(S_1 \cap S_2) \\
&= p\pi_r(S_1 \cup S_2) + p\pi_r(S_1 \cap S_2),
\end{aligned}
$$

where the inequality follows from the fact that $\pi(\cdot)$ is a submodular function. This means, $p\pi_r(\cdot)$ is a submodular function. □

Algorithm 1:

1 **for** $r = 1$ to $n + 1$ **do**
2   Construct the set B_r and the auxiliary function $p\pi_r(\cdot)$ defined on all the subset of $J \setminus B_r$ defined as above.
3   Set $S_r := \arg\min_{S: S \subseteq J \setminus B_r} p\pi_j(S)$, and $R_r := B_r \cup S_r$. Schedule all jobs in $A_r (= J \setminus R_r)$ on the single machine using the earliest release date rule.
4   Set $Z_r = r_{r-1} + p(A_r) + \pi(R_r)$ be the objective value of (A_r, R_r), where $r_0 = 0$.
5 Set $(A, R) = \arg\min_{(A_r, R_r)} Z_r$, and output (A, R).

Theorem 2. (A, R) *can be found in polynomial time satisfying that its objective value is no more than* $2Z^*$, *where* Z^* *is the optimal value.*

Proof. Let (A^*, R^*) be the optimal schedule in J, and its objective value is

$$Z^* = C(A^*) + \pi(R^*),$$

where $C(A^*)$ is the makespan of the jobs in A^*. Let $r^* = \max\{j | j \in A^*\} + 1$, then we have

$$Z^* \geq r_{r^*-1} + \pi(B_{r^*}), \tag{2}$$

where r_{r^*-1} is the maximum release date of the jobs in A^*, and $B_{r^*} = \{j \in J | j \geq r^*\}$.

Since $S_{r^*} := \arg\min_{S: S \subseteq J \setminus B_{r^*}} p\pi_j(S)$, we have $p\pi_j(S_{r^*}) \leq p\pi_j(R^* \setminus B_{r^*})$, and

$$\begin{aligned}
p(A_{r^*}) + \pi(S_{r^*}) &= p((J \setminus B_{r^*}) \setminus S_{r^*}) + \pi(S_{r^*}) \\
&\leq p((J \setminus B_{r^*}) \setminus (R^* \setminus B_{r^*})) + \pi(R^* \setminus B_{r^*}) \\
&= p(A^*) + \pi(R^* \setminus B_{r^*}) \\
&\leq Z^*.
\end{aligned} \tag{3}$$

Then the objective value of (A, R) is

$$\begin{aligned}
Z \leq Z_{r^*} &= r_{r^*-1} + p(A_{r^*}) + \pi(R_{r^*}) \\
&\leq r_{r^*-1} + p(A_{r^*}) + \pi(S_{r^*}) + \pi(B_{R^*}) \\
&\leq 2Z^*,
\end{aligned}$$

where the first second inequality follows from the fact that $\pi(\cdot)$ is a submodular function and $\pi(\emptyset) = 0$, and the third inequality follows from inequalities (2) and (3).

By Lemma 1, for any $r \in \{1, 2, 3, \ldots, n + 1\}$, the set S_r can be found in polynomial-time [4]. Thus, Algorithm 1 can be implemented in polynomial time. $\square$

Note that the following lemma is important which is used in the online case.

Lemma 2. *If job n is in set A generated by Algorithm 1, then there exists an optimal schedule (A^*, R^*) satisfying*

$$n \in A^*.$$

Proof. If job n is in set A generated by Algorithm 1, then $(A, R) = (A_{n+1}, R_{n+1})$; otherwise, since job n is in set B_r for any $r \in \{1, 2, \ldots, n\}$, then n is in R_r for any $r \in \{1, 2, \ldots, n\}$. We have

$$p(J \setminus R) + \pi(R) \le p(J \setminus (S \cup R)) + \pi(S \cup R), \; \forall S \subseteq J \tag{4}$$

by $B_{n+1} = \emptyset$ and $R = R_{n+1} = S_{n+1} = \arg\min_{S:S \subseteq J \setminus B_{n+1}} p\pi_j(S)$.

Assume that any optimal schedule satisfying that job n is rejected, and let (A^*, R^*) be an optimal schedule in J satisfying $n \notin A^*$, and its objective value is Z^*. Then, we have

$$\begin{aligned} Z^* &= C(A^*) + \pi(R^*) \\ &< C(A^* \cup A) + \pi(R^* \setminus A) \\ &\le \max\{C(A^*), r_n\} + p(A \setminus A^*) + \pi(R^* \setminus A), \end{aligned} \tag{5}$$

where $C(A^*)$ is the makespan of the jobs in A^*, the first inequality follows from the assumption and $n \in A$.

If $r_n \le C(A^*)$, by rearranging inequality (5), we have

$$\begin{aligned} p(A \setminus A^*) &> \pi(R^*) - \pi(R^* \setminus A) \\ &= \pi(R^*) - \pi(R^* \setminus (J \setminus R)) \\ &= \pi(R^*) - \pi(R^* \cap R) \\ &\ge \pi(R^* \cup R) - \pi(R). \end{aligned}$$

where the last inequality follows from the diminishing marginal value $\pi(\cdot)$ by definition (1). Then, we have

$$\begin{aligned} p(J \setminus R) + \pi(R) &= p(J \setminus (R^* \cup R)) + p(R^* \setminus R) + \pi(R) \\ &= p(J \setminus (R^* \cup R)) + p(A \setminus A^*) + \pi(R) \\ &> p(J \setminus (R^* \cup R)) + \pi(R^* \cup R), \end{aligned}$$

which contradicts the inequality (4).

Otherwise, $r_n > C(A^*)$, let $r^* = \max_{j:j \in A^*} +1$. we have

$$r_{r^*} \le C(A^*) < r_n$$

and

$$r_{r^*} + \pi(R^*) \le Z^* \le r_n + p(A \setminus A^*) + \pi(R^* \setminus A),$$

where the second inequality follows from inequality (5). These statements imply that

$$
\begin{aligned}
0 > r_n - r_{r^*} &\geq \pi(R^*) - p(A \setminus A^*) - \pi(R^* \setminus A) \\
&= \pi(R^*) - \pi(R^* \setminus (J \setminus R)) - p((J \setminus R) \setminus (J \setminus R^*)) \\
&= \pi(R^*) - \pi(R^* \cap R) - p(R^* \setminus (R^* \cap R)) \\
&\geq \pi(R^* \cup R) - \pi(R) - p(R^* \setminus (R^* \cap R)) \\
&= p(J \setminus (R^* \cup R)) + \pi(R^* \cup R) - p(J \setminus (R)) - \pi(R),
\end{aligned}
$$

where the last inequality follows the diminishing marginal value $\pi(\cdot)$ by definition (1). This implies that

$$
p(J \setminus R) + \pi(R) > p(J \setminus (R^* \cup R)) + \pi(R^* \cup R),
$$

which contradicts the inequality (4).

Therefore, the lemma holds.

3.2   The On-line Algorithm

In this subsection, we present an on-line algorithm with a competitive ratio 3 for $1|r, on\text{-}line, reject|C_{max} + \pi(R)$. In particular, we prove that the competitive ratio of this on-line algorithm is 2 if all the jobs have the same release date.

Given a set of a sequence jobs, $J = \{1, 2, \ldots, n\}$, for each $j \in \{1, 2, \ldots, n\}$, let $J_j = \{1, \ldots, j\}$ be the set of the first j jobs in J.

For instance J_j, we can find a feasible schedule (A_j, R_j) in polynomial-time using Algorithm 1. The objective value of (A_j, R_j) satisfies $Z_j \leq Z_j^*$ by Theorem 2, where Z_j^* is the optimal value for instance J_j.

Next, we provide the detailed the on-line algorithm in Algorithm 2 below.

Algorithm 2:

---

1 Initially, set the makspan $C = 0$, $j = 1$, $A = \emptyset$ and $R = \emptyset$.
2 Assume that a job j arrives.
3 Using Algorithm 1, find the feasible solution (A_j, R_j) for instance J_j.
4 **if** $j \in R_j$ **then**
5 | reject job j, and set $R := R \cup \{j\}$

6 **else**
7 | process job j on the machine at time $\max\{C, r_j\}$, and set
   | $C := \max\{C, r_j\} + p_j$ and $A := A \cup \{j\}$.

8 If no new job arrives, stop and output the current schedule (A, R) and its value $Z = C + \pi(R)$; otherwise, set $j := j + 1$ and go to **2**.

Lemma 3. *Let C_j be the makespan generated Algorithm 2 for instance J_j, then we have*

$$
C_j \leq Z_j^*, \quad \forall j \in \{1, 2, \ldots, n\}.
$$

Proof. Our proof is by mathematical induction. For the instance $J_1 = \{1\}$, let (A_1, R_1) be the feasible schedule generated by Algorithm 1, then we have either $R_1 = \{1\}$ or $A_1 = \{1\}$. If $R_1 = \{1\}$, then job 1 is rejected by Algorithm 2, and $C_1 = 0 \leq Z_1^*$; otherwise, $A_1 = \{1\}$, then $(\{1\}, \emptyset)$ is an optimal schedule by Lemma 2 and $Z_1^* = r_1 + p_1$. job 1 is processed on the machine by Algorithm 2 and $C_1 = r_1 + p_i = Z_1^*$. Thus, we have $C_1 \leq Z_1^*$.

Then, assume that $C_j \leq Z_j^*$ holds for each $j < k$. We consider $j = k$. It is obvious that

$$Z_{k-1}^* \leq Z_k^*.$$

If $k \in R_k$, then job k is rejected by Algorithm 2, and

$$C_k = C_{k-1} \leq Z_{k-1}^* \leq Z_k^*.$$

Otherwise, $k \in A_k$, there exists an optimal schedule (A_k^*, R_k^*) for instance J_k satisfying $k \in A_k^*$. Thus, we have

$$\begin{aligned}
Z_k^* &= C(A_k^*) + \pi(R_k^*) \\
&= \max\{C(A_k^* \setminus \{k\}), r_k\} + p_k + \pi(R_k^*) \\
&\geq C(A_k^* \setminus \{k\}) + p_k + \pi(R_k^*) \\
&\geq Z_{k-1}^* + p_k,
\end{aligned} \qquad (6)$$

where $C(A_k^*)$ is the makespan of job set A^* using the earliest release date rule, and the last inequality follows because $(A_k^* \setminus \{J_k\}, R_k^*)$ is a feasible schedule for instance J_{k-1} and Z_{k-1}^* is the optimal objective value for instance J_{k-1}.

Since job $k \in A_k$, job k is processed on the machine and

$$C_k = \max\{C_{k-1}, r_k\} + p_k.$$

If $C_{k-1} \leq r_k$, by $k \in A_k^*$, we have $C_k = r_k + p_k \leq C(A_k^*) \leq Z_k^*$; Otherwise, $C^{k-1} > r_k$, then we have

$$C^k = C^{k-1} + p_k \leq Z_{k-1}^* + p_k \leq Z_k^*,$$

where the first inequality follows from by the assumptions and the second follows from inequality (6).

Therefore, we have $C^k \leq Z_k^*$ and the lemma holds.

Lemma 4. *Let $R(j)$ be the rejected set generated Algorithm 2 for instance J_j, then we have*

$$\pi(R(j)) \leq 2Z_j^* \ \forall j \in \{1, 2, \ldots, n\}.$$

Proof. For any $j = 1, 2, \ldots, n$, let (A_j, R_j) be the feasible solution generated by Algorithm 1, then its objective value is

$$Z_j = r_{r_j-1} + p(A_j) + \pi(R_j) \leq 2Z_j^*, \qquad (7)$$

where $r_j - 1$ is the number of loops to generate (A_j, R_j), and the inequality follows from Theorem 2.

If $R(j) \setminus R_j = \emptyset$, we have

$$\pi(R(j)) \leq \pi(R_j) \leq Z_j \leq 2Z_j^*,$$

where the first inequality follows from that $\pi(\cdot)$ is monotone, and the second inequality follows from inequality (7).

Otherwise, $R(j) \setminus R_j \neq \emptyset$, let

$$k = \arg \max_{j' \in R(j) \setminus R_j} j',$$

be the maximum job in $R(j) \setminus R_j$. Then, $k \in A_j$, we have

$$r_k \leq r_{r_j - 1}, \tag{8}$$

by Algorithm 1.

Let (A_k, R_k) be the feasible schedule for instance J_k generated by Algorithm 1, and Algorithm 1 needs to find a feasible schedule $(A_{k,r}, R_{k,r})$ for each $r \in \{1, 2, \ldots, k+1\}$ to find (A_k, R_k), i.e.,

$$(A_k, R_k) = \arg \min_{(A_{k,r}, R_{k,r})} Z_{k,j},$$

where $Z_{k,r} = r_{r-1} + p(A_{k,r}) + \pi(R_{k,r})$. We define $\hat{r} = \arg \min_r (r_{r-1} + p(A_{k,r}) + \pi(R_{k,r})$, i.e., $(A_k, R_k) = (A_{k,\hat{r}}, R_{k,\hat{r}})$. Thus, the objetive value of (A_k, R_k) is $r_{\hat{r}-1} + p(A_k) + \pi(R_k) \leq r_k + p(J_k \setminus (R_j \cap R_k)) + \pi(R_j \cap R_k)$, and

$$r_k - r_{\hat{r}-1}$$
$$\geq p(A_k) - p(J_k \setminus (R_j \cap R_k)) + \pi(R_k) - \pi(R_j \cap R_k)$$
$$\geq \pi(R_k \cup R_j) - \pi(R_j) - p(R_k \setminus R_j),$$

where the last inequality follows from the diminishing marginal value $\pi(\cdot)$ by definition (1). Thus, we have

$$\pi(R_k \cup R_j) \leq Z_j.$$

If $R(j) \setminus (R_k \cup R_j) = \emptyset$, we have

$$\pi(R(j)) \leq \pi(R_k \cup R_j) \leq Z_j \leq 2Z_j^*,$$

where the first inequality follows from that $\pi(\cdot)$ is monotone, and the last inequality follows from inequality (7). Thus, the lemma holds; otherwise, let $R(j) = R_k \cup R_j$, repeat the above process up to $|R^j \setminus (R_k \cup R_j)| (\leq n)$ times, we can obtain that the inequality $R(j) \setminus (R_k \cup R_j) = \emptyset$ follows.

Theorem 3. *The online Algorithm 2 is a 3-competitive for any instance of the* $1|r, on\text{-}line, reject|C_{max} + \pi(R)$ *and the bound is tight.*

Proof. By Lemma 3 and Lemma 4, for any instance J_j, the objective value of the schedule generated by Algorithm 2 is

$$C_j + \pi(R(j)) \leq 3Z_j^*.$$

To show that the bound is tight, consider the job list $J = \{1, 2\}$, where $r_1 = 0$, $p_1 = P$, $r_2 = P$ and $p_2 = 1$ (P is a large constant). The polymatroid function $\pi(\cdot)$ is defined as follows: $\pi(\emptyset) = 0$, $\pi(\{1\}) = P + 1$, $\pi(\{2\}) = 2P - 1$, $\pi(\{1, 2\}) = 2P$.

When the first job 1 arrives, job $1 \in A_1$ by the Algorithm 1, and job 1 is accepted and processed on the machine by the Algorithm 2. Then, the second job 2 arrives, job $2 \in R_2$ by the Algorithm 1, and job 2 is rejected by the Algorithm 2. Therefore, we have $Z = r_1 + p_1 + \pi(\{2\}) = 0 + P + 2P - 1 = 3P - 1$, and $Z^* = r_1 + p_1 + p_2 = P + 1$. Thus, we have

$$\frac{Z}{Z^*} = \frac{3P - 1}{P + 1} \longrightarrow 3, \text{ when } P \to \infty.$$

4   Conclusion

In this paper, we consider the on-line single machine scheduling problem with release dates and submodular rejection penalties. We prove that there is no on-line algorithm with a constant competitive ratio if the penalty submodular function is nonmonotone, and present an on-line algorithm with a competitive ratio of 3 for the problem when the penalty submodular function is monotone.

The topic could be further studied in the following ways. It is challenging to either find a greater lower bound or design an on-line algorithm with a better competitive ratio. In the real world, we always know all the information about all the jobs that have the same release date, thus the on-line-over-time version of this problem is worth considering. Moreover, algorithm for the on-line single machine vector scheduling problem with release dates and submodular rejection penalties could be further developed.

Acknowledgements. The work is supported in part by the National Natural Science Foundation of China [No. 12071417], and National Key R&D Program of China [No. 2020AAA 0105200].

Conflict of Interest. The authors declare that they have no conflict of interest.

References

1. Bartal, Y., Leonardi, S., Marchetti-Spaccamela, A., Sgall, J., Stougie, L.: Multiprocessor scheduling with rejection. SIAM J. Discrete Math. **13**(1), 64–78 (2000)
2. Dai, B., Li, W.: Vector scheduling with rejection on two machines. Int. J. Comput. Math. **97**(12), 2507–2515 (2020)

3. Epstein, L., Noga, J., Woeginger, G.: On-line scheduling of unit time jobs with rejection: minimizing the total completion time. Oper. Res. Lett. **30**(6), 415–420 (2002)
4. Fleischer, L., Iwata, S.: A push-relabel framework for submodular function minimization and applications to parametric optimization. Discrete Appl. Math. **131**(2), 311–322 (2003)
5. Li, W., Cui, Q.: Vector scheduling with rejection on a single machine. 4OR **16**, 95–104 (2018)
6. Li, W., Li, J., Zhang, X., Chen, Z.: Penalty cost constrained identical parallel machine scheduling problem. Theor. Comput. Sci. **607**, 181–192 (2015)
7. Liu, X., Li, W.: Approximation algorithm for the single machine scheduling problem with release dates and submodular rejection penalty. Mathematics **8**(1), 133 (2020)
8. Liu, X., Li, W.: Approximation algorithms for the multiprocessor scheduling with submodular penalties. Optimiz. Lett. **15**, 2165–2180 (2021)
9. Liu, X., Li, W., Zhu, Y.: Single machine vector scheduling with general penalties. Mathematics **9**(16), 1965 (2021)
10. Liu, X., Dai, H., Li, S., Li, W.: k-prize-collecting minimum power cover problem with submodular penalties on a plane (in chinese). Sci. Sin. Inform. **52**(6), 947 (2022)
11. Lu, L., Ng, C., Zhang, L.: Optimal algorithms for single-machine scheduling with rejection to minimize the Makespan. Int. J. Prod. Econ. **130**(2), 153–158 (2011)
12. Ou, J., Zhong, X., Wang, G.: An improved heuristic for parallel machine scheduling with rejection. Eur. J. Oper. Res. **241**(3), 653–661 (2015)
13. Seiden, S.: Preemptive multiprocessor scheduling with rejection. Theor. Comput. Sci. **262**(1–2), 437–458 (2001)
14. Zhang, L., Lu, L., Yuan, J.: Single machine scheduling with release dates and rejection. Eur. J. Oper. Res. **198**(3), 975–978 (2009)
15. Zhang, X., Xu, D., Du, D., Wu, C.: Approximation algorithms for precedence-constrained identical machine scheduling with rejection. J. Combinat. Optimiz. **35**, 318–330 (2018)

The Optimal Dynamic Rationing Policy in the Stock-Rationing Queue

Quan-Lin Li[1], Yi-Meng Li[2], Jing-Yu Ma[3]([⊠]) [iD], and Heng-Li Liu[2]

[1] School of Economics and Management,
Beijing University of Technology, Beijing 100124, China
[2] School of Economics and Management,
Yanshan University, Qinhuangdao 066004, China
[3] Bussiness School, Xuzhou University of Technology, Xuzhou 221018, China
mjy0501@126.com

Abstract. In this paper, we study a stock-rationing queue with two demand classes by means of the sensitivity-based optimization, and develop a complete algebraic solution for the optimal dynamic rationing policy. To do this, we establish a policy-based birth-death process to show that the optimal dynamic rationing policy must be of transformational threshold type. Based on this finding, we can refine three sufficient conditions under each of which the optimal dynamic rationing policy is of threshold type (i.e., critical rationing level). Crucially, we characterize the monotonicity and optimality of the long-run average profit of this system, and establish some new structural properties of the optimal dynamic rationing policy by observing any given reference policy. Finally, we use numerical examples to verify computability of our theoretical results. We believe that the methodology and results developed in this paper can shed light on the study of stock-rationing queue and open a series of potentially promising research.

Keywords: Stock-rationing queue · Inventory rationing · Dynamic rationing policy · Sensitivity-based optimization · Markov decision process

1 Introduction

In this paper, we consider a stock-rationing queueing problem of a warehouse with one type of products and two classes of demands, which may be viewed as coming from retailers with two different priority levels. Now, such a stock-rationing warehouse system becomes more and more important in many large cities under the current COVID-19 environment. For example, Beijing has seven

Supported by the National Natural Science Foundation of China under grant No. 71932002.

J.-Y. Ma and Q.-L. Li—Contributed to the work equally and should be regarded as co-first authors.

super-large warehouses, which always supply various daily necessities (e.g., vegetables, meat, eggs, seafood) to more than 40 million people every day. In the warehouses, each type of daily necessities are supplied by lots of different companies in China and other countries, which lead to that the successive supply stream of each type of products can be well described as a Poisson process. In addition, the two retailers may be regarded as a large supermarket group and another community retail store group. Typically, the large supermarket group has a higher supply priority than the community retail store group. When the COVID-19 at Beijing is at a serious warning, the stock-rationing management of the warehouses and their optimal rationing policy play key roles in strengthening the effective management of the warehouses such that every family at Beijing can have a very comprehensive life guarantee.

From the perspective of practical applications, such a stock-rationing queueing problem with multiple demand classes can always be encountered in many different real areas, assemble-to-order systems, make-to-stock queues and multi-echelon inventory systems by Ha [6]; manufacturing by Zhao et al. [21]; airline by Wang et al. [19]; rental business by Altug and Ceryan [1] and Jain et al. [9]; health care by Moosa and Luyckx [15] and Baron et al. [2]; and so forth. All the studies above show that the stock-rationing queues with multiple demand classes are not only necessary and important in many practical applications, but also have their own theoretical interest. In the stock-rationing queueing systems, the rationing policies always assign different supply priorities to multiple classes of demands. In the early literature, the so-called *critical rationing level* was imagined intuitively, and its existence was further proved by Veinott Jr [18] and Topkis [17]. Thus designing and optimizing the critical rationing levels become a basic issue of inventory rationing across multiple demand classes.

So far some research has applied the Markov decision processes (MDPs) to discuss inventory rationing (and stock-rationing queues) across multiple demand classes by means of the submodular (or supermodular) technique, among which important examples include Ha [6–8], Gayon et al. [5], Benjaafar and ElHafsi [3] and Nadar et al. [14]. To this end, it is a key that the structural properties of the optimal rationing policy need to be identified by using a set of structured value functions that are preserved under an optimal operator. Based on this finding, the optimal rationing policy of the inventory rationing across multiple demand classes can be further analyzed by means of the structural properties. In many more general cases, it is not easy and even very difficult to set up the structural properties of the optimal rationing policy. For this reason, some stronger model assumptions have to be further added. The purpose of improving the applicability of the MDPs motivates us to propose a new algebraic method to find a complete algebraic solution to the optimal rationing policy by means of the sensitivity-based optimization which is proposed by Cao [4]. To the best of our knowledge, this paper is the first to apply the sensitivity-based optimization to the study of stock-rationing queues.

Based on the above analysis, we summarize the main contributions of this paper as follows:

(1) A complete algebraic solution: This paper develops a complete algebraic solution to the optimal dynamic rationing policy of the stock-rationing queue by means of the sensitivity-based optimization, and shows that the optimal dynamic rationing policy must be of transformational threshold type, which can lead to refining three sufficient conditions under each of which the optimal dynamic rationing policy is of threshold type. It is worthwhile to note that our transformational threshold type results are sharper than the bang-bang control given in Ma et al. [12,13] and Xia et al. [20]. Therefore, our algebraic method provides a new way of optimality proofs when comparing to the frequently-used submodular (or supermodular) technique of MDPs.

(2) Two different methods can sufficiently support each other: Note that our algebraic method sets up a complete algebraic solution to the optimal dynamic rationing policy, thus it can provide not only a necessary complement of policy spatial structural integrity but also a new way of optimality proof when comparing to the frequently-used submodular (or supermodular) technique of MDPs. Since our algebraic method and the submodular (or supermodular) technique are all important parts of the MDPs (the former is to use the poisson equations, while the latter is to apply the optimality equation), it is clear that the two different methods will sufficiently support each other in the study of stock-rationing queues (and rationing inventory) with two demand classes.

The remainder of this paper is organized as follows. Section 2 gives model description for the stock-rationing queue with two demand classes. Section 3 establishes an optimization problem to find the optimal dynamic rationing policy, in which we set up a policy-based birth-death process and introduce a key reward function. Section 4 discusses the monotonicity and optimality of the long-run average profit of this system, and finds the optimal dynamic rationing policy under three different areas of the penalty cost. Finally, some concluding remarks are given in Sect. 5.

2   Model Description

In this section, we describe a stock-rationing queue, in which a single class of products are supplied to stock at a warehouse, and the two classes of demands come from two retailers with different priorities. In addition, we provide system structure, operational mode and mathematical notations.

A Stock-Rationing Queue: The warehouse has the maximal capacity N to stock a single class of products, and needs to pay a holding cost C_1 per product per unit time. There are two classes of demands to order the products, in which the demands of Class 1 have a higher priority than that of Class 2, such that the demands of Class 1 can be satisfied in any non-zero inventory, while the demands of Class 2 may be either satisfied or refused based on the inventory level of products. Figure 1 depicts a simple physical system to understand the stock-rationing queue.

The Supply Process: The supply stream of the products to the warehouse is a Poisson process with arrival rate λ, where the price of per product is C_3 paid

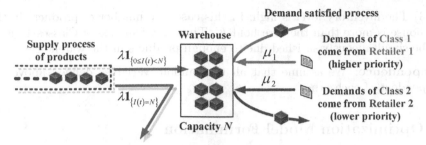

I(t): The total number of products in the stock-rationing queue at time *t*

Fig. 1. A stock-rationing queue with two demand classes.

by the warehouse to the external product supplier. If the warehouse is full, then any new arriving product has to be lost. In this case, the warehouse will have an opportunity cost C_4 per product rejected into the warehouse.

The Service Processes: The service times provided by the warehouse to satisfy the demands of Classes 1 and 2 are i.i.d. and exponential with service rates μ_1 and μ_2, respectively. The service disciplines for the two classes of demands are all First Come First Serve (FCFS). The warehouse can obtain the service price R when one product is sold to Retailer 1 or 2. Note that each demand of Class 1 or 2 is satisfied by one product every time.

The Stock-Rationing Rule: For the two classes of demands, each demand of Class 1 can always be satisfied in any non-zero inventory, while for satisfying the demands of Class 2, we need to consider three different cases as follows:

(1) The inventory level is zero. In this case, any new arriving demand has to be rejected immediately. This leads to the lost sales cost $C_{2,1}$ (resp. $C_{2,2}$) per unit time for any lost demand of Class 1 (resp. 2). We assume that $C_{2,1} > C_{2,2}$, which is used to guarantee the higher priority service for the demands of Class 1 when comparing to the lower priority for the demands of Class 2.

(2) The inventory level is low. In this case, the number of products in the warehouse is not more than a key threshold K, where the threshold K is subjectively designed by means of some real experience. Note that the demands of Class 1 have a higher priority to receive the products than the demands of Class 2. Thus the warehouse will not provide any product to satisfy the demands of Class 2 under an equal service condition if the number of products in the warehouse is not more than K. Otherwise, such a service priority is violated (i.e., the demands of Class 2 are satisfied from a low stock), so that the warehouse must pay a penalty cost P per product supplied to the demands of Class 2 at a low stock. Note that the penalty cost P measures different priority levels to supply the products between the two classes of demands.

(3) The inventory level is high. In this case, the number of products in the warehouse is more than the threshold K. Thus the demands of Classes 1 and 2 can be simultaneously satisfied due to enough products in the warehouse.

Independence: We assume that all the random variables defined above are independent of each other.

3　Optimization Model Formulation

In this section, we establish an optimization problem to find the optimal dynamic rationing policy in the stock-rationing queue. To do this, we set up a policy-based birth-death process and introduce a key reward function.

3.1　The States and Policies

To study the stock-rationing queue with two demand classes, we define both 'states' and 'policies' to express stochastic dynamics of the stock-rationing queue.

Let $I(t)$ be the number of products in the warehouse at time t. Then it is regarded as the state of this system at time t. Obviously, all the cases of State $I(t)$ form a state space as follows:

$$\Omega = \{0, 1, 2, \ldots, N\}.$$

Also, State $i \in \Omega$ is regarded as an inventory level of this system.

From the states, some policies are defined with a little bit more complicated. Let d_i be a policy related to State $i \in \Omega$, and it expresses whether or not the warehouse prefers to supply some products to the demands of Class 2 when the inventory level is not more than the threshold K for $0 < K \leq N$. Thus, we have

$$d_i = \begin{cases} 0, & i = 0, \\ 0 \text{ or } 1, & i = 1, 2, \ldots, K, \\ 1, & i = K+1, K+2, \ldots, N, \end{cases} \tag{1}$$

where $d_i = 0$ and 1 represents that the warehouse rejects and satisfies the demands of Class 2, respectively. Obviously, not only does the policy d_i depend on State $i \in \Omega$, but also it is controlled by the threshold K. Of course, for a special case, if $K = N$, then $d_i \in \{0, 1\}$ for $1 \leq i \leq N$.

Corresponding to each state in Ω, we define a time-homogeneous policy of the stock-rationing queue as

$$\mathbf{d} = (d_0; d_1, d_2, \ldots, d_K; d_{K+1}, d_{K+2}, \ldots, d_N).$$

It follows from (1) that

$$\mathbf{d} = (0; d_1, d_2, \ldots, d_K; 1, 1, \ldots, 1). \tag{2}$$

Thus the rationing policy $\mathbf{d}$ depends on $d_i \in \{0,1\}$, which is related to State i for $1 \leq i \leq K$. Let all the possible policies of the stock-rationing queue, given in (2), form a policy space as follows:

$$\mathcal{D} = \{\mathbf{d} : \mathbf{d} = (0; d_1, d_2, \ldots, d_K; 1, 1, \ldots, 1), d_i \in \{0,1\}, 1 \leq i \leq K\}.$$

Let $I^{(\mathbf{d})}(t)$ be the state of the stock-rationing queue at time t under any given rationing policy $\mathbf{d} \in \mathcal{D}$. It is easy to check that $\{I^{(\mathbf{d})}(t) : t \geq 0\}$ is a policy-based birth-death process. Based on this, the infinitesimal generator is given by

$$\mathbf{B}^{(\mathbf{d})} = \begin{pmatrix} -\lambda & \lambda & & & & & \\ v(d_1) & -[\lambda + v(d_1)] & \lambda & & & & \\ & \ddots & \ddots & \ddots & & & \\ & & v(d_K) & -[\lambda + v(d_K)] & \lambda & & \\ & & & v(1) & -[\lambda + v(1)] & \lambda & \\ & & & & \ddots & \ddots & \ddots \\ & & & & & v(1) & -[\lambda + v(1)] & \lambda \\ & & & & & & v(1) & -v(1) \end{pmatrix}, \quad (3)$$

where $v(d_i) = \mu_1 + d_i \mu_2$ for $i = 1, 2, \ldots, K$, and $v(1) = \mu_1 + \mu_2$. It is clear that $v(d_i) > 0$ for $i = 1, 2, \ldots, K$. Thus the policy-based birth-death process $\mathbf{B}^{(\mathbf{d})}$ must be irreducible, aperiodic and positive recurrent for any given policy $\mathbf{d} \in \mathcal{D}$.

3.2   The Stationary Probability Vector

From the infinitesimal generator, we write the stationary probability vector of the policy-based birth-death process $\{I^{(\mathbf{d})}(t) : t \geq 0\}$ as

$$\pi^{(\mathbf{d})} = \left(\pi^{(\mathbf{d})}(0); \pi^{(\mathbf{d})}(1), \ldots, \pi^{(\mathbf{d})}(K); \pi^{(\mathbf{d})}(K+1), \ldots, \pi^{(\mathbf{d})}(N)\right). \quad (4)$$

Obviously, the stationary probability vector $\pi^{(\mathbf{d})}$ is the unique solution to the system of linear equations: $\pi^{(\mathbf{d})}\mathbf{B}^{(\mathbf{d})} = \mathbf{0}$ and $\pi^{(\mathbf{d})}\mathbf{e} = 1$, where $\mathbf{e}$ is a column vector of ones with a suitable dimension. We write

$$\xi_0 = 1, \qquad\qquad\qquad i = 0,$$

$$\xi_i^{(\mathbf{d})} = \begin{cases} \dfrac{\lambda^i}{\prod\limits_{j=1}^{i} v(d_j)}, & i = 1, 2, \ldots, K, \\[4mm] \dfrac{\lambda^i}{(\mu_1 + \mu_2)^{i-K} \prod\limits_{j=1}^{K} v(d_j)}, & i = K+1, K+2, \ldots, N, \end{cases} \quad (5)$$

and

$$h^{(\mathbf{d})} = 1 + \sum_{i=1}^{N} \xi_i^{(\mathbf{d})}.$$

It follows from Li [10] that

$$\pi^{(\mathbf{d})}(i) = \begin{cases} \dfrac{1}{h^{(\mathbf{d})}}, & i = 0, \\[3mm] \dfrac{1}{h^{(\mathbf{d})}} \xi_i^{(\mathbf{d})}, & i = 1, 2, \ldots, N. \end{cases} \quad (6)$$

3.3   The Reward Function

By using the policy-based birth-death process $\mathbf{B}^{(\mathbf{d})}$, now we define a key reward function in the stock-rationing queue. It is seen from that the reward function with respect to both states and policies is defined as a profit rate (i.e., the total system revenue minus the total system cost per unit time). By observing the impact of rationing policy $\mathbf{d}$ on the profit rate, the reward function at State i under rationing policy $\mathbf{d}$ is given by

$$f^{(\mathbf{d})}(i) = R\left(\mu_1 1_{\{i>0\}} + \mu_2 d_i\right) - C_1 i - C_{2,1}\mu_1 1_{\{i=0\}} - C_{2,2}\mu_2\left(1 - d_i\right)$$
$$- C_3\lambda 1_{\{i<N\}} - C_4\lambda 1_{\{i=N\}} - P\mu_2 d_i 1_{\{1\leq i\leq K\}}, \qquad (7)$$

where $1_{\{.\}}$ represents the indicator function whose value is one when the event occurs; otherwise it is zero. By using the indicator function, satisfying and rejecting the demands of Class 1 are expressed as $1_{\{i>0\}}$ and $1_{\{i=0\}}$, respectively; the external products enter or are lost by the warehouse according to $1_{\{i<N\}}$ and $1_{\{i=N\}}$, respectively; and a penalty cost paid by the warehouse is denoted by $1_{\{1\leq i\leq K\}}$ due to that the warehouse supplies the products to the demands of Class 2 at a low stock.

Based on the above analysis, we define an $(N+1)$-dimensional column vector composed of the elements $f(0)$, $f^{(\mathbf{d})}(i)$ for $1 \leq i \leq K$, and $f(j)$ for $K+1 \leq j \leq N$ as follows:

$$\mathbf{f}^{(\mathbf{d})} = \left(f(0); f^{(\mathbf{d})}(1), f^{(\mathbf{d})}(2), \ldots, f^{(\mathbf{d})}(K); f(K+1), f(K+2), \ldots, f(N)\right)^T.$$
$$(8)$$

Now, we consider the long-run average profit of the stock-rationing queue (or the continuous-time policy-based birth-death process $\{I^{(\mathbf{d})}(t) : t \geq 0\}$) under any given rationing policy $\mathbf{d}$. Let

$$\eta^{\mathbf{d}} = \lim_{T\to\infty} E\left\{\frac{1}{T}\int_0^T f^{(\mathbf{d})}\left(I^{(\mathbf{d})}(t)\right) dt\right\}.$$

Then

$$\eta^{\mathbf{d}} = \pi^{(\mathbf{d})}\mathbf{f}^{(\mathbf{d})},$$

where $\pi^{(\mathbf{d})}$ and $\mathbf{f}^{(\mathbf{d})}$ are given by (4) and (8), respectively.

Based on this, our objective is to find an optimal dynamic rationing policy $\mathbf{d}^*$ such that the long-run average profit $\eta^{\mathbf{d}}$ is maximal, that is,

$$\mathbf{d}^* = \arg\max_{\mathbf{d}\in\mathcal{D}}\left\{\eta^{\mathbf{d}}\right\}. \qquad (9)$$

3.4   The Perturbation Realization Factor

For any given rationing policy $\mathbf{d} \in \mathcal{D}$, it follows from Cao [4] that for the continuous-time policy-based birth-death process $\{I^{(\mathbf{d})}(t), t \geq 0\}$, we define the performance potential as

$$g^{(\mathbf{d})}(i) = E\left\{\int_0^{+\infty}\left[f^{(\mathbf{d})}\left(I^{(\mathbf{d})}(t)\right) - \eta^{\mathbf{d}}\right] dt \,\middle|\, I^{(\mathbf{d})}(0) = i\right\}, \qquad (10)$$

which quantifies the contribution of the initial State i to the long-run average profit of the stock-rationing queue. Furthermore, we define a perturbation realization factor as

$$G^{(\mathbf{d})}(i) \overset{\text{def}}{=} g^{(\mathbf{d})}(i-1) - g^{(\mathbf{d})}(i), i = 1, 2, \ldots, N, \tag{11}$$

which quantifies the difference among two adjacent performance potentials $g^{(\mathbf{d})}(i)$ and $g^{(\mathbf{d})}(i-1)$, and measures the effect on the long-run average profit of the stock-rationing queue when the system state is changed from $i-1$ to i.

By using the policy-based Poisson equation in Ma et al. [12] , we can derive a new system of linear equations, which can be used to directly express the perturbation realization factor $G^{(\mathbf{d})}(i)$ for $i = 1, 2, \ldots, N$, as follows:

Theorem 1. *For any given rationing policy* $\mathbf{d}$*, the perturbation realization factor* $G^{(\mathbf{d})}(i)$ *is given by*
    (a) for $1 \leq i \leq K$*,*

$$G^{(\mathbf{d})}(i) = \lambda^{-i}\left[f(0) - \eta^{\mathbf{d}}\right]\prod_{k=1}^{i-1} v(d_k) + \sum_{r=1}^{i-1} \lambda^{r-i}\left[f^{(\mathbf{d})}(r) - \eta^{\mathbf{d}}\right]\prod_{k=r+1}^{i-1} v(d_k);$$

(b) for $K+1 \leq i \leq N$*,*

$$G^{(\mathbf{d})}(i) = \lambda^{-i}\left[f(0) - \eta^{\mathbf{d}}\right]\prod_{k=1}^{K} v(d_k)\left[v(1)\right]^{i-K-1}$$

$$+ \sum_{r=1}^{K-1} \lambda^{r-K}\left[f^{(\mathbf{d})}(r) - \eta^{\mathbf{d}}\right]\prod_{k=r+1}^{K} v(d_k)$$

$$+ \sum_{r=K}^{i-1} \lambda^{r-i}\left[f(r) - \eta^{\mathbf{d}}\right]\left[v(1)\right]^{i-r-2}.$$

3.5   The Penalty Cost

When the inventory level is low, if the service priority is violated (i.e., the demands of Class 2 are served at a low stock), then the warehouse has to pay the penalty cost P for each product supplied to the demands of Class 2.

For our later discussion, we will see that $G^{(\mathbf{d})}(i) + b$ where $b = R + C_{2,2} - P$ plays a fundamental role in the performance optimization of the stock-rationing queue and the sign directly determines the selection of decision actions. Based on this, we study the influence of the penalty cost P on the sign of $G^{(\mathbf{d})}(i) + b$. From $G^{(\mathbf{d})}(i) + b = 0$, we have

$$P \left\{ 1 + \lambda^{-i} \left[A_0 - F^{(\mathbf{d})} \right] \prod_{k=1}^{i-1} v\left(d_k\right) + \sum_{r=1}^{i-1} \lambda^{r-i} \left[A_r^{(\mathbf{d})} - F^{(\mathbf{d})} \right] \prod_{k=r+1}^{i-1} v\left(d_k\right) \right\}$$

$$= R + C_{2,2} + \lambda^{-i} \left[B_0 - D^{(\mathbf{d})} \right] \prod_{k=1}^{i-1} v\left(d_k\right) + \sum_{r=1}^{i-1} \lambda^{r-i} \left[B_r^{(\mathbf{d})} - D^{(\mathbf{d})} \right] \prod_{k=r+1}^{i-1} v\left(d_k\right),$$

$$\tag{12}$$

thus, the unique solution of the penalty cost P to the linear equation (12) is given by

$$\mathfrak{P}_i^{(\mathbf{d})} = \frac{R + C_{2,2} + \lambda^{-i} \left[B_0 - D^{(\mathbf{d})} \right] \prod_{k=1}^{i-1} v\left(d_k\right) + \sum_{r=1}^{i-1} \lambda^{r-i} \left[B_r^{(\mathbf{d})} - D^{(\mathbf{d})} \right] \prod_{k=r+1}^{i-1} v\left(d_k\right)}{1 + \lambda^{-i} \left[A_0 - F^{(\mathbf{d})} \right] \prod_{k=1}^{i-1} v\left(d_k\right) + \sum_{r=1}^{i-1} \lambda^{r-i} \left[A_r^{(\mathbf{d})} - F^{(\mathbf{d})} \right] \prod_{k=r+1}^{i-1} v\left(d_k\right)}.$$

$$\tag{13}$$

It's easy to see that if $\mathfrak{P}_i^{(\mathbf{d})} > 0$ and $0 \leq P \leq \mathfrak{P}_i^{(\mathbf{d})}$, then $G^{(\mathbf{d})}\left(i\right) + b \geq 0$; while if $P \geq \mathfrak{P}_i^{(\mathbf{d})}$, then $G^{(\mathbf{d})}\left(i\right) + b \leq 0$. Note that the equality can hold only if $P = \mathfrak{P}_i^{(\mathbf{d})}$.

4 Monotonicity and Optimality

In this section, we analyze the optimal dynamic rationing policy in the three different areas of the penalty cost $P \geq P_H\left(\mathbf{d}\right)$; $P_L\left(\mathbf{d}\right) > 0$ and $0 < P \leq P_L\left(\mathbf{d}\right)$; and $P_L\left(\mathbf{d}\right) < P < P_H\left(\mathbf{d}\right)$. For the third area: $P_L\left(\mathbf{d}\right) < P < P_H\left(\mathbf{d}\right)$, the optimal dynamic rationing policy may not be of threshold type but it must be of transformational threshold type.

The following lemma provides a useful equation for the difference $\eta^{\mathbf{d}'} - \eta^{\mathbf{d}}$ corresponding to any two different policies $\mathbf{d}, \mathbf{d}' \in \mathcal{D}$, called performance difference equation. See Cao [4] and Ma et al. [12] for more details.

Lemma 1. *For any two policies* $\mathbf{d}, \mathbf{d}' \in \mathcal{D}$*, we have*

$$\eta^{\mathbf{d}'} - \eta^{\mathbf{d}} = \pi^{(\mathbf{d}')} \left[\left(\mathbf{B}^{(\mathbf{d}')} - \mathbf{B}^{(\mathbf{d})} \right) \mathbf{g}^{(\mathbf{d})} + \left(\mathbf{f}^{(\mathbf{d}')} - \mathbf{f}^{(\mathbf{d})} \right) \right]. \tag{14}$$

To find the optimal rationing policy $\mathbf{d}^*$, we define two rationing policies $\mathbf{d}$ and $\mathbf{d}'$ with an interrelated structure at Position i as follows:

$$\mathbf{d} = \left(0; d_1, d_2, \ldots, d_{i-1}, \underline{d_i}, d_{i+1}, \ldots, d_K; 1, 1, \ldots, 1 \right),$$
$$\mathbf{d}' = \left(0; d_1, d_2, \ldots, d_{i-1}, \underline{d_i'}, d_{i+1}, \ldots, d_K; 1, 1, \ldots, 1 \right),$$

where $d_i', d_i \in \{0,1\}$ with $d_i' \neq d_i$. Clearly, if the two rationing policies $\mathbf{d}$ and $\mathbf{d}'$ have an interrelated structure at Position i, then only the difference between the two rationing policies $\mathbf{d}$ and $\mathbf{d}'$ is at their ith elements: d_i and d_i'.

Lemma 2. *For the two rationing policies* **d** *and* **d'** *with an interrelated structure at Position* i: d_i *and* d'_i, *we have*

$$\eta^{\mathbf{d}'} - \eta^{\mathbf{d}} = \mu_2 \pi^{(\mathbf{d}')}(i)(d'_i - d_i)\left[G^{(\mathbf{d})}(i) + b\right]. \tag{15}$$

In the stock-rationing queue, we define two critical values related to the penalty cost P as

$$P_H(\mathbf{d}) = \max_{\mathbf{d} \in \mathcal{D}}\left\{0, \mathfrak{P}_1^{(\mathbf{d})}, \mathfrak{P}_2^{(\mathbf{d})}, \ldots, \mathfrak{P}_K^{(\mathbf{d})}\right\} \tag{16}$$

and

$$P_L(\mathbf{d}) = \min_{\mathbf{d} \in \mathcal{D}}\left\{\mathfrak{P}_1^{(\mathbf{d})}, \mathfrak{P}_2^{(\mathbf{d})}, \ldots, \mathfrak{P}_K^{(\mathbf{d})}\right\}. \tag{17}$$

The following proposition uses the two critical values $P_H(\mathbf{d})$ and $P_L(\mathbf{d})$, together with the penalty cost P, to provide some sufficient conditions under which the function $G^{(\mathbf{d})}(i) + b$ is either non-positive or non-negative.

Proposition 1. *(1) If* $P \geq P_H(\mathbf{d})$ *for any given rationing policy* $\mathbf{d} \in \mathcal{D}$, *then for each* $i = 1, 2, \ldots, K$,
$$G^{(\mathbf{d})}(i) + b \leq 0.$$

(2) If $P_L(\mathbf{d}) > 0$ *and* $0 \leq P \leq P_L(\mathbf{d})$ *for any given rationing policy* $\mathbf{d} \in \mathcal{D}$, *then for each* $i = 1, 2, \ldots, K$,

$$G^{(\mathbf{d})}(i) + b \geq 0.$$

The following theorems provide the optimal dynamic rationing policy of the stock-rationing queue in two different areas, see the proof in Li et al. [11].

Theorem 2. *If* $P \geq P_H(\mathbf{d})$ *for any given rationing policy* $\mathbf{d}$, *then the optimal dynamic rationing policy of the stock-rationing queue is given by*

$$\mathbf{d}^* = (0; 0, 0, \ldots, 0; 1, 1, \ldots, 1).$$

This shows that if the penalty cost is higher with $P \geq P_H(\mathbf{d})$ *for any given rationing policy* $\mathbf{d}$, *then the warehouse can not supply any product to the demands of Class* 2.

Theorem 3. *If* $P_L(\mathbf{d}) > 0$ *and* $0 \leq P \leq P_L(\mathbf{d})$ *for any given rationing policy* $\mathbf{d}$, *then the optimal dynamic rationing policy of the stock-rationing queue is given by*

$$\mathbf{d}^* = (0; 1, 1, \ldots, 1; 1, 1, \ldots, 1).$$

This shows that if the penalty cost is lower with $P_L(\mathbf{d}) > 0$ *and* $0 \leq P \leq P_L(\mathbf{d})$, *then the warehouse would like to supply the products to the demands of Class* 2.

In what follows, we discuss the third area of the penalty cost: $P_L(\mathbf{d}) < P < P_H(\mathbf{d})$. Note that the analysis for this area is a little complicated. To this end, we propose a new algebraic method to find the optimal dynamic rationing policy of the stock-rationing queue.

Based on (13), we introduce a convention: If $\mathfrak{P}_{n-1}^{(\mathbf{d})} < \mathfrak{P}_n^{(\mathbf{d})} = \mathfrak{P}_{n+1}^{(\mathbf{d})} = \cdots = \mathfrak{P}_{n+i}^{(\mathbf{d})} = c$ and $\mathfrak{P}_{n-1}^{(\mathbf{d})} < P \le c$, then we write

$$\mathfrak{P}_{n-1}^{(\mathbf{d})} < P \le \mathfrak{P}_n^{(\mathbf{d})} = \mathfrak{P}_{n+1}^{(\mathbf{d})} = \cdots = \mathfrak{P}_{n+i}^{(\mathbf{d})},$$

that is, the penalty cost P is written in front of all the equal elements in the sequence $\left\{\mathfrak{P}_k^{(\mathbf{d})} : n \le k \le n+i\right\}$.

For the sequence $\left\{\mathfrak{P}_k^{(\mathbf{d})} : 1 \le k \le K\right\}$, we set up a new permutation from the smallest to the largest as follows:

$$\mathfrak{P}_{i_1}^{(\mathbf{d})} \le \mathfrak{P}_{i_2}^{(\mathbf{d})} \le \cdots \le \mathfrak{P}_{i_{K-1}}^{(\mathbf{d})} \le \mathfrak{P}_{i_K}^{(\mathbf{d})},$$

it is clear that $\mathfrak{P}_{i_1}^{(\mathbf{d})} = P_L(\mathbf{d})$ and $\mathfrak{P}_{i_K}^{(\mathbf{d})} = P_H(\mathbf{d})$. For convenience of description, for the incremental sequence $\left\{\mathfrak{P}_{i_j}^{(\mathbf{d})} : 1 \le j \le K\right\}$, we write its subscript vector as $(i_1, i_2, \ldots, i_K)$. Note that the subscript vector $(i_1, i_2, \ldots, i_K)$ depends on rationing policy $\mathbf{d}$.

The following lemma shows how the penalty cost P is distributed in the sequence $\left\{\mathfrak{P}_k^{(\mathbf{d})} : 1 \le k \le K\right\}$.

Lemma 3. If $P_L(\mathbf{d}) < P < P_H(\mathbf{d})$ for any given rationing policy $\mathbf{d}$, then there exists the minimal positive integer $n_0 \in \{1, 2, \ldots, K\}$ such that either

$$\mathfrak{P}_{i_{n_0}}^{(\mathbf{d})} < P = \mathfrak{P}_{i_{n_0+1}}^{(\mathbf{d})}$$

or

$$\mathfrak{P}_{i_{n_0}}^{(\mathbf{d})} < P < \mathfrak{P}_{i_{n_0+1}}^{(\mathbf{d})}.$$

Now, our task is to develop a new method for finding the optimal dynamic rationing policy by means of the two useful information: (a) The incremental sequence

$$P_L(\mathbf{d}) = \mathfrak{P}_{i_1}^{(\mathbf{d})} \le \mathfrak{P}_{i_2}^{(\mathbf{d})} \le \cdots \le \mathfrak{P}_{i_{K-1}}^{(\mathbf{d})} \le \mathfrak{P}_{i_K}^{(\mathbf{d})} = P_H(\mathbf{d});$$

and (b) the penalty cost P has a fixed position: $\mathfrak{P}_{i_{n_0}}^{(\mathbf{d})} < P \le \mathfrak{P}_{i_{n_0+1}}^{(\mathbf{d})}$, where n_0 is the minimal positive integer in the set $\{1, 2, \ldots, K\}$.

In what follows we discuss two different cases: A simple case with

$$P_L(\mathbf{d}) = \mathfrak{P}_1^{(\mathbf{d})} \le \mathfrak{P}_2^{(\mathbf{d})} \le \cdots \le \mathfrak{P}_{K-1}^{(\mathbf{d})} \le \mathfrak{P}_K^{(\mathbf{d})} = P_H(\mathbf{d}) \tag{18}$$

and a general case with

$$P_L(\mathbf{d}) = \mathfrak{P}_{i_1}^{(\mathbf{d})} \le \mathfrak{P}_{i_2}^{(\mathbf{d})} \le \cdots \le \mathfrak{P}_{i_{K-1}}^{(\mathbf{d})} \le \mathfrak{P}_{i_K}^{(\mathbf{d})} = P_H(\mathbf{d}). \tag{19}$$

Case One: A Simple Case

In case of (18), the subscript vector is expressed as $\{1, 2, \ldots, K\}$ depending on the rationing policy $\mathbf{d}$.

If $P_L(\mathbf{d}) < P < P_H(\mathbf{d})$ for any given rationing policy $\mathbf{d}$, then there exists the minimal positive integer $n_0 \in \{1, 2, \ldots, K-1, K\}$ such that

$$P_L(\mathbf{d}) = \mathfrak{P}_1^{(\mathbf{d})} \leq \cdots \leq \mathfrak{P}_{n_0-1}^{(\mathbf{d})} < P \leq \mathfrak{P}_{n_0}^{(\mathbf{d})} \leq \cdots \leq \mathfrak{P}_K^{(\mathbf{d})} = P_H(\mathbf{d}).$$

Based on this, we take two different sets

$$\Lambda_1 = \left\{ \mathfrak{P}_1^{(\mathbf{d})}, \mathfrak{P}_2^{(\mathbf{d})}, \ldots, \mathfrak{P}_{n_0-1}^{(\mathbf{d})} \right\}$$

and

$$\Lambda_2 = \left\{ \mathfrak{P}_{n_0}^{(\mathbf{d})}, \mathfrak{P}_{n_0+1}^{(\mathbf{d})}, \ldots, \mathfrak{P}_K^{(\mathbf{d})} \right\}.$$

By using the two sets Λ_1 and Λ_2, we write

$$\overline{P}_H(\mathbf{d}; 1 \to n_0 - 1) = \max_{1 \leq i \leq n_0-1} \left\{ \mathfrak{P}_i^{(\mathbf{d})} \right\}$$

and

$$\overline{P}_L(\mathbf{d}; n_0 \to K) = \min_{n_0 \leq j \leq K} \left\{ \mathfrak{P}_j^{(\mathbf{d})} \right\}.$$

It is clear that $\overline{P}_H(\mathbf{d}; 1 \to n_0 - 1) = \mathfrak{P}_{n_0-1}^{(\mathbf{d})}$ and $\overline{P}_L(\mathbf{d}; n_0 \to K) = \mathfrak{P}_{n_0}^{(\mathbf{d})}$.

For this simple case, the following theorem finds the optimal dynamic rationing policy is of threshold type.

Theorem 4. *For the simple case with $P_L(\mathbf{d}) < P < P_H(\mathbf{d})$ for any given rationing policy $\mathbf{d}$, if there exists the minimal positive integer $n_0 \in \{1, 2, \ldots, K\}$ such that*

$$P_L(\mathbf{d}) = \mathfrak{P}_1^{(\mathbf{d})} \leq \cdots \leq \mathfrak{P}_{n_0-1}^{(\mathbf{d})} < P \leq \mathfrak{P}_{n_0}^{(\mathbf{d})} \leq \cdots \leq \mathfrak{P}_K^{(\mathbf{d})} = P_H(\mathbf{d}), \qquad (20)$$

then the optimal dynamic rationing policy is given by

$$\mathbf{d}^* = \left(0; \underbrace{0, 0, \ldots, 0}_{n_0-1 \ zeros}, \underbrace{1, 1, \ldots, 1}_{K-n_0+1 \ ones}; 1, 1, \ldots, 1 \right).$$

Case Two: A General Case

For the incremental sequence $\left\{ \mathfrak{P}_{i_j}^{(\mathbf{d})} : j = 1, 2, \ldots, K \right\}$, we write its subscript vector as $(i_1, i_2, \ldots, i_K)$, which depends on rationing policy $\mathbf{d}$. In the general case, we assume that $(i_1, i_2, \ldots, i_K) \neq (1, 2, \ldots, K)$.

If $P_L(\mathbf{d}) < P < P_H(\mathbf{d})$ for any given rationing policy $\mathbf{d}$, then there exists the minimal positive integer $n_0 \in \{1, 2, \ldots, K\}$ such that

$$P_L(\mathbf{d}) = \mathfrak{P}_{i_1}^{(\mathbf{d})} \leq \cdots \leq \mathfrak{P}_{i_{n_0-1}}^{(\mathbf{d})} < P \leq \mathfrak{P}_{i_{n_0}}^{(\mathbf{d})} \leq \cdots \leq \mathfrak{P}_{i_K}^{(\mathbf{d})} = P_H(\mathbf{d}).$$

Based on this, we take two sets

$$\Lambda_1^G = \left\{ \mathfrak{P}_{i_1}^{(\mathbf{d})}, \mathfrak{P}_{i_2}^{(\mathbf{d})}, \ldots, \mathfrak{P}_{i_{n_0-1}}^{(\mathbf{d})} \right\}$$

and

$$\Lambda_2^G = \left\{ \mathfrak{P}_{i_{n_0}}^{(\mathbf{d})}, \mathfrak{P}_{i_{n_0+1}}^{(\mathbf{d})}, \ldots, \mathfrak{P}_{i_K}^{(\mathbf{d})} \right\}.$$

For the two sets Λ_1^G and Λ_2^G, we write

$$\overline{P}_H^G (\mathbf{d}; 1 \to n_0 - 1) = \max_{1 \le k \le n_0 - 1} \left\{ \mathfrak{P}_{i_k}^{(\mathbf{d})} \right\}$$

and

$$\overline{P}_L^G (\mathbf{d}; n_0 \to K) = \min_{n_0 \le k \le K} \left\{ \mathfrak{P}_{i_k}^{(\mathbf{d})} \right\},$$

It is clear that $\overline{P}_H^G (\mathbf{d}; 1 \to n_0 - 1) = \mathfrak{P}_{i_{n_0-1}}^{(\mathbf{d})}$ and $\overline{P}_L^G (\mathbf{d}; n_0 \to K) = \mathfrak{P}_{i_{n_0}}^{(\mathbf{d})}$.

Corresponding to the subscript vector of the incremental sequence $\left\{ \mathfrak{P}_{i_k}^{(\mathbf{d})} : 1 \le k \le K \right\}$, we transfer rationing policy

$$\mathbf{d} = (0; d_1, d_2, \ldots, d_{n_0-1}, d_{n_0}, d_{n_0+1}, \ldots, d_K; 1, 1, \ldots, 1)$$

into a new transformational rationing policy

$$\mathbf{d} \, (\text{Transfer}) = \left(0; d_{i_1}, d_{i_2}, \ldots, d_{i_{n_0-1}}, d_{i_{n_0}}, d_{i_{n_0+1}}, \ldots, d_{i_K}; 1, 1, \ldots, 1 \right).$$

Therefore, a transformation of the optimal dynamic rationing policy $\mathbf{d}^*$ is

$$(1, 2, \ldots, K - 1, K) \Rightarrow (i_1, i_2, \ldots, i_{K-1}, i_K);$$

and an inverse transformation of the optimal transformational dynamic rationing policy $\mathbf{d}^*$ (Transfer) is

$$(i_1, i_2, \ldots, i_{K-1}, i_K) \Rightarrow (1, 2, \ldots, K - 1, K).$$

For the general case, the following theorem finds the optimal dynamic rationing policy, which may not be of threshold type, but it must be of transformational threshold type.

Theorem 5. *For the general case with $P_L (\mathbf{d}) < P < P_H (\mathbf{d})$ for any given rationing policy $\mathbf{d}$, if there exists the minimal positive integer $n_0 \in \{1, 2, \ldots, K\}$ such that*

$$P_L (\mathbf{d}) = \mathfrak{P}_{i_1}^{(\mathbf{d})} \le \cdots \le \mathfrak{P}_{i_{n_0-1}}^{(\mathbf{d})} < P \le \mathfrak{P}_{i_{n_0}}^{(\mathbf{d})} \le \cdots \le \mathfrak{P}_{i_K}^{(\mathbf{d})} = P_H (\mathbf{d}),$$

then the optimal transformational dynamic rationing policy is given by

$$\mathbf{d}^* \, (\text{Transfer}) = \left(0; \underbrace{0, 0, \ldots, 0}_{n_0-1 \ zeros}, \ \underbrace{1, 1, \ldots, 1}_{K-n_0+1 \ ones} \, ; 1, 1, \ldots, 1 \right).$$

The following theorem provides a useful summarization for Theorems 2 to 5, and shows that we obtain a complete algebraic solution to the optimal dynamic rationing policy of the stock-rationing queue.

Theorem 6. *For the stock-rationing queue with two demand classes, there must exist an optimal transformational dynamic rationing policy*

$$\mathbf{d}^* \left(\textit{Transfer} \right) = \left(0; \underbrace{0, 0, \ldots, 0}_{n_0 - 1 \ zeros}, \underbrace{1, 1, \ldots, 1}_{K - n_0 \ ones}; 1, 1, \ldots, 1 \right).$$

Based on this finding, we can achieve the following two useful results:

(a) The optimal dynamic rationing policy $\mathbf{d}^*$ is of critical rationing level (i.e., threshold type) under each of the three conditions: (i) $P \geq P_H (\mathbf{d})$ for any given rationing policy $\mathbf{d}$; (ii) $P_L (\mathbf{d}) > 0$ and $0 \leq P \leq P_L (\mathbf{d})$ for any given rationing policy $\mathbf{d}$; and (iii) $P_L (\mathbf{d}) < P < P_H (\mathbf{d})$ with the subscript vector $(1, 2, \ldots, K)$ depending on the rationing policy $\mathbf{d}$.

(b) The optimal dynamic rationing policy is not of critical rationing level (i.e., threshold type) if $P_L (\mathbf{d}) < P < P_H (\mathbf{d})$ with the subscript vector $(i_1, i_2, \ldots, i_K) \neq (1, 2, \ldots, K)$ depending on rationing policy $\mathbf{d}$.

In the remainder of this section, we provide two numerical examples to verify computability of our theoretical results and analyze how the optimal long-run average profit of the stock-rationing queue depends on some key system parameters.

In the numerical examples, we take some common parameters: $C_1 = 1, C_{2,1} = 4, C_{2,2} = 1, C_3 = 5, C_4 = 1; R = 15, N = 100; \mu_1 = 30, \mu_2 = 40; K = 5, 6, 10$.

(a) A higher penalty cost. Let $P = 10$ and $\mathbf{d}^* = (0; 0, 0, \ldots, 0; 1, 1, \ldots, 1)$. We discuss how the optimal long-run average profit $\eta^{\mathbf{d}^*}$ depends on λ for $\lambda \in (47, 50)$.

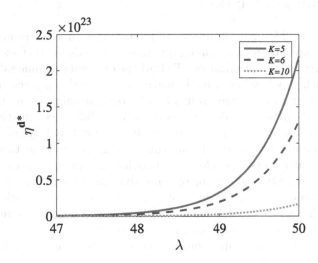

Fig. 2. $\eta^{\mathbf{d}^*}$ vs. λ under three different thresholds K with a higher penalty cost.

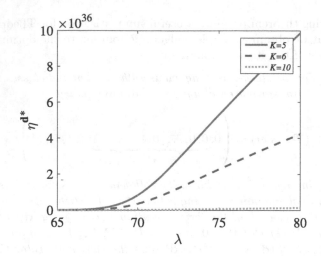

Fig. 3. $\eta^{\mathbf{d}^*}$ vs. λ under three different thresholds K with a lower penalty cost.

From Fig. 2, it is seen that the optimal long-run average profit $\eta^{\mathbf{d}^*}$ increases as λ increases. In addition, with the threshold K increases, the optimal long-run average profit $\eta^{\mathbf{d}^*}$ increases less slowly as λ increases.

(b) A lower penalty cost. Let $P = 0.1$ and $\mathbf{d}^* = (0; 1, 1, \ldots, 1; 1, 1, \ldots, 1)$. We discuss how the optimal long-run average profit $\eta^{\mathbf{d}^*}$ depends on λ for $\lambda \in (65, 80)$. From Fig. 3, it is seen that the optimal long-run average profit $\eta^{\mathbf{d}^*}$ increases as λ increases. In addition, with the threshold K increases, the optimal long-run average profit $\eta^{\mathbf{d}^*}$ increases less slowly as λ increases.

5   Concluding Remarks

In this paper, we highlight intuitive understanding on the optimal dynamic rationing policy of the stock-rationing queue with two demand classes by means of the sensitivity-based optimization. To find the optimal dynamic rationing policy, we establish a policy-based birth-death process and a key reward function such that the long-run average profit of the stock-rationing queue is expressed explicitly. Based on this, we derive a performance difference equation between any two rationing policies such that we can find the optimal dynamic rationing policy and compute the maximal long-run average profit from three different areas of the penalty costs. Therefore, we provide an algebraic method to set up a complete algebraic solution to the optimal dynamic rationing policy. We show that the optimal dynamic rationing policy must be of transformational threshold type, which leads to refining three simple sufficient conditions under each of which the optimal dynamic rationing policy is of threshold type.

Along such a line, there are a number of interesting directions for potential future research, for example:

- Extending to the stock-rationing queues with multiple demand classes, multiple types of products, backorders, batch order, batch production, and so on;
- analyzing non-Poisson input, such as Markovian arrival processes (MAPs); and/or non-exponential service times, e.g., the PH distributions;
- discussing how the long-run average profit can be influenced by some concave or convex reward functions;
- studying individual or social optimization for stock-rationing queues from a perspective of game theory by means of the sensitivity-based optimization.

References

1. Altug, M.S., Ceryan, O.: Optimal dynamic allocation of rental and sales inventory for fashion apparel products. IISE Trans. **54**(6), 603–617 (2021)
2. Baron, O., Lu, T., Wang, J.: Priority, capacity rationing, and ambulance diversion in emergency departments. SSRN. 3387439 (2019)
3. Benjaafar, S., ElHafsi, M.: Production and inventory control of a single product assemble-to-order system with multiple customer classes. Manage. Sci. **52**(12), 1896–1912 (2006)
4. Cao, X.R.: Stochastic Learning and Optimization–A Sensitivity-based Approach. Springer, New York (2007). https://doi.org/10.1007/978-0-387-69082-7
5. Gayon, J.P., De Vericourt, F., Karaesmen, F.: Stock rationing in an $M/E_k/1$ multiclass make-to-stock queue with backorders. IIE Trans. **41**(12), 1096–1109 (2009)
6. Ha, A.Y.: Inventory rationing in a make-to-stock production system with several demand classes and lost sales. Manage. Sci. **43**(8), 1093–1103 (1997)
7. Ha, A.Y.: Stock-rationing policy for a make-to-stock production system with two priority classes and backordering. Nav. Res. Log. **44**(5), 457–472 (1997)
8. Ha, A.Y.: Stock rationing in an $M/E_k/1$ make-to-stock queue. Manage. Sci. **46**(1), 77–87 (2000)
9. Jain, A., Moinzadeh, K., Dumrongsiri, A.: Priority allocation in a rental model with decreasing demand. M. Som-Manuf. Serv. Op. **17**(2), 236–248 (2015)
10. Li, Q.L.: Constructive Computation in Stochastic Models with Applications: The RG-Factorizations. Springer (2010). https://doi.org/10.1007/978-3-642-11492-2
11. Li, Q.L., Li, Y.M., Ma, J.Y., Liu, H.L.: A complete algebraic solution to the optimal dynamic rationing policy in the stock-rationing queue with two demand classes. arXiv: 1908.09295 (2019)
12. Ma, J.Y., Xia, L., Li, Q.L.: Optimal energy-efficient policies for data centers through sensitivity-based optimization. Discrete Event Dyn. S. **29**(4), 567–606 (2019)
13. Ma, J.Y., Li, Q.L., Xia, L.: Optimal asynchronous dynamic policies in energy-efficient data centers. Systems **10**(2), 27 (2022)
14. Nadar, E., Akan, M., Scheller-Wolf, A.: Optimal structural results for assemble-to-order generalized M-systems. Oper. Res. **62**(3), 571–579 (2014)
15. Moosa, M.R., Luyckx, V.A.: The realities of rationing in health care. Nat. Rev. Nephrol. **17**(7), 435–436 (2021)
16. Puterman, M.L.: Markov Decision Processes: Discrete Stochastic Dynamic Programming. John Wiley & Sons, Hoboken (2014)
17. Topkis, D.M.: Optimal ordering and rationing policies in a nonstationary dynamic inventory model with n demand classes. Manage. Sci. **15**(3), 160–176 (1968)

18. Veinott, A.F., Jr.: Optimal policy in a dynamic, single product, nonstationary inventory model with several demand classes. Oper. Res. **13**(5), 761–778 (1965)
19. Wang, R., Qin, Y., Sun, H.: Research on location selection strategy for airlines spare parts central warehouse based on METRIC. Comput. Intel. Neurosc. **2021**, 1–16 (2021)
20. Xia, L., Zhang, Z.G., Li, Q.L.: A c/μ-rule for job assignment in heterogeneous group-server queues. Prod. Oper. Manag. **31**(3), 1191–1215 (2021)
21. Zhao, H., Deshpande, V., Ryan, J.K.: Inventory sharing and rationing in decentralized dealer networks. Manage. Sci. **51**(4), 531–547 (2005)

The Constrained Parallel-Machine Scheduling Problem with Divisible Processing Times and Penalties

Jianping Li[1]($\boxtimes$), Runtao Xie[1], Junran Lichen[2], Guojun Hu[1], Pengxiang Pan[1], and Ping Yang[1]

[1] Department of Mathematics, Yunnan University, East Outer Ring South Road, University Town, Chenggong District, Kunming 650504, People's Republic of China
jianping@ynu.edu.cn, xieruntao7@163.com, huguojun@mail.ynu.edu.cn, panpx@outlook.com, yp1573395725@hotmail.com
[2] School of Mathematics and Physics, Beijing University of Chemical Technology, No. 15, North Third Ring East Road, Chaoyang District, Beijing 100029, People's Republic of China
J.R.Lichen@buct.edu.cn

Abstract. We consider the constrained parallel-machine scheduling problem with divisible processing times and penalties (the CPS-DTP problem, for short). Specifically, given a set $M = \{a_1, a_2, \ldots, a_m\}$ of m identical machines, and a set $J = \{b_1, b_2, \ldots, b_n\}$ of n jobs, each job $b_j \in J$ has a processing time $p_j \in Z^+$ and a penalty $e_j \in Z^+$, and the job processing times are divisible, *i.e.*, either $p_i | p_j$ or $p_j | p_i$ for any two different jobs b_i and b_j in J. Each job b_j is either executed in processing time p_j with which we schedule this job on one of m machines, or rejected with its penalty e_j that we must pay for, it is asked to determine a subset $A \subseteq J$ such that each job $b_j \in A$ has to be scheduled only on one of m machines and each job $b_j \in J \backslash A$ has to be rejected. We consider three versions of the CPS-DTP problem, respectively. (1) The constrained parallel-machine scheduling problem with divisible processing times and total penalties (the CPS-DTTP problem, for short) is asked to determine a subset $A \subseteq J$ to satisfy the constraint mentioned-above, the objective is to minimize the makespan of the schedule T for accepted jobs in A plus the value of total penalties of the rejected jobs in $J \backslash A$; (2) The constrained parallel-machine scheduling problem with divisible processing times and maximum penalty (the CPS-DTMP problem, for short) is asked to determine a subset $A \subseteq J$ to satisfy the constraint mentioned-above, the objective is to minimize the makespan of the schedule T for accepted jobs in A plus the maximum penalty paid for rejected jobs in $J \backslash A$; (3) The constrained parallel-machine scheduling problem with divisible processing times and bounded penalty (the CPS-DTBP problem, for short) is asked to determine a subset $A \subseteq J$ to

This paper is fully supported by the National Natural Science Foundation of China [Nos.11861075,12101593] and Project for Innovation Team (Cultivation) of Yunnan Province [No.202005AE160006]. Junran Lichen is also supported by Fundamental Research Funds for the Central Universities (buctrc202219), and Jianping Li is also supported by Project of Yunling Scholars Training of Yunnan Province.

Q. Ni and W. Wu (Eds.): AAIM 2022, LNCS 13513, pp. 83–95, 2022.
https://doi.org/10.1007/978-3-031-16081-3_8

satisfy the constraint mentioned-above and the value of total penalties of the rejected jobs in $J\backslash A$ is no more than a given bound, the objective is to minimize the makespan of the schedule T for accepted jobs in A.

In this paper, we design an exact algorithm in pseudo-polynomial time to solve the CPS-DTTP problem, an exact algorithm in strongly polynomial time to solve the CPS-DTMP problem and an exact algorithm in polynomial time to solve the CPS-DTBP problem, respectively.

Keywords: Combinatorial optimization · Scheduling · Divisible processing times · Penalties · Exact algorithms

1   Introduction

The classical scheduling problem [6] is defined as follows. Given a set $M = \{a_1, a_2, \ldots, a_m\}$ of m parallel machines and a set $J = \{b_1, b_2, \ldots, b_n\}$ of n jobs, where each job $b_j \in J$ has a processing time p_j, it is asked to determine a schedule T such that each job b_j has to be executed only on one of the machines, the objective is to minimize the makespan, *i.e.*, the maximum completion time taken over all machines. The classical scheduling problem has many applications in some domains, for example, personnel scheduling, computer systems, engineering design and transportation scheduling. Following the convention of Lawler et al. [9], we denote the classical scheduling problem by $P \parallel C_{\max}$.

It is well-known that the $P \parallel C_{\max}$ problem is *NP*-hard [10]. Graham [6] in 1966 proposed the classical list scheduling algorithm (the LS algorithm, for short) which is to arrange all jobs on the machines and assign the job which is next processed to the currently finishing fastest machine, and the approximation ratio of the LS algorithm is $2 - \frac{1}{m}$. Faigle [5] in 1989 proved that the LS algorithm is an optimal algorithm for the $P \parallel C_{\max}$ problem when $m = 2$ or $m = 3$. And Graham [7] in 1969 proposed another famous longest processing time algorithm (the LPT algorithm, for short) for the off-line version of the $P\|C_{max}$ problem which sort the jobs according to their processing times in non-increasing order, and then using the LS algorithm. The approximation ratio of the LPT algorithm is $\frac{4}{3} - \frac{1}{3m}$, which is significantly better than the LS algorithm. Coffman [2] in 1978 presented the MULTIFIT approximation algorithm for the $P \parallel C_{\max}$ problem according to the corresponding relationship between the bin packing problem and this problem, and then proved that the upper bound of the MULTIFIT algorithm does not exceed 1.22 in the worst case.

However, in many practical cases, processing all jobs may cause the high cost when some processing times of jobs are very large. So as a survey written by Shabtay et al. [12], in such cases, the firm may wish to reject these jobs to keep the completion of orders, though they need to pay some rejection penalties. Bartal et al. [1] in 1996 first considered the parallel-machine scheduling problem with rejection penalties (the PS-P problem, for short), which is a generalization of the $P \parallel C_{\max}$ problem, and it is defined as follows. Given a set $M = \{a_1, a_2, \ldots, a_m\}$ of m identical machines and a set $J = \{b_1, b_2, \ldots, b_n\}$ of n jobs, each job $b_j \in J$

has a processing time p_j and a penalty e_j. It is asked to determine a subset $A \subseteq J$ such that each job $b_j \in A$ has to be scheduled only on one of m machines and each job $b_j \in J \backslash A$ has to be rejected. The objective is to minimize the makespan of the schedule T for accepted jobs in A plus the value of total penalties of the rejected jobs in $J \backslash A$. For the PS-P problem, Bartal et al. [1] designed a fully polynomial-time approximation scheme (FPTAS, for short) for fixed m and a polynomial-time approximation scheme (PTAS, for short) for arbitrary m. In addition, they presented a $(2 - \frac{1}{m})$-approximation algorithm with running time $O(n \log n)$ for arbitrary m.

Zhang et al. [15] in 2009 introduced the parallel-machine scheduling problem with bounded penalty (the PS-BP problem, for short), which is a generalization of the PS-P problem. In addition, in order to satisfy the constraint of the PS-P problem, the PS-BP problem is asked to satisfy the constraint that the value of total penalties of the rejected jobs in $J \backslash A$ is no more than B, the objective is to minimize the makespan. Zhang et al. [15] designed an exact pseudo-polynomial-time dynamic programming algorithm for the PS-BP problem and an FPTAS when m is a constant. Li et al. [11] in 2014 designed a strongly polynomial-time 2-approximation algorithm and a PTAS for the PS-BP problem, and then presented a FPTAS for the case where the number of machines is a fixed constant, which improved previous best running time to $O(\frac{1}{\epsilon^{2m+3}} + mn^2)$.

Yue et al. [14] in 2019 proposed the parallel-machine scheduling problem with divisible processing times (the PS-DT problem, for short), which is a special version of the $P \parallel C_{\max}$ problem. In addition to satisfying the constraint of the $P \parallel C_{\max}$ problem, the PS-DT problem is asked to satisfy the constraint that the job processing times are divisible, $i.e.$, either $p_i | p_j$ or $p_j | p_i$, for any two different jobs b_i and b_j in J. For convenience, we denote this case by $(p_i, p_j) = \min\{p_i, p_j\}$. The objective is to minimize the makespan. Divisible processing times are of interest because they arise naturally in certain applications, such as memory allocation in computer systems, where device capacities and block sizes are commonly restricted to powers of 2 [8]. For the PS-DT problem, Yue et al. [14] designed a polynomial-time algorithm that can exactly solve the PS-DT problem in time $O(n \log n)$.

Zheng et al. [16] in 2018 introduced the parallel-machine scheduling problem with penalties under special conditions (the PS-PSC problem, for short), which is a special version of the PS-P problem. In addition to satisfying the constraint of the PS-P problem, the PS-PSC problem is asked to satisfy the constraint that $e_i/p_i \geq e_j/p_j$ for any two different jobs b_i and b_j in J whenever $p_i \geq p_j$, the objective is to minimize the makespan plus the value of total penalties of the rejected jobs. Zheng et al. [16] proved that the PS-PSC problem is solvable in polynomial time, and then presented an efficient polynomial-time algorithm to solve this problem.

Yue [13] in 2020 considered the parallel-machine scheduling problem with bounded penalty under special conditions (the PS-BPSC problem, for short), which is a special version of the PS-BP problem. In addition to satisfying the constraint of the PS-BP problem, the PS-PSC problem is asked to satisfy the constraint that each job $b_j \in J$ has a processing time $p_j \in 2^{Z^+}$ and a penalty $e_j \in \{1, 2\}$, the objective is to minimize the makespan plus the value of total penalties of the rejected jobs in $J \backslash A$. Yue [13] designed an exact algorithm in polynomial time that can exactly solve the PS-BPSC problem .

As the results mentioned-above, when more conditions about processing times and penalties are satisfied, we can find that the PS-P problem and its variants are solvable in polynomial time. But the conditions so far are strictly, it is natural to ask whether there exist some more general conditions that can keep polynomial-time solvability of those problems. Motivated by the problems mention-above, we consider the following three problems. Specifically, given a set $M = \{a_1, a_2, \ldots, a_m\}$ of m identical machines, and a set $J = \{b_1, b_2, \ldots, b_n\}$ of n jobs, each job $b_j \in J$ has a processing time $p_j \in Z^+$ and a penalty $e_j \in Z^+$, where the job processing times are divisible, $i.e.$, either $p_i | p_j$ or $p_j | p_i$, for any two different jobs b_i and b_j in J. Each of these n jobs is either processed on a machine and not allowed to be interrupted in the process, or rejected with its penalty that we must pay for, and each machine can only process one job at a time. It is asked to determine a subset $A \subseteq J$ such that each job $b_j \in A$ has to be scheduled only on one of m machines and each job $b_j \in J \backslash A$ has to be rejected. We consider three versions of this constrained parallel-machine scheduling problem with divisible processing times and penalties (the CPS-DTP problem, for short), $i.e.$, (1) The constrained parallel-machine scheduling problem with divisible processing times and total penalties (the CPS-DTTP problem, for short) is asked to determine a subset $A \subseteq J$ to satisfy the constraint mentioned-above, the objective is to minimize the makespan of the schedule T for accepted jobs in A plus the value of total penalties of the rejected jobs in $J \backslash A$, $i.e.$, $\min\{C_{\max}(A) + \sum_{b_j \in J \backslash A} e_j \mid A \subseteq J\}$ (2) The constrained parallel-machine scheduling problem with divisible processing times and maximum penalty (the problem CPS-DTMP, for short) is asked to determine a subset $A \subseteq J$ to satisfy the constraint mentioned-above, the objective is to minimize the makespan of the schedule T for the accepted jobs in A plus maximum penalty paid for the rejected jobs in $J \backslash A$, $i.e.$, $\min\{C_{\max}(A) + \max\{e_j | b_j \in J \backslash A\} \mid A \subseteq J\}$ (3) The constrained parallel-machine scheduling problem with divisible processing times and bounded penalty (the CPS-DTBP problem, for short) is asked to determine a subset $A \subseteq J$ of to satisfy the constraint mentioned-above and the value of total penalties of the rejected jobs in $J \backslash A$ is no more than a given bound B, $i.e.$, $\sum_{b_j \in J \backslash A} e_j \leq B$, the objective is to minimize the makespan of the schedule T for accepted jobs in A , $i.e.$, $\min\{C_{\max}(A) \mid A \subseteq J\}$.

For convenience, by following the convention of Lawler et al. [9], we denote these three versions of the CPS-DTP problem by $P \mid (p_i, p_j) = \min\{p_i, p_j\}, rej \mid C_{\max} + \sum_{b_j \in J \backslash A} e_j$, $P \mid (p_i, p_j) = \min\{p_i, p_j\}, rej \mid C_{\max} + \max\{e_j | b_j \in J \backslash A\}$ and $P \mid (p_i, p_j) = \min\{p_i, p_j\}, \sum_{b_j \in J \backslash A} e_j \leq B \mid C_{\max}$, respectively.

As far as what we have known, the PS-P problem [1,4] and the PS-BP problem [11,13,15] have been studied for many years. However, to our best knowledge, the three problems mentioned-above have not been considered in the literature. Our contribution in this paper is to design three exact algorithms to solve the CPS-DTTP problem, the CPS-DTMP problem and the CPS-DTBP problem, respectively.

The paper is organized as follows. In Sect. 2, we introduce some notations and terminologies, and then present fundamental lemmas to ensure the correctness of our exact algorithms; In Sect. 3, we present an exact algorithm to solve the CPS-DTTP problem, and this algorithm runs in time $O((n \log n + nm)C)$, where n is the number of jobs, m is the number of machines and C is the optimal value for the PS-DT problem; In Sect. 4, we design an exact algorithm to solve the CPS-DTMP problem, and this algorithm runs in time $O(n^2 \log n)$; In Sect. 5, we provide an exact algorithm to solve the CPS-DTBP problem, and that algorithm runs in time $O((n \log n + nm) \log C)$; In Sect. 6, we provide our conclusion and further research.

2   Terminologies and Fundamental Lemmas

In order to clearly present our exact algorithms to solve the CPS-DTTP problem, the CPS-DTMP problem and the CPS-DTBP, respectively, we restate a special version of classical scheduling problem and a special version of the multiple knapsack problem in the sequel. And we provide some fundamental lemmas as follows.

The parallel-machine scheduling problem with divisible processing times (the PS-DT problem, for short) is defined as follows.

Definition 1. [14] *Given a set $M = \{a_1, a_2, \ldots, a_m\}$ of m identical machines, and a set $J = \{b_1, b_2, \ldots, b_n\}$ of n jobs, each job $b_j \in J$ has a processing time $p_j \in Z^+$, where the job processing times are divisible, i.e., either $p_i | p_j$ or $p_j | p_i$, for any two different jobs b_i and b_j in J. Each machine can only process one job at a time, and the job is not allowed to be interrupted in the processing. It is asked to determine a schedule T such that each job $b_j \in J$ has to be processed on one of the machines. The objective is to minimize the makespan, i.e., $\min_T \{C_{\max}(T)\}$.*

For convenience, we use $I = (M, J, p, e)$ to denote an instance of the CPS-DTTP problem and the CPS-DTMP, use $I' = (M, J, p, e; B)$ to denote an instance of the CPS-DTBP problem in the sequel. For any subset $X \subseteq J$, we define $e(X) = \sum_{b_j \in J} e_j$. Now we can construct an instance $\tau(I) = (M, J, p)$ of the PS-DT problem with a processing time function $p: J \to Z^+$.

We have known that Yue [14] in 2019 proved the LPT algorithm [7] can solve the PS-DT problem in polynomial time, which is restated as follows

Lemma 1. [14] *There is an exact polynomial algorithm, denoted by the $\mathcal{A}_{LPT}$ algorithm [7], to solve the PS-DT problem, and this algorithm runs in time $O(n \log n)$, where n is the number of jobs with divisible processing times.*

The multiple knapsack problem with divisible item sizes (the MKP-DS problem, for short) is defined as follows.

Definition 2. [3] *Given a set N of m knapsacks with same capacity limitation L and a set $Y = \{y_1, \ldots, y_n\}$ of n items, each item $y_j \in Y$ has its size $s_j \in Z^+$ and value $v_j \in R^+$, where item sizes are divisible, i.e., either $s_i | s_j$ or $s_j | s_i$, for any two different items y_i and y_j in Y. It is asked to determine an assignment $(Y_1, Y_2, \ldots, Y_m; Y_0)$ of items, satisfying $Y = Y_0 \cup (\cup_{i=1}^m Y_i)$, $Y_0 \cap Y_k = \phi$ for each $k \in \{1, 2, \ldots, m\}$, and $Y_i \cap Y_j = \phi$ for two different integers $i, j \in \{1, 2, \ldots, m\}$, such that the items in $Y_i \subseteq Y$ $(i = 1, \ldots, n)$ can be put into the ith knapsack under the constraint that the total size of items in Y_i does not exceed L, the objective is to maximize the summation of values of assigned items, i.e., $\max\{\sum_{y_j \in Y'} v_j \mid Y' = \cup_{i=1}^m Y_i\}$.*

Similarly, we use $Q = (N, Y; s, v; L)$ to denote an instance of the MKP-DS problem in the sequel. For any subset $Y' \subseteq Y$, we define $v(Y') = \sum_{y_j \in Y'} v_j$.

We have known that Detti [3] in 2009 presented an exact algorithm in polynomial time to solve the MKP-DS problem, which is restated as follows

Lemma 2. [3] *There is an exact polynomial-time algorithm, denoted by the $\mathcal{A}_{Detti}$ algorithm, to solve the MKP-DS problem, and this algorithm runs in time $O(n \log n + nm)$, where m is the number of knapsacks and n is the number of items.*

Given an instance I of either the CPS-DTMP problem or the CPS-DTBP problem, we can construct an instance $\rho_k(I)$ of the MKP-DS problem in the following ways. Specifically, given an instance $I = (M, J, p, e)$ and a positive integer k, we construct an instance $\rho_k(I) = (N, Y; s, v; k)$ of the MKP-DS problem, i.e., N is the same set of m knapsacks with same capacity limitation k, each item $y_j \in Y$ has its size $s_j \in Z^+$ and value $v_j \in R^+$, where $s_j = p_j$ and $v_j = e_j$ for each job $b_j \in J$.

3   An Exact Algorithm to Solve the CPS-DTTP Problem

In this section, we consider the constrained parallel-machine scheduling problem with divisible processing times and total penalties (the CPS-DTTP problem), and we plan to find a schedule T and a subset $A \subseteq J$ such that the value $C_{\max}(T) + e(J \backslash A)$ is minimized.

Our algorithm, denoted by the CPS-DTTP algorithm, to solve the CPS-DTTP problem is described in details as follows.

Algorithm: CPS-DTTP
INPUT: An instance $I = (M, J, p, e)$ of the CPS-DTTP problem;
OUTPUT: A schedule T_{k_0} and a value $k_0 + e(J \backslash A_{k_0})$.
Begin
Step 1. Using an input instance I to construct an instance $\tau(I) = (M, J, p)$ of the PS-DT problem, we execute the $\mathcal{A}_{LPT}$ algorithm [7] on the instance $\tau(I)$ to produce an optimal solution and its optimal value C for the PS-DT problem;

Step 2. For $k = 1$ to C do:

   (2.1) Using an input instance I of the CPS-DTTP problem and a positive integer k, we construct an instance $\rho_k(I) = (N, Y; s, v; k)$ of the MKP-DS problem mentioned in Sect. 2;

   (2.2) Executing the $\mathcal{A}_{Detti}$ algorithm [3] on the instance $\rho_k(I)$ of the MKP-DS problem, we determine a schedule $T_k = (S_{1_k}, S_{2_k}, \ldots, S_{m_k}; R_k)$, and we execute all jobs in $A_k = \cup_{i=1}^m S_{i_k}$ on these m machines such that the value $e(A_k)$ is maximized, equivalently, that $e(J \backslash A_k)$ is minimized;

Step 3. Compute $e(J)$, and denote $e(J \backslash A_0) = e(J)$, $T_0 = (\phi, \phi, \ldots, \phi; J)$

Step 4. Determine a schedule $T_{k_0} \in \{T_0, T_1, T_2, \ldots, T_C\}$, satisfying the following
$$C_{\max}(T_{k_0}) + e(J \backslash A_{k_0}) = \min\{C_{\max}(T_k) + e(J \backslash A_k) \mid k = 0, 1, 2, \ldots, C\};$$

Step 5. Output "the schedule T_{k_0} and the value $C_{\max}(T_{k_0}) + e(J \backslash A_{k_0})$".

End

Using the CPS-DTTP algorithm, we can determine the following

Theorem 1. *The CPS-DTTP algorithm is an exact algorithm to solve the CPS-DTTP problem, and this algorithm runs in $O((n \log n + nm)C)$ time, where n is the number of jobs, m is the number of machines and C is the optimal value for an instance $\tau(I) = (M, J, p)$ of the PS-DT problem.*

Proof. Suppose that there is an optimal solution for an instance $I = (M, J, p, e)$ of the CPS-DTTP problem, *i.e.*, there is an optimal schedule $T^* = (S_1^*, S_2^*, \ldots, S_m^*; R^*)$ and an optimal value $V^* = C_{\max}(T^*) + e(J \backslash A^*)$, where $A^* = \cup_{i=1}^m S_i^*$. For the same instance I, the CPS-DTTP algorithm obtains a schedule $T_{k_0} = (S_{1k_0}, S_{2k_0}, \ldots, S_{mk_0}; R_{k_0})$ and the output value is $V_0 = C_{\max}(T_{k_0}) + e(J \backslash A_{k_0})$.

For convenience, we denote $k^* = C_{\max}(T^*)$. According to Lemma 2, we can obtain that the $\mathcal{A}_{Detti}$ algorithm [3] at Step 2 produces a subset $A_{k^*} \subseteq J$ such that the jobs in A_{k^*} are all process on one of the m machines, and the sum of the penalties of all processed jobs is the largest in all other subsets of J, *i.e.*, $e(A_{k^*}) = \max\{e(A) \mid$ the jobs in a subset $A(\subseteq X)$ are all process on m machines, and the makespan is no more than $k^*\}$, implying that $e(A_{k^*}) \geq e(A)$, where $A(\subseteq J)$ is a set of jobs that can be processed on m machines and whose makespan does not exceed k^*.

Since $e(J) = e(A_{k^*}) + e(J \backslash A_{k^*}) = e(A) + e(J \backslash A)$ for each subset $A (\subseteq J)$ mentioned-above (including the subset A^* and the subset A_{k^*} because of the definition $k^* = C_{\max}(T^*)$), using the fact $e(A_{k^*}) \geq e(A)$, we have the following $e(J) - e(A_{k^*}) \leq e(J) - e(A)$, implying that $e(J \backslash A_{k^*}) \leq e(J \backslash A^*)$ whenever $A = A^*$.

Now, we obtain the following
$$V_0 = C_{\max}(T_{k_0}) + e(J \backslash A_{k_0}) = \min\{k + e(J \backslash A_k) \mid k = 0, 1, 2, ..., C\}$$
$$\leq k^* + e(J \backslash A_{k^*}) \leq k^* + e(J \backslash A^*) = C_{\max}(T^*) + e(J \backslash A^*) = V^*$$

where the second inequality comes from the facts $0 \leq k^* \leq C$, the third inequality comes from the fact $e(J \backslash A_{k^*}) \leq e(J \backslash A^*)$ and the fourth equality comes from the definition $k^* = C_{\max}(A^*)$.

Thus, we have $V_0 = V^*$ by the minimality of the optimal solution $A^* \subseteq J$ for an instance I of the CPS-DTTP problem, implying that the schedule T_{k_0} and the value V_0 produced by the CPS-DTTP algorithm is also an optimal solution for an instance I of the CPS-DTTP problem.

The complexity of the CPS-DTTP algorithm can be determined as follows. (1) Step 1 needs at most time $O(n \log n)$ to compute C ; (2) For each $k \in \{0, 1, 2, \ldots, C\}$, the $\mathcal{A}_{Detti}$ algorithm [3] needs time $O(n \log n + nm)$ to find a schedule T_k such that all jobs in the subset A_j which obtain from T_k can be processed on m machines and that $e(A_k)$ is maximized, implying that Step 2 needs at most time $O((n \log n + nm)C)$ to execute C iterations; (3) Step 3 needs at most time $O(n)$ to compute $e(J)$; (4) Step 4 needs at most time $O(C)$ to find a a schedule $T_{k_0} \in \{T_0, T_1, T_2, \ldots, T_C\}$ with minimum value $C_{\max}(T_{k_0}) + e(J \setminus A_{k_0})$. Hence, the CPS-DTTP algorithm needs total time $O((n \log n + nm)C)$.

4   An Exact Algorithm to Solve the CPS-DTMP Problem

In this section, we consider the constrained parallel-machine scheduling problem with divisible processing times and minimum penalty (the CPS-DTMP problem), and we plan to find a schedule T such that the value $C_{\max}(T) + \max\{e_j | b_i \in J \setminus A\}$ is minimized.

Our algorithm, denoted by the CPS-DTMP algorithm, to solve the CPS-DTMP problem is described in details as follows.

Algorithm: CPS-DTMP
INPUT: An instance $I = (M, J, p, e)$ of the CPS-DTMP problem;
OUTPUT: A schedule T_{j_0} and a value $C_{\max}(T_{j_0}) + e_{j_0}$.
Begin
Step 1. We construct a fabricated job b_0, and denote $e_0 = \min\{e_j - 1 | b_j \in J\}$;
Step 2. For $j = 0$ to n do:
    (2.1) Reject b_j, and denote $A_j = \{b_t \in J | e_t > e_j\}$;
    (2.2) Using an input instance I, we construct another instance $I_j = (M, A_j, p, e)$ of the CPS-DTMP problem;
    (2.3) Using the instance I_j of the CPS-DTMP problem, we construct an instance $\tau(I_j) = (M, A_j, p)$ of the PS-DT problem mentioned in Section 2;
    (2.4) Executing the $\mathcal{A}_{LPT}$ algorithm [7] on the instance $\tau(I_j)$ of the PS-DT problem, determine an optimal schedule $T_j = (S_{1j}, S_{2j}, \ldots, S_{mj}; R_j)$ and the makespan $C_{\max}(T_j)$, where $A_j = \cup_{i=1}^m S_{ij}$ and $R_j = \phi$;
Step 3. Determine a schedule $T_{j_0} \in \{T_0, T_1, T_2, \ldots, T_n\}$, satisfying the following
        $C_{\max}(T_{j_0}) + e_{j_0} = \min\{C_{\max}(T_j) + e_j \mid j = 0, 1, 2, \ldots, n\}$,
    and denote $T_{j_0} = (S_{1j_0}, S_{2j_0}, \ldots, S_{mj_0}; R')$, where $R' = J \setminus \cup_{i=1}^m S_{ij_0}$;
Step 4. Output "the schedule T_{j_0} and the value $C_{\max}(T_{j_0}) + e_{j_0}$".
End

Using the CPS-DTMP algorithm, we can determine the following

Theorem 2. *The CPS-DTMP algorithm is an exact algorithm to solve the CPS-DTMP problem, and this algorithm runs in $O(n^2 \log n)$ time, where n is the number of jobs.*

Proof. Suppose that there is an optimal solution for an instance $I = (M, J, p, e)$ of the CPS-DTMP problem, *i.e.*, there is an optimal schedule $T^* = (S_1^*, S_2^*, \dots, S_m^*; R^*)$ and an optimal value $V^* = C_{\max}(T^*) + e_{j^*}$, where $A^* = \cup_{i=1}^{m} S_i^*$. For the same instance I, the CPS-DTMP algorithm obtains a schedule $T_{j_0} = (S_{1j_0}, S_{2j_0}, \dots, S_{mj_0}; R_{j_0})$ and a value $V_0 = C_{\max}(T_{j_0}) + e_{j_0}$.

Without loss of generality, we assume that the job b_{j^*} is the job with the largest penalty in subset $J \backslash A^*$, then the new job subset is $A_{j^*} = \{b_t | e_t > e_{j^*}\} = A^*$. Using the $\mathcal{A}_{LPT}$ algorithm [7], according to Lemma 1, we can get a schedule T_{j^*} which process all the jobs in A_{j^*} on one of the m machines such that the makespan is minimum, then there is $C_{\max}(T_{j^*}) \leq C_{\max}(T_{j^*}')$, where the processing set is A_{j^*} in every schedule T_{j^*}' and the maximum penalty of the job in the rejection set is e_{j^*}. This shows that $C_{\max}(T_{j^*}) \leq C_{\max}(T^*)$ whenever $T_{j^*}' = T^*$.

Now, we obtain the following

$$V_0 = C_{\max}(T_{j_0}) + e_{j_0} = \min \{C_{\max}(T_j) + e_j | j = 1, 2, \dots, n\}$$

$$\leq C_{\max}(T_{j^*}) + e_{j^*} \leq C_{\max}(T^*) + e_{j^*} = V^*$$

where the first inequality comes from the facts $1 \leq j^* \leq n$, the second inequality comes from the fact $C_{\max}(T_{j^*}) \leq C_{\max}(T^*)$.

Thus we have $V_0 = V^*$, by the minimality of the optimal solution for an instance I of the CPS-DTMP problem, implying that the schedule T_{j_0} and the value V_0 produced by the CPS-DTMP algorithm is also an optimal solution for an instance I of the CPS-DTMP problem.

The complexity of the CPS-DTMP algorithm can be determined as follows. (1) Step 1 needs at most time $O(n)$; (2) For each $j \in \{1, 2, \dots, n\}$, the $\mathcal{A}_{LPT}$ algorithm [7] needs time $O(n \log n)$ to find a schedule T_j such that all jobs in the subset A_j which obtain from T_j can be processed on m machines and that $C_{\max}(T_j)$ is minimal, implying that Step 2 needs at most time $O(n^2 \log n)$ to execute n iterations; (3) Step 3 needs at most time $O(n)$ to find a a schedule $T_{j_0} \in \{T_0, T_1, T_2, \dots, T_n\}$ with minimum value $C_{\max}(T_{j_0}) + e_{j_0}$. Hence, the CPS-DTMP algorithm needs total time $O(n^2 \log n)$.

5   An Exact Algorithm to Solve the CPS-DTBP Problem

In this section, we consider the constrained parallel-machine scheduling problem with divisible processing times and bounded penalty (the CPS-DTBP problem), and we plan to find a schedule T such that the value of total penalties of rejected jobs is no more than a given bound B, the objective is minimize the makespan.

Our algorithm, denoted by the CPS-DTBP algorithm, to solve the CPS-DTBP problem is presented as follows.

Algorithm: CPS-DTBP
INPUT: An instance $I = (M, J, p, e; B)$ of the CPS-DTBP problem;
OUTPUT: A schedule T_{k_0} and a value k_0.
Begin
Step 1. Using an input instance I, we construct an instance $\tau(I) = (M, J, p)$ of
the PS-DT problem, and executing the $\mathcal{A}_{LPT}$ algorithm [7] on the instance
$\tau(I)$, we produce the optimal solution and its optimal value C;
Step 2. If $(e(J) \leq B)$ then
           Output "the schedule $T_{k_0} = (\phi, \phi, \ldots, \phi; J)$, and the value $k_0 = 0$",
  Stop;
    Else
        Denote $H = 1$ and $H' = C$;
Step 3. Set $k = \lceil (H + H')/2 \rceil$;
Step 4. We do the following two steps
    (4.1) Using an input instance I of the CPS-DTBP problem and a positive
integer k, we construct an instance $\rho_k(I) = (N, Y; s, v; k)$ of the MKP-DS
problem mentioned in Sect. 2;
    (4.2) Using the $\mathcal{A}_{Detti}$ algorithm [3] on the instance $\rho_k(I)$ of the MKP-
DS problem, we determine a schedule $T_k = (S_{1_k}, S_{2_k}, \ldots, S_{m_k}; R_k)$ to execute
all jobs in $A_k = \cup_{i=1}^{m} S_{i_k}$ on these m machines such that the value $e(A_k)$ is
maximized, equivalently, that $e(J \backslash A_k)$ is minimized;
Step 5. If $(e(J \backslash A_k) \leq B)$ then
      Set $H' = k$;
    Else
      Set $H = k$;
Step 6. If $(H' - H > 1)$ then Go to Step 3;
Step 7. Output "the schedule T_{k_0} and the value $k_0 = H'$".
End

Using the CPS-DTBP algorithm, we can determine the following

Theorem 3. *The CPS-DTBP algorithm is an exact algorithm to solve the CPS-
DTBP problem, and this algorithm runs in time $O((n \log n + nm) \log C)$, where
n is the number of jobs, m is the number of machines and C is the optimal value
for an instance $\tau(I) = (M, J, p)$ of the PS-DT problem.*

Proof. Whenever the case $e(J) \leq B$ happens, we can easily use the CPS-DTBP
algorithm to obtain an optimal solution , *i.e.*, reject all jobs in J in this case.

In the sequel arguments, we should consider the case $e(J) > B$. Suppose that
there is an optimal solution for an instance $I = (M, J, p, e; B)$ of the CPS-
DTBP problem, *i.e.*, there is an optimal schedule $T^* = (S_1^*, S_2^*, \ldots, S_m^*; R^*)$
and an optimal value $V^* = C_{\max}(T^*)$. For the same instance I, the CPS-DTBP
algorithm obtains a value k_0 and a schedule $T_{k_0} = (S_{1k_0}, S_{2k_0}, \ldots, S_{mk_0}; R_{k_0})$.

In Step 5, we determine that if $e(J \backslash A_k) \leq B$, assign the value of k to the
upper bound, otherwise assign the value of k to the lower bound. That is to say,
when k is the upper bound, we have $e(J \backslash A_k) \leq B$, and when k is the lower

bound, we have $e(J\backslash A_k)>B$. In Step 6, we can find that the penalty of the jobs are all positive integers, so the upper and lower bounds are also both positive integers. When the difference between the upper and lower bounds is equal to 1 and k_0 is the upper bound at this time, we can obtain that $e(J\backslash A_{k_0}) \leq B$ satisfying the constraints, and for k_0-1, we have $e(J\backslash A_{k_0-1})>B$ which does not satisfy the constraints, implying that k_0 is the smallest positive integer that satisfies the constraints $e(J\backslash A_k) \leq B$, i.e., $k_0 = \min\{k \in \{1,2,\ldots,C\}|e(J\backslash A_k) \leq B\}$.

For convenience ,when we denote $k^* = C_{\max}(T^*)$, using Lemma 2, we can obtain that the $\mathcal{A}_{Detti}$ algorithm [3] at Step 4 produces a subset $A_{k^*} \subseteq J$ such that the sum of the penalties of all processed jobs in A_{k^*} is the largest among all subsets of J, implying that $e(A_{k^*}) \geq e(A^*)$.

Now, we obtain the following

$$e(J\backslash A_{k^*}) = e(J) - e(A_{k^*}) \leq e(J) - e(A^*) = e(J\backslash A^*) \leq B$$

where the first inequality comes from the facts $e(A_{k^*}) \geq e(A^*)$, the second inequality comes from the facts that the optimal solution is also a feasible solution, implying $e(J\backslash A_{k^*}) \leq B$.

Due to the minimality of k_0 in the algorithm, i.e., $k_0 = \min\{k \in \{1,2,\ldots,C\}|e(J\backslash A_k) \leq B\}$, we have $k_0 = C_{\max}(T_{k_0}) \leq k^* = C_{\max}(T^*)$, and by the minimality of the optimal solution k^* for an instance I of the CPS-DTBP problem, we have $k_0 = k^*$, implying that the schedule T_{k_0} and the value k_0 produced by the CPS-DTBP algorithm is also an optimal solution for an instance I' of the CPS-DTBP problem.

The complexity of the CPS-DTBP algorithm can be determined as follows. (1) Step 1 needs at most time $O(n \log n)$ to compute C ; (2) Step 2 needs at most time $O(n)$ to compute $e(J)$; (3) Step 3 needs at most time $O(1)$ to compute k; (4)Step 4 needs at most time $O(n \log n + nm)$to find a schedule T_k for some k by using the $\mathcal{A}_{Detti}$ algorithm [3]; (5) Step 5 needs at most time $O(1)$ to determine the size relationship between $e(J\backslash A_k)$ and B and assign k to H or H'; (6) Step 6 determine the difference between H and H', if the difference is 1, we can find the value of k_0, otherwise it returns to Step 3 and uses the binary search algorithm to iterate, so that Step 6 needs at most time $O((n \log n + nm) \log C)$ to execute $\log C$ iterations. Hence, the CPS-DTBP algorithm needs total time $O((n \log n + nm) \log C)$.

6   Conclusion and Further Work

In this paper, we consider the constrained parallel-machine scheduling problem with divisible processing times and total penalties (the CPS-DTTP problem), the constrained parallel-machine scheduling problem with divisible processing times and maximum penalty (the CPS-DTMP problem) and the constrained parallel-machine scheduling problem with divisible processing times and bounded penalty (the CPS-DTBP problem), respectively. We obtain the following three main results.

(1) We design an exact algorithm in time $O((n \log n + nm)C)$ to solve the the CPS-DTTP problem, where n is the number of jobs with divisible sizes, m is the number of machines and C is the optimal value for an instance $\tau(I) = (M, J, p)$ of the PS-DT problem;

(2) We present an exact algorithm in time $O(n^2 \log n)$ to solve the CPS-DTMP problem;

(3) We provide an exact algorithm in time $O((n \log n + nm) \log C)$ to solve the CPS-DTBP problem, where C is the optimal value for an instance $\tau(I) = (M, J, p)$ of the PS-DT problem.

In further work, we shall study other versions of the parallel-machine scheduling problem with divisible job sizes and penalties, and we shall try to design other exact algorithm in lower running time to resolve CPS-DTTP problem.

References

1. Bartal, Y., Leonardi, S., Marchetti-Spaccamela, A., Sgall, J., Stougie, L.: Multiprocessor scheduling with rejection. SIAM J. Dis. Math. **13**, 64–78 (2000)
2. Coffman, E.G., Jr., Garey, M.R., Johnson, D.S.: An application of bin-packing to multiprocessor scheduling. SIAM J. Comput. **7**(1), 1–17 (1978)
3. Detti, P.: A polynomial algorithm for the multiple knapsack problem with divisible item sizes. Inf. Process. Lett. **109**(11), 582–584 (2009)
4. Dósa, G., He, Y.: Bin packing problems with rejection penalties and their dual problems. Inf. Comput. **204**(5), 795–815 (2006)
5. Faigle, U., Kern, W., Turn, G.: On the performance of on-line algorithms for partition problems. Acta Cybern. **9**, 107–119 (1989)
6. Graham, R.L.: Bounds for centain multiprocessing anomalies. Bell Syst. Tech. J. **45**, 1563–1581 (1966)
7. Graham, R.L.: Bounds on multiprocessing timing anomalies. SIAM J. Appl. Math. **17**, 416–429 (1969)
8. Knuth, D.E.: Foundamental Algorithms, vol. 1, 2nd ed. Addison-Wesley, Reading (1973)
9. Lawler, E.L., Lenstra, J.K., Rinnooy Kan, A.H.G., Shmoys, D.B.: Sequencing and scheduling: algorithms and complexity. Handb. Oper. Res. Manag. Sci. **4**, 445–522 (1993)
10. Lenstra, J.K., Kan, A.R., Brucker, P.: Complexity of machine scheduling problems. Ann. Discrete Mach. **1**, 343–362 (1977)
11. Li, W., Li, J., Zhang, X.,Chen, Z.: Penalty cost constrained identical parallel machine scheduling problem. Theor. Comput. Sci. **607**, 181–192 (2015)
12. Shabtay, D., Gaspar, N., Kaspi, M.: A survey on offline scheduling with rejection. J. Sched. **16**(1), 3–28 (2013)
13. Yue, X.: Parallel machine scheduling problem with rejection cost under special conditions. Kunming University of Science and Technology, China (2020)
14. Yue, X., Gao, J., Chen, Z.: A polynomial time algorithm for scheduling on processing time constraints. In: ACM International Conference Proceeding Series, 2019 the 9th International Conference on Communication and Network Security, pp. 109–113 (2019)

15. Zhang, Y., Ren, J., Wang, C.: Scheduling with rejection to minimize the makespan. In: Du, D.-Z., Hu, X., Pardalos, P.M. (eds.) COCOA 2009. LNCS, vol. 5573, pp. 411–420. Springer, Heidelberg (2009). https://doi.org/10.1007/978-3-642-02026-1_39

16. Zheng, S., Yue, X., Chen, Z.: Parallel machine scheduling with rejection under special conditions. In: ACM International Conference Proceeding Series, 2018 the 8th International Conference on Communication and Network Security, pp. 139–143 (2018)

Obnoxious Facility Location Games
with Candidate Locations

Ling Gai[1]([✉]) [ID], Mengpei Liang[1] [ID], and Chenhao Wang[2,3] [ID]

[1] Glorious Sun School of Business and Management,
Donghua University, Shanghai 200051, China
lgai@dhu.edu.cn, liangmengpei@mail.dhu.edu.cn
[2] Advanced Institute of Natural Sciences, Beijing Normal University,
Zhuhai 519087, China
[3] BNU-HKBU United International College, Zhuhai 519087, China
chenhwang@bnu.edu.cn

Abstract. We study obnoxious facility location games with facility candidate locations. For obnoxious single facility location games under social utility objective, we present a group strategy-proof mechanism with approximation ratio of 3. Then we prove the ratio is tight by giving a corresponding lower bound instance. This is also proved to be the best possible mechanism. For obnoxious two-facility location games with facility candidate locations, we study the heterogeneous facility case in this paper. We design a group strategy-proof mechanism and prove that the approximation ratio is 2. We also prove that the problem lower bound is $\frac{3}{2}$.

Keywords: Obnoxious facility location game · Mechanism design · Strategyproof · Approximation ratio

1 Introduction

The classical facility location problem is a consideration of where to place one or more facilities to serve the agent and achieve the goal of maximizing utility, i.e., a trade-off between service and benefit. For the facility location game, the location of an agent is private information and needs to be reported by the agent. The agents, being rational in the game, will try to maximize their utilities by misreporting so that the facility is closer to them. Thus, we are more concerned about how to design a *strategy-proof* mechanism to incentivize agents to report their positions truthfully while ensuring a relatively good facility location solution. The concept of *strategy-proof approximation* mechanism design without money was first introduced by Procaccia and Tennenholtz [10]. Following their work, this branch of study has received a great deal of attention. Chan et al. [2] gave a thorough and comprehensive survey for it.

In the general setting, most studies focused on the case that facilities can be placed anywhere (e.g., see [1,3,8,9,16]). However, in practice, most of the facilities can only be built in fixed areas or given locations. For example, the garbage disposal plants or landfills are limited to particular places based on the factors of geographical position and wind direction. A gas station is generally not located in the center of residential community. In this paper, we study the facility location games where the locations can only be selected from a given candidate set, and the facilities should not be built at the same place.

In the following we present a brief review on the results of facility location games with candidate locations. Tang et al. [12] studied single facility and two-facility location games for social cost objective and maximum cost objective. For the single-facility problem under the maximum-cost objective, they gave a deterministic 3-approximation group strategy-proof mechanism, and proved that no deterministic (or randomized) strategy-proof mechanism can have an approximation ratio better than 3 (or 2). For the two-facility problem, they gave an anonymous deterministic group strategy-proof mechanism that is $(2n - 3)$-approximation for the total-cost objective, and 3-approximation for the maximum-cost objective. Walsh [14] analyzed six different objectives and showed that limiting the location of a facility makes the problem harder to approximate. Thang [13] assumed an agent could control multiple locations. He designed a 3-approximation randomized strategy-proof mechanism and a deterministic group strategy-proof mechanism with $2n + 1$ approximation ratio for social cost objective. Feldman et al. [5] studied three types of candidate selection mechanisms (single candidate, ranking and location mechanisms), and they gave the relationships among them. Serafino and Ventre [11] and Kanellopoulos et al. [7] studied a different version of facility location games. They assumed the positions of agents to be common knowledge, while every agent has a private preference over the given facility candidates. So the cost of each agent is defined to be the distance to the set of facilities he is interested in, rather than accessing the nearest facility. Dokow et al. [4] analyzed the location game on discrete unweighted graph, where the agents and the facility are restricted to vertices only. They gave a characterization of strategy-proof mechanisms on lines and on sufficiently large cycles. Filimonov and Meir [6] gave a characterization of strategy-proof mechanisms on discrete trees.

To the best of our knowledge, there has been no analysis on obnoxious facility location game with candidate locations. In this paper, the obnoxious facility location game on a line is studied. We assume there are n agents distributed in the interval $[0, 1]$, and the candidate locations are given as a set $M \subseteq [a, b], a, b \in \Re$. We will design strategy-proof mechanisms, and analyze their approximation ratios. When multiple facilities are to be located, we consider the heterogeneous facilities case. [2, 15] presented some motivations in real life for this problem. The agent's utility is the sum of its distances to all the facilities. And for obnoxious heterogeneous two-facility location games with minimum distance requirement, [15] designed group strategy-proof mechanism with approximate ratio in $(1, 2]$. When we consider two facilities with a distance limit of 0, the approximation ratio of their mechanism can be converted to 2.

Our Results. For the obnoxious single facility location game under social utility objective, we prove the cases where an optimal group strategy-proof mechanism exists. Then we present a group strategy-proof mechanism with approximation ratio of 3 for other cases. This is the best possible strategy-proof mechanism for this problem. For the obnoxious heterogeneous two-facility location game, we propose an optimal group strategy-proof mechanism for the case where $[a, b]$ has no intersection with $(0, 1)$ and a 2-approximation mechanism for the case with intersection, where a and b are the leftmost and rightmost point of the facility candidate locations, respectively. We also prove the ratio is tight by giving a corresponding lower bound instance. The problem lower bound is proved to be $\frac{3}{2}$. A summary of our results is shown in Table 1.

Table 1. Main results

Facility	Candidate location	Approximation ratio	Lower bound
Single facility	$\frac{a+b}{2} \leq 0$ or $\frac{a+b}{2} \geq 1$	Optimal	/
	$0 < \frac{a+b}{2} < 1$	3	3
Heterogeneous Two-facility	$[a, b] \subseteq (-\infty, 0]$ or $[a, b] \subseteq [1, +\infty)$	Optimal	/
	Other cases	2	$\frac{3}{2}$

2   Definition and Terminology

Let $N = \{1, 2, ..., n\}$ be the set of agents. They are in an interval $[0, 1]$ and their location set is denoted as X. Let M be the set of candidate locations for the facility, $M \subseteq [a, b]$. That is, the leftmost candidate point is a and the rightmost candidate point is b, $a, b \in \Re$. The distance between any two points $x, y \in \Re$ is $d(x, y) = |x - y|$. We denote the location profile reported by the agents as $\mathbf{x} = (x_1, x_2, ..., x_n)$. A deterministic mechanism f based on the agents' location profile $\mathbf{x}$, outputs k facility locations $\mathbf{y} = (y_1, ..., y_k) \in M^k$.

When $k = 1$, a single obnoxious facility is to be located. Assume the facility location to be $\mathbf{y} = f(\mathbf{x}) = y$, the utility of agent i is his distance to the facility y. That is,

$$u(x_i, \mathbf{y}) = u(x_i, y) = d(x_i, y).$$

If $k = 2$, two obnoxious facilities are to be located. Suppose the locations output by the mechanism are $\mathbf{y} = f(\mathbf{x}) = \{y_1, y_2\}$. Then for the heterogeneous facilities case, the utility is defined to be the sum of distances to both facilities [15]

$$u(x_i, \mathbf{y}) = d(x_i, y_1) + d(x_i, y_2).$$

The social utility of a mechanism $f(x)$ with respect to x is denoted as the sum of utilities of n agents,

$$SU(\mathbf{x}, \mathbf{y}) = \sum_{i \in N} u(x_i, \mathbf{y}).$$

A mechanism f is *strategy-proof* if no agent can acquire more utility from misreporting. Specifically, assume that an agent $i \in N$ misreports its location profile x_i as x_i', then

$$u(x_i, f(x_i', \mathbf{x}_{-i})) \leq u(x_i, f(x_i, \mathbf{x}_{-i}))$$

A mechanism f is *group strategy-proof* if there exists at least one agent in group F who cannot benefit from misreporting. That is, there exists an agent $i \in F \subseteq N$ such that

$$u(x_i, f(\mathbf{x}_F', \mathbf{x}_{-F})) \leq u(x_i, f(\mathbf{x}_F, \mathbf{x}_{-F}))$$

Given an instance $\mathbf{c}$, let $OPT(\mathbf{c})$ be the optimum social objective value, and $f(\mathbf{c})$ be the objective value of mechanism f. We say mechanism f has an approximate ratio β if for any instance $\mathbf{c}$ there exists a number β such that $\frac{OPT(\mathbf{c})}{f(\mathbf{c})} \leq \beta$.

3 Obnoxious Single Facility Location Game

In this section, we study the obnoxious single facility location game with the objective of maximizing social utility. Specifically, given an agent location set $X \subseteq [0, 1]$, one facility is to be selected from candidate location set M which belongs to interval $[a, b]$, $a, b \in \Re$. The agent utility is his distance to the facility output by the mechanism, and the social utility is the sum of agents' utility.

It is easy to see that an optimal facility location solution may not guarantee strategy-proofness. For example, there are two agents with $x_1 = \frac{1}{3}$, $x_2 = \frac{3}{5}$, and the facility candidate location set is $M = \{0, 1\}$. Because $\sum_{i=1}^{2} d(x_i, 1) > \sum_{i=1}^{2} d(x_i, 0)$, the optimal facility location is $y^* = 1$. However, if x_2 misreports its location to be 1, the mechanism returns the facility location $y^* = 0$ and x_2 can benefit from misreporting.

In the following, a group strategy-proof mechanism will be presented. The performance analysis of this mechanism will be carried out based on the position of interval $[a, b]$ and its relationship with agent location set X.

Mechanism 1. Given an agent location profile $\mathbf{x}$ in the interval $[0, 1]$ and a facility candidate location set $M \subseteq [a, b]$. Let n_1 be the number of agents with $x_i \leq \frac{a+b}{2}$ and n_2 be the number of agents with $x_i > \frac{a+b}{2}$. If $n_1 \geq n_2$, return the rightmost facility candidate point b; otherwise, return the leftmost facility candidate point a.

Theorem 1. *For the obnoxious single facility location game, Mechanism 1 is group strategy-proof. It is optimal if $[a, b] \cap X = \emptyset$, or $[a, b] \cap X \neq \emptyset$, $\frac{a+b}{2} \leq 0$ or $[a, b] \cap X \neq \emptyset$, $\frac{a+b}{2} \geq 1$; and is 3-approximate if $[a, b] \cap X \neq \emptyset$ and $0 < \frac{a+b}{2} < 1$.*

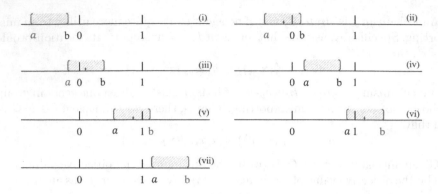

Fig. 1. Seven possible positions of interval $[a, b]$ comparing to $[0, 1]$, $[a, b] \cap X = \emptyset$.

Proof. As shown in Fig. 1, there is no intersection between $[a, b]$ and interval $(0, 1)$. For the cases (i)(ii)(iii), a will be selected as the output facility location since $n_1 < n_2$. And point a is the farthest facility candidate point for all agents. So it is optimal and no agents have the motivation to misreport their positions. Cases (v)(vi)(vii) are with the similar analysis.

For case (iv) with $a > 0, b < 1, 0 < \frac{a+b}{2} < 1$. Let N_1 be the set of the agents with $x_i \in [0, \frac{a+b}{2}]$, N_2 be the set of the agents with $x_i \in (\frac{a+b}{2}, 1]$, and $i \in N$. Suppose $n_1 \geq n_2$, then Mechanism 1 outputs $y = b$, the rightmost candidate location. For any $a < y' < y$, we have,

$$\sum_{i=1}^{n} d(x_i, y) = \sum_{i \in N_1} d(x_i, y') + \sum_{i \in N_1} d(y', y) + \sum_{i \in N_2} d(x_i, y') - \sum_{i \in N_2} d(y', y)$$

$$= \sum_{i=1}^{n} d(x_i, y') + (n_1 - n_2) \cdot d(y', y)$$

$$\geq \sum_{i=1}^{n} d(x_i, y'),$$

where the last inequality holds because $n_1 \geq n_2$, implying the optimality of y.

Therefore, Mechanism 1 is optimal. Next, we analyse the group strategy-proofness of Mechanism 1. Let $F \in N$ be a coalition. We need to prove that at least one agent in F does not benefit from lying. Without loss of generality, we assume that $n_1 \geq n_2$, and the mechanism outputs the facility location profile $\mathbf{y} = b$. If the output of the mechanism is to be changed, i.e., the facility is located on point a, at least one agent with $x_i \leq \frac{a+b}{2}$ misreports his/her location to $x_i' > \frac{a+b}{2}$. However, $u(x_i, f(\mathbf{x}_F', \mathbf{x}_{-F})) = d(x_i, a) \leq u(x_i, f(\mathbf{x}_F, \mathbf{x}_{-F})) = d(x_i, b)$, implying no increase in the utility of agent i.

When there are agents with positions between $[a, b]$, while the midpoint $\frac{a+b}{2}$ is outside of $[0, 1]$, Mechanism 1 is still group strategy-proof and optimal. Suppose $\frac{a+b}{2} \leq 0$, then Mechanism 1 returns $y = a$, which is the farthest facility candidate location for all agents. Therefore, Mechanism 1 is optimal.

When $[a,b] \cap X \neq \emptyset$, $0 < \frac{a+b}{2} < 1$, suppose $n_1 \geq n_2$, the optimal location is $\mathbf{y}^* = y^*$ and Mechanism 1 returns $y = b$. Since the distance from the agent in N_1 to point y is at least $(y - \frac{a+b}{2})$. Then,

$$SU(\mathbf{x}, \mathbf{y}) = \sum_{i \in N} u(x_i, \mathbf{y}) = \sum_{i \in N} d(x_i, y) \geq n_1 \cdot (y - \frac{a+b}{2}).$$

Let $D_1 = SU(\mathbf{x}, \mathbf{y}) - n_1 \cdot (y - \frac{a+b}{2})$, we have

$$D_1 = \sum_{x_i \in [0, \frac{a+b}{2}]} [d(x_i, y) - d(\frac{a+b}{2}, y)] + \sum_{x_i \in (\frac{a+b}{2}, 1]} d(x_i, y)$$

$$= \sum_{x_i \in [0, \frac{a+b}{2}]} [d(x_i, \frac{a+b}{2})] + \sum_{x_i \in (\frac{a+b}{2}, 1]} d(x_i, y).$$

Consider a new location profile $\mathbf{x}'$, in which there are n_1 agents at point $\frac{a+b}{2}$ and n_2 agents are at point y (when $y > 1$, n_2 agents at point 1). Since y^* is to the left of $y = b$ and the distance to y is at most $(b - a)$. We have

$$SU(\mathbf{x}', \mathbf{y}^*) = n_1 \cdot d(\frac{a+b}{2}, y^*) + n_2 \cdot d(y^*, y)$$

$$\leq n_1 \cdot (y - \frac{a+b}{2}) + 2n_2 \cdot (y - \frac{a+b}{2}) \leq 3n_1 \cdot (y - \frac{a+b}{2}).$$

Let D_2 be the difference of $SU(\mathbf{x}, \mathbf{y}^*)$ and $SU(\mathbf{x}', \mathbf{y}^*)$. Then,

$$D_2 = SU(\mathbf{x}, \mathbf{y}^*) - SU(\mathbf{x}', \mathbf{y}^*)$$

$$= \sum_{x_i \in [0, \frac{a+b}{2}]} [d(x_i, y^*) - d(\frac{a+b}{2}, y^*)] + \sum_{x_i \in (\frac{a+b}{2}, 1]} [d(x_i, y^*) - d(y, y^*)].$$

Thus,

$$D_1 - D_2 = \sum_{x_i \in [0, \frac{a+b}{2}]} [d(x_i, \frac{a+b}{2}) + d(\frac{a+b}{2}, y^*) - d(x_i, y^*)]$$

$$+ \sum_{x_i \in (\frac{a+b}{2}, 1]} [d(x_i, y) + d(y, y^*) - d(x_i, y^*)].$$

By the triangle inequality, it is easy to get that $D_1 - D_2 \geq 0$. (When $y > 1$, $D_1 - D_2 \geq 0$ still holds.) In summary,

$$\frac{OPT(\mathbf{x})}{SU(\mathbf{x}, \mathbf{y})} = \frac{SU(\mathbf{x}, \mathbf{y}^*)}{SU(\mathbf{x}, \mathbf{y})} \leq \frac{SU(\mathbf{x}, \mathbf{y}^*) - D_1}{SU(\mathbf{x}, \mathbf{y}) - D_1} \leq \frac{SU(\mathbf{x}, \mathbf{y}^*) - D_2}{SU(\mathbf{x}, \mathbf{y}) - D_1}$$

$$= \frac{SU(\mathbf{x}', \mathbf{y}^*)}{n_1 \cdot (y - \frac{a+b}{2})} \leq \frac{3n_1 \cdot (y - \frac{a+b}{2})}{n_1 \cdot (y - \frac{a+b}{2})} = 3.$$

$\square$

It is obvious that the approximation ratio of 3 for Mechanism 1 is tight. Given the agents' location profile $\mathbf{x} = (\frac{1}{2}, \frac{3}{4})$ and candidate location profile $(\frac{1}{4}, \frac{3}{4})$, the optimal social utility is $\frac{3}{4}$ with the facility locating at point $\frac{1}{4}$, and the social utility of Mechanism 1 is $\frac{1}{4}$ with the facility at point $\frac{3}{4}$.

Note that in [3], Cheng et al. pointed out that for any deterministic strategy-proof mechanism selecting one of the endpoints as the facility location, 3 is the best possible approximation ratio when the agents are located on a path. This conclusion applies in our case with facility candidate location constraint, too. We present the following theorem.

Theorem 2. *For the obnoxious single facility location game with candidate location, no deterministic strategy-proof mechanism can have an approximation ratio better than 3 if $[a, b] \cap X \neq \phi$ and $0 < \frac{a+b}{2} < 1$.*

Proof. Suppose f is a deterministic mechanism, the output of the mechanism f is $f(\mathbf{x}) = y$. Consider the profile $\mathbf{x} = (x_1, x_2) = (\frac{a+b}{2} - \varepsilon, \frac{a+b}{2} + \varepsilon), 0 < \varepsilon < 1$. Assume a location set of facility candidates $M = \{a, b\}, 0 \leq a < b \leq 1$. We can see that this profile satisfies the constraint of $[a, b] \cap X \neq \phi$ and $0 < \frac{a+b}{2} < 1$.

The possible facility location of mechanism f could be a or b. If $y = b$, the utility of agent 2 is $\frac{b-a}{2} - \varepsilon$. Consider the profile $\mathbf{x}' = (x_1, x_2) = (\frac{a+b}{2} - \varepsilon, b)$. By the strategy-proofness, $u(x_2, y') = |y' - x_2| \leq \frac{b-a}{2} - \varepsilon$. So we have $y' = b$. Then the social utility of f is at most $\frac{b-a}{2} + \varepsilon$ while the optimal social utility is $\frac{3(b-a)}{2} - \varepsilon$. Thus the approximation ratio is at least 3 when ε tends to 0. Therefore, the general lower bound for the approximation ratio is 3.   □

4   Obnoxious Heterogeneous Two-facility Location Game

In this section, two heterogeneous facilities are to be located inside the candidate location set $M \subseteq [a, b], a, b \in \Re$. Agents are still located in the interval $[0, 1]$. Each of them tries to maximize the total distance from two facilities, that is, for agent i, his utility is $d(x_i, y_1) + d(x_i, y_2)$. Denote the optimal location profile as $\mathbf{y}^* = (y_1^*, y_2^*)$ and the output of our mechanism as (y_1, y_2). Without loss of generality, suppose $y_1^* < y_2^*$ and $y_1 < y_2$.

Mechanism 2. Given a location profile $\mathbf{x}$ in the interval $[0, 1]$ and a location set of facility candidates $M \subseteq [a, b]$. The two facilities at the two candidate positions which are farthest from point 0.

Theorem 3. *For the obnoxious heterogeneous two-facility location game, Mechanism 2 is a group strategy-proof optimal mechanism if $[a, b] \subseteq (-\infty, 0]$ or $[a, b] \subseteq [1, +\infty)$.*

Proof. The group strategy-proof of Mechanism 2 is easy to verify since the output of Mechanism 2 does not depend on agents' report. The conclusion of optimal is trivial since Mechanism 2 ensures the selected facilities being the farthest from each agent.   □

We then carry out our study by considering other possible positions of interval $[a, b]$.

Mechanism 3. (Endpoints Mechanism). Select two endpoints of the facility candidate locations, i.e., point a and point b.

Theorem 4. *Mechanism 3 is group strategy-proof for the obnoxious heterogeneous two-facility location game under the maximizing social utility objective. When $[a, b] \subseteq [0, 1]$, or $a < 0$ and $b > 1$, or $a < 0$ and $0 < b \leq 1$, or $0 \leq a < 1$ and $b > 1$, the approximation ratio is 2.*

Proof. No agent misreporting will affect the output of the mechanism, i.e., the output of the mechanism is fixed. So, Mechanism 3 is group strategy-proof.

Four cases are shown in Fig. 2. In either case, the approximate ratio of Mechanism 3 is 2. For any agent in the interval $[0, 1]$, it is either farthest from a or b. For each agent, assume he is farther from point a, his utility from any mechanism is at most $2|x_i - a|$. Since our mechanism outputs two endpoints, it must output point a, so the utility of agent from our mechanism is at least $|x_i - a|$. Thus, the social utility of Mechanism 3 is at least half of the maximum social utility when all the agents are considered. Therefore, the approximate ratio of Mechanism 3 is 2.   □

As for the lower bound of Mechanism 3, consider the following example. Suppose all the agents are at the point ε. The facility candidate locations are points 0, 1 and $1 - \varepsilon$. Mechanism 3 outputs two locations $a = 0$, $b = 1$, and the social utility is $n(1 - 0)$. The optimal solution outputs the two rightmost points, point 1 and point $1 - \varepsilon$. The social utility is $n(2 - 3\varepsilon)$. So we have $\frac{n(2-3\varepsilon)}{n(1-0)} \to 2$, when $\varepsilon \to 0$. Therefore, the analysis of the approximation ratio is tight.

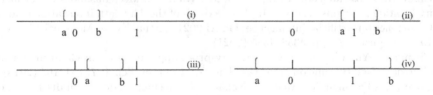

Fig. 2. Four possible positions of interval $[a, b]$ comparing to $[0, 1]$.

Theorem 5. *For the obnoxious heterogeneous two-facility location game with candidate location, no deterministic strategy-proof mechanism f can have an approximation ratio better than $\frac{3}{2}$ if $[a, b] \subseteq [0, 1]$, or $a < 0$ and $b > 1$, or $a < 0$ and $0 < b \leq 1$, or $0 \leq a < 1$ and $b > 1$, under the social utility objective.*

Proof. Suppose f is a strategy-proof mechanism with approximation ratio better than $\frac{3}{2}$ for heterogeneous two-facility location. Assume w.l.o.g. that f builds at least one facility on the rightmost two candidates. In this case, the utility of agent 2 is at most $\frac{1}{2} + \varepsilon + 1 - (\frac{1}{2} + \varepsilon) = 1$. However, if agent 2 misreports her location as 1, then by the approximation ratio, f must locate two facilities at 0 and ε, and thus the utility of agent 2 increases to $\frac{1}{2} + \varepsilon + \frac{1}{2} = 1 + \varepsilon$, giving a contradiction to the strategyproofness.   □

5   Conclusion

This paper initiates the studies on obnoxious facility location games with candidate locations. We design deterministic strategy-proof mechanisms and analyze the approximation ratio comparing with the optimal solution. We consider both the single facility location game and the two-facility location game. In the single facility game, the agent utility is defined as the distance to that facility; In the two-facility game, the agent utility is defined to be the sum of distances to both facilities for the heterogeneous case. The social utility is the sum of agents' utility.

We give a 3-approximation group strategy-proof mechanism for the obnoxious single facility location game with candidate locations. The ratio is proved to be tight. For the obnoxious heterogeneous two-facility game with candidate locations, we propose a 2-approximation mechanism besides an optimal group strategy-proof mechanism under different cases of $[a, b]$. We also proved that for obnoxious heterogeneous two-facility location game with candidate locations, no strategy-proof mechanism could get better approximation ratio less than $\frac{3}{2}$.

In the future work we will study the mechanism design and analysis on obnoxious homogeneous two-facility game with candidate locations. The cases when agents are distributed on a tree or general graphs will also be considered.

References

1. Anastasiadis, E., Deligkas, A.: Heterogeneous facility location games. In: Proceedings of the 17th International Conference on Autonomous Agents and MultiAgent Systems, pp. 623–631 (2018)
2. Chan, H., Filos-Ratsikas, A., Li, B., Li, M., Wang, C.: Mechanism design for facility location problems: a survey. In: Proceedings of the Thirtieth International Joint Conference on Artificial Intelligence, IJCAI 2021, Virtual Event/Montreal, Canada, 19–27 August 2021, pp. 4356–4365 (2021)
3. Cheng, Y., Yu, W., Zhang, G.: Strategy-proof approximation mechanisms for an obnoxious facility game on networks. Theoret. Comput. Sci. **497**, 154–163 (2013)
4. Dokow, E., Feldman, M., Meir, R., Nehama, I.: Mechanism design on discrete lines and cycles. In: Proceedings of the 13th ACM Conference on Electronic Commerce, pp. 423–440 (2012)
5. Feldman, M., Fiat, A., Golomb, I.: On voting and facility location. In: Proceedings of the 2016 ACM Conference on Economics and Computation, pp. 269–286 (2016)
6. Filimonov, A., Meir, R.: Strategyproof facility location mechanisms on discrete trees. In: AAMAS 2021: 20th International Conference on Autonomous Agents and Multiagent Systems, Virtual Event, United Kingdom, 3–7 May 2021, pp. 510–518 (2021)
7. Kanellopoulos, P., Voudouris, A.A., Zhang, R.: On discrete truthful heterogeneous two-facility location. arXiv preprint arXiv:2109.04234 (2021)
8. Meir, R.: Strategyproof facility location for three agents on a circle. In: International Symposium on Algorithmic Game Theory, pp. 18–33 (2019)
9. Nehama, I., Todo, T., Yokoo, M.: Manipulation-resistant false-name-proof facility location mechanisms for complex graphs. Auton. Agent. Multi-Agent Syst. **36**(1), 1–58 (2022). https://doi.org/10.1007/s10458-021-09535-5

10. Procaccia, A.D., Tennenholtz, M.: Approximate mechanism design without money. In: Proceedings of the 10th ACM Conference on Electronic Commerce, pp. 177–186 (2009)
11. Serafino, P., Ventre, C.: Heterogeneous facility location without money. Theoret. Comput. Sci. **636**, 27–46 (2016)
12. Tang, Z., Wang, C., Zhang, M., Zhao, Y.: Mechanism design for facility location games with candidate locations. In: International Conference on Combinatorial Optimization and Applications, pp. 440–452 (2020)
13. Thang, N.K.: On (group) strategy-proof mechanisms without payment for facility location games. In: International Workshop on Internet and Network Economics, pp. 531–538 (2010)
14. Walsh, T.: Strategy proof mechanisms for facility location at limited locations. In: Pacific Rim International Conference on Artificial Intelligence, pp. 113–124 (2021)
15. Xu, X., Li, B., Li, M., Duan, L.: Two-facility location games with minimum distance requirement. J. Artif. Intell. Res. **70**, 719–756 (2021)
16. Zou, S., Li, M.: Facility location games with dual preference. In: Proceedings of the 2015 international Conference on Autonomous Agents and Multiagent Systems, pp. 615–623 (2015)

Nonlinear Combinatorial Optimization

Streaming Adaptive Submodular Maximization

Shaojie Tang[1]([✉]) [iD] and Jing Yuan[2] [iD]

[1] Naveen Jindal School of Management,
University of Texas at Dallas, Richardson, USA
`shaojie.tang@utdallas.edu`
[2] Department of Computer Science, University of North Texas, Denton, USA

Abstract. Many sequential decision making problems can be formulated as an adaptive submodular maximization problem. However, most of existing studies in this field focus on pool-based setting, where one can pick items in any order, and there have been few studies for the stream-based setting where items arrive in an arbitrary order and one must immediately decide whether to select an item or not upon its arrival. In this paper, we introduce a new class of utility functions, semi-policywise submodular functions. We develop a series of effective algorithms to maximize a semi-policywise submodular function under the stream-based setting.

1 Introduction

Many machine learning and artificial intelligence tasks can be formulated as an adaptive sequential decision making problem. The goal of such a problem is to sequentially select a group of items, each selection is based on the past, in order to maximize some give utility function. It has been shown that in a wide range of applications, including active learning [4] and adaptive viral marketing [12], their utility functions satisfy the property of adaptive submodularity [4], a natural diminishing returns property under the adaptive setting. Several effective solutions have been developed for maximizing an adaptive submodular function subject to various practical constraints. For example, [4] developed a simple adaptive greedy policy that achieves a $1 - 1/e$ approximation ratio for maximizing an adaptive monotone and adaptive submodular function subject to a cardinality constraint. Recently, [9] extends the aforementioned studies to the non-monotone setting and they propose a $1/e$ approximated solution for maximizing a non-monotone adaptive submodular function subject to a cardinality constraint. In the same work, they develop a faster algorithm whose running time is linear in the number of items. [10] develops the first constant approximation algorithms subject to more general constraints such as knapsack constraint and k-system constraint.

We note that most of existing studies focus on the pool-based setting where one is allowed to select items in any order. In this paper, we tackle this problem under the stream-based setting. Under our setting, items arrive one by one in an online fashion where the order of arrivals is decided by the adversary. Upon the arrival of an item, one must decide immediately whether to select that item

Q. Ni and W. Wu (Eds.): AAIM 2022, LNCS 13513, pp. 109–120, 2022.
https://doi.org/10.1007/978-3-031-16081-3_10

or not. If this item is selected, then we are able to observe its realized state; otherwise, we skip this item and wait for the next item. Our goal is to adaptively select a group items in order to maximize the expected utility subject to a knapsack constraint. For solving this problem, we introduce the concept of *semi-policywise submodularity*, which is another adaptive extension of the classical notation of submodularity. We show that this property can be found in many real world applications such as active learning and adaptive viral marketing. We develop a series of simple adaptive policies for this problem and prove that if the utility function is semi-policywise submodular, then our policies achieve constant approximation ratios against the optimal pool-based policy. In particular, for a single cardinality constraint, we develop a stream-based policy that achieves an approximation ratio of $\frac{1-1/e}{4}$. For a general knapsack constraint, we develop a stream-based policy that achieves an approximation ratio of $\frac{1-1/e}{16}$.

2   Related Work

Stream-Based Submodular Optimization. Non-adaptive submodular maximization under the stream-based setting has been extensively studied. For example, [1] develop the first efficient non-adaptive streaming algorithm SieveStreaming that achieves a $1/2 - \epsilon$ approximation ratio against the optimum solution. Their algorithm requires only a single pass through the data, and memory independent of data size. [6] develop an enhanced streaming algorithm which requires less memory than SieveStreaming. Very recently, [7] propose a new algorithm that works well under the assumption that a single function evaluation is very expensive. [3] extend the previous studies from the non-adaptive setting to the adaptive setting. They develop constant factor approximation solutions for their problem. However, they assume that items arrive in a random order, which is a large difference from our adversarial arrival model. Our work is also related to submodular prophet inequalities [2,8]. Although they also consider an adversarial arrival model, their setting is different from ours in that 1. they assume items are independent and 2. they are allowed to observe an item's state before selecting it.

Adaptive Submodular Maximization. [4] introduce the concept of adaptive submodularity that extends the notation of submodularity from sets to policies. They develop a simple adaptive greedy policy that achieves a $1 - 1/e$ approximation ratio if the function is adaptive monotone and adaptive submodular. When the utility function is non-monotone, [9] show that a randomized greedy policy achieves a $1/e$ approximation ratio subject to a cardinality constraint. Very recently, they generalize their previous study and develop the first constant approximation algorithms subject to more general constraints such as knapsack constraint and k-system constraint [10]. Other variants of adaptive submodular maximization have been studied in [11,13–15].

3 Preliminaries

3.1 Items

We consider a set E of n items. Each items $e \in E$ belongs to a random state $\Phi(e) \in O$ where O represents the set of all possible states. Denote by ϕ a *realization* of Φ, i.e., for each $e \in E$, $\phi(e)$ is a realization of $\Phi(e)$. In the application of experimental design, an item e represents a test, such as the blood pressure, and $\Phi(e)$ is the result of the test, such as, *high*. We assume that there is a known prior probability distribution $p(\phi) = \Pr(\Phi = \phi)$ over realizations ϕ. The distribution p completely factorizes if realizations are independent. However, we consider a general setting where the realizations are dependent. For any subset of items $S \subseteq E$, we use $\psi : S \to O$ to represent a *partial realization* and $\mathrm{dom}(\psi) = S$ is called the *domain* of ψ. For any pair of a partial realization ψ and a realization ϕ, we say ϕ is consistent with ψ, denoted $\phi \sim \psi$, if they are equal everywhere in $\mathrm{dom}(\psi)$. For any two partial realizations ψ and ψ', we say that ψ is a *subrealization* of ψ', and denoted by $\psi \subseteq \psi'$, if $\mathrm{dom}(\psi) \subseteq \mathrm{dom}(\psi')$ and they are consistent in $\mathrm{dom}(\psi)$. In addition, each item $e \in E$ has a cost $c(e)$. For any $S \subseteq E$, let $c(S) = \sum_{e \in S} c(e)$ denote the total cost of S.

3.2 Policies

In the stream-based setting, we assume that items arrive one by one in an adversarial order σ. A policy has to make an irrevocable decision on whether to select an item or not when an item arrives. If an item is selected, then we are able to observe its realized state; otherwise, we can not reveal its realized state. Formally, a *stream-based* policy is a partial mapping that maps a pair of partial realizations ψ and an item e to some distribution of $\{0,1\}$: $\pi : 2^E \times O^E \times E \to \mathcal{P}(\{0,1\})$, specifying whether to select the arriving item e based on the current observation ψ. For example, assume that the current observation is ψ and the newly arrived item is e, then $\pi(\psi, e) = 1$ (resp. $\pi(\psi, e) = 0$) indicates that π selects (res. does not select) e.

Assume that there is a utility function $f : 2^{E \times O} \to \mathbb{R}_{\geq 0}$ which is defined over items and states. Letting $E(\pi, \phi, \sigma)$ denote the subset of items selected by a stream-based policy π conditioned on a realization ϕ and a sequence of arrivals σ, the expected utility $f_{avg}(\pi)$ of a stream-based policy π conditioned on a sequence of arrivals σ can be written as

$$\mathbb{E}[f_{avg}(\pi) \mid \sigma] = \mathbb{E}_{\Phi \sim p, \Pi}[f(E(\pi, \Phi, \sigma), \Phi)]$$

where the expectation is taken over all possible realizations Φ and the internal randomness of the policy π.

We next introduce the concept of policy concatenation which will be used in our proofs.

Definition 1 (Policy Concatenation). *Given two policies π and π', let $\pi@\pi'$ denote a policy that runs π first, and then runs π', ignoring the observation obtained from running π.*

Pool-Based Policy. When analyzing the performance of our stream-based policy, we compare our policy against the optimal *pool-based* policy which is allowed to select items in any order. Note that any stream-based policy can be viewed as a special case of pool-based policy, hence, an optimal pool-based policy can not perform worse than any optimal stream-based policy. By abuse of notation, we still use π to represent a pool-based policy. Formally, a pool-based policy can be encoded as a partial mapping π that maps partial realizations ψ to some distribution of E: $\pi : 2^E \times O^E \to \mathcal{P}'(E)$. Intuitively, $\pi(\psi)$ specifies which item to select next based on the current observation ψ. Letting $E(\pi, \phi)$ denote the subset of items selected by a pool-based policy π conditioned on a realization ϕ, the expected utility $f_{avg}(\pi)$ of a pool-based policy π can be written as

$$f_{avg}(\pi) = \mathbb{E}_{\Phi \sim p, \Pi}[f(E(\pi, \Phi), \Phi)]$$

where the expectation is taken over all possible realizations Φ and the internal randomness of the policy π. Note that if π is a pool-based policy, then for any sequence of arrivals σ, $f_{avg}(\pi) = \mathbb{E}[f_{avg}(\pi) \mid \sigma]$. This is because the output of a pool-based policy does not depend on the sequence of arrivals.

3.3 Problem Formulation and Additional Notations

Our objective is to find an stream-based policy that maximizes the worst-case expected utility subject to a budget constraint B, i.e.,

$$\max_{\pi \in \Omega^s} \min_{\sigma} \mathbb{E}[f_{avg}(\pi) \mid \sigma] \tag{1}$$

where $\Omega^s = \{\pi \mid \forall \phi, \sigma' : c(E(\pi, \Phi, \sigma')) \leq B\}$ represents a set of all feasible stream-based policies subject to a knapsack constraint (c, B). That is, a feasible policy must satisfy the budget constraint under all possible realizations and sequences of arrivals.

We next introduce some additional notations and important assumptions in order to facilitate our study.

Definition 2 (Conditional Expected Marginal Utility of an Item).
Given a utility function $f : 2^{E \times O} \to \mathbb{R}_{\geq 0}$, the conditional expected marginal utility $\Delta(e \mid \psi)$ of an item e on top of a partial realization ψ is

$$\Delta(e \mid \psi) = \mathbb{E}_{\Phi}[f(S \cup \{e\}, \Phi) - f(S, \Phi) \mid \Phi \sim \psi] \tag{2}$$

where the expectation is taken over Φ with respect to $p(\phi \mid \psi) = \Pr(\Phi = \phi \mid \Phi \sim \psi)$.

Definition 3. *[4][Adaptive Submodularity and Monotonicity] A function $f : 2^{E \times O} \to \mathbb{R}_{\geq 0}$ is adaptive submodular with respect to a prior $p(\phi)$ if for any two partial realization ψ and ψ' such that $\psi \subseteq \psi'$ and any item $e \in E \setminus \mathrm{dom}(\psi')$,*

$$\Delta(e \mid \psi) \geq \Delta(e \mid \psi') \tag{3}$$

Moreover, if $f : 2^{E \times O} \to \mathbb{R}_{\geq 0}$ is adaptive monotone with respect to a prior $p(\phi)$, then we have $\Delta(e \mid \psi) \geq 0$ for any partial realization ψ and any item $e \in E \setminus \mathrm{dom}(\psi)$.

Definition 4 (Conditional Expected Marginal Utility of a Pool-Based Policy). *Given a utility function* $f : 2^{E \times O} \to \mathbb{R}_{\geq 0}$, *the conditional expected marginal utility* $\Delta(\pi \mid \psi)$ *of a pool-based policy* π *on top of partial realization* ψ *is*

$$\Delta(\pi \mid \psi) = \mathbb{E}_{\Phi, \Pi}[f(E(\pi, \Phi), \Phi) - f(\mathrm{dom}(\psi), \Phi) \mid \Phi \sim \psi]$$

where the expectation is taken over Φ *with respect to* $p(\phi \mid \psi) = \Pr(\Phi = \phi \mid \Phi \sim \psi)$ *and the internal randomness of* π.

We next introduce a new class of stochastic functions.

Definition 5 (Semi-policywise Submodularity). *A function* $f : 2^{E \times O} \to \mathbb{R}_{\geq 0}$ *is semi-policywise submodular with respect to a prior* $p(\phi)$ *and a knapsack constraint* (c, B) *if for any partial realization* ψ,

$$f_{avg}(\pi^*) \geq \max_{\pi \in \Omega^p} \Delta(\pi \mid \psi) \tag{4}$$

where Ω^p *denotes the set of all possible pool-based policies subject to a knapsack constraint* (c, B), *i.e.,* $\Omega^p = \{\pi \mid \forall \phi, c(E(\pi, \phi)) \leq B\}$, *and*

$$\pi^* \in \arg\max_{\pi \in \Omega^p} f_{avg}(\pi)$$

represents an optimal pool-based policy subject to (c, B).

In the rest of this paper, we always assume that our utility function $f : 2^{E \times O} \to \mathbb{R}_{\geq 0}$ is adaptive monotone, adaptive submodular and semi-policywise submodular with respect to a prior $p(\phi)$ and a knapsack constraint (c, B). In appendix, we show that this type of function can be found in a variety of important real world applications. All missing materials are moved to appendix.

4   Uniform Cost

We first study the case when all items have uniform costs, i.e., $\forall e \in E, c(e) = 1$. Without loss of generality, assume B is some positive integer. To solve this problem, we extend the non-adaptive solution in [1] to the adaptive setting.

4.1   Algorithm Design

Recall that $\pi^* \in \arg\max_{\pi \in \Omega^p} f_{avg}(\pi)$ represents an optimal pool-based policy subject to a budget constraint B, suppose we can estimate $f_{avg}(\pi^*)$ approximately, i.e., we know a value v such that $\beta \cdot f_{avg}(\pi^*) \geq v \geq \alpha \cdot f_{avg}(\pi^*)$ for some $\alpha \in [0, 1]$ and $\beta \in [1, 2]$. Our policy, called Online Adaptive Policy π^c, starts with an empty set $S = \emptyset$. In each subsequent iteration i, after observing an arriving item $\sigma(i)$, π^c adds $\sigma(i)$ to S if the marginal value of $\sigma(i)$ on top of the current partial realization ψ_t is at least $\frac{v}{2B}$; otherwise, it skips $\sigma(i)$. This process iterates until there are no more arriving items or it reaches the cardinality constraint. A detailed description of π^c is listed in Algorithm 1.

Algorithm 1. Online Adaptive Policy π^c

1: $S = \emptyset; i = 1; t = 1; \psi_1 = \emptyset.$
2: **while** $i \leq n$ and $|S| < B$ **do**
3:    **if** $\Delta(\sigma(i) \mid \psi_t) \geq \frac{v}{2B}$ **then**
4:       $S \leftarrow S \cup \{\sigma(i)\}; \psi_{t+1} \leftarrow \psi_t \cup \{(\sigma(i), \Phi(\sigma(i)))\}; t \leftarrow t + 1;$
5:    $i = i + 1;$
6: **return** S

4.2 Performance Analysis

We present the main result of this section in the following theorem.

Theorem 1. *Assuming that we know a value v such that $\beta \cdot f_{avg}(\pi^*) \geq v \geq \alpha \cdot f_{avg}(\pi^*)$ for some $\beta \in [1, 2]$ and $\alpha \in [0, 1]$, we have $\mathbb{E}[f_{avg}(\pi^c) \mid \sigma] \geq \min\{\frac{\alpha}{4}, \frac{2-\beta}{4}\} f_{avg}(\pi^*)$ for any sequence of arrivals σ.*

4.3 Offline Estimation of $f_{avg}(\pi^*)$

Recall that the design of π^c requires that we know a good approximation of $f_{avg}(\pi^*)$. We next explain how to obtain such an estimation. It is well known that a simple greedy pool-based policy π^g (which is outlined in Algorithm 2) provides a $(1-1/e)$ approximation for the pool-based adaptive submodular maximization problem subject to a cardinality constraint [4], i.e., $f_{avg}(\pi^g) \geq (1-1/e)f_{avg}(\pi^*)$. Hence, $f_{avg}(\pi^g)$ is a good approximation of $f_{avg}(\pi^*)$. In particular, if we set $v = f_{avg}(\pi^g)$, then we have $f_{avg}(\pi^*) \geq v \geq (1-1/e)f_{avg}(\pi^*)$. This, together with Theorem 2, implies that π^c achieves a $\frac{1-1/e}{4}$ approximation ratio against π^*. One can estimate the value of $f_{avg}(\pi^g)$ by simulating π^g on every possible realization ϕ to obtain $E(\pi^g, \phi)$ and letting $f_{avg}(\pi^g) = \sum_\phi p(\phi) f(E(\pi^g, \phi), \phi)$. When the number of possible realizations is large, one can sample a set of realizations according to $p(\phi)$ then run the simulation. Although obtaining a good estimation of $f_{avg}(\pi^g)$ may be time consuming, this only needs to be done once in an offline manner. Thus, it does not contribute to the running time of the online implementation of π^c.

Algorithm 2. Offline Adaptive Greedy Policy π^g

1: $S = \emptyset; t = 1; \psi_1 = \emptyset.$
2: **while** $t \leq B$ **do**
3:    let $e' = \arg\max_{e \in E} \Delta(e \mid \psi_t);$
4:    $S \leftarrow S \cup \{e'\}; \psi_{t+1} \leftarrow \psi_t \cup \{(e', \Phi(e'))\}; t \leftarrow t + 1;$
5: **return** S

---

5 Nonuniform Cost

We next study the general case when items have nonuniform costs.

5.1   Algorithm Design

Algorithm 3. Online Adaptive Policy with Nonuniform Cost π^k

1: $S = \emptyset; t = 1; i = 1; \psi_1 = \emptyset$.
2: **while** $i \leq n$ **do**
3:     **if** $\frac{\Delta(\sigma(i)|\psi_t)}{c(\sigma(i))} \geq \frac{v}{2B}$ **then**
4:         **if** $\sum_{e \in S} c(e) + c(\sigma(i)) > B$ **then**
5:             break;
6:         **else**
7:             $S \leftarrow S \cup \{\sigma(i)\}; \psi_{t+1} \leftarrow \psi_t \cup \{(\sigma(i), \Phi(\sigma(i)))\}; t \leftarrow t + 1;$
8:         $i = i + 1$
9: **return** S

Suppose we can estimate $f_{avg}(\pi^*)$ approximately, i.e., we know a value v such that $\beta \cdot f_{avg}(\pi^*) \geq v \geq \alpha \cdot f_{avg}(\pi^*)$ for some $\alpha \in [0,1]$ and $\beta \in [1,2]$. For each $e \in E$, let $f(e)$ denote $\mathbb{E}_\Phi[f(\{e\}, \Phi)]$ for short. Our policy randomly selects a solution from $\{e^*\}$ and π^k with equal probability, where $e^* = \arg\max_{e \in E} f(e)$ is the best singleton and π^k, which is called Online Adaptive Policy with Nonuniform Cost, is a density-greedy policy. Hence, the expected utility of our policy is $(f(e^*) + \mathbb{E}[f_{avg}(\pi^k) \mid \sigma])/2$ for any given sequence of arrivals σ. We next explain the design of π^k. π^k starts with an empty set $S = \emptyset$. In each subsequent iteration i, after observing an arriving item $\sigma(i)$, it adds $\sigma(i)$ to S if the marginal value per unit budget of $\sigma(i)$ on top of the current realization ψ_t is at least $\frac{v}{2B}$, i.e., $\frac{\Delta(\sigma(i)|\psi_t)}{c(\sigma(i))} \geq \frac{v}{2B}$, and adding $\sigma(i)$ to S does not violate the budget constraint; otherwise, if $\frac{\Delta(\sigma(i)|\psi_t)}{c(\sigma(i))} < \frac{v}{2B}$, π^k skips $\sigma(i)$. This process iterates until there are no more arriving items or it reaches the first item (excluded) that violates the budget constraint. A detailed description of π^k is listed in Algorithm 3.

5.2   Performance Analysis

Before presenting the main theorem, we first introduce a technical lemma.

Lemma 1. *Assuming that we know a value v such that $\beta \cdot f_{avg}(\pi^*) \geq v \geq \alpha \cdot f_{avg}(\pi^*)$ for some $\alpha \in [0,1]$ and $\beta \in [1,2]$, we have $\max\{f(e^*), \mathbb{E}[f_{avg}(\pi^k) \mid \sigma]\} \geq \min\{\frac{\alpha}{4}, \frac{2-\beta}{4}\} f_{avg}(\pi^*)$ for any sequence of arrivals σ.*

Proof: We first introduce an auxiliary policy π^{k+} that follows the same procedure of π^k except that π^{k+} is allowed to add the first item that violates the budget constraint. Although π^{k+} is not necessarily feasible, we next show that the expected utility $\mathbb{E}[f_{avg}(\pi^{k+}) \mid \sigma]$ of π^{k+} is upper bounded by $\max\{f(e^*), \mathbb{E}[f_{avg}(\pi^k) \mid \sigma]\}$ for any sequence of arrivals σ, i.e., $\mathbb{E}[f_{avg}(\pi^{k+}) \mid \sigma] \leq \max\{f(e^*), \mathbb{E}[f_{avg}(\pi^k) \mid \sigma]\}$.

Proposition 1. *For any sequence of arrivals σ,*

$$\mathbb{E}[f_{avg}(\pi^{k+}) \mid \sigma] \leq \max\{f(e^*), \mathbb{E}[f_{avg}(\pi^k) \mid \sigma]\}$$

Proposition 1, whose proof is deferred to appendix, implies that to prove this lemma, it suffices to show that $\mathbb{E}[f_{avg}(\pi^{k+}) \mid \sigma] \geq \min\{\frac{\alpha}{4}, \frac{2-\beta}{4}\} f_{avg}(\pi^*)$. The rest of the analysis is devoted to proving this inequality for any fixed sequence of arrivals σ. We use $\lambda = \{\psi_1^\lambda, \psi_2^\lambda, \psi_3^\lambda, \cdots, \psi_{z^\lambda}^\lambda\}$ to denote a fixed run of π^{k+}, where ψ_t^λ is the partial realization of the first t selected items and z^λ is the total number of selected items under λ. Let $U = \{\lambda \mid \Pr[\lambda] > 0\}$ represent all possible stories of running π^{k+}, U^+ represent those stories where π^{k+} meets or violates the budget, i.e., $U^+ = \{\lambda \in U \mid c(\mathrm{dom}(\psi_{z^\lambda}^\lambda)) \geq B\}$, and U^- represent those stories where π^{k+} does not use up the budget, i.e., $U^- = \{\lambda \in U \mid c(\mathrm{dom}(\psi_{z^\lambda}^\lambda)) < B\}$. Therefore, $U = U^+ \cup U^-$. For each λ and $t \in [z^\lambda]$, let e_t^λ denote the t-th selected item under λ. Define $\psi_0^\lambda = \emptyset$ for any λ. Using the above notations, we can represent $\mathbb{E}[f_{avg}(\pi^{k+}) \mid \sigma]$ as follows:

$$\mathbb{E}[f_{avg}(\pi^{k+}) \mid \sigma] = \sum_{\lambda \in U} \Pr[\lambda] \left(\sum_{t \in [z^\lambda]} \Delta(e_t^\lambda \mid \psi_{t-1}^\lambda) \right) \tag{5}$$

$$= \underbrace{\sum_{\lambda \in U^+} \Pr[\lambda] \left(\sum_{t \in [z^\lambda]} \Delta(e_t^\lambda \mid \psi_{t-1}^\lambda) \right)}_{I} + \sum_{\lambda \in U^-} \Pr[\lambda] \left(\sum_{t \in [z^\lambda]} \Delta(e_t^\lambda \mid \psi_{t-1}^\lambda) \right) \tag{6}$$

Then we consider two cases. We first consider the case when $\sum_{\lambda \in U^+} \Pr[\lambda] \geq 1/2$ and show that the value of I is lower bounded by $\frac{\alpha}{4} f_{avg}(\pi^*)$. According to the definition of U^+, we have $\sum_{t \in [z^\lambda]} c(e_t^\lambda) \geq B$ for any $\lambda \in U^+$. Moreover, recall that for all $t \in [z^\lambda]$, $\frac{\Delta(e_t^\lambda \mid \psi_{t-1}^\lambda)}{c(e_t^\lambda)} \geq \frac{v}{2B}$ due to the design of our algorithm. Therefore, for any $\lambda \in U^+$,

$$\sum_{t \in [z^\lambda]} \Delta(e_t^\lambda \mid \psi_{t-1}^\lambda) \geq \frac{v}{2B} \times B = \frac{v}{2} \tag{7}$$

Because we assume that $\sum_{\lambda \in U^+} \Pr[\lambda] \geq 1/2$, we have

$$\sum_{\lambda \in U^+} \Pr[\lambda] \left(\sum_{t \in [z^\lambda]} \Delta(e_t^\lambda \mid \psi_{t-1}^\lambda) \right) \geq \left(\sum_{\lambda \in U^+} \Pr[\lambda] \right) \times \frac{v}{2} \geq \frac{v}{4} \geq \frac{\alpha}{4} f_{avg}(\pi^*) \tag{8}$$

The first inequality is due to (7) and the third inequality is due to the assumption that $v \geq \alpha \cdot f_{avg}(\pi^*)$. We conclude that the value of I (and thus $\mathbb{E}[f_{avg}(\pi^{k+}) \mid \sigma]$) is no less than $\frac{\alpha}{4} f_{avg}(\pi^*)$, i.e.,

$$\mathbb{E}[f_{avg}(\pi^{k+}) \mid \sigma] \geq \frac{\alpha}{4} f_{avg}(\pi^*) \tag{9}$$

We next consider the case when $\sum_{\lambda \in U^+} \Pr[\lambda] < 1/2$. We show that under this case,

$$\mathbb{E}[f_{avg}(\pi^{k+}) \mid \sigma] \geq \frac{2-\beta}{4} f_{avg}(\pi^*) \tag{10}$$

Because $f : 2^{E \times O} \to \mathbb{R}_{\geq 0}$ is adaptive monotone, we have $\mathbb{E}[f_{avg}(\pi^{k+} @ \pi^*) \mid \sigma] \geq f_{avg}(\pi^*)$. To prove (10), it suffices to show that

$$\mathbb{E}[f_{avg}(\pi^{k+}) \mid \sigma] \geq \frac{2 - \beta}{4} \mathbb{E}[f_{avg}(\pi^{k+} @ \pi^*) \mid \sigma]$$

Observe that we can represent the gap between $f_{avg}(\pi^{k+} @ \pi^*)$ and $f_{avg}(\pi^*)$ conditioned on σ as follows:

$$\mathbb{E}[f_{avg}(\pi^{k+} @ \pi^*) - f_{avg}(\pi^{k+}) \mid \sigma] = \sum_{\lambda \in U} \Pr[\lambda] \left(\sum_{t \in [z^\lambda]} \Delta(\pi^* \mid \psi_{z^\lambda}^\lambda) \right) \quad (11)$$

$$= \underbrace{\sum_{\lambda \in U^+} \Pr[\lambda] \left(\sum_{t \in [z^\lambda]} \Delta(\pi^* \mid \psi_{z^\lambda}^\lambda) \right)}_{II} + \underbrace{\sum_{\lambda \in U^-} \Pr[\lambda] \left(\sum_{t \in [z^\lambda]} \Delta(\pi^* \mid \psi_{z^\lambda}^\lambda) \right)}_{III} \quad (12)$$

Because $f : 2^{E \times O} \to \mathbb{R}_{\geq 0}$ is semi-policywise submodular with respect to $p(\phi)$ and (c, B), we have $\max_{\pi \in \Omega^p} \Delta(\pi \mid \psi_{z^\lambda}^\lambda) \leq f_{avg}(\pi^*)$. Moreover, because $\Delta(\pi^* \mid \psi_{z^\lambda}^\lambda) \leq \max_{\pi \in \Omega^p} \Delta(\pi \mid \psi_{z^\lambda}^\lambda)$, we have

$$\Delta(\pi^* \mid \psi_{z^\lambda}^\lambda) \leq f_{avg}(\pi^*) \quad (13)$$

It follows that

$$II = \sum_{\lambda \in U^+} \Pr[\lambda] \left(\sum_{t \in [z^\lambda]} \Delta(\pi^* \mid \psi_{z^\lambda}^\lambda) \right) \leq \left(\sum_{\lambda \in U^+} \Pr[\lambda] \right) f_{avg}(\pi^*) \quad (14)$$

Next, we show that III is upper bounded by $\left(\sum_{\lambda \in U^-} \Pr[\lambda] \right) \frac{\beta}{2} f_{avg}(\pi^*)$. For any $\psi_{z^\lambda}^\lambda$, we number all items $e \in E$ by decreasing ratio $\frac{\Delta(e \mid \psi_{z^\lambda}^\lambda)}{c(e)}$, i.e., $e(1) \in \arg\max_{e \in E} \frac{\Delta(e \mid \psi_{z^\lambda}^\lambda)}{c(e)}$. Let $l = \min\{i \in \mathbb{N} \mid \sum_{j=1}^{i} c(e(i)) \geq B\}$. Define $D(\psi_{z^\lambda}^\lambda) = \{e(i) \in E \mid i \in [l]\}$ as the set containing the first l items. Intuitively, $D(\psi_{z^\lambda}^\lambda)$ represents a set of *best-looking* items conditional on $\psi_{z^\lambda}^\lambda$. Consider any $e \in D(\psi_{z^\lambda}^\lambda)$, assuming e is the i-th item in $D(\psi_{z^\lambda}^\lambda)$, let

$$x(e, \psi_{z^\lambda}^\lambda) = \min\{1, \frac{B - \sum_{s \in \cup_{j \in [i-1]} \{e(j)\}} c(s)}{c(e)}\}$$

where $\cup_{j \in [i-1]} \{e(j)\}$ represents the first $i - 1$ items in $D(\psi_{z^\lambda}^\lambda)$.

In analogy to Lemma 1 of [5],

$$\sum_{e \in D(\psi_{z^\lambda}^\lambda)} x(e, \psi_{z^\lambda}^\lambda) \Delta(e \mid \psi_{z^\lambda}^\lambda) \geq \Delta(\pi^* \mid \psi_{z^\lambda}^\lambda) \quad (15)$$

Note that for every $\lambda \in U^-$, we have $\sum_{t \in [z^\lambda]} c(e_t^\lambda) < B$, that is, π^{k+} does not use up the budget under λ. This, together with the design of π^{k+}, indicates that for any $e \in E$, its benefit-to-cost ratio on top of $\psi_{z^\lambda}^\lambda$ is less than $\frac{v}{2B}$, i.e., $\frac{\Delta(e|\psi_{z^\lambda}^\lambda)}{c(e)} < \frac{v}{2B}$. Therefore,

$$\sum_{e \in D(\psi_{z^\lambda}^\lambda)} x(e, \psi) \Delta(e \mid \psi_{z^\lambda}^\lambda) \leq B \times \frac{v}{2B} = \frac{v}{2} \tag{16}$$

(15) and (16) imply that

$$\Delta(\pi^* \mid \psi_{z^\lambda}^\lambda) \leq \frac{v}{2} \tag{17}$$

We next provide an upper bound of III,

$$III = \sum_{\lambda \in U^-} \Pr[\lambda](\sum_{t \in [z^\lambda]} \Delta(\pi^* \mid \psi_{z^\lambda}^\lambda)) \leq (\sum_{\lambda \in U^-} \Pr[\lambda])\frac{v}{2} \tag{18}$$

$$\leq (\sum_{\lambda \in U^-} \Pr[\lambda])\frac{\beta}{2} f_{avg}(\pi^*) \tag{19}$$

where the first inequality is due to (17) and the second inequality is due to $v \leq \beta \cdot f_{avg}(\pi^*)$.

Now we are in position to bound the value of $\mathbb{E}[f_{avg}(\pi^{k+}@\pi^*) - f_{avg}(\pi^{k+}) \mid \sigma]$,

$$\mathbb{E}[f_{avg}(\pi^{k+}@\pi^*) - f_{avg}(\pi^{k+}) \mid \sigma] = II + III \tag{20}$$

$$\leq (\sum_{\lambda \in U^+} \Pr[\lambda])f_{avg}(\pi^*) + (\sum_{\lambda \in U^-} \Pr[\lambda])\frac{\beta}{2} f_{avg}(\pi^*) \tag{21}$$

$$\leq \frac{1}{2} f_{avg}(\pi^*) + \frac{1}{2} \times \frac{\beta}{2} f_{avg}(\pi^*) \tag{22}$$

$$= \frac{2+\beta}{4} f_{avg}(\pi^*) \tag{23}$$

The first inequality is due to (14) and (19). The second inequality is due to $\sum_{\lambda \in U^+} \Pr[\lambda] + \sum_{\lambda \in U^-} \Pr[\lambda] = 1$ and the assumptions that $\sum_{\lambda \in U^+} \Pr[\lambda] < 1/2$ and $\beta \in [1, 2]$. Because $\mathbb{E}[f_{avg}(\pi^{k+}@\pi^*) \mid \sigma] \geq \mathbb{E}[f_{avg}(\pi^*) \mid \sigma]$, which is due to $f : 2^{E \times O} \to \mathbb{R}_{\geq 0}$ is adaptive monotone, we have

$$\mathbb{E}[f_{avg}(\pi^*) - f_{avg}(\pi^{k+}) \mid \sigma] \leq \mathbb{E}[f_{avg}(\pi^{k+}@\pi^*) - f_{avg}(\pi^{k+}) \mid \sigma] \tag{24}$$

$$\leq \frac{2+\beta}{4} f_{avg}(\pi^*) \tag{25}$$

where the second inequality is due to (23). This, together with the fact that $\mathbb{E}[f_{avg}(\pi^*) \mid \sigma] = f_{avg}(\pi^*)$, i.e., the optimal pool-based policy is not dependent on the sequence of arrivals, implies (10).

Combining the above two cases ((9) and (10)), we have

$$\mathbb{E}[f_{avg}(\pi^{k+}) \mid \sigma] \geq \min\{\frac{\alpha}{4}, \frac{2-\beta}{4}\} f_{avg}(\pi^*) \tag{26}$$

This, together with Proposition 1, immediately concludes this lemma. □

Recall that our final policy randomly picks a solution from $\{e^*\}$ and π^k with equal probability, thus, its expected utility is $\frac{f(e^*)+\mathbb{E}[f_{avg}(\pi^k)|\sigma]}{2}$ which is lower bounded by $\frac{\max\{f(e^*),\mathbb{E}[f_{avg}(\pi^k)|\sigma]\}}{2}$. This, together with Lemma 1, implies the following main theorem.

Theorem 2. *If we randomly pick a solution from $\{e^*\}$ and π^k with equal probability, then it achieves a $\min\{\frac{\alpha}{8}, \frac{2-\beta}{8}\}$ approximation ratio against the optimal pool-based policy π^*.*

5.3 Offline Estimation of $f_{avg}(\pi^*)$

Algorithm 4. Offline Greedy Policy with Nonuniform Cost π^{gn}

1: $S = \emptyset; t = 1; \psi_1 = \emptyset$.
2: **while** $t \leq B$ **do**
3:     let $e' = \arg\max_{e \in E} \frac{\Delta(e|\psi_t)}{c(e)}$;
4:     **if** $\sum_{e \in S} c(e) + c(e') > B$ **then**
5:         break;
6:     $S \leftarrow S \cup \{e'\}; \psi_{t+1} \leftarrow \psi_t \cup \{(e', \Phi(e'))\}; t \leftarrow t + 1$;
7: **return** S

To complete the design of π^k, we next explain how to estimate the utility of the optimal pool-based policy $f_{avg}(\pi^*)$. It has been shown that the better solution between $\{e^*\}$ and a pool-based density-greedy policy π^{gn} (Algorithm 4) achieves a $(1 - 1/e)/2$ approximation for the pool-based adaptive submodular maximization problem subject to a knapsack constraint [16], i.e., $\max\{f_{avg}(\pi^{gn}), f(e^*)\} \geq \frac{1-1/e}{2} f_{avg}(\pi^*)$. If we set $v = \max\{f_{avg}(\pi^{gn}), f(e^*)\}$ in π^k, then we have $\alpha = (1 - 1/e)/2$ and $\beta = 1$. This, together with Theorem 2, implies that π^c achieves a $\frac{1-1/e}{16}$ approximation ratio against π^*. One can estimate the value of $f_{avg}(\pi^{gn})$ by simulating π^{gn} on every possible realization ϕ to obtain $E(\pi^{gn}, \phi)$ and letting $f_{avg}(\pi^{gn}) = \sum_\phi p(\phi)f(E(\pi^{gn}, \phi), \phi)$. To estimate the value of $f(e^*)$, one can compute the value of $f(e)$ using $f(e) = \sum_\phi p(\phi)f(\{e\}, \phi)$ for all $e \in E$, then return the best result as $f(e^*)$.

References

1. Badanidiyuru, A., Mirzasoleiman, B., Karbasi, A., Krause, A.: Streaming submodular maximization: massive data summarization on the fly. In: Proceedings of the 20th ACM SIGKDD International Conference on Knowledge Discovery and Data Mining, pp. 671–680 (2014)
2. Chekuri, C., Livanos, V.: On submodular prophet inequalities and correlation gap. arXiv preprint arXiv:2107.03662 (2021)
3. Fujii, K., Kashima, H.: Budgeted stream-based active learning via adaptive submodular maximization. In: NIPS, vol. 16, pp. 514–522 (2016)

4. Golovin, D., Krause, A.: Adaptive submodularity: theory and applications in active learning and stochastic optimization. J. Artifi. Intell. Res. **42**, 427–486 (2011)
5. Gotovos, A., Karbasi, A., Krause, A.: Non-monotone adaptive submodular maximization. In: Twenty-Fourth International Joint Conference on Artificial Intelligence (2015)
6. Kazemi, E., Mitrovic, M., Zadimoghaddam, M., Lattanzi, S., Karbasi, A.: Submodular streaming in all its glory: Tight approximation, minimum memory and low adaptive complexity. In: International Conference on Machine Learning, pp. 3311–3320. PMLR (2019)
7. Kuhnle, A.: Quick streaming algorithms for maximization of monotone submodular functions in linear time. In: International Conference on Artificial Intelligence and Statistics, pp. 1360–1368. PMLR (2021)
8. Rubinstein, A., Singla, S.: Combinatorial prophet inequalities. In: Proceedings of the Twenty-Eighth Annual ACM-SIAM Symposium on Discrete Algorithms, pp. 1671–1687. SIAM (2017)
9. Tang, S.: Beyond pointwise submodularity: non-monotone adaptive submodular maximization in linear time. Theoret. Comput. Sci. **850**, 249–261 (2021)
10. Tang, S.: Beyond pointwise submodularity: Non-monotone adaptive submodular maximization subject to knapsack and k-system constraints. In: 4th international Conference on Modelling, Computation and Optimization in Information Systems and Management Sciences (2021)
11. Tang, S.: Robust adaptive submodular maximization. CoRR abs/2107.11333 (2021). https://arxiv.org/abs/2107.11333
12. Tang, S., Yuan, J.: Influence maximization with partial feedback. Oper. Res. Lett. **48**(1), 24–28 (2020)
13. Tang, S., Yuan, J.: Adaptive regularized submodular maximization. In: 32nd International Symposium on Algorithms and Computation (ISAAC 2021). Schloss Dagstuhl-Leibniz-Zentrum für Informatik (2021)
14. Tang, S., Yuan, J.: Non-monotone adaptive submodular meta-learning. In: SIAM Conference on Applied and Computational Discrete Algorithms (ACDA 2021), pp. 57–65. SIAM (2021)
15. Tang, S., Yuan, J.: Optimal sampling gaps for adaptive submodular maximization. In: AAAI (2022)
16. Yuan, J., Tang, S.J.: Adaptive discount allocation in social networks. In: Proceedings of the 18th ACM International Symposium on Mobile Ad Hoc Networking and Computing, pp. 1–10 (2017)

Constrained Stochastic Submodular Maximization with State-Dependent Costs

Shaojie Tang$^{(\boxtimes)}$ (iD)

Naveen Jindal School of Management, University of Texas at Dallas, Texas, USA
shaojie.tang@utdallas.edu

Abstract. In this paper, we study the constrained stochastic submodular maximization problem with state-dependent costs. The input of our problem is a set of items whose states (i.e., the marginal contribution and the cost of an item) are drawn from a known probability distribution. The only way to know the realized state of an item is to select that item. We consider two constraints, i.e., *inner* and *outer* constraints. Recall that each item has a state-dependent cost, and the inner constraint states that the total *realized* cost of all selected items must not exceed a give budget. Thus, inner constraint is state-dependent. The outer constraint, on the other hand, is state-independent. It can be represented as a downward-closed family of sets of selected items regardless of their states. Our objective is to maximize the objective function subject to both inner and outer constraints. Under the assumption that larger cost indicates larger "utility", we present a constant approximate solution to this problem.

1 Introduction

In this paper, we study a novel constrained stochastic submodular maximization problem. We follow the framework developed in [4] and introduce the state-dependent item costs into the classic stochastic submodular maximization problem. The input of our problem is a set of items, each item has a random state which is drawn from a known probability distribution. The marginal contribution and the cost of an item are dependent on its actual state. The utility function is a mapping from sets of items and their states to a real number. We must select an item before observing its actual state. Our objective is to sequentially select a group of items to maximize the objective function. We must obey two constraints, namely, inner and outer constraints, through the selection process. The inner constraint requires that the total *realized* cost of all selected items must not exceed a given budget B. Thus, the inner constraint is state-dependent. The outer constraint is represented as a downward-closed family of sets of selected items regardless of their states. Thus, the outer constraint is state-independent. Under the assumption that the cost of an item is larger if it is in a "better" state, we present a constant approximate solution.

Our model is general enough to capture many real-world applications. Here we discuss three examples.

Q. Ni and W. Wu (Eds.): AAIM 2022, LNCS 13513, pp. 121–132, 2022.
https://doi.org/10.1007/978-3-031-16081-3_11

Adaptive Coupon Allocation. The objective of this problem [7] is to distribute coupons to a group of up to k seed users such that those who redeem the coupon can generate the largest cascade of influence. The state, which is stochastic, of a user is her decision on whether or not to redeem the coupon. In this case, it is uncertain in advance how many users redeem the coupon and help to promote the product. Our framework can capture this scenario by treating k as a outer constraint and treating the budget on the total value of redeemed coupons as a inner constraint.

Recruiting Crowd Workers. Crowdsourcing is an effective way of obtaining information from a large group of workers. A typical crowdsourcing process can be described as follows: the task-owner sequentially hires up to k workers to work on a set of similar tasks. Each worker reports her results to the task-owner after completing her task. The state of a worker is the quality of the results returned by her. It is clearly reasonable to assume that the actual amount of reward paid to a particular worker depends on her state, which can only be observed after she delivers the task. By treating k as a outer constraint and treating the budget on the total value of total payments as a inner constraint, our objective is to maximize the overall quality of the completed tasks subject to outer and inner constraints.

Recommendation. In the context of product or news recommendation, our objective is to recommend a group of items to a customer in order to maximize some utility function which often satisfies the diminishing marginal return property. The performance of a recommended item depends on many random factors such as the customer's preferences. For example, after receiving a recommended article, the customer decides to skip or read it in a probabilistic manner and she must spend her own resource such as time and money on reading a recommended article. Our framework can capture this scenario by treating the customer's decision on a particular article as the state of that article. Hence, the performance, as well as the cost, of a recommended article is determined by its state.

Related Works. Stochastic submodular maximization has been extensively studied in the literature [5,8,9,11]. While most of existing works assume that the cost of each item is deterministic and pre-known, we consider state-dependent item costs. Our problem reduces to the stochastic knapsack problem [6] when considering linear objective function. Recently, [4,10] extended the previous study to the stochastic submodular maximization problem, however, their model does not incorporate outer constraints. Hence, our study can be considered as an extension of [4]. Our work is also closely related to submodular probing problem [1] where they assume each item has binary states, our model allows each item to have multiple states.

2   Preliminaries and Problem Formulation

Lattice-submodular functions Let $I = \{1, 2, \cdots, n\}$ be a set of items and $[B] = \{1, 2, \cdots, B\}$ be a set of states. We further define $[0; B] = \{0, 1, 2, \cdots, B\}$. Given two vectors $u, v \in [0; B]^I$, $u \leq v$ means that $u(i) \leq v(i)$ for all $i \in I$. Define

$(u \lor v)(i) = \max\{u(i), v(i)\}$ and $(u \land v)(i) = \min\{u(i), v(i)\}$. For $i \in I$, define $\mathbf{1}_i$ as the vector that has a 1 in the i-th coordinate and 0 in all other coordinates. A function $f : [0; B]^I \rightarrow \mathbb{R}_+$ is called *monotone* if $f(u) \leq f(v)$ holds for any $u, v \in [0; B]^I$ such that $u \leq v$, and f is called *lattice submodular* if $f(u \lor s\mathbf{1}_i) - f(u) \geq f(v \lor s\mathbf{1}_i) - f(v)$ holding for any $u, v \in [0; B]^I$ such that $u \leq v$, $s \in [0; B]$, $i \in I$.

Items and States. We use a vector $\phi \in [B]^I$ to denote a *realization* where for each item $i \in I$, $\phi(i) \in [B]$ denotes the state of i under realization ϕ. We assume that there is a known prior probability distribution p_i over realizations for each item i, i.e., $p_i = \{\Pr[\phi(i) = s] : s \in [B]\}$. The states of all items are decided independently at random, i.e., ϕ is drawn randomly from the product distribution $p = \prod_{i \in I} p_i$. For each item $i \in I$ and state $s \in [B]$, let $c_i(s)$ denote the cost of i when its state is s. We made the following assumption.

Assumption 1. *For all $i \in I$ and $s, s' \in [B]$ such that $s \geq s'$, we have $c_i(s) \geq c_i(s')$, i.e., the cost of an item is larger if it is in a "better" state.*

Adaptive Policy and Problem Formulation. Formally, a policy π is a function that specifies which item to select next based on the observations made so far. Consider any $S \subseteq I$ and any realization ϕ, we use ϕ_S to denote a vector in $[0; B]^I$ such that for each $i \in I$, set $\phi_S(i) = \phi(i)$ if $i \in S$, and $\phi_S(i) = 0$ otherwise. The utility of S conditioned on ϕ is $f(\phi_S)$ where $f : [0; B]^I \rightarrow \mathbb{R}_+$ is a monotone and lattice-submodular function. Consider an arbitrary policy π, for each ϕ, let $I(\pi, \phi)$ denote the set of items selected by π conditional on ϕ^1. Let Φ denote a random realization, the expected utility of π is written as

$$f_{avg}(\pi) = \mathbb{E}_{\Phi \sim p}[f(\Phi_{I(\pi,\Phi)})] \qquad (1)$$

Moreover, for any subset of items $S \subseteq I$, define $\overline{f}(S) = \mathbb{E}_{\Phi \sim p}[f(\Phi_S)]$ as the expected utility of S with respect to the distribution p.

Definition 1. *We say a policy π is feasible if it satisfies both outer and inner constraints:*

1. *(Inner Constraint) For all ϕ, we have $\sum_{i \in I(\pi, \phi)} c_i(\phi(i)) \leq C$.*
2. *(Outer Constraint) For all ϕ, we have $I(\pi, \phi) \in \mathcal{I}^{out}$, where $\mathcal{I}^{out}$ is a downward-closed family of sets of items.*

Our goal is to identify the best feasible policy that maximizes its expected utility.

$$\max_{\pi} f_{avg}(\pi) \text{ subject to } \forall \phi : \sum_{i \in I(\pi, \phi)} c_i(\phi(i)) \leq C; I(\pi, \phi) \in \mathcal{I}^{out}.$$

Following the framework developed in [3,4], our algorithm is composed of two phases, a continuous optimization phase and a rounding phase. We first

---

[1] For simplicity, we only consider deterministic policy. However, all results can be easily extended to random policies.

solve a continuous optimization problem and obtain a fractional solution. In the rounding phase, we convert the continuous solution to a feasible adaptive policy that obeys inner and outer constraints. We first explain the continuous optimization phase.

3   Continuous Optimization Phase

We present our solution based on the concept of "time". We assume that each item i is associated with a random processing time $c_i(\phi(i))$. Hence, if an item i is selected at time t, we must wait until $t + c_i(\phi(i))$ to select the next item. We treat the budget C as the time limit. We can not select an item i at slot t if the processing of i may finish after a time limit C, i.e., i can not be selected at time t if there exists some state $s \in [B]$ such that $t + c_i(s) > C$. We define a variable $x(i, t)$ for each item i and slot t, and it indicates whether an item i is selected at t. Let $\overline{x} \in \mathbb{R}_+^I$ denote a vector defined by $\overline{x}(i) = \sum_{t \in [C - c_i(B)]} x(i, t)$. We next introduce the multilinear extension $F(\overline{x})$ of $\overline{f}$.

$$F(\overline{x}) = \sum_{U \subseteq I} \prod_{i \in U} \overline{x}(i) \prod_{i \notin U} (1 - \overline{x}(i)) \overline{f}(U)$$

Let $P_{\mathcal{I}^{out}} \subseteq [0, 1]^I$ denote a polytope that is a relaxation for $\mathcal{I}^{out} \subseteq 2^I$, i.e., $P_{\mathcal{I}^{out}} = \text{conv}\{\mathbf{1}_S \mid S \in \mathcal{I}^{out}\}$. For any $d \in [0, 1]$, let $d \cdot P_{\mathcal{I}^{out}} = \{d \cdot \overline{x} \mid \overline{x} \in P_{\mathcal{I}^{out}}\}$. Now we are ready to introduce our continuous optimization problem as follows:

P1: *Maximize $F(\overline{x})$*
subject to:

$$\begin{cases} \forall i \in I : \overline{x}(i) = \sum_{t \in [C - c_i(B)]} x(i, t) \\ \forall i \in I : \overline{x}(i) \leq 1 \\ \overline{x} \in P_{\mathcal{I}^{out}} \\ \forall t \in [C] : \sum_{i \in I} \mathbb{E}[\min\{c_i(\phi(i)), t\}] \sum_{t' \in [t]} x(i, t') \leq 2t \quad (C1) \end{cases}$$

In constraint (C1), the expectation $\mathbb{E}[\min\{c_i(\phi(i)), t\}]$ is taken with respect to p, i.e., $\mathbb{E}[\min\{c_i(\phi(i)), t\}] = \sum_{s=1}^B p_i(s) \min\{c_i(s), t\}$. Note that the formulation of **P1** involves $\Omega(n \times C)$ variables, which makes our algorithm pseudo polynomial. However, we can apply the technique used in [6] to convert the algorithm into a polynomial-time algorithm at the expense of weakening the approximation ratio by a constant factor. If $P_{\mathcal{I}^{out}}$ is a solvable polytope, we can adopt the stochastic continuous greedy algorithm developed in [2] to solve **P1**. Their algorithm involves two controlling parameters: stopping time $l \in [0, 1]$ and step size δ. Their original analysis can be easily extended to show that for a stopping time $l \in [0, 1]$, the algorithm outputs a solution x such that $\overline{x} \in l \cdot P_{\mathcal{I}^{out}}$, $\forall t \in [C] : \sum_{i \in I} \mathbb{E}[\min\{c_i(\phi(i)), t\}] \sum_{t' \in [t]} x(i, t') \leq l \cdot 2t$, and $F(\overline{x}) \geq (1 - e^{-l} - O(n^3 \delta)) f_{avg}(\pi^{opt})$ where π^{opt} denotes the optimal policy of our original problem. The following lemma follows immediately from the above observation.

Lemma 1. *Let π^{opt} denote the optimal policy of our original problem. Assume $P_{\mathcal{I}^{out}}$ is a solvable polytope, if we apply the stochastic continuous greedy algorithm with stopping time $l \in [0,1]$ and step size $\delta = o(n^{-3})$ to solve* **P1***, then the algorithm outputs a solution x such that $\overline{x} \in l \cdot P_{\mathcal{I}^{out}}, \forall t \in [C] : \sum_{i \in I} \mathbb{E}[\min\{c_i(\phi(i)), t\}] \sum_{t' \in [t]} x(i, t') \leq l \cdot 2t$ and $F(\overline{x}) \geq (1 - e^{-l} - o(1)) f_{avg}(\pi^{opt})$.*

4 Rounding Phase

In this section, we introduce an effective rounding approach that converts the continuous solution to an adaptive policy. Before explaining the rounding phase, we first introduce two important concepts: (β, γ)-*balanced contention resolution scheme* [3] and α-*contention resolution scheme* [4].

4.1 Contention Resolution Scheme

(β, γ)-balanced contention resolution scheme $((\beta, \gamma)$-balanced CRS) is a general framework designed for maximizing set-submodular functions. In [4], the authors extend this concept to the lattice-submodular functions by introducing α-contention resolution scheme (α-CRS).

We first introduce the concept of (β, γ)-balanced CRS.

Definition 2 ((β, γ)-balanced CRS). *Given a vector $\overline{z} \in \beta \cdot P_{\mathcal{I}^{out}}$, let R denote a random set of I obtained by including each item $i \in I$ independently with probability $\overline{z}(i)$. A (β, γ)-balanced CRS with regards to $\overline{z}$ is a mapping $\chi : 2^I \to \mathcal{I}^{out}$ such that $\Pr[i \in \chi(R) | i \in R] \geq \gamma$, where the probability considers two sources of randomness: one is the randomness in choosing R, and the other source is the randomness in the execution of χ. A (β, γ)-balanced CRS is said to be monotone if for any two sets R, R' such that $R \subseteq R'$, the following inequality holds: $\Pr[i \in \chi(R) | i \in R] \geq \Pr[i \in \chi(R') | i \in R']$.*

We next introduce the concept of α-contention resolution scheme. Consider a probability distribution $q : I \times [0; B] \to [0, 1]$. Let $v \in [0; B]^I$ denote a random vector such that, for each $i \in I$, the value of $v(i)$ is set to $j \in [0; B]$ independently with probability $q(i, j)$.

Definition 3 (α-CRS). *Let $\mathcal{F} \subseteq [0; B]^I$ be a downward-closed subset of $[0; B]^I$, that is, $u \leq v \in \mathcal{F}$ implies $u \in \mathcal{F}$, and let $\alpha \in [0, 1]$. An α-contention resolution scheme (α-CRS) with regards to q is a mapping $\psi : [0; B]^I \to \mathcal{F}$ that satisfies the following two conditions:*

- *For each $i \in I$, $\psi(v)(i) \in \{0, v(i)\}$;*
- *For each $i \in I$ and each $j \in [B]$, we have $\Pr[\psi(v)(i) = j | v(i) = j] \geq \alpha$, where the probability considers two sources of randomness: one is the randomness in choosing v, and the other source is the randomness in the execution of ψ.*

An α-CRS ψ is said to be monotone if, for each $u, v \in [0; B]^I$ such that $u(i) = v(i)$ and $u \leq v$, we have $\Pr[\psi(u)(i) = u(i)] \geq \Pr[\psi(v)(i) = v(i)]$, where the probability here considers only the randomness in the execution of ψ.

In the context of maximizing set-submodular functions, Lemma 1.6 in [3] states that one can combine contention resolution schemes for different constraints. We next follow a similar proof of theirs to show that this result also holds for lattice-submodular functions.

Lemma 2. *Let $\mathcal{F} = \bigcap_{t=1}^{k} \mathcal{F}^t$ denote the intersection of several different subsets of $[0; B]^I$ where for each $t \in [k]$, $\mathcal{F}^t \subseteq [0; B]^I$ is a downward-closed subset of $[0; B]^I$. Suppose each $\mathcal{F}^t$ has a monotone α_t-CRS with regards to q. Then $\mathcal{F}$ has a monotone $\prod_{t=1}^{k} \alpha_t$-CRS with regards to q.*

Proof: We assume $k = 2$ for simplicity; the general statement can be proved by induction. Given a vector $v \in [0; B]^I$, for each $t \in \{1, 2\}$, assume that we can apply a monotone α_t-CRS ψ^t separately to obtain $\psi^t(v)$. Then we define a mapping $\psi : [0; B]^I \to \mathcal{F}$ such that

$$\text{for each } i \in I : \psi(v)(i) = \begin{cases} v(i) & \text{if for all } t \in \{1, 2\}, \psi^t(v)(i) = v(i) \\ 0 & \text{otherwise} \end{cases}$$

We next show that ψ is a monotone $\alpha_1 \alpha_2$-CRS with regards to q. Conditioned on v, the value of $\psi^1(v)$, $\psi^2(v)$ are independent, which means that

$$\Pr[\psi(v)(i) = v(i)] = \Pr[\psi^1(v)(i) = v(i) \& \psi^2(v)(i) = v(i)] \tag{2}$$
$$= \Pr[\psi^1(v)(i) = v(i)] \Pr[\psi^2(v)(i) = v(i)] \tag{3}$$

Taking an expectation over v conditioned on $v(i) = j$, we get

$$\Pr[\psi(v)(i) = j | v(i) = j] = \mathbb{E}_{v \sim q}[\Pr[\psi(v)(i) = j] | v(i) = j] \tag{4}$$
$$= \mathbb{E}_{v \sim q}[\Pr[\psi^1(v)(i) = j \& \psi^2(v)(i) = j] | v(i) = j] \tag{5}$$
$$= \mathbb{E}_{v \sim q}[\Pr[\psi^1(v)(i) = j] \Pr[\psi^2(v)(i) = j] | v(i) = j] \tag{6}$$

Due to both ψ^1 and ψ^2 are monotone, we have both $\Pr[\psi^1(v)(i) = j | v]$ and $\Pr[\psi^2(v)(i) = j | v]$ are non-increasing function of v on the product space of vectors that satisfy $v(i) = j$. By the FKG inequality, we have

$$\mathbb{E}_{v \sim q}[\Pr[\psi^1(v)(i) = j] \Pr[\psi^2(v)(i) = j] | v(i) = j] \tag{7}$$
$$\geq \mathbb{E}_{v \sim q}[\Pr[\psi^1(v)(i) = j] | v(i) = j] \mathbb{E}_{v \sim q}[\Pr[\psi^2(v)(i) = j] | v(i) = j] \tag{8}$$
$$= \Pr[\psi^1(v)(i) = j | v(i) = j] \Pr[\psi^2(v)(i) = j | v(i) = j] \tag{9}$$

(6) and (9) imply that

$$\Pr[\psi(v)(i) = j | v(i) = j] \tag{10}$$
$$\geq \Pr[\psi^1(v)(i) = j | v(i) = j] \Pr[\psi^2(v)(i) = j | v(i) = j] \tag{11}$$
$$\geq \alpha_1 \alpha_2 \tag{12}$$

The second inequality is due to the assumption that ψ^1 is an α_1-CRS with regards to q and ψ^2 is an α_2-CRS with regards to q.

We next prove the monotonicity of ψ. For each $u, v \in [0; B]^I$ such that $u(i) = v(i)$ and $u \leq v$, we have

$$\Pr[\psi(u)(i) = u(i)] = \Pr[\psi^1(u)(i) = u(i)] \Pr[\psi^2(u)(i) = u(i)] \tag{13}$$
$$\geq \Pr[\psi^1(v)(i) = u(i)] \Pr[\psi^2(v)(i) = u(i)] \tag{14}$$
$$= \Pr[\psi(v)(i) = v(i)] \tag{15}$$

The inequality is due to the assumption that both ψ^1 and ψ^2 are monotone. $\square$

4.2   Algorithm Design

Algorithm 1. Inner and Outer Constrained Adaptive Policy π^{io}

1: $A = \emptyset; i = 1; j = 1$.
2: compute a solution y for **P1** by the stochastic continuous greedy algorithm with stopping time $l = \min\{\beta, 1/4\}$ and step size $\delta = o(n^{-3})$
3: **for** $i \in I$ **do**
4:     add i to R^{io} with probability $\overline{y}(i)$
5: apply an outer constraint-specific monotone (β, γ)-balanced CRS χ^{io} to R^{io} to obtain a subset of items $\chi^{io}(R^{io})$ which satisfies the outer constraint
6: **for** $i \in \chi^{io}(R^{io})$ **do**
7:     sample a number t from $[C - c_i(B)]$ with probability $y(i,t)/\overline{y}(i)$
8:     $t^{io}(i) \leftarrow t$
9: $\sigma^{io} \leftarrow$ sequence of items in $\chi^{io}(R^{io})$ sorted in a nondecreasing order of $t^{io}(i)$, breaking ties with the least index tie breaking rule
10: **for** $i \in I$ **do**
11:     **if** $C' \leq t^{io}(\sigma_i^{io})$ **then**
12:         select σ_i^{io} and observe $\phi(\sigma_i^{io})$
13:         $C' = C' + c_{\sigma_i^{io}}(\phi(\sigma_i^{io}))$

Assume $P_{\mathcal{I}^{out}}$ is a solvable polytope and there exists a monotone (β, γ)-balanced CRS for $\mathcal{I}^{out}$. Now we are ready the present the design of our *Inner and Outer Constrained Adaptive Policy* π^{io} (Algorithm 1). Our policy is composed of three steps:

1. Compute a solution y for **P1** by the stochastic continuous greedy algorithm with stopping time $l = \min\{\beta, 1/4\}$ and step size $\delta = o(n^{-3})$.
2. Generate a random set R^{io} by including each item i with probability $\overline{y}(i)$. Then apply a monotone (β, γ)-balanced CRS χ^{io} to obtain a subset of items $\chi^{io}(R^{io})$ which satisfies the outer constraint.
3. Sample a number t from $[C - c_i(B)]$ with probability $y(i,t)/\overline{y}(i)$ for each $i \in \chi^{io}(R^{io})$. Let σ^{io} denote a sequence of items in $\phi^{io}(R^{io})$ sorted in a nondecreasing order of $t^{io}(i)$.

(a) Add σ_1^{io} to the solution and observe $\phi(\sigma_1^{io})$.
(b) Starting with $i = 2$. If $\sum_{i\in[i-1]} c_{\sigma_i^{io}}(\phi(\sigma_i^{io})) \leq t^{io}(\sigma_i^{io})$, add σ_i^{io} to the solution and observe $\phi(\sigma_i^{io})$; otherwise, set $c_{\sigma_i^{io}}(\phi(\sigma_i^{io})) = 0$. Repeat this step with the next item $i \leftarrow i + 1$. This process continues until all items from σ^{io} have been visited.

4.3 Performance Analysis

This section is devoted to proving the approximation ratio of π^{io}. Recall that y is obtained from solving problem **P1** using the stochastic continuous greedy algorithm with stopping time $l = \min\{\beta, 1/4\}$ and step size $\delta = o(n^{-3})$. Then Lemma 1 implies the following Corollary.

Corollary 1. *Let π^{opt} denote the optimal policy of our original problem. Assume $P_{\mathcal{I}^{out}}$ is a solvable polytope, if we apply the stochastic continuous greedy algorithm with stopping time $l = \min\{\beta, 1/4\}$ and step size $\delta = o(n^{-3})$ to solve **P1**, then the algorithm outputs a solution y such that $\overline{y} \in \min\{\beta, 1/4\} \cdot P_{\mathcal{I}^{out}}$, $\forall t \in [C] : \sum_{i\in I} \mathbb{E}[\min\{c_i(\phi(i)), t\}] \sum_{t'\in[t]} y(i, t') \leq \min\{\beta, 1/4\} \cdot 2t$ and $F(\overline{y}) \geq (1 - e^{-\min\{\beta, 1/4\}} - o(1)) f_{avg}(\pi^{opt})$.*

We next focus on proving that if there exists a monotone (β, γ)-balanced CRS for $\mathcal{I}^{out}$, then $f_{avg}(\pi^{io}) \geq (1 - \min\{2\beta, 1/2\})\gamma F(\overline{y})$. This together with Corollary 1 implies that $f_{avg}(\pi^{io}) \geq (1 - \min\{2\beta, 1/2\})\gamma(1 - e^{-\min\{\beta, 1/4\}} - o(1)) f_{avg}(\pi^{opt})$.

Consider a random vector $v \in [0; B]^I$ such that, for each $i \in I$, $v(i)$ is determined independently as $j \in [0; B]$ with probability

$$h(i, j) = p_i(j) \cdot \overline{y}(i) \tag{16}$$

where we define $p_i(0) = \frac{1-\overline{y}(i)}{\overline{y}(i)}$ for each $i \in I$. Let $R(v) = \{i | i \in I \text{ and } v(i) \neq 0\}$. For the purpose of analyzing the performance of Algorithm 1, we introduce three mapping functions: ψ^a, ψ^b, and ψ^c.

- **Design of ψ^a.** We apply a monotone (β, γ)-balanced CRS χ^{io} used in Algorithm 1 to $R(v)$ and obtain a set $\chi^{io}(R(v)) \subseteq R(v)$. We set $\psi^a(v)(i) = 0$ for all $i \in I \setminus \chi^{io}(R(v))$, and set $\psi^a(v)(i) = v(i)$ for all $i \in \chi^{io}(R(v))$.
- **Design of ψ^b.** Sample a *starting time* $t(i)$ from $[C - c_i(B)]$ with probability $y(i, t)/\overline{y}(i)$ for each $i \in R(v)$. Let $\sigma(v)$ denote the sequence of items in $R(v)$ sorted in a nondecreasing order of $t(i)$, breaking ties with the least index tie breaking rule. Let $\sigma(v)_{\leq t(i)}$ denote the sequence of items whose starting time is no later than $t(i)$. For each $i \in I \setminus R(v)$, we set $\psi^b(v)(i) = 0$. For each $i \in R(v)$, we set $\psi^b(v)(i) = v(i)$, if

$$\sum_{\substack{i'\in\sigma(v)_{\leq t(i)}\setminus\{i\}}}^{i-1} c_{i'}(v(i')) \leq t(i) \tag{17}$$

and set $\psi^b(v)(i) = 0$ otherwise.

- **Design of ψ^c.** The third mapping function ψ^c takes the intersection of ψ^a and ψ^b: First apply ψ^a and ψ^b to v separately to obtain $\psi^a(v)$ and $\psi^b(v)$, then generate $\psi^c(v)$ as follows.

$$\text{For all } i \in I : \psi^c(v)(i) = \begin{cases} v(i) & \text{if } \psi^a(v)(i) = v(i) \text{ and } \psi^b(v)(i) = v(i) \\ 0 & \text{otherwise} \end{cases}$$

Before presenting the main theorem of this paper, we first provide several technical lemmas. The first four lemmas are used to lower bound the expected utility of $\psi^c(v)$ with regards to h. The fifth lemma shows that the expected utility of our policy is lower bounded by the expected utility of $\psi^c(v)$ with regards to h. Combing these two results, we are able to derive a lower bound on the expected utility of our policy.

Lemma 3. ψ^a *is a monotone γ-CRS with regards to h.*

Proof: We first prove that ψ^a is a γ-CRS with regards to h. Recall that we set $\psi^a(v)(i) = 0$ for all $i \in I \setminus S$, and set $\psi^a(v)(i) = v(i)$ for all $i \in \chi^{io}(R(v))$. It follows that for all $i \in I$ and all $j \in [B]$, we have

$$\Pr[\psi^a(v)(i) = j | v(i) = j] = \Pr[i \in \chi^{io}(R(v)) | i \in R(v)] \tag{18}$$

By the definition of monotone (β, γ)-balanced CRS, we have $\Pr[i \in \chi^{io}(R(v)) | i \in R(v)] \geq \gamma$. Hence, $\Pr[\psi^a(v)(i) = j | v(i) = j] \geq \gamma$. Next we prove that ψ^a is monotone. Consider any two vectors $u, v \in [0; B]^I$ such that $u(i) = v(i)$ and $u \leq v$, we have $R(u) \subseteq R(v)$. Based on the definition of monotone (β, γ)-balanced CRS, we have $\Pr[i \in \chi^{io}(R(v)) | i \in R(u)] \geq \Pr[i \in \chi^{io}(R(v)) | i \in R(v)]$. Together with (18), we have $\Pr[\psi^a(u)(i) = j | v(i) = j] \geq \Pr[\psi^a(v)(i) = j | v(i) = j]$ for all $i \in I$ and $j \in [B]$. $\qquad\square$

Lemma 4. ψ^b *is a monotone $(1 - \min\{2\beta, 1/2\})$-CRS with regards to h.*

Proof: The monotonicity of ψ^b follows from Lemma 3 in [4]. We next focus on proving that ψ^b is a $(1 - \min\{2\beta, 1/2\})$-CRS. Consider any $i \in I$ and $k \in [C]$, we have

$$\Pr[\psi^b(v)(i) = v(i) \mid i \in R(v), t(i) = k] \tag{19}$$

$$= \Pr[\sum_{i' \in \sigma(v)_{\leq k} \setminus \{i\}} c_{i'}(v(i')) \leq k \mid i \in R(v), t(i) = k] \tag{20}$$

$$= \Pr[\sum_{i' \in \sigma(v)_{\leq k} \setminus \{i\}} \min\{c_{i'}(v(i')), k\} \leq k \mid i \in R(v), t(i) = k] \tag{21}$$

where the probability considers two sources of randomness: one is the randomness in choosing v, and the other source is the randomness in the generation of t. Because the event that $\sum_{i' \in \sigma(v)_{\leq k} \setminus \{i\}} \min\{c_{i'}(v(i')), k\} \leq k$ is independent of the event that $i \in R(v)$ and $t(i) = k$, we have

$$\Pr[\sum_{i' \in \sigma(v)_{\leq k} \setminus \{i\}} \min\{c_{i'}(v(i')), k\} \leq k \mid i \in R(v), t(i) = k] \qquad (22)$$

$$= \Pr[\sum_{i' \in \sigma(v)_{\leq k} \setminus \{i\}} \min\{c_{i'}(v(i')), k\} \leq k] \qquad (23)$$

We next provide a lower bound of $\Pr[\sum_{i' \in \sigma(v)_{\leq k} \setminus \{i\}} \min\{c_{i'}(v(i')), k\} \leq k]$. Observe that

$$\mathbb{E}[\sum_{i' \in \sigma(v)_{\leq k} \setminus \{i\}} \min\{c_{i'}(v(i')), k\}] \qquad (24)$$

$$= \sum_{i' \in I \setminus \{i\}} \mathbb{E}[\min\{c_{i'}(v(i')), k\}] \Pr[t(i') \leq k] \qquad (25)$$

$$= \sum_{i' \in I \setminus \{i\}} \mathbb{E}[\min\{c_{i'}(v(i')), k\}] \sum_{t \in [k]} y(i', t) \qquad (26)$$

$$\leq \sum_{i' \in I} \mathbb{E}[\min\{c_{i'}(v(i')), k\}] \sum_{t \in [k]} y(i', t) \qquad (27)$$

$$\leq \min\{\beta, 1/4\} 2k = \min\{2\beta, 1/2\} k \qquad (28)$$

The second inequality is due to $t(i)$ from $[C - c_i(B)]$ with probability $y(i, t)/\overline{y}(i)$ for each $i \in R(v)$. The second inequality is due to Corollary 1 and the fact that y is obtained from solving problem **P1** using the stochastic continuous greedy algorithm with stopping time $l = \min\{\beta, 1/4\}$ and step size $\delta = o(n^{-3})$. Hence, $\Pr[\sum_{i' \in \sigma(v)_{\leq k} \setminus \{i\}} \min\{c_{i'}(v(i')), k\} > k] < \min\{2\beta, 1/2\}$ due to Markov inequality. It follows that

$$\Pr[\sum_{i' \in \sigma(v)_{\leq k} \setminus \{i\}} \min\{c_{i'}(v(i')), k\} \leq k] \qquad (29)$$

$$= 1 - \Pr[\sum_{i' \in \sigma(v)_{\leq k} \setminus \{i\}} \min\{c_{i'}(v(i')), k\} > k] \qquad (30)$$

$$> 1 - \min\{2\beta, 1/2\} \qquad (31)$$

$\square$

The following lemma follows from Lemma 3, Lemma 4 and Lemma 2.

Lemma 5. ψ^c is a monotone $(1 - \min\{2\beta, 1/2\})\gamma$-CRS with regards to h.

Theorem 4 in [4] states that if ψ^c is a monotone α-CRS with respect to h, then $\mathbb{E}_{v \sim h}[f(\psi^c(v))] \geq \alpha F(\overline{y})$. This together with Lemma 5 implies the following lemma.

Lemma 6. $\mathbb{E}_{v \sim h}[f(\psi^c(v))] \geq (1 - \min\{2\beta, 1/2\})\gamma F(\overline{y})$.

We next show that the expected utility of π^{io} is bounded by $\mathbb{E}_{v \sim h}[f(\psi^c(v))]$ from below.

Lemma 7. $f_{avg}(\pi^{io}) \geq \mathbb{E}_{v \sim h}[f(\psi^c(v))]$.

Proof: Recall that for any $S \subseteq I$ and any realization ϕ, we use ϕ_S to denote a vector in $[0; B]^I$ such that $\phi_S(i) = \phi(i)$ if $i \in S$, and $\phi_S(i) = 0$ otherwise. Let Φ_S denote a random realization of S. As specified in Algorithm 1, R^{io} is a random set that is obtained by including each item $i \in I$ independently with probability $\overline{y}(i)$. Thus, $\Phi_{R^{io}} \in [0; B]^I$ can be considered as a random vector such that, for each $i \in I$, $\Phi_{R^{io}}(i)$ is determined independently as $j \in [0; B]$ with probability

$$h(i, j) = p_i(j) \cdot \overline{y}(i) \tag{32}$$

where we define $p_i(0) = \frac{1 - \overline{y}(i)}{\overline{y}(i)}$ for each $i \in I$. Note that the probability considers two sources of randomness: one is the randomness in choosing R^{io}, and the other source is the randomness of realization Φ.

Now consider a fixed realization $v \in [0; B]^I$ of $\Phi_{R^{io}}$. Recall that in the design of ψ^a, we define $R(v) = \{i | i \in I \text{ and } v(i) \neq 0\}$. It is easy to verify that R^{io} coincides with $R(v)$. For purpose of analysis, we further assume that $\chi^{io}(R^{io})$ coincides with $\chi^{io}(R(v))$. Moreover, for each $i \in \chi^{io}(R^{io})$, we assume that $t(i)$ (the sampled starting time of i in the implementation of ψ^b) coincides with $t^{io}(i)$ (the sampled starting time of i in Algorithm 1). This assumption indicates that σ^{io} (a sorted sequence of items as specified in Algorithm 1) is a subsequence of $\sigma(v)$ (a sorted sequence of items as specified in the design of ψ^b) due to $\chi^{io}(R^{io}) \subseteq R^{io}$.

We next define a new mapping function ψ^{io} as follows:

$$\text{For all } i \in I : \psi^{io}(v)(i) = \begin{cases} v(i) & \text{if } i \text{ is selected by } \pi^{io} \text{ conditional on } v \\ 0 & \text{otherwise} \end{cases}$$

Note that for $i \in I$ to be selected by π^{io} conditional on v, it must satisfy $i \in \chi^{io}(R^{io})$ as well as the condition defined in Line 11 of Algorithm 1 which can be written as

$$\sum_{i' \in \sigma^{io}_{\leq t(i)} \setminus \{i\} : i' \text{ is selected by } \pi^{io}} c_{i'}(v(i')) \leq t(i) \tag{33}$$

where $\sigma^{io}_{\leq t(i)}$ denotes the subsequence of σ^{io} by including all items whose starting time is no later than $t(i)$. It is easy to verify that $f_{avg}(\pi^{io}) = \mathbb{E}_{v \sim h}[f(\psi^{io}(v))]$. We next focus on proving that $\mathbb{E}_{v \sim h}[f(\psi^{io}(v))] \geq \mathbb{E}_{v \sim h}[f(\psi^c(v))]$.

According to the design of ψ^c, for each $i \in I$, $\psi^c(v)(i) = v(i)$ if and only if $i \in \chi^{io}(R(v))$ and condition (17) is satisfied. Given that σ^{io} is a subsequence of $\sigma(v)$, condition (17) is stronger than the condition (33). This together with the fact that $\chi^{io}(R^{io}) = \chi^{io}(R(v))$ implies that for each $i \in I$, $\psi^c(v)(i) = v(i)$ implies that $\psi^{io}(v)(i) = v(i)$. Hence, we have $\psi^{io}(v) \geq \psi^c(v)$, which implies that $f(\psi^{io}(v)) \geq f(\psi^c(v))$. It follows that $\mathbb{E}_{v \sim h}[f(\psi^{io}(v))] \geq \mathbb{E}_{v \sim h}[f(\psi^c(v))]$. This finishes the proof of this lemma due to $f_{avg}(\pi^{io}) = \mathbb{E}_{v \sim h}[f(\psi^{io}(v))]$. $\qquad \square$

Corollary 1, Lemma 6, and Lemma 7 imply the following main theorem.

Theorem 1. *Assume* $P_{\mathcal{I}^{out}}$ *is a solvable polytope and there exists a monotone* (β, γ)-*balanced CRS for* $\mathcal{I}^{out}$, $f_{avg}(\pi^{io}) \geq (1 - \min\{2\beta, 1/2\})\gamma(1 - e^{-\min\{\beta, 1/4\}} - o(1))f_{avg}(\pi^{opt})$.

4.4   Completing the Last Piece of the Puzzle: Discussion on β and γ

As the approximation ratio of π^{io} is depending on the values of β and γ, we next discuss some practical outer constraints under which β and γ are well defined. In [3], they present monotone (β, γ)-balanced CRSs for a wide range of practical constraints including (multiple) matroid constraints, knapsack constraints, and their intersections. We can use their results as subroutines in Algorithm 1 to handle a variety of outer constraints. For example, if $\mathcal{I}^{out}$ is the intersection of a fixed number of knapsack constraints, there exists a $(1 - \epsilon, 1 - \epsilon)$-balanced CRS. If $\mathcal{I}^{out}$ is induced by a matroid constraint, there exists a $(b, \frac{1-e^{-b}}{b})$-balanced CRS for any $b \in (0, 1]$.

References

1. Adamczyk, M., Sviridenko, M., Ward, J.: Submodular stochastic probing on matroids. Math. Oper. Res. **41**(3), 1022–1038 (2016)
2. Asadpour, A., Nazerzadeh, H.: Maximizing stochastic monotone submodular functions. Manage. Sci. **62**(8), 2374–2391 (2016)
3. Chekuri, C., Vondrák, J., Zenklusen, R.: Submodular function maximization via the multilinear relaxation and contention resolution schemes. SIAM J. Comput. **43**(6), 1831–1879 (2014)
4. Fukunaga, T., Konishi, T., Fujita, S., Kawarabayashi, K.I.: Stochastic submodular maximization with performance-dependent item costs. In: Proceedings of the AAAI Conference on Artificial Intelligence, vol. 33, pp. 1485–1494 (2019)
5. Golovin, D., Krause, A.: Adaptive submodularity: theory and applications in active learning and stochastic optimization. J. Artif. Intell. Res. **42**, 427–486 (2011)
6. Gupta, A., Krishnaswamy, R., Molinaro, M., Ravi, R.: Approximation algorithms for correlated knapsacks and non-martingale bandits. In: 2011 IEEE 52nd Annual Symposium on Foundations of Computer Science, pp. 827–836. IEEE (2011)
7. Tang, S.: Stochastic coupon probing in social networks. In: Proceedings of the 27th ACM International Conference on Information and Knowledge Management, pp. 1023–1031 (2018)
8. Tang, S.: Beyond pointwise submodularity: non-monotone adaptive submodular maximization in linear time. Theoret. Comput. Sci. **850**, 249–261 (2021)
9. Tang, S.: Beyond pointwise submodularity: non-monotone adaptive submodular maximization subject to Knapsack and k-system constraints. In: Le Thi, H.A., Pham Dinh, T., Le, H.M. (eds.) MCO 2021. LNNS, vol. 363, pp. 16–27. Springer, Cham (2022). https://doi.org/10.1007/978-3-030-92666-3_2
10. Tang, S.: Stochastic submodular probing with state-dependent costs. In: Wu, W., Du, H. (eds.) AAIM 2021. LNCS, vol. 13153, pp. 170–178. Springer, Cham (2021). https://doi.org/10.1007/978-3-030-93176-6_15
11. Tang, S., Yuan, J.: Influence maximization with partial feedback. Oper. Res. Lett. **48**(1), 24–28 (2020)

Bicriteria Algorithms for Maximizing the Difference Between Submodular Function and Linear Function Under Noise

Mengxue Geng[1], Shufang Gong[1(✉)], Bin Liu[1], and Weili Wu[2]

[1] School of Mathematical Sciences, Ocean University of China,
Qingdao 266100, People's Republic of China
shufanggong@stu.ouc.edu.cn
[2] School of Computer Science, University of Texas at Dallas,
Richardson, TX 75080, USA

Abstract. Submodular optimization is an essential problem in many fields due to its diminishing marginal benefit. This property of submodular function plays an important role in many applications. In recent years, the problem of maximizing a non-negative monotone submodular function minus a linear function under various constraints has gradually emerged and is widely used in many practical scenarios such as team formation and recommendation. In this paper, We focus on maximizing a non-negative monotone normalized submodular function minus a linear function under ϵ–multiplicative noise and the result is similar in the case of ϵ-additive noise. Many previous studies were conducted in a noiseless environment, here we consider optimization of this problem in a noisy environment for the first time. In addition, our study will be conducted under two situations, that is, the cardinality constraint and the matroid constraint. Based on these two situations, we propose two bicriteria approximation algorithms respectively and all these algorithms can obtain good results.

Keywords: Submodular function · Linear function · Noise · Bicriteria algorithm

1 Introduction

Due to the good property of the submodular function, which is the property of diminishing marginal benefit, the optimization problem about it has also become a hot topic in recent years. Many researches and literatures show that this property plays an important role in many applications, especially in artificial intelligence [20], social welfare [21], machine learning [14] fields showing extremely

This work was supported in part by the National Natural Science Foundation of China (11971447, 11871442), and the Fundamental Research Funds for the Central Universities.

strong applicability, specifically in recommendation systems [11], influence max-imization in social networks [12], sensor settings [6] and so on. The submodular maximization problem subject to various constraints has been widely studied [5,18]. With the emergence of practical application scenarios such as team for-mation [15], recommender systems [11] and social work [13], we began to investi-gate the maximization of the non-negative monotone submodular function $f(S)$ minus the cost function $c(S)$, i.e., c is a linear function.

In this model, we observe that the objective function $h(S) = f(S) - c(S)$ is submodular but not necessarily non-negative and monotone. However, exist-ing studies suggest that maximizing a potentially negative submodular func-tion is possible without a multiplicative approximation factor [15]. Therefore, the approximation guarantee cannot be expressed in the form of a traditional approximation ratio. In this paper, we use the bicriteria approximation algorithm proposed by [2], that is the output solution S satisfies

$$f(S) - c(S) \geq \mu f(OPT) - \nu c(OPT),$$

in which $0 \leq \mu \leq 1$. In recent years, the relevant work under this model has been performed in a noise-free environment. However, in many practical scenarios, it is difficult or costly to get a specific value of function f under the noise model, but it is easy to get a value of F such as using Neural Net(NN) training [7]. Thus, we can only obtain a noisy evaluation function F of f. In the discussion as follows, we consider this problem under the ϵ-multiplicative noise and the proof is similar in the case of ϵ-additive noise. Generally the type of noise can be regarded as multiplicative noise and additive noise [19], i.e.,

$$(1 - \epsilon)f(X) \leq F(X) \leq (1 + \epsilon)f(X).$$

and

$$f(X) - \epsilon \leq F(X) \leq f(X) + \epsilon,$$

Thus, the problem becomes how to find the approximate solution with the help of noisy evaluation function.

For the set S obtained after each iteration, we assume that there are two oracles that can be used to calculate the value of $F(S)$ and $c(S)$.

2 Related Work

The study of maximizing $f - c$ under noise-free model has been studied as follows. For the cardinality constraint, Harshaw et al. [10] combined standard greedy algorithms with distortion techniques to solve the problem, where f is a γ-weakly submodular. They obtained a solution set satisfying $(1 - \frac{1}{e} - \varepsilon, 1)$-bicriteria approximation ratio when f is submodular. Then, Nikolakaki et al. [15] provided a simple greedy algorithm with $(\frac{1}{2}, 1)$-bicriteria approximation ratio. Meanwhile, the same approximation ratio is obtained under the online unconstrained in [15]. In addition, [15] also showed that a solution set S satisfying $(\frac{3-\sqrt{5}}{2}, 1)$-bicriteria approximation ratio under streaming model with the cardinality constraint. For the matroid constraint, Sviridenko et al. [17] provided a stochastic algorithm to solve this problem, and then got a solution S satisfying $(1 - \frac{1}{e}, 1)$-bicriteria approximation ratio. However, the time complexity of this algorithm is very bad. Then, Feldman [4] improved it by using a continuous distortion greedy algorithm to obtain the same approximation ratio. Recently, Nikolakaki et al. [15] proposed a algorithm based on the standard greedy algorithm yields a solution set Q satisfying $(\frac{1}{2}, 1)$-bicriteria approximation ratio.

For submodular maximization problems in noisy environments, Horel et al. [9] first introduced the concept of approximate submodular, they proved that the greedy algorithm can achieve $(1 - e^{-1} - O(\delta))$-approximation ratio for maximizing non-negative monotone submodular function with the cardinality constraint, when $\varepsilon \leq \frac{1}{k}$ and $\delta = \varepsilon k$. Gölz [8] studied the submodular maximization with the P-matroids constraint under the ϵ-multiplicative noise model, and obtained $(P + 1 + \frac{4\epsilon k}{1-\epsilon})^{-1}$-approximation ratio, in which k is the size of the maximum feasible set of the P-matroids. In addition, there have been many results for the study of maximization problem under the streaming model. Yang et al. [22] developed two streaming thresholding algorithms for maximizing streaming submodular with the cardinality constraint under the ϵ-multiplicative noise and ϵ-additive noise. If the parameter $\epsilon \to 0$, their algorithms all had $\frac{2}{k}$-approximation to the optimal solution. For k-submodular maximization with streaming model under the cardinality constraint, Nguyen and Thai [16] proposed two streaming algorithm, one is deterministic and the other is random. Both algorithms provided $O((1 - \epsilon)^{-2}\epsilon B)$-approximation ratio when f has monotonicity as well as $O((1 - \epsilon)^{-3}\epsilon B)$-approximation ratio when f is non-monotone.

To the best of our knowledge, although there have been many studies on maximization $f - c$ where f is an non-negative monotone normalized submodular function and c is the sum of the costs in noise-free environments, there is a lack of research in noisy environments. Thus, this is the first paper to study the $f - c$ maximization problem in noisy environment. We focus on this problem under the ϵ-multiplicative noise in this paper because the solution of the problem under the ϵ-additive noise model is similar to the ϵ-multiplicative noise model. Under the noise model, we cannot get the exact value of the function f, but use certain method such as Neural Net (NN) training can easily get the noise version F [7]. Then we can use the approximate version to obtain an approximate solution.

The rest of this paper is constructed as follows. In Sect. 3, we introduce some basic definitions. In Sect. 4, we develop the bicriteria algorithm under noise with cardinality constraint. In Sect. 5, we consider the problem with matroid constraint. The last section summarizes the full paper.

3 Preliminaries

In this paper, we study the problem of maximizing the difference between a non-negative monotone submodular function and a linear function in a noisy environment. Under this premise, we cannot know the oracle in which calculates the value of function f. Therefore among all the algorithms, we use the noise version function F to obtain an approximation ratio. Recall that a set function $f: 2^N \to \mathbf{R}^+$ is *non-negative* if $f(A) \geq 0$ for any subset $A \subseteq N$. A function f is called *monotone* if $f(A) \geq f(B)$ whenever $A \subseteq B$. In addition, f is *normalized* if $f(\emptyset) = 0$. For the arbitrary elements $e \in N$ and subsets $A \subseteq N$, the marginal benefit of j in A can be expressed as

$$f_A(j) := f(A \cup \{j\}) - f(A).$$

It is clear that when j is in A there is $f_A(j) = 0$. Similarly, represented by $f_A(B) := f(A \cup B) - f(A)$ the marginal benefit of B in A. In addition, we claim that a function f is *submodular* function if it satisfies

$$f(A) + f(B) \geq f(A \cup B) + f(A \cap B),$$

for any subsets $A \subseteq N$, $B \subseteq N$. It also has an equivalent definition, as shown as follows

$$f(A \cup \{j\}) - f(A) \geq f(B \cup \{j\}) - f(B),$$

for any $A \subseteq B$ and $j \in N \setminus B$.

In addition, the noisy function F satisfies normalization rule and is not necessarily submodular.

Definition 1 [19]. *For some $\epsilon > 0$, we say set function $F: 2^N \to \mathbf{R}$ is an ϵ-multiplicative noise oracle of submodular function f if*

$$(1 - \epsilon)f(X) \leq F(X) \leq (1 + \epsilon)f(X), \quad \forall X \subseteq N.$$

In reality, the maximization problem of a potential negative submodular function is not approximable and may not have a constant multiplicative approximation factor. Thus, unlike the traditional approximation ratio representation, we use a weaker approximation concept to evaluate the degree of approximation of the solution. We therefore give the definition of the bicriteria approximation algorithm.

Definition 2 [2]. *An algorithm is called a (μ, ν)-bicriteria approximation algorithm if the output solution of it satisfying $f(S) - c(S) \geq \mu f(O) - \nu c(O)$, where O is the optimal solution to $\max_{S \subseteq N}\{f(S) - c(S)\}$ and $0 \leq \mu \leq 1$.*

4   The Bicriteria Algorithm Under Noise with the Cardinality Constraint

In this section, we study the problem of maximizing $f(S) - c(S)$ under noise with the cardinality constraint in which f is a non-negative normalized monotone submodular function and c is the sum of the costs of all elements in the solution, i.e., a non-negative linear function. Then, we present a bicriteria algorithm with a surrogate objective function of $F(S) - xc(S)$, where the value of x can be

---

Algorithm 1. The bicriteria algorithm for $f - c$ under noise with the cardinality constraint

---

Input: Given a ground set G, noisy function F, $x = 2$, a non-negative linear function c and cardinality constraint $k \in \mathbf{N}_+$.
Output: A solution subset Q
 1: Initially set $Q := \emptyset$
 2: **for** $i = 1, 2, \cdots, k$ **do**
 3:     select $e_i = \arg\max_{e \in G \backslash Q}\{F_Q(e) - xc(e)\}$
 4:     **if** $F_Q(e_i) - xc(e_i) > 0$ **then**
 5:         update $Q := Q \cup \{e_i\}$
 6:     **else**
 7:         break
 8:     **end if**
 9: **end for**
10: **return** Q

---

determined later. Suppose that in each iteration, when element e_i is added, the current set becomes $Q^{(i)} = \{e_1, e_2, \cdots, e_i\}$, and the final output solution is Q. Let $O = \arg\max_{|S| \leq k}\{f(S) - c(S)\}$, and the surrogate objective function is $\hat{h}(S) = F(S) - xc(S)$.

Before the concrete proof, we sort the elements in $Q \cup O$: $e_1, e_2, \cdots, e_{|Q \cup O|}$, such that $e_i = \arg\max_{e \in (Q \cup O) \backslash \{e_1, \cdots, e_{i-1}\}} \hat{h}_{\{e_1, \cdots, e_{i-1}\}}(e_i)$. According to the execution rules of the Algorithm 1, in this order, the first $|Q|$ elements are the output solution of this algorithm and their order is the same as the order in which this algorithm adds them to the set Q i.e., $Q = S^{(|Q|)}$. In addition, the first i elements in this order are denoted as $S^{(i)} = \{e_1, \cdots, e_i\}$, where $i = 1, 2, \cdots, |Q \cup O|$.

Lemma 1. *Let $|O| = l \leqslant k$, then we have*

$$
\begin{aligned}
\hat{h}(S^{(l)}) \geqslant &\ \frac{1 - \epsilon}{2}\hat{h}(S^{(l)} \cup O) - \frac{2\epsilon k(1 + \epsilon)}{1 - \epsilon}F(S^{(l)} \cup O) - \frac{1 + \epsilon}{2}xc(O) \\
&+ \frac{1 - \epsilon}{2}x\big(c(S^{(l)} \cup O) - c(S^{(l)})\big).
\end{aligned}
\tag{1}
$$

Proof. Consider the elements in $Q \cup O$ after sorting. By definition, $S^{(l)}$ denotes the first $l = |O|$ elements in $Q \cup O$, it is clearly that $l = |S^{(l)}| = |O|$. Then we can construct a bijection $\sigma : O \to S^{(l)}$, which satisfies condition: for each $i \leq l$, $\sigma^{-1}(e_i) = e_j$ for some index $i \leq j$.

From the construction of the bijection, we know, for $i = 1, 2, \cdots, l$

$$
\hat{h}_{S^{(i-1)}}(e_i) \geq \hat{h}_{S^{(i-1)}}(\sigma^{-1}(e_i)).
$$

For each $i = 1, \cdots, l$, let $O^{(i)} = \sigma^{-1}(S^{(i)})$, $O^{(i)} = O^{(i-1)} \cup \sigma^{-1}(e_i)$, then we have

$$
\begin{aligned}
\hat{h}_{S^{(i-1)}}(\sigma^{-1}(e_i)) &= \hat{h}(\sigma^{-1}(e_i) \cup S^{(i-1)}) - \hat{h}(S^{(i-1)}) \\
&= F(\sigma^{-1}(e_i) \cup S^{(i-1)}) - xc(\sigma^{-1}(e_i) \cup S^{(i-1)}) \\
&\quad - F(S^{(i-1)}) + xc(S^{(i-1)}) \\
&\geq (1 - \epsilon)f(\sigma^{-1}(e_i) \cup S^{(i-1)}) - xc(\sigma^{-1}(e_i)) \\
&\quad - (1 + \epsilon)f(S^{(i-1)}) \\
&\geq (1 - \epsilon)\Big(f(\sigma^{-1}(e_i) \cup S^{(i-1)}) - f(S^{(i-1)})\Big) \\
&\quad - 2\epsilon f(S^{(i-1)}) - xc(\sigma^{-1}(e_i)),
\end{aligned}
$$

where the first inequality used the definition of function $F(S)$ and the inequality:

$$
c(\sigma^{-1}(e_i) \cup S^{(i-1)}) - c(S^{(i-1)}) \leq c(\sigma^{-1}(e_i)).
$$

Based on the submodularity of function f, we can get

$$
\hat{h}_{S^{(i-1)}}(\sigma^{-1}(e_i)) \geq (1 - \epsilon)f_{S^{(l)} \cup O^{(i-1)}}(\sigma^{-1}(e_i)) - 2\epsilon f(S^{(i-1)}) - xc(\sigma^{-1}(e_i)).
$$

By the definition of F, we obtain

$$
\begin{aligned}
\hat{h}_{S^{(i-1)}}(\sigma^{-1}(e_i)) &\geq (1 - \epsilon)\Big(\frac{F(S^{(l)} \cup O^{(i)})}{1 + \epsilon} - \frac{F(S^{(l)} \cup O^{(i-1)})}{1 - \epsilon}\Big) \\
&\quad - 2\epsilon f(S^{(i-1)}) - xc(\sigma^{-1}(e_i)) \\
&\geq \frac{1 - \epsilon}{1 + \epsilon}\Big(F(S^{(l)} \cup O^{(i)}) - F(S^{(l)} \cup O^{(i-1)})\Big) \\
&\quad - \frac{2\epsilon}{1 + \epsilon}F(S^{(l)} \cup O^{(i-1)}) - 2\epsilon f(S^{(l)} \cup O^{(i-1)}) - xc(\sigma^{-1}(e_i)) \\
&\geq \frac{1 - \epsilon}{1 + \epsilon}\Big(F(S^{(l)} \cup O^{(i)}) - xc(S^{(l)} \cup O^{(i)}) - F(S^{(l)} \cup O^{(i-1)}) \\
&\quad + xc(S^{(l)} \cup O^{(i-1)})\Big) - 4\epsilon f(S^{(l)} \cup O^{(i-1)}) - xc(\sigma^{-1}(e_i)) \\
&\quad + \frac{1 - \epsilon}{1 + \epsilon}x\Big(c(S^{(l)} \cup O^{(i)}) - c(S^{(l)} \cup O^{(i-1)})\Big).
\end{aligned}
$$

Therefore, we have

$$
\begin{aligned}
\hat{h}_{S^{(i-1)}}(e_i) &\geq \frac{1 - \epsilon}{1 + \epsilon}\Big(\hat{h}(S^{(l)} \cup O^{(i)}) - \hat{h}(S^{(l)} \cup O^{(i-1)})\Big) - 4\epsilon f(S^{(l)} \cup O^{(i-1)}) \\
&\quad - xc(\sigma^{-1}(e_i)) + \frac{1 - \epsilon}{1 + \epsilon}x\Big(c(S^{(l)} \cup O^{(i)}) - c(S^{(l)} \cup O^{(i-1)})\Big).
\end{aligned}
$$

Summing up all $i = 1, \cdots, l$, since the monotonicity of function f and the definition of function $F(S)$, we can get

$$
\begin{aligned}
\hat{h}(S^{(l)}) &\geq \hat{h}(S^{(l)}) - \hat{h}(\emptyset) \geq \frac{1 - \epsilon}{1 + \epsilon}\Big(\hat{h}(S^{(l)} \cup O) - \hat{h}(S^{(l)})\Big) - \frac{4\epsilon k}{1 - \epsilon}F(S^{(l)} \cup O) \\
&\quad - xc(O) + \frac{1 - \epsilon}{1 + \epsilon}x\Big(c(S^{(l)} \cup O) - c(S^{(l)})\Big),
\end{aligned}
$$

where $O^{(l)} = \sigma^{-1}(S^{(l)}) = O$ and $\hat{h}(\emptyset) = F(\emptyset) \geq 0$. Rewriting the inequality, we obtain

$$\hat{h}(S^{(l)}) \geq \frac{1-\epsilon}{2}\hat{h}(S^{(l)} \cup O) - \frac{2\epsilon k(1+\epsilon)}{1-\epsilon}F(S^{(l)} \cup O)$$
$$- \frac{1+\epsilon}{2}xc(O) + \frac{1-\epsilon}{2}x\Big(c(S^{(l)} \cup O) - c(S^{(l)})\Big).$$

Thus we complete this Lemma.

From Algorithm 1, we can divide the output solution into two cases: *Case 1*, $|Q| \geq l$; and *Case 2*, $|Q| < l$. Considering these two cases respectively, we can get the following Lemmas.

Lemma 2. *When $|Q| \geq l$, Algorithm 1 outputs a feasible solution Q such that*

$$f(Q) - c(Q) \geq \frac{\alpha}{2(1+\epsilon)}f(O) - c(O), \tag{2}$$

where $\alpha = (1 - 4k)\epsilon^2 - (2 + 4k)\epsilon + 1$ and $x = 2$.

Proof. In this case, it is clearly $S^{(l)} \subseteq Q$.

According to the execution rules of Algorithm 1, for each elements $e_i \in Q$, we know $\hat{h}_{Q^{(i-1)}}(e_i) \geq 0$, thus

$$\hat{h}(Q) - \hat{h}(S^{(l)}) = \sum_{i=l+1}^{|Q|} \hat{h}_{S^{(i-1)}}(e_i) \geq 0.$$

Combining the inequality (1), let $\alpha = (1 - 4k)\epsilon^2 - (2 + 4k)\epsilon + 1$, we have

$$\hat{h}(Q) \geq \hat{h}(S^{(l)}) \geq \frac{\alpha}{2(1-\epsilon)}F(S^{(l)} \cup O) - \frac{1-\epsilon}{2}xc(S^{(l)}) - \frac{1+\epsilon}{2}xc(O).$$

As function c is non-negative linear function, it holds that $c(S^{(l)}) \leq c(Q)$. Then

$$\hat{h}(Q) \geq \frac{\alpha}{2}f(S^{(l)} \cup O) - \frac{1-\epsilon}{2}xc(Q) - \frac{1+\epsilon}{2}xc(O)$$
$$\geq \frac{\alpha}{2}f(O) - \frac{1-\epsilon}{2}xc(Q) - \frac{1+\epsilon}{2}xc(O),$$

where the first inequality holds by the property of F and the second inequality follows by the monotonicity of f.

By the definition of $\hat{h}$, we obtain

$$F(Q) - xc(Q) \geq \frac{\alpha}{2}f(O) - \frac{1-\epsilon}{2}xc(Q) - \frac{1+\epsilon}{2}xc(O).$$

Since $(1 - \epsilon)f(X) \leq F(X) \leq (1 + \epsilon)f(X)$, rearranging this inequality, we have

$$f(Q) - \frac{x}{2}c(Q) \geq \frac{\alpha}{2(1+\epsilon)}f(O) - \frac{x}{2}c(O).$$

In order to obtain the approximate ratio that meets the conditions, we only need to make following inequality established

$$-\frac{x}{2}c(O) + \frac{x}{2}c(Q) - c(Q) \geq -c(O),$$

Obviously, the above formula holds when $x = 2$. Thus Lemma 2 holds, where $\alpha = (1 - 4k)\epsilon^2 - (2 + 4k)\epsilon + 1$.

Lemma 3. *When $|Q| < l$, Algorithm 1 outputs a feasible solution Q such that*

$$f(Q) - c(Q) \geq \frac{\beta}{2(1+\epsilon)}f(O) - \frac{2}{1-\epsilon}c(O), \tag{3}$$

where $\beta = (1 - 4k)\epsilon^2 - (2 + 12k)\epsilon + 1$ and $x = 2$.

Based on Lemma 2 and Lemma 3, there is the final bicriteria ratio in Theorem 1.

Theorem 1. *The bicriteria ratio for Algorithm 1 is $(\frac{\beta}{2(1+\epsilon)}, \frac{2}{1-\epsilon})$, where $\beta = (1 - 4k)\epsilon^2 - (2 + 12k)\epsilon + 1$. If the parameter $\epsilon \to 0$, we then have $(\frac{1}{2}, 2)$-approximation.*

5   The Bicriteria Algorithm Under Noise with the Matroid Constraint

In this section, we describe the problem of maximizing a non-negative monotone normalized submodular function f minus a non-negative linear function c under ϵ-multiplicative noise model with the matroid constraint. Before analyzing the specific approximation guarantee of Algorithm 2, we firstly introduce the definition of matroid $\mathcal{M} = (E, \mathcal{F})$ by [3].

Definition 3 [3]. *Assume a finite set E is the ground set and $\mathcal{F}$ is a family of subsets of E, then a pair $(E, \mathcal{F})$ is called a matroid if both E and $\mathcal{F}$ satisfy the following conditions:*

(1) $\emptyset \in \mathcal{F}$;
(2) if $A \subseteq B$ and $B \in \mathcal{F}$, then $A \in \mathcal{F}$;
(3) if $A, B \in \mathcal{F}$ and $|B| > |A|$, then there exists $e \in B \backslash A$ such that $A \cup \{e\} \in \mathcal{F}$.

Further, we also need to use the following Lemma about the relationship between the two independent sets in the matroid $\mathcal{M} = (E, \mathcal{F})$, which is a natural result provided by [1].

Lemma 4 [1]. *Let X and Y be two independent sets in the matroid $\mathcal{M} = (E, \mathcal{F})$ such that $|X| = |Y|$. There is a bijection $\psi : X \backslash Y \to Y \backslash X$ such that $(Y \backslash \psi(e)) \cup \{e\}$ is also an independent set for each element $e \in X \backslash Y$ and $\psi(e)) \in Y \backslash X$.*

Algorithm 2. The bicriteria algorithm for $f - c$ under noise with the matroid constraint

Input: Given a ground set E, noisy function F, $x = 2 + 2\epsilon + 4\epsilon p$, $0 \leq \epsilon < 1$, a non-negative linear function c and matroid $\mathcal{M} = (E, \mathcal{F})$.
Output: A solution subset S
1: Initially set $S := \emptyset$, $H := E$
2: **for** $i = 1, 2, \cdots$ **do**
3:    **if** $H = \emptyset$ **then**
4:        break
5:    **else**
6:        select $e_i = \arg\max_{e \in H}\{F(e|Q) - xc(e)\}$
7:        **if** $F(e_i|Q) - xc(e_i) > 0$ **then**
8:            update $S := S \cup \{e_i\}$
9:        **else**
10:            break
11:        **end if**
12:    **end if**
13:    delete all the elements in H such that $S \cup \{e\} \notin \mathcal{F}$
14: **end for**
15: **return** S

Suppose that the output solution of Algorithm 2 is S and the element added to S to in the i-th iteration is e_i. Then the current set becomes $S_i = \{e_1, \cdots, e_i\}$. Let $O = \arg\max_{S \subseteq E}\{f(S) - c(S), S \subseteq \mathcal{F}\}$ and p is the rank of the matroid $\mathcal{M}$(the size of the largest independent set). The surrogate objective function is $\hat{h}(S) = F(S) - xc(S)$, where the value of x can be determined later.

Considering the final result of the Algorithm 2, let $H^{'}$ be the set of elements remaining in E at the end of Algorithm 2. Then we consider the elements of the set O in two parts, one for the elements that are in both O and $H^{'}$, which is called O_1 and the other for the remaining elements in O, which is called O_2.

On the one hand, through the property of matroid, it is easy to prove that the number of elements in S is not less than the number of elements in O_2 by the converse method which can refer to the [15]. Let $|O_2| = q$, we analyse the approximation ratio by the following Lemmas.

Lemma 5. *Let $S_q = \{e_1, \cdots, e_q\}$, it holds that*

$$(1 - \epsilon)f(S_q \cup O_2) - 2\epsilon p f(S) - xc(O_2) \leq 2f(S_q) - xc(S_q). \qquad (4)$$

On the other hand, there is also a relationship between S and O_1.

Lemma 6. *It holds that*

$$(1 - \epsilon)f(S \cup O_1) - xc(O_1) \leq (1 - \epsilon + 2\epsilon p)f(S). \qquad (5)$$

By the above two Lemmas, we get Theorem 2 as follows.

142   M. Geng et al.

Theorem 2. *When $x = 2 + 2\epsilon + 4\epsilon p$, Algorithm 2 returns a set $S \in \mathcal{F}$ such that*

$$f(S) - c(S) \geq \frac{1 - \epsilon}{2 + 2\epsilon + 4\epsilon p} f(O) - c(O), \tag{6}$$

where p is the rank of matroid $\mathcal{F}$. If the parameter $\epsilon \to 0$, the approximate ratio is $(\frac{1}{2}, 1)$.

6 Conclusions

In this paper, we firstly study the problem of maximizing the difference between the submodular function and linear function in a noisy environment under two constraints such as cardinality constraint and matroid constraint respectively. The target of our study is the noise function F instead of the traditional submodular function f. As our major contribution, we present two bicriteria algorithms under ϵ-multiplicative noise. With the cardinality constraint, Algorithm 1 gives the $(\frac{1+\epsilon^2-2\epsilon-12\epsilon k-4\epsilon^2 k}{2(1+\epsilon)}, \frac{2}{1-\epsilon})$-bicriteria approximation ratio; With the matroid constraint, the $(\frac{1-\epsilon}{2+2\epsilon+4\epsilon p}, 1)$-bicriteria approximation ratio is obtained. If the parameter $\epsilon \to 0$, we then have $(\frac{1}{2}, 2)$-approximation and $(\frac{1}{2}, 1)$-approximation to their respective problems.

References

1. Brualdi, R.A.: Comments on bases in dependence structures. Bull. Aust. Math. Soc. **1**(2), 161–167 (1969)
2. Du, D., Li, Y., Xiu, N., Xu, D.: Simultaneous approximation of multi-criteria submodular function maximization. J. Oper. Res. Soc. China **2**(3), 271–290 (2014)
3. Edmonds, J.: Submodular functions, matroids, and certain polyhedra. In: Jünger, M., Reinelt, G., Rinaldi, G. (eds.) Combinatorial Optimization — Eureka, You Shrink! LNCS, vol. 2570, pp. 11–26. Springer, Heidelberg (2003). https://doi.org/10.1007/3-540-36478-1_2
4. Feldman, M.: Guess free maximization of submodular and linear sums. In: Friggstad, Z., Sack, J.-R., Salavatipour, M.R. (eds.) WADS 2019. LNCS, vol. 11646, pp. 380–394. Springer, Cham (2019). https://doi.org/10.1007/978-3-030-24766-9_28
5. Feige, U., Mirrokni, V.S., Vondrák, J.: Maximizing non-monotone submodular functions. SIAM J. Comput. **40**(4), 1133–1153 (2011)
6. Guestrin, C., Krause, A., Singh, A.P.: Near-optimal sensor placements in gaussian processes. In: 22th International Conference on Machine Learning, pp. 265–272. PMLR (2005)
7. Goldberger, J., Ben-Reuven, E.: Training deep neural-networks using a noise adaptation layer (2016)
8. Gölz, P., Procaccia, A.D.: Migration as submodular optimization. In: 33th AAAI Conference on Artificial Intelligence, pp. 549–556. AAAI Press, Palo Alto, California USA (2019)
9. Horel, T., Singer, Y.: Maximization of approximately submodular functions. In: Advances in Neural Information Processing Systems, vol. 29 (2016)

10. Harshaw, C., Feldman, M., Ward, J., et al.: Submodular maximization beyond non-negativity: guarantees, fast algorithms, and applications. In: 36th International Conference on Machine Learning, pp. 2634–2643. PMLR (2019)
11. Kazemi, E., Minaee, S., Feldman, M., et al.: Regularized submodular maximization at scale. In: 38th International Conference on Machine Learning, pp. 5356–5366 (2021)
12. Kempe, D., Kleinberg, J., Tardos, É.: Maximizing the spread of influence through a social network. In: 9th ACM SIGKDD International Conference on Knowledge Discovery and Data Mining, pp. 137–146. Association for Computing Machinery, New York (2003)
13. Leskovec, J., Krause, A., Guestrin, C., et al.: Cost-effective outbreak detection in networks. In: 13th ACM SIGKDD International Conference on Knowledge Discovery and Data Mining, pp. 420–429. Association for Computing Machinery, New York (2007)
14. Mirzasoleiman, B., Badanidiyuru, A., Karbasi, A.: Fast constrained submodular maximization: personalized data summarization. In: 33th International Conference on Machine Learning, pp. 1358–1367. PMLR (2016)
15. Nikolakaki, S.M., Ene, A., Terzi, E.: An efficient framework for balancing submodularity and cost. In: 27th ACM SIGKDD Conference on Knowledge Discovery and Data Mining, pp. 1256–1266. Association for Computing Machinery, New York (2021)
16. Nguyen, L., Thai, M.T.: Streaming k-submodular maximization under noise subject to size constraint. In: 37th International Conference on Machine Learning, pp. 7338–7347. PMLR (2020)
17. Sviridenko, M., Vondrák, J., Ward, J.: Optimal approximation for submodular and supermodular optimization with bounded curvature. Math. Oper. Res. **42**(4), 1197–1218 (2017)
18. Sviridenko, M.: A note on maximizing a submodular set function subject to a knapsack constraint. Oper. Res. Lett. **32**(1), 41–43 (2004)
19. Singer, Y., Vondrák, J.: Information-theoretic lower bounds for convex optimization with erroneous oracles. In: Advances in Neural Information Processing Systems, vol. 28 (2015)
20. Shahaf, D., Horvitz, E.: Generalized task markets for human and machine computation. In: 24th AAAI Conference on Artificial Intelligence, pp. 986–993. AAAI Press, Palo Alto, California USA (2010)
21. Vondrák, J.: Optimal approximation for the submodular welfare problem in the value oracle model. In: 14th Annual ACM Symposium on Theory of Computing, pp. 67–74. Association for Computing Machinery, New York (2008)
22. Yang, R., Xu, D., Cheng, Y., et al.: Streaming submodular maximization under noises. In: 39th International Conference on Distributed Computing Systems, pp. 3348–357. IEEE, Dallas (2019)

Monotone k-Submodular Knapsack Maximization: An Analysis of the Greedy+Singleton Algorithm

Jingwen Chen[1], Zhongzheng Tang[2], and Chenhao Wang[1,3(✉)]

[1] BNU-HKBU United International College, Zhuhai, China
[2] Beijing University of Posts and Telecommunications, Beijing, China
[3] Beijing Normal University, Zhuhai, China
chenhwang@bnu.edu.cn

Abstract. This paper studies the problem of maximizing a non-negative monotone k-submodular function. A k-submodular function is a generalization of a submodular function, where the input consists of k disjoint subsets, instead of a single subset. For the problem under a knapsack constraint, we consider the algorithm that returns the better solution between the single element of highest value and the result of the fully greedy algorithm, to which we refer as Greedy+Singleton, and prove an approximation ratio $\frac{1}{4}(1 - \frac{1}{e}) \approx 0.158$. Though this ratio is strictly smaller than the best known factor for this problem, Greedy+Singleton is simple, fast, and of special interests. Our experiments demonstrates that the algorithm performs well in terms of the solution quality.

Keywords: k-submodular · Knapsack · Approximation ratio

1 Introduction

We investigate k-submodular functions in this paper, which generalize submodular functions in a natural way. The input of a k-submodular function consists of k disjoint subsets of a finite nonempty set V, instead of a single subset of V in a submodular function. Submodular and bisubmodular functions are included in our setting as the special cases $k = 1$ and $k = 2$ respectively.

The k-submodular maximization problem has been widely studied due to its broad applications, e.g., influence maximization with k types of topics or rumors, and sensor placement with k types of sensors, etc. In addition to the unconstrained setting [22], the k-submodular maximization problem is also investigated in various constrained setting, such as cardinality constraints [14], matroid constraints [16], and knapsack constraints [20]. This paper will focus on the knapsack constraints.

Let us define k-submodular functions formally. Given a finite nonempty set V with $|V| = n$, let $(k + 1)^V := \{(X_1, \ldots, X_k) \mid X_i \subseteq V \ \forall i \in [k], X_i \cap X_j = \varnothing \ \forall i \neq j\}$ be the family of k disjoint sets, where $[k] := \{1, \ldots, k\}$.

Q. Ni and W. Wu (Eds.): AAIM 2022, LNCS 13513, pp. 144–155, 2022.
https://doi.org/10.1007/978-3-031-16081-3_13

Definition 1 (k-submodularity [8]). *A function* $f : (k+1)^V \to \mathbb{R}$ *is called* k-submodular, *if for any* $\mathbf{x} = (X_1, \ldots, X_k)$ *and* $\mathbf{y} = (Y_1, \ldots, Y_k)$ *in* $(k+1)^V$, *we have*

$$f(\mathbf{x}) + f(\mathbf{y}) \geq f(\mathbf{x} \sqcup \mathbf{y}) + f(\mathbf{x} \sqcap \mathbf{y}),$$

where

$$\mathbf{x} \sqcup \mathbf{y} := \left(X_1 \cup Y_1 \backslash (\bigcup_{i \neq 1} X_i \cup Y_i), \ldots, X_k \cup Y_k \backslash (\bigcup_{i \neq k} X_i \cup Y_i) \right),$$

$$\mathbf{x} \sqcap \mathbf{y} := (X_1 \cap Y_1, \ldots, X_k \cap Y_k).$$

Roughly speaking, k-submodularity captures the property that, if we choose exactly one set $X_a \in \{X_1, \ldots, X_k\}$ that an element a can belong to for each $a \in V$, then the resulting function is submodular. As we know, the submodularity is equivalent to the property of *diminishing marginal returns*, and we shall see that this property also play an important role in k-submodular functions. However, in k-submodular functions, we must specify not only which element we are adding to the solution, but also which set in the partition it is being added to.

Our Contributions

In this paper, we study the maximization problem of a non-negative monotone k-submodular function under a knapsack constraint, that is, each item $a \in V$ has a cost c_a, and the total cost of selected items must not exceed a given budget $B \in \mathbb{R}_+$. A natural heuristic for knapsack problem is **Greedy**, which maintains a feasible solution, and at each step adds to this solution an item that maximizes the marginal value per unit of weight (i.e., marginal density), until there is no longer budget. Unfortunately, this greedy algorithm is well known to have unbounded approximations, even for linear objectives.

We consider the **Greedy+Singleton** algorithm that returns the better solution between **Greedy** and **Singleton**, where **Singleton** returns a solution that contains a single item of highest value among all dimensions. We prove that the **Greedy+Singleton** algorithm is $(1 - \frac{1}{e})/4$-approximation.

This approximation ratio is not as good as those in the literature. For the k-submodular knapsack maximization, Tang et al. [20] provide a $(1 - \frac{1}{e})/2$-approximation algorithm within $O(n^4 k^3)$ queries, which combines **Singleton** with a greedy algorithm that completes all feasible solutions of size 2 greedily. This approximation ratio has been improved to $\frac{1}{2} - \epsilon$ by Wang and Zhou [21] using multilinear extension techniques.

Nevertheless, we highlight the novelty and special interests of our contribution. From the perspective of time complexity, while both algorithms in [20] and [21] are not very efficient, our algorithm **Greedy+Singleton** takes only $O(n^2 k)$ queries. From the perspective of theoretical interests, special attention was given to **Greedy+Singleton** for submodular functions: the approximation ratio is improved from 0.387 to 0.393 and to 0.427 (see more details in the part of Related Work). Hence, our approximation ratio $(1 - \frac{1}{e})/4$ is the first step for k-submodular functions, and may inspire follow-up studies to improve this ratio by more careful analysis.

Related Work
We discuss the related work in three parts: submodular knapsack maximization, **Greedy+Singleton** for submodular functions, and k-submodular maximization.

Submodular Knapsack. For monotone submodular maximization under a knapsack constraint, Sviridenko [18] presents a greedy $(1 - \frac{1}{e})$-approximation algorithm with $O(n^5)$ queries, which enumerates all feasible sets of size no more than 3 and then expands each set of size 3 greedily by the marginal density. This is the best possible approximation ratio among polynomial-time algorithms. Faster algorithms with $(1 - \frac{1}{e} - \epsilon)$-approximation exist [2], but the time is exponential to $\frac{1}{\epsilon}$. Yaroslavtsev et al. [23] presented a Greedy+Max algorithm that is a $\frac{1}{2}$-approximation with query complexity $O(\tilde{K}n)$, where $\tilde{K}$ is an upper bound on the number of elements in any feasible solution. Huang et al. [6,7] considered this problem in a streaming setting.

Greedy+Singleton for Submodular Functions. The algorithm was first suggested in [10] for coverage functions, and adapted to monotone submodular function in [12]. Both works stated an approximation guarantee of $(1-e^{-0.5}) \approx 0.393$, though the proofs in both works were flawed. Tang et al. [19] establish a correct proof for the $(1 - e^{-0.5})$-approximation, improving upon an earlier approximation guarantee $\frac{e-1}{2e-1} \approx 0.387$ [1]. Feldman, Nutov and Shoham showed that the approximation ratio of the algorithm is within $[0.427, 0.462]$ [3]. More recently, Kulik, Schwartz and Shachnai [11] present an improved upper bound of 0.42945, which combined with the result of [3], limits the approximation ratio of **Greedy+Singleton** to the narrow interval $[0.427, 0.42945]$.

k-submodular Maximization. The k-submodular functions were proposed by Huber and Kolmogorov [8], which express the submodularity on choosing k disjoint sets of elements instead of a single set, and recently become a popular subject of research [4,5,13,17]. For unconstrained non-monotone k-submodular maximization, Ward and Živný [22] first proposed an approximation algorithm, and the ratio was later improved by Iwata et al. [9] to $\frac{1}{2}$, which is more recently improved to $\frac{k^2+1}{2k^2+1}$ by Oshima [15]. For unconstrained *monotone k*-submodular maximization, Ward and Živný [22] proved that a greedy algorithm is $\frac{1}{2}$-approximaion, and later, Iwata et al. [9] proposed a randomized $\frac{k}{2k-1}$-approximation algorithm, which is asymptotically tight.

The maximization of a monotone k-submodular function is also studied under various constraints. Under a total size constraint (i.e., $\cup_{i \in [k]}|X_i| \leq B$ for an integer budget B), Ohsaka and Yoshida [14] proposed a $\frac{1}{2}$-approximation algorithm, and under individual size constraints (i.e., $|X_i| \leq B_i \ \forall i \in [k]$ with integers B_i), they proposed a $\frac{1}{3}$-approximation algorithm. Under a matroid constraint on the union of the sets, Sakaue [16] proposed a $\frac{1}{2}$-approximation algorithm. Under a knapsack constraint, Tang et al. [20] provided a $(1 - \frac{1}{e})/2$-approximation algorithm that combines **Singleton** with a greedy algorithm that completes all feasible solutions of size 2 greedily, and later Wang and Zhou [21] improved it to $\frac{1}{2} - \epsilon$ using multilinear extension techniques.

2   Preliminaries

Denote $\mathbf{x} \preceq \mathbf{y}$, if $\mathbf{x} = (X_1, \ldots, X_k)$ and $\mathbf{y} = (Y_1, \ldots, Y_k)$ with $X_i \subseteq Y_i$ for each $i \in [k]$. Define the marginal gain when adding item a to the i-th set of $\mathbf{x}$ to be

$$\Delta_{a,i}(\mathbf{x}) := f(X_1, \ldots, X_{i-1}, X_i \cup \{a\}, X_{i+1}, \ldots, X_k) - f(\mathbf{x}).$$

A k-submodular function f clearly satisfies the *orthant submodularity*, that is,

$$\Delta_{a,i}f(\mathbf{x}) \geq \Delta_{a,i}f(\mathbf{y}), \quad \text{for any } \mathbf{x}, \mathbf{y} \in (k+1)^V \text{ with } \mathbf{x} \preceq \mathbf{y}, a \notin \cup_{j \in [k]} Y_j, i \in [k].$$

A function $f : (k+1)^V \to \mathbb{R}$ is called *monotone*, if $f(\mathbf{x}) \leq f(\mathbf{y})$ for any $\mathbf{x} \preceq \mathbf{y}$. Ward and Živný [22] shows that when monotonicity holds, f is k-submodular if and only if it is orthant submodular.

For notational ease, we express the family of solutions $(k+1)^V$ in an alternative way:

$$\mathcal{S} = \left\{ \cup_{j=1}^t \{(a_j, i_j)\} \mid t \in [n] \cup \{0\}, a_j \in V, i_j \in [k] \right\}.$$

That is, $\mathcal{S}$ is the family of the sets of item-index pairs, where an item-index pair (a, i) indicates that item $a \in V$ is assigned to the i-th set. Any k-disjoint set $\mathbf{x} = (X_1, \ldots, X_k) \in (k+1)^V$ uniquely corresponds to an item-index pairs set $S \in \mathcal{S}$, such that $(a_j, i_j) \in S$ if and only if $a_j \in X_{i_j}$. From now on, we rewrite $f(\mathbf{x})$ as $f(S)$ with a slight abuse of notation, and thus $\Delta_{a,i}(S)$ means the marginal gain $f(S \cup \{(a, i)\}) - f(S)$. For any $S \in \mathcal{S}$, we define $U(S) := \{a \in V \mid \exists i \in [k] \text{ s.t. } (a, i) \in S\}$ to be the set of items included, and the *size* of S is $|S| = |U(S)|$. In this paper, let f be a non-negative, monotone, k-submodular function. We further assume that $f(\varnothing) = 0$, which is without loss of generality because otherwise we can redefine $f(S) := f(S) - f(\varnothing)$ for all $S \in \mathcal{S}$.

We first introduce an important lemma.

Lemma 1 ([20]). *For any $S, S' \in \mathcal{S}$ with $S \subseteq S'$, we have*

$$f(S') - f(S) \leq \sum_{(a,i) \in S' \setminus S} \Delta_{a,i}(S).$$

The following proposition from Ward and Živný [22] says that unconstrained Greedy (see Algorithm 1) is $\frac{1}{2}$-approximation for maximizing f without any constraint. This algorithm considers items in an arbitrary order, and assigns each item the *best* index that brings the largest marginal gain. We will use this algorithm as a subroutine in the analysis in Sect. 3.

Proposition 1 ([22]). *Let $T \in \mathcal{S}$ be a solution that maximizes f in the unconstrained setting, and $S \in \mathcal{S}$ be the solution returned by Unconstrained Greedy. Then $f(T) \leq 2 \cdot f(S)$.*

Algorithm 1. Unconstrained Greedy

Input: Set $V = \{1, 2, \ldots, n\}$, monotone k-submodular function f
Output: A solution $S \in \mathcal{S}$
1: $S \leftarrow \varnothing$
2: **for** $a = 1$ to n **do**
3: $\quad i_a \leftarrow \arg\max_{i \in [k]} \Delta_{a,i}(S)$
4: $\quad S \leftarrow S \cup \{(a, i_a)\}$
5: **end for**
6: **return** S

3 Approximations of Greedy+Singleton

We consider the problem of maximizing f under a knapsack constraint, that is, each item $a \in V$ has a cost c_a, and the total cost of selected items must not exceed a given budget $B \in \mathbb{R}_+$. For any solution $S \in \mathcal{S}$, define $c(S)$ to be the total cost of all items in S. Algorithm 2 is the procedure of **Greedy+Singleton**.

In the remainder of this section, we prove an approximation ratio $\frac{1}{4}(1 - \frac{1}{e})$, following the framework provided by Khuller et al. [10] for the budgeted maximum coverage problem, which can derive a $\frac{1}{2}(1 - \frac{1}{e})$ approximation for the submodular knapsack maximization.

Algorithm 2. Greedy+Singleton

1: Let $S^* \in \arg\max\limits_{S: |S|=1, c(S) \leq B} f(S)$ be a singleton solution giving the largest value.
2: $S^0 \leftarrow \varnothing$, $V^0 \leftarrow V$
3: **for** t from 1 to n **do**
4: $\quad$ Let $(a_t, i_t) = \arg\max\limits_{a \in V^{t-1}, i \in [k]} \frac{\Delta_{a,i}(S^{t-1})}{c_a}$ be the pair that maximizes the marginal density
5: $\quad$ **if** $c(S^{t-1}) + c_{a_t} \leq B$ **then**
6: $\quad\quad S^t = S^{t-1} \cup \{(a_t, i_t)\}$
7: $\quad$ **else**
8: $\quad\quad S^t = S^{t-1}$
9: $\quad$ **end if**
10: $\quad V^t = V^{t-1} \setminus \{a_t\}$
11: **end for**
12: $S^* \leftarrow S^n$ if $f(S^n) > f(S^*)$
13: **return** S^*

Let OPT be the optimal solution, and $f(OPT)$ be the optimal value. Let $l + 1$ be the first time when Algorithm 2 does not add an item in $U(OPT)$ to the current solution, due to the budget constraint. We can further assume that $l + 1$ is the first step t for which $S^t = S^{t-1}$. This assumption is without loss of generality, because if it happens earlier for some $t' < l + 1$, then $a_{t'}$ does not belong to the optimal solution T, nor the approximate solution we are interested in; thus, we can remove $a_{t'}$ from the ground set V, without affecting the analysis, the optimal solution T, and the approximate solution returned by the algorithm.

For each $t = 1,\ldots,l$, we define $G_t = S_t$ to be the solution after the t-th iteration, and define $G_{l+1} = S_l \cup \{(a_{l+1}, i_{l+1})\}$ to be the solution obtained by adding (a_{l+1}, i_{l+1}) to S_l though violating the budget. The following lemma bounds the marginal gain in every iteration.

Lemma 2. *For each $t = 1,\ldots,l+1$, we have*

$$f(G_t) - f(G_{t-1}) \geq \frac{c_{a_t}}{2B}\left(f(OPT) - 2f(G_{t-1})\right)$$

Proof. We first consider the items in $U(OPT)$. We run the subroutine Algorithm 1 on ground set $U(OPT)$ with respect to an order of items in which the first l items are $a_1, a_2, \ldots, a_l$. Let $\overline{OPT}$ be the solution returned by this subroutine, and by Proposition 1, $f(OPT) \leq 2 \cdot f(\overline{OPT})$. Noting that a_t is the item added by Algorithm 2 in the t-th iteration for every $t = 1,\ldots,l$, we have $S_t \subseteq \overline{OPT}$. Then by Lemma 1, we have

$$\begin{aligned}
f(OPT) &\leq 2 \cdot f(\overline{OPT}) \\
&\leq 2 \cdot f(G_{t-1}) + 2 \sum_{(a,i)\in\overline{OPT}\backslash G_{t-1}} \Delta_{a,i}(G_{t-1}) \\
&\leq 2 \cdot f(G_{t-1}) + 2B \cdot \frac{\Delta_{a_t,i_t}(G_{t-1})}{c_{a_t}} \\
&= 2 \cdot f(G_{t-1}) + 2B \cdot \frac{f(G_t) - f(G_{t-1})}{c_{a_t}},
\end{aligned}$$

where the last inequality follows from the facts that the marginal density is maximized in each iteration and the capacity remained is at most B. Then immediately we have $f(G_t) - f(G_{t-1}) \geq \frac{c_{a_t}}{2B}\left(f(OPT) - 2f(G_{t-1})\right)$. $\square$

Lemma 3. *For each $t = 1,\ldots,l+1$, we have*

$$f(G_t) \geq (1 - x_t) \cdot f(OPT),$$

where $x_1 = 1 - \frac{c_{a_1}}{B}$ and $x_t = (1 - \frac{c_{a_t}}{B})x_{t-1} + \frac{c_{a_t}}{2B}$.

Proof. We prove it by induction. Firstly, when $t = 1$, clearly we have $f(S_1) \geq \frac{c_{a_1}}{B}f(OPT)$. Suppose the statement holds for iterations $1, 2, \ldots, t-1$. We show that it also holds for iteration t:

$$f(G_t) = f(G_{t-1}) + f(G_t) - f(G_{t-1})$$
$$\geq f(G_{t-1}) + \frac{c_{a_t}}{2B}(f(OPT) - 2f(G_{t-1}))$$
$$= (1 - \frac{c_{a_t}}{B})f(G_{t-1}) + \frac{c_{a_t}}{2B}f(OPT)$$
$$\geq (1 - \frac{c_{a_t}}{B})(1 - x_{t-1}) \cdot f(OPT) + \frac{c_{a_t}}{2B}f(OPT)$$
$$= (1 - \frac{c_{a_t}}{2B} - x_{t-1} + \frac{c_{a_t}}{B}x_{t-1})f(OPT)$$
$$= [1 - ((1 - \frac{c_{a_t}}{B})x_{t-1} + \frac{c_{a_t}}{2B})]f(OPT),$$

as desired.   □

It is not hard to see that the recurrence relation $x_t = (1 - \frac{c_{a_t}}{B})x_{t-1} + \frac{c_{a_t}}{2B}$ with initial state $x_1 = 1 - \frac{c_{a_1}}{B}$ can be written as

$$x_t - \frac{1}{2} = (1 - \frac{c_{a_t}}{B})x_{t-1} - \frac{1}{2}(1 - \frac{c_{a_t}}{B})$$
$$= (1 - \frac{c_{a_t}}{B})(x_{t-1} - \frac{1}{2}).$$

Hence, we can easily get a general formula

$$x_t = (\frac{1}{2} - \frac{c_{a_1}}{B})\prod_{j=2}^{t}(1 - \frac{c_{a_j}}{B}) + \frac{1}{2}.$$

Now we are ready to prove our main theorem.

Theorem 1. *Greedy+Singleton* *achieves an approximation ratio of* $\frac{1}{4}(1 - \frac{1}{e}) \approx 0.158$ *for k-submodular knapsack maximization with* $O(n^2k)$ *queries.*

Proof. By Lemma 3, we have

$$f(G_{l+1}) \geq (1 - x_{l+1}) \cdot f(OPT)$$
$$= \left(\frac{1}{2} - (\frac{1}{2} - \frac{c_1}{B})\prod_{j=2}^{l+1}(1 - \frac{c_j}{B})\right) \cdot f(OPT)$$
$$\geq \left(\frac{1}{2} - \frac{1}{2}(1 - \frac{c_1}{B})\prod_{j=2}^{l+1}(1 - \frac{c_j}{B})\right) \cdot f(OPT) \qquad (1)$$
$$\geq \left(\frac{1}{2} - \frac{1}{2}\prod_{j=1}^{l+1}(1 - \frac{c_j}{c(G_{l+1})})\right) \cdot f(OPT)$$
$$\geq \left(\frac{1}{2} - \frac{1}{2} \cdot (1 - \frac{1}{l+1})^{l+1}\right) \cdot f(OPT)$$
$$\geq \left(\frac{1}{2} - \frac{1}{2e}\right) \cdot f(OPT). \qquad (2)$$

Note that inequality (1) follows from the fact that adding the item a_{l+1} to G_l violates the budget B, and thus

$$c(G_{l+1}) = c(G_l) + c_{l+1} > B.$$

By (2), we know that

$$f(G_l) + \max_{(a,i)} \Delta_{a,i}(G_l) \geq \left(\frac{1}{2} - \frac{1}{2e}\right) \cdot f(OPT).$$

The LHS is no more than $f(G_l)$ plus the maximum profit of a single item, i.e., the outcome of **Singleton**, say (a^*, i^*). Therefore, the better solution between G_l and $\{(a^*, i^*)\}$ is has a value

$$\max\{f(G_l), f(\{(a^*, i^*)\})\} \geq \frac{1}{2}\left(\frac{1}{2} - \frac{1}{2e}\right) \cdot f(OPT),$$

Since $G_l = S_l$ is a part of the solution returned by **Greedy+Singleton** when **Greedy** performs better than **Singleton**, it establishes an approximation ratio $\frac{1}{4}(1 - \frac{1}{e})$.   □

4   Experiments

In this section, we evaluate the performance of algorithms empirically on synthetic dataset. We conducted experiments on Visual Studio 2020 with Intel(R) Core(TM) i5-10500 (3.10 GHz) and 16 GB of main memory. We implemented all algorithms in C++. We apply the algorithms to the problem of influence maximization of several topics in social networks.

In the information diffusion model, called the k-*topic independent cascade* (k-IC) model [14], suppose that there are k kinds of topic (or rumors) which can spread independently in a social network. Let $G = (V, E)$ be a social network with an edge probability $p_{u,v}^i$ for each edge $(u, v) \in E$, representing the probability of spread the i-th topic from vertex u to v. Suppose we are given some seeds $S \in (k+1)^V$, the infection process about the i-th topic starts by activating all seed nodes with topic i, independently from other topics. And then their spread conforms to the following rule: When a node u is activated in step t for the first time, it has a single chance to activate each of its neighbor nodes who are current inactive, with a probability $p_{u,v}^i$ of successfully activating its neighbor v. If u succeeds, v becomes active in step $t + 1$ with the i-th topic, and then v will continue to infect according to this rule in the next round. If the infection is not successful, then u will not try again to infect v. The diffusion process will stop when there is no more chance to activate any node.

The *influence spread* function $f : (k+1)^V \to \mathbb{R}_+$ in the k-IC model is defined as the expected total number of vertices who are eventually activated in one of the k topics given seeds $S \in (k+1)^V$. That is, $f(S) = \mathbb{E}[|\bigcup_{i \in [k]} A_i(S)|]$, where $A_i(S)$ is a random variable representing the set of activated vertices with the i-th topic. Suppose each vertex $v \in V$ has a cost $c_v \in \mathbb{R}_+$. Given a directed graph

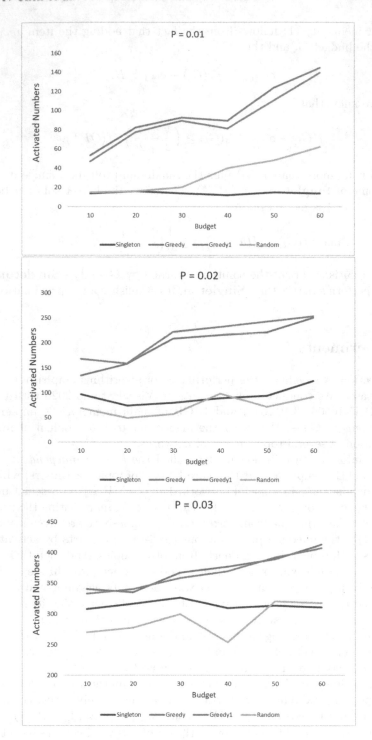

Fig. 1. The experiments results of the algorithms in the k-IC model, for a range of budget from 10 to 60.

$G = (V, E)$, edge probabilities $p_{u,v}^i$ for all $(u, v) \in [E], i \in [k])$, and a budget B, the problem is to select seeds $S \in (k + 1)^V$ that maximizes $f(S)$ subject to a knapsack constraint $\sum_{v \in U(S)} c_v \leq B$. This influence spread function f is monotone k-submodular [14].

In the experiment, we generate a synthetic dataset in the Erdős-Rényi random graph model with 500 vertices. Each edge is included in the graph with probability $p = 0.01, 0.02, 0.03$, independently from every other edge. Let the number of topics be $k = 5$. The cost of each vertex is a random number following the uniform distribution $Uni(0, 5)$. We compare the performances of the following algorithms by changing the value of the budget.

- **Singleton**: a single node as well as it topic that maximizes the total number of activated nodes.
- **Greedy**: at each step we add to the seeds set a feasible node as well as its topic that maximizes the marginal density. The process ends when there is no more budget for any further node.
- **Greedy+Singleton** (Algorithm 2): the better solution between **Singleton** and **Greedy**.
- **Greedy1**: we initialize the set maintained in **Greedy** to be the output of **Singleton**, and then expand this set greedily.
- **Random**: at each step we add to the seeds set a random feasible node as well as a random topic, until there is no more feasible node.

The experiments results (see Fig. 1) show that for every instance, **Greedy** performs strictly better than **Singleton**, and thus **Greedy+Singleton** is exactly the same to **Greedy**. When $p = 0.01$, **Greedy** is better than **Greedy1**, while when $p = 0.02$, **Greedy1** is better than **Greedy**. When $p = 0.03$, the two algorithms have a similar performance. In all cases, **Greedy** and **Greedy1** are much better than **Singleton** and **Random**.

5 Conclusion

We have provided an approximation ratio $\frac{1}{4}(1 - \frac{1}{e})$ of the **Greedy+Singleton** algorithm for the problem of maximizing a monotone k-submodular function under a knapsack constraint. An immediate open question is whether this ratio is tight with respect to the algorithm? The answer seems to be negative, because the framework of the proof is a generalization of that in [10], which gives a very loose analysis of **Greedy+Singleton** for the submodular knapsack problem. Thus, it would be interesting to explore what approximation this algorithm can achieve. Further, another future direction is to look at other simple and fast algorithms for k-submodular maximization, or multiple knapsack constraints.

Acknowledgements. This work is partially supported by Artificial Intelligence and Data Science Research Hub, BNU-HKBU United International College (UIC), No. 2020KSYS007, and by a grant from UIC (No. UICR0400025-21). Zhongzheng Tang is supported by National Natural Science Foundation of China under Grant No. 12101069 and Innovation Foundation of BUPT for Youth (No. 500422309). Chenhao Wang is supported by a grant from UIC (No. UICR0700036-22).

References

1. Cohen, R., Katzir, L.: The generalized maximum coverage problem. Inf. Process. Lett. **108**(1), 15–22 (2008)
2. Ene, A., Nguyen, H.L.: A nearly-linear time algorithm for submodular maximization with a knapsack constraint. In: Proceedings of the 46th International Colloquium on Automata, Languages, and Programming (ICALP) (2019)
3. Feldman, M., Nutov, Z., Shoham, E.: Practical budgeted submodular maximization. arXiv preprint arXiv:2007.04937 (2020)
4. Gridchyn, I., Kolmogorov, V.: Potts model, parametric maxflow and k-submodular functions. In: Proceedings of the IEEE International Conference on Computer Vision (ICCV), pp. 2320–2327 (2013)
5. Hirai, H., Iwamasa, Y.: On k-submodular relaxation. SIAM J. Discret. Math. **30**(3), 1726–1736 (2016)
6. Huang, C.C., Kakimura, N.: Improved streaming algorithms for maximizing monotone submodular functions under a knapsack constraint. Algorithmica **83**(3), 879–902 (2021)
7. Huang, C.C., Kakimura, N., Yoshida, Y.: Streaming algorithms for maximizing monotone submodular functions under a Knapsack constraint. In: Approximation, Randomization, and Combinatorial Optimization. Algorithms and Techniques (APPROX/RANDOM) (2017)
8. Huber, A., Kolmogorov, V.: Towards minimizing k-submodular functions. In: Mahjoub, A.R., Markakis, V., Milis, I., Paschos, V.T. (eds.) ISCO 2012. LNCS, vol. 7422, pp. 451–462. Springer, Heidelberg (2012). https://doi.org/10.1007/978-3-642-32147-4_40
9. Iwata, S., Tanigawa, S.I., Yoshida, Y.: Improved approximation algorithms for k-submodular function maximization. In: Proceedings of the 27th Annual ACM-SIAM Symposium on Discrete Algorithms (SODA), pp. 404–413 (2016)
10. Khuller, S., Moss, A., Naor, J.S.: The budgeted maximum coverage problem. Inf. Process. Lett. **70**(1), 39–45 (1999)
11. Kulik, A., Schwartz, R., Shachnai, H.: A refined analysis of submodular greedy. Oper. Res. Lett. **49**(4), 507–514 (2021)
12. Lin, H., Bilmes, J.: Multi-document summarization via budgeted maximization of submodular functions. In: Human Language Technologies: The 2010 Annual Conference of the North American Chapter of the Association for Computational Linguistics, pp. 912–920 (2010)
13. Nguyen, L., Thai, M.T.: Streaming k-submodular maximization under noise subject to size constraint. In: Proceedings of the 37th International Conference on Machine Learning (ICML), pp. 7338–7347. PMLR (2020)
14. Ohsaka, N., Yoshida, Y.: Monotone k-submodular function maximization with size constraints. In: Proceedings of the 28th International Conference on Neural Information Processing Systems (NeurIPS), vol. 1, pp. 694–702 (2015)

15. Oshima, H.: Improved randomized algorithm for k-submodular function maximization. SIAM J. Discret. Math. **35**(1), 1–22 (2021)
16. Sakaue, S.: On maximizing a monotone k-submodular function subject to a matroid constraint. Discret. Optim. **23**, 105–113 (2017)
17. Soma, T.: No-regret algorithms for online k-submodular maximization. In: Proceedings of the 22nd International Conference on Artificial Intelligence and Statistics (AISTATS), pp. 1205–1214. PMLR (2019)
18. Sviridenko, M.: A note on maximizing a submodular set function subject to a Knapsack constraint. Oper. Res. Lett. **32**(1), 41–43 (2004)
19. Tang, J., Tang, X., Lim, A., Han, K., Li, C., Yuan, J.: Revisiting modified greedy algorithm for monotone submodular maximization with a Knapsack constraint. Proc. ACM Measure. Anal. Comput. Syst. **5**(1), 1–22 (2021)
20. Tang, Z., Wang, C., Chan, H.: On maximizing a monotone k-submodular function under a knapsack constraint. Oper. Res. Lett. **50**(1), 28–31 (2022)
21. Wang, B., Zhou, H.: Multilinear extension of k-submodular functions. arXiv preprint arXiv:2107.07103 (2021)
22. Ward, J., Živný, S.: Maximizing k-submodular functions and beyond. ACM Trans. Algorithms **12**(4), 1–26 (2016)
23. Yaroslavtsev, G., Zhou, S., Avdiukhin, D.: "Bring your own greedy"+ max: near-optimal 1/2-approximations for submodular knapsack. In: International Conference on Artificial Intelligence and Statistics, pp. 3263–3274. PMLR (2020)

Guarantees for Maximization of k-Submodular Functions with a Knapsack and a Matroid Constraint

Kemin Yu, Min Li, Yang Zhou, and Qian Liu[✉]

School of Mathematics and Statistics, Shandong Normal University,
Jinan 250014, People's Republic of China
lq_qsh@163.com, {liminemily,zhouyang}@sdnu.edu.cn

Abstract. A k-submodular function is a generalization of a submodular function, whose definition domain is the collection of k disjoint subsets. In our paper, we apply a greedy and local search technique to obtain a $\frac{1}{6}(1-e^{-2})$-approximate algorithm for the problem of maximizing a k-submodular function subject to the intersection of a knapsack constraint and a matroid constraint. Furthermore, we use a special analytical method to improve the approximation ratio to $\frac{1}{3}(1-e^{-3})$, when the k-submodular function is monotone.

Keywords: k-submodularity · Knapsack constraint · Matroid constraint · Approximation algorithm

1 Introduction

Consider a ground set G composed of n elements and $k \in N_+$, we define $(k+1)^G$ as the family of k disjoint subset $(X_1, \ldots, X_k)$, where $X_i \subseteq G, \forall i \in [k]$ and $X_i \cap X_j = \emptyset$, $\forall i \neq j$. A function $f : (k+1)^G \to R$ is said to be k-submodular [7], if

$$f(\boldsymbol{x}) + f(\boldsymbol{y}) \geq f(\boldsymbol{x} \sqcup \boldsymbol{y}) + f(\boldsymbol{x} \sqcap \boldsymbol{y}),$$

for any $\boldsymbol{x} = (X_1, \ldots, X_k)$ and $\boldsymbol{y} = (Y_1, \ldots, Y_k)$ in $(k+1)^G$, where

$$\boldsymbol{x} \sqcup \boldsymbol{y} := (X_1 \cup Y_1 \setminus (\bigcup_{i \neq 1} X_i \cup Y_i), \ldots, X_k \cup Y_k \setminus (\bigcup_{i \neq k} X_i \cup Y_i)),$$

$$\boldsymbol{x} \sqcap \boldsymbol{y} := (X_1 \cap Y_1, \ldots, X_k \cap Y_k).$$

Obviously, it is a submodular function for $k = 1$.

As early as 1978, Nemhauser et al. [11] studied the monotone submodular maximization problem subject to cardinality constraints and obtained a greedy $(1 - 1/e)$-approximation algorithm. Many scholars extended submodular maximization to different constraints and design approximate algorithms, see [1–6, 10, 17, 20]. Among them,

Supported by National Science Foundation of China (No. 12001335) and Natural Science Foundation of Shandong Province of China (Nos. ZR2019PA004, ZR2020MA029, ZR2021MA100).

Q. Ni and W. Wu (Eds.): AAIM 2022, LNCS 13513, pp. 156–167, 2022.
https://doi.org/10.1007/978-3-031-16081-3_14

knapsack constraint and matroid constraint are mainly concerned, and most of the algorithms can achieve the tight $1 - 1/e$ approximation ratio. However, under the intersection constraint of a knapsack and a matroid, we have not found that the algorithm can achieve $1 - 1/e$-approximation, since the loss of rounding is difficult to avoid. Recently, by combining greedy and local search techniques, Sarpatwa et al. [16] contributed an algorithm for reaching $\frac{1-e^{-2}}{2}$-approximation ratio.

In recent years, k-submodular maximization problem has been widely concerned and studied. There have been many research results. For k-submodular maximization without constraint, Ward and Zivny [19] gave a deterministic greedy algorithm, whose approximate ratio reached $1/3$, and a randomized greedy algorithm whose approximate ratio is $\frac{1}{1+a}$, where $a = \max\{1, \sqrt{\frac{k-1}{4}}\}$. Iwata et al. [8] improved the approximation ratio to $1/2$. Later, [14] contributed an algorithm with ratio $\frac{k^2+1}{2k^2+1}$. Under the monotonicity assumption, Ward and Zivny [19] gave a 1/2 approximation algorithm and Iwata et al. [8] improved the approximation ratio to $k/(2k-1)$, which is asymptotically tight. There are also many results for nonnegative monotone k-submodular maximization with constraints. In 2015, Ohsaka and Yoshida [13] designed a 1/2-approximation algorithm for a total size constraint. Sakaue [15] presented a 1/2-approximation algorithm with a matroid constraint. And for monotone k-submodular maximization subject to a knapsack constraint, Tang et al. [18] proposed an algorithm of $\frac{1-1/e}{2}$ approximate ratio. Liu et al. [9] design a combinatorial approximation algorithm for monotone k-submodular maximization subject to a knapsack and a matroid constraint and obtained a $\frac{1}{4}(1 - e^{-2})$ approximate ratio.

In this paper, we consider the k-submodular maximization subject to a knapsack and a matroid constraint, and do some work on the basis of the algorithm given by Liu et al. [9]. The main contributions of this paper are as follows:

- We extend the algorithm for k-submodular maximization problem with a knapsack and a matroid constraint to nonmonotone case, and achieve a $\frac{1}{6}(1 - e^{-2})$ approximate ratio, based on the pairwise monotone property.
- We improve the approximate ratio from $\frac{1}{4}(1 - e^{-2})$ in [9] to $\frac{1}{3}(1 - e^{-3})$ under the monotonicity assumption. In the theoretical analysis of the algorithm, we no longer rely on the results of the greedy algorithm for the unconstrained k-submodular maximization problem, and use the properties of k-submodular function to get the new result.

We organize our paper as follows. In Sect. 2, we first introduce the k-submdodular function and some corresponding results, then present the k-submodular maximization problem with a knapsack and a matroid constraint. We present our results for non-monotone case in Sect. 3. In Sect. 4, we show our theoretical analysis for monotone case.

2 Preliminaries

2.1 k-Submodular Function

For any two k disjoint subsets $\boldsymbol{x} = (X_1, \ldots, X_k)$ and $\boldsymbol{y} = (Y_1, \ldots, Y_k)$ in $(k+1)^G$, we need to introduce a remove operation and a partial order, i.e.

$$\boldsymbol{x} \setminus \boldsymbol{y} := (X_1 \setminus Y_1, \ldots, X_k \setminus Y_k),$$

$$x \preceq y, \text{ if } X_i \subseteq Y_i, \forall i \in [k].$$

Define $\emptyset := (\emptyset, \ldots, \emptyset) \in (k+1)^G$ and $(v, i) \in (k+1)^G$ such that $X_i = \{v\}$ and $X_j = \emptyset$ for $\forall j \in [k]$ with $j \neq i$. Refer $U(x) = \bigcup_{i=1}^{k} X_i$. For $v \notin U(x)$, we use $f_x((v, i)) = f(x \sqcup (v, i)) - f(x)$ to represent the marginal gain of f. A function f is said to be pairwise monotone if $f_x((v, i)) + f_x((v, j)) \geq 0$ for any $i \neq j \in [k]$ holds. In addition, we call that the function f is orthant submodular, if $f_x((v, i)) \geq f_y((v, i))$ holds, for any $x \preceq y$. According to the above definition, we have the equivalent definition and property of the k-submodular function as follows.

Definition 1 [19]. *A function $f : (k + 1)^G \rightarrow R$ is k-submodular iff it is pairwise monotone and orthant submodular.*

Lemma 1 [18]. *Given a k-submodular f, we have*

$$f(y) - f(x) \leq \sum_{(v,i) \preceq y \backslash x} f_x((v, i)),$$

for any $x \preceq y$.

Check the definition of k-submodular, we have the lemma as follows.

Lemma 2. *Given a k-submodular f, we set $g(x) = f(x \sqcup (v, i))$: $(k + 1)^{G \backslash v} \rightarrow R$, then $g(x)$ is k-submodular.*

2.2 k-Submodular Maximization with a Knapsack and a Matroid Constraint

We define $\mathcal{L} \subseteq 2^G$ as the family of subsets of G. A pair $(G, \mathcal{L})$ is called as an independence system if $(\mathcal{M}1)$ and $(\mathcal{M}2)$ holds. And if $(\mathcal{M}3)$ also holds, the independence system $(G, \mathcal{L})$ is a matroid.

Definition 2. *Given a pair $\mathcal{M} = (G, \mathcal{L})$, where $\mathcal{L} \subseteq 2^G$. We call $\mathcal{M}$ is a matroid if the following holds:*

$(\mathcal{M}1)$: *$\emptyset \in \mathcal{L}$.*
$(\mathcal{M}2)$: *for any subset $A \in \mathcal{L}$, $B \subseteq A$ indicates $B \in \mathcal{L}$.*
$(\mathcal{M}3)$: *for any two subset $A, B \in \mathcal{L}$, $| A | > | B |$ indicates that there exists a point $v \in A \backslash B$, such that $B \cup \{v\} \in \mathcal{L}$.*

Given a subset $A \in \mathcal{L}$ and a pair of points (a, b), where $a \in A \cup \{\emptyset\}$ and $b \in G \backslash A$, we refer the pair (a, b) as a swap(a, b) if $A \backslash \{a\} \cup \{b\} \in \mathcal{L}$. It means that only some special points pair called swap can guarantee that $A \backslash \{a\} \cup \{b\} \in \mathcal{L}$ is still an independent set.

We highlight that the next lemma ensures that a swap(a, b) must exist between the optimal solution $\boldsymbol{x}^*$ and the current solution $\boldsymbol{x}^t$ in the later analysis. Consider the support set of the current solution $U(\boldsymbol{x}^t)$ as $A \in \mathcal{L}$ and $U(\boldsymbol{x}^*)$ as $B \in \mathcal{L}$. We will consider finding a special kind of swap$(y(b), b)$ of $U(\boldsymbol{x}^t)$, where $b \in U(\boldsymbol{x}^*)\backslash U(\boldsymbol{x}^t)$ and $y(b) \in U(\boldsymbol{x}^t)\backslash U(\boldsymbol{x}^*) \cup \{\emptyset\}$.

Lemma 3 [16]. *Assume two sets $A, B \in \mathcal{L}$, then we can construct a mapping y : $B\backslash A \rightarrow (A\backslash B) \cup \{\emptyset\}$, where every point $b \in B\backslash A$ satisfies $(A\backslash\{y(b)\}) \cup \{b\} \in \mathcal{L}$, and $a \in A\backslash B$ satisfies $|y^{-1}(a)| \leq 1$.*

Consider every point v in G, we give it a weight $w_v \geq 0$ and a total upper bound B. In the following, we assume that w_v is an integer, because we can always change all w_v and B proportionally without losing generality. The two constraints reduce the domain of candidate solutions, so we can only find some solutions $\boldsymbol{x} \in (k+1)^G$ such that the sum of weight w_v of all points v in $U(\boldsymbol{x})$ is less than B and $U(\boldsymbol{x})$ is an independent set. Define $w_{\boldsymbol{x}} = \sum_{v \in U(\boldsymbol{x})} w_v$. The problem can be written as

$$\max_{\boldsymbol{x} \in (k+1)^G} \{f(\boldsymbol{x}) \mid w_{\boldsymbol{x}} \leq B \text{ and } U(\boldsymbol{x}) \in \mathcal{L}\}. \tag{1}$$

In addition, in the later proof, we need to use the following lemma.

Lemma 4 [11]. *Given two fixed $P, D \in N_+$ and a sequence of numbers $\gamma_i \in R_+$, where $i \in [P]$, then we have*

$$\frac{\sum_{i=1}^{P} \gamma_i}{\min_{t \in [P]}(\sum_{i=1}^{t-1} \gamma_i + D\gamma_t)} \tag{2}$$
$$\geq 1 - (1 - \frac{1}{D})^P \geq 1 - e^{-P/D}.$$

2.3 Algorithm

Before giving the algorithm to solve problem (1), we firstly introduce a greedy algorithm for unconstrained k-submodular by [19]. We know that a k-submodular function f is pairwise monotone due to Definition 1, that is, $f_{\boldsymbol{x}}((v, i)) + f_{\boldsymbol{x}}((v, j)) \geq 0$ for any $i \neq j \in [k]$. It means that for a fixed $\boldsymbol{x} \in (k+1)^G$ and $v \in G\backslash U(\boldsymbol{x})$, there are no two positions $i \neq j \in [k]$ such that $f_{\boldsymbol{x}}((v, i)) < 0$ and $f_{\boldsymbol{x}}((v, j)) < 0$ both hold. So we can always find a position $i \in [k]$ such that $f_{\boldsymbol{x}}((v, i)) \geq 0$ for any $v \in G\backslash U(\boldsymbol{x})$. Therefore, for every current solution $\boldsymbol{x}^t$ in the Algorithm 1, we add $v \in G\backslash U(\boldsymbol{x}^t)$ with a greedy position i_j until all points $v \in G$ are added to $U(\boldsymbol{x}^t)$.

Then we give an algorithm inspired by [16] and [18] for problems (1) called MK-KM abbreviated as maximizing k-submodular function with a knapsack constraint and a matroid constraint. Let's highlight some important nodes. Firstly, we select three elements with the largest marginal return from the optimal solution $\boldsymbol{x}^*$ by enumerating. Second, for every current solution $\boldsymbol{x}^t \in \mathcal{L}$ and the optimal solution $\boldsymbol{x}^* \in \mathcal{L}$, we can always find a swap$(y(b), b)$ satisfying $y(b) \in \boldsymbol{x}^t\backslash\boldsymbol{x}^*$ and $b \in \boldsymbol{x}^*\backslash\boldsymbol{x}^t$ by Lemma 3. But we always choose a swap(a, b) with the highest marginal profit density $\rho(a, b)$. In the

---

Algorithm 1. Greedy Algorithm (f, G)

Require: A function $f : (k + 1)^G \rightarrow R_+$ and a set $G = [n]$
Ensure: A k-disjoint set $\boldsymbol{x} \in (k + 1)^G$
  1: $\boldsymbol{x} \leftarrow (\emptyset, \ldots, \emptyset)$
  2: **for** $j = 1$ to n **do**
  3:     $i_j \leftarrow \arg\max_{i \in [k]} f_{\boldsymbol{x}}((v, i))$
  4:     $\boldsymbol{x} \leftarrow \boldsymbol{x} \sqcup (v, i_j)$
  5: **end for**
  6: **return** $\boldsymbol{x}$

---

line 9 of MK-KM, we reorder the $U(\boldsymbol{x}^t)$ after the operation of swap(a, b) and ensure $\boldsymbol{x}^0 \preceq \boldsymbol{x}^t$. Considering the order of each element in $(U(\boldsymbol{x}^{t-1} \setminus \boldsymbol{x}^0) \setminus \{a\}) \cup \{b\})$ as it is added to current solution in MK-KM, we add them to Greddy Algorithm in the same order. Last but not least, only when $\boldsymbol{x}^t$ is updated, S will be regenerated in line 5. Otherwise, MK-KM will continue to pick and remove the next swap in the loop from 6 to 13. So MK-KM will break the loop when $S = \emptyset$ in line 6.

Algorithm 2. MK-KM (G, B, M)

Require: A function $f : (k + 1)^G \rightarrow R_+$, a budget $B \in R_+$ and a matroid $(G, \mathcal{L})$
Ensure: A k-disjoint set $\boldsymbol{x} \in (k + 1)^G$ satisfying $w_{\boldsymbol{x}} \leq B$ and $U(\boldsymbol{x}) \in \mathcal{L}$
  1: Let $\boldsymbol{x}^\alpha \in \arg\max\limits_{|U(\boldsymbol{x})|=1, \boldsymbol{x} \preceq \boldsymbol{x}^*} f(\boldsymbol{x})$, $\boldsymbol{x}^\beta \in \arg\max\limits_{|U(\boldsymbol{x})|=2, \boldsymbol{x}^\alpha \preceq \boldsymbol{x} \preceq \boldsymbol{x}^*} f(\boldsymbol{x})$
     $\boldsymbol{x}^\gamma \in \arg\max\limits_{|U(\boldsymbol{x})|=3, \boldsymbol{x}^\beta \preceq \boldsymbol{x} \preceq \boldsymbol{x}^*} f(\boldsymbol{x})$ and $t = 0$
  2: $\boldsymbol{x}^t \leftarrow \boldsymbol{x}^\gamma$ and $switch = false$
  3: **while** $switch = false$ **do**
  4:     $switch = true$
  5:     Generate a collection of all swaps $S = S(U(\boldsymbol{x}^t \backslash \boldsymbol{x}^0))$
  6:     **while** $switch = true$ and $S \neq \emptyset$ **do**
  7:         Pick a swap (a, b) from S with a maximum value of $\rho(a, b) =$
             $\max_{j \in [k]} \frac{f((\boldsymbol{x}^t \backslash (a,i)) \sqcup (b,j)) - f(\boldsymbol{x}^t)}{w_b}$
  8:         **if** $\rho(a, b) > 0$ and $w_{\boldsymbol{x}} - w_a + c_b \leq B$ **then**
  9:             $\widetilde{\boldsymbol{x}}^t \leftarrow$ **Greedy Algorithm** for $f(\widetilde{\boldsymbol{x}}^t \sqcup \boldsymbol{x}^0)$ over $(U(\boldsymbol{x}^t \setminus \boldsymbol{x}^0) \setminus \{a\}) \cup \{b\}$
 10:             $\boldsymbol{x}^{t+1} = \widetilde{\boldsymbol{x}}^t \sqcup \boldsymbol{x}^0$
 11:             $w_{\boldsymbol{x}^{t+1}} = w_{\boldsymbol{x}^t} - w_a + w_b$
 12:             $switch = false$
 13:         **end if**
 14:         $S = S \setminus \{(a, b)\}$
 15:     **end while**
 16: **end while**
 17: **return** $\boldsymbol{x}$

We modify MK-KM and give MK-KM' algorithm for problem (1) with monotonicity. MK-KM' selects two elements with the largest marginal return from the optimal solution x^* by enumerating. This modification reduces the running time.

Algorithm 3. MK-KM' (G, B, M)

Require: A function $f : (k+1)^G \to R_+$, a budget $B \in R_+$ and a matroid $(G, \mathcal{L})$
Ensure: A k-disjoint set $x \in (k+1)^G$ satisfying $w_x \le B$ and $U(x) \in \mathcal{L}$
1: Let $x^\alpha \in \arg \max\limits_{|U(x)|=1, x \preceq x^*} f(x)$, $x^\beta \in \arg \max\limits_{|U(x)|=2, x^\alpha \preceq x \preceq x^*} f(x)$, and $t = 0$
2: $x^t \leftarrow x^\beta$ and $switch = false$
3: **while** $switch = false$ **do**
4:     $switch = true$
5:     Generate a collection of all swaps $S = S(U(x^t \backslash x^0))$
6:     **while** $switch = true$ and $S \ne \emptyset$ **do**
7:         Pick a swap (a, b) from S with a maximum value of $\rho(a, b) = \max_{j \in [k]} \frac{f((x^t \backslash (a,i)) \sqcup (b,j)) - f(x^t)}{w_b}$
8:         **if** $\rho(a, b) > 0$ and $w_x - w_a + c_b \le B$ **then**
9:             $\tilde{x}^t \leftarrow$ **Greedy Algorithm** for $f(\tilde{x}^t \sqcup x^0)$ over $(U(x^t \backslash x^0) \backslash \{a\}) \cup \{b\}$
10:            $x^{t+1} = \tilde{x}^t \sqcup x^0$
11:            $w_{x^{t+1}} = w_{x^t} - w_a + w_b$
12:            $switch = false$
13:        **end if**
14:        $S = S \backslash \{(a, b)\}$
15:    **end while**
16: **end while**
17: **return** x

In order to pave the way for analysis of Sect. 4, we consider the process of the current solution x^t generated by $x^0 \sqcup \tilde{x}^t$. We carefully define $\tilde{x}_j^t$ as the current solution of each iteration of the greedy algorithm of the 9th line, where $j \in \{1, \ldots, |U(x^t) - 2|\}$ for every fixed t. Define $(v_j, i_j) = \tilde{x}_j^t \backslash \tilde{x}_{j-1}^t$ in Greedy Algorithm.

For the convenience of writing, we define $x_j^t = \tilde{x}_j^t \sqcup x^0$. Then immediately $(v_j, i_j) = (x_j^t \backslash x_{j-1}^t) = ((\tilde{x}_j^t \sqcup x^0) \backslash (\tilde{x}_{j-1}^t \sqcup x^0))$ holds. For each fixed iteration step t, there are a string of iteration steps $j \in \{1, \ldots, |U(x^t) - 2|\}$ for the nested greedy algorithm.

3   Analysis for Non-monotone k-submodular Maximization with a Knapsack Constraint and a Matroid Constraint

In this section, we will draw support from the nested greedy algorithm to solve problem (1). For nonnegative, non-monotone and unconstrained k-submodular, we need the following conclusions. Lemma 5 comes from Proposition 2.1 in [8]. If there exists a solution achieving the optimal value, we can construct an optimal solution containing all points of ground set. Therefore, for unconstrained k-submodular maximization, we only analyze the optimal solution which is the partition of ground set of Algorithm 1. And Lemma 6 ensures that we can obtain a 1/3-approximate greedy solution in the nested greedy Algorithm 1 by using $(U(x^t \setminus x^0) \setminus \{a\}) \cup \{b\}$ as ground set G, where $OPT_f(G)$ is the optimal value of unconstrained k-submodular f maximization over G.

Lemma 5 [8]. *For maximizing a non-monotone k-submodular f over a set G, there exists a partition of G achieving the optimal value.*

Lemma 6 [19]. *For maximizing a non-monotone k-submodular f over a set G, by greedy algorithm, we can get a solution x such that $U(x) = G$ and $3f(x) \geq OPT_f(G)$.*

Drawing support from the nested greedy algorithm, we reorder each iterative solution of MK-KM and analyze the approximate ratio in two cases.

Theorem 1. *Applying MK-KM algorithm to problem (1), we can obtain a $\frac{1}{6}(1 - e^{-2})$-approximate ratio.*

Proof. Using Lemma 3 between the iterative solution x^t of MK-KM and the optimal solution x^*, there exists swap $(y(b), b)$ satisfying $y(b) \in (U(x^t) \setminus U(x^*)) \cup \{\emptyset\}$ and $b \in U(x^*) \setminus U(x^t)$.

For any iteration step t, we construct a solution $\hat{x}^t$. Considering all $(b, i) \preceq x^* \setminus x^t$, we add them to x^t and get $\hat{x}^t$. Note that $x^0 \preceq x^t \preceq \hat{x}^t$ and $U(\hat{x}^t) = U(x^*) \cup U(x^t)$.

Due to Lemma 5, there exists an optimal solution containing all points in ground set G. And by Lemma 2, we know that $f(x \sqcup x^0)$ is a k-submodular over $U(\hat{x}^t) \setminus U(x^0)$. So we define that $OPT_{f(x \sqcup x^0)}(U(\hat{x}^t) \setminus U(x^0))$ is the optimal value of $f(x \sqcup x^0)$ over $U(\hat{x}^t \setminus x^0)$. Using Lemma 6 for each x^t in MK-KM, we always have

$$
\begin{aligned}
OPT_{f(x \sqcup x^0)}&(U(\hat{x}^t) \setminus U(x^0)) \\
&\leq 3f(\hat{x}^t) \\
&\leq 3f(x^t) + 3 \sum_{(b,i) \preceq \hat{x}^t \setminus x^t} [f(x^t \sqcup (b, i)) - f(x^t)] \\
&\leq 3f(x^t) + 3 \sum_{(b,i) \preceq \hat{x}^t \setminus x^t} [f((x^t \setminus (y(b), j)) \sqcup (b, i)) - f((x^t \setminus (y(b), j)))].
\end{aligned} \tag{3}
$$

The first inequality is due to Lemma 6. And the second is due to Lemma 1. By orthant submodularity, we get the third inequality. Recall that MK-KM breaks all loops when $S = \emptyset$ in line 6. It implies that we cannot find a qualified swap(a, b) to update the output solution x. We only consider swaps$(y(b), b)$ in $S(U(x \setminus x^0))$ related to $b \in U(x^*) \setminus U(x)$ instead of all candidate swaps(a, b). Now we use this construction method to analyze the algorithm in two cases.

Case 1: Consider a very special case that every swap$(y(b), b)$ was rejected just due to $\rho(y(b), b) \leq 0$ instead of knapsack constraint.

Applying formula (3) for the output solution $\boldsymbol{x}$ and constructed solution $\hat{\boldsymbol{x}}$, we get

$$
\begin{aligned}
f(\boldsymbol{x}^*) &\\
&\le OPT_{f(\boldsymbol{x} \sqcup \boldsymbol{x}^0)}(U(\hat{\boldsymbol{x}}) \backslash U(\boldsymbol{x}^0)) \\
&\le 3f(\boldsymbol{x}) + 3 \sum_{(b,i) \preceq \hat{\boldsymbol{x}} \backslash \boldsymbol{x}} [f((\boldsymbol{x} \backslash (y(b), j)) \sqcup (b, i)) - f((\boldsymbol{x} \backslash (y(b), j))].
\end{aligned}
\tag{4}
$$

Since $\rho(y(b), b) \le 0$, we have

$$
f((\boldsymbol{x} \backslash (y(b), j)) \sqcup (b, i)) \le f(\boldsymbol{x})
\tag{5}
$$

for all $(b, i) \preceq \hat{\boldsymbol{x}} \backslash \boldsymbol{x}$. We define $\{(y(b), j)\}_{b \in U(\hat{\boldsymbol{x}}^t \backslash \boldsymbol{x}^t) \backslash \{\emptyset\}} = \{(y_1, j_1), \dots, (y_K, j_K)\}$, then we get

$$
\begin{aligned}
&\sum_{(b,i) \preceq \hat{\boldsymbol{x}} \backslash \boldsymbol{x}} [f(\boldsymbol{x}) - f((\boldsymbol{x} \backslash (y(b), j))] \\
&\le \sum_{l=1}^{K} [f((\boldsymbol{x} \backslash ((y_1, j_1) \sqcup \cdots \sqcup (y_K, j_K))) \sqcup ((y_1, j_1) \sqcup \cdots \sqcup (y_l, j_l))) \\
&\quad - f((\boldsymbol{x} \backslash ((y_1, j_1) \sqcup \cdots \sqcup (y_K, j_K))) \sqcup ((y_1, j_1) \sqcup \cdots \sqcup (y_{l-1}, j_{l-1})))] \\
&= f(\boldsymbol{x}) - f(\boldsymbol{x} \backslash ((y_1, j_1) \sqcup \cdots \sqcup (y_K, j_K))) \\
&\le f(\boldsymbol{x}).
\end{aligned}
\tag{6}
$$

The first inequality is due to orthant submodularity. Because f is nonnegative, the second inequality holds. So we can get

$$
f(\boldsymbol{x}^*) \le 6f(\boldsymbol{x}).
\tag{7}
$$

Therefore, we find a $1/6$-approximate solution in Case 1.

Case 2: Consider the opposite of Case 1 that there exists at least one swap$(y(b), b)$ satisfying $w_{\boldsymbol{x}} - w_{y(b)} + w_b > B$.

Assume a special iteration step t^*. For the first time, there appears a swap $(y(b_*), b_*)$ in $S(U(\boldsymbol{x}^{t^*} \backslash \boldsymbol{x}^0))$ such that $w_{\boldsymbol{x}^{t^*}} - w_{y(b_*)} + w_{b_*} > B$, where $b_* \in U(\boldsymbol{x}^*) \backslash U(\boldsymbol{x}^{t^*})$ and $y(b_*) \in (U(\boldsymbol{x}^{t^*}) \backslash U(\boldsymbol{x}^*)) \cup \{\emptyset\}$.

Although this swap$(y(b_*), b_*)$ violates the knapsack constraint, we use it to construct a solution $(\boldsymbol{x}^{t^*} \backslash (y(b_*), j_{y(b_*)})) \sqcup (b_*, i_{b_*})$. By orthant submodularity, pairwise monotonicity and the greedy choice of $\boldsymbol{x}^\alpha$, $\boldsymbol{x}^\beta$ and $\boldsymbol{x}^\gamma$, we have

$$
f((\boldsymbol{x}^{t^*} \backslash (y(b_*), j_{y(b_*)})) \sqcup (b_*, i_{b_*})) - f(\boldsymbol{x}^{t^*}) \le \frac{2}{3} f(\boldsymbol{x}^0).
\tag{8}
$$

The detailed process of proof is shown in the Appendix. By Lemma 2, we know that $g(\boldsymbol{x}) = f(\boldsymbol{x}) - f(\boldsymbol{x}^0)$ is a k-submodular function. Then applying formula (3) for the current solution $\boldsymbol{x}^t$ and constructed solution $\hat{\boldsymbol{x}}^t$, we can get

$$
g(\boldsymbol{x}^*) \le 6[g(\boldsymbol{x}^t) + \frac{(B - w_{\boldsymbol{x}^0})}{2} \rho_{t+1}].
\tag{9}
$$

for all $t \in \{1, \ldots, t^*\}$. The detailed process of proof is shown in the Appendix. We introduce a construction method inspired by K. K. Sarpatwar [16]. Its details are still in the Appendix. Due to the construction method, we can get

$$\frac{g((\boldsymbol{x}^{t^*} \setminus (y(b_*), j_{y(b_*)})) \sqcup (b_*, i_{b_*}))}{g(\boldsymbol{x}^*)} \geq \frac{1}{6}(1 - e^{-2}). \tag{10}$$

Then, combing (8) and (10), we have

$$
\begin{aligned}
&f(\boldsymbol{x}^{t^*}) \\
&= f(\boldsymbol{x}^0) + g(\boldsymbol{x}^{t^*}) \\
&= f(\boldsymbol{x}^0) + g((\boldsymbol{x}^{t^*} \setminus (y(b_*), j_{y(b_*)})) \sqcup (b_*, i_{b_*})) \\
&\quad - [g((\boldsymbol{x}^{t^*} \setminus (y(b_*), j_{y(b_*)})) \sqcup (b_*, i_{b_*}))) - g(\boldsymbol{x}^{t^*})] \\
&= f(\boldsymbol{x}^0) + g((\boldsymbol{x}^{t^*} \setminus (y(b_*), j_{y(b_*)})) \sqcup (b_*, i_{b_*})) \\
&\quad - [f((\boldsymbol{x}^{t^*} \setminus (y(b_*), j_{y(b_*)})) \sqcup (b_*, i_{b_*}))) - f(\boldsymbol{x}^{t^*})] \\
&\geq f(\boldsymbol{x}^0) + \frac{1}{6}(1 - e^{-2})g(\boldsymbol{x}^*) - \frac{2}{3}f(\boldsymbol{x}^0) \\
&\geq \frac{1}{6}(1 - e^{-2})f(\boldsymbol{x}^*).
\end{aligned}
\tag{11}
$$

Therefore, we have a $\frac{1}{6}(1 - e^{-2})$-approximate solution $\boldsymbol{x}^{t^*}$ for MK-KM.

4 Analysis for Monotone k-Submodular Maximization with a Knapsack and a Matroid Constraint

A function f is said to be monotone, if $f(\boldsymbol{x}) \leq f(\boldsymbol{y})$ for any $\boldsymbol{x} \preceq \boldsymbol{y}$. It is easy to see that f must be pairwise monotone if f is monotone. Therefore, a monotone function $f : (k+1)^G \to R$ is k-submodular if and only it is orthant submodular. In this section, we introduce a special construction method inspired by Lan N. Nguyen [12], and obtain a better approximate ratio by MK-KM' algorithm.

For a fixed iteration t, recall that $(v_j, i_j) = \boldsymbol{x}_j^t \backslash \boldsymbol{x}_{j-1}^t$. Define $(v_j, i_*) \preceq \boldsymbol{x}^*$. We construct two sequences $\{o_{j-1/2}\}$ and $\{o_j\}$ such that $o_{j-1/2} = (\boldsymbol{x}^* \sqcup \boldsymbol{x}_j^t) \sqcup \boldsymbol{x}_{j-1}^t$ and $o_j = (\boldsymbol{x}^* \sqcup \boldsymbol{x}_j^t) \sqcup \boldsymbol{x}_j^t$, where $j \in \{1, \ldots, |U(\boldsymbol{x}^t)| - 2\}$ and $o_{j=0} = \boldsymbol{x}^*$.

Note that $\boldsymbol{x}_{j-1}^t \preceq \boldsymbol{x}_j^t \preceq o_j$ and $o_{j-1/2} \preceq o_j$. By Lemma 2, we know that $g(\boldsymbol{x}) = f(\boldsymbol{x}) - f(\boldsymbol{x}^0)$ is a monotone k-submodular function. Then for any $j \in \{1, \ldots, |U(\boldsymbol{x}^t)| - 2\}$, we have

$$g(o_{j-1}) - g(o_j) \leq g(o_{j-1}) - g(o_{j-1/2}) \leq g(\boldsymbol{x}_j^t) - g(\boldsymbol{x}_{j-1}^t). \tag{12}$$

The first inequality is due to monotonicity and $o_{j-1/2} \preceq o_j$. When $v_j \notin U(\boldsymbol{x}^*)$ or $v_j \in U(\boldsymbol{x}^*)$ with $i_j = i_*$, we have $g(o_{j-1}) - g(o_{j-1/2}) \leq 0$ by monotonicity. When $v_j \in U(\boldsymbol{x}^*)$ and $i_j \neq i_*$, we have $g(o_{j-1}) - g(o_{j-1/2}) \geq 0$. Using orthant submodularity, we get the following inequality.

$$g(o_{j-1}) - g(o_{j-1/2}) \leq g(\boldsymbol{x}_{j-1}^t \sqcup (v_j, i_*)) - g(\boldsymbol{x}_{j-1}^t) \tag{13}$$

Then by greedy choice, the inequality (12) holds.

Theorem 2. *According to MK-KM' algorithm, a $\frac{1}{3}(1 - e^{-3})$-approximate solution of problem (1) can be obtained, if f is monotone.*

Proof. Similarly to Theorem 1, we analyze the algorithm in two cases. When we get the output solution x, there is not any qualified swap (a, b) to update x. We only consider swaps$(y(b), b)$ in $S(U(x \backslash x^0))$ related to $b \in U(x^*) \backslash U(x)$ instead of all candidate swaps(a, b).

Case 1: Consider a very special case that every swap$(y(b), b)$ was rejected just due to $\rho(y(b), b) \le 0$ instead of knapsack constraint.

For the optimal solution x^* and the output solution x, we construct two sequences $\{o_{j-1/2}\}$ and $\{o_j\}$, where $j \in \{1, \ldots, |U(x)| - 2\}$. Sum (12) for j from 1 to $(|U(x)| - 2)$, we have

$$
g(x^*) - g(o_{|U(x)|-2}) = \sum_{j=1}^{|U(x)|-2} [g(o_{j-1}) - g(o_j)]
$$

$$
\le \sum_{j=1}^{|U(x)|-2} [g(x_j) - g(x_{j-1})] \tag{14}
$$

$$
= g(x).
$$

Using Lemma 1, orthant submodularity and $\rho(y(b), b) \le 0$, we get

$$
g(x^*) \le g(o_{|U(x)|-2}) + g(x)
$$

$$
\le g(x) + \sum_{(b,i) \preceq (o_{|U(x)|-2} \backslash x)} [g(x \sqcup (b, i)) - g(x)] + g(x)
$$

$$
\le 2g(x) + \sum_{(b,i) \preceq (o_{|U(x)|-2} \backslash x)} [g((x \backslash (y(b), j)) \sqcup (b, i)) - g(x \backslash (y(b), j))] \tag{15}
$$

$$
\le 2g(x) + \sum_{(b,i) \preceq (o_{|U(x)|-2} \backslash x)} [g(x) - g(x \backslash (y(b), j))].
$$

Let $\{(y(b), j)\}_{b \in U(o_{|U(x)|} \backslash x) \backslash \{\emptyset\}} = \{(y_1, j_1), \ldots, (y_K, j_K)\}$, then we have

$$
g(x^*) \le 2g(x) + \sum_{l=1}^{K} [g((y_1, j_1) \sqcup \cdots \sqcup (y_l, j_l)) - g((y_1, j_1) \sqcup \cdots \sqcup (y_{l-1}, j_{l-1}))]
$$

$$
\le 2g(x) + \sum_{l=1}^{K} g((y_1, j_1) \sqcup \cdots \sqcup (y_K, j_K))
$$

$$
\le 3g(x).
$$

$$\tag{16}$$

Therefore,

$$
f(x^*) \le 3f(x) - 2f(x^0) \le 3f(x). \tag{17}
$$

We obtain 1/3-approximate ratio in case 1.

Case 2: Consider the opposite of case 1 that there exists at least one swap$(y(b), b)$ satisfying $w_x - w_{y(b)} + w_b > B$.

For the first time, there appears a swap $(y(b_*), b_*)$ in $S(U(\boldsymbol{x}^{t^*}\backslash\boldsymbol{x}^0))$ such that $w_{\boldsymbol{x}^{t^*}} - w_{y(b_*)} + w_{b_*} > B$, where $b_* \in U(\boldsymbol{x}^*)\backslash U(\boldsymbol{x}^{t^*})$ and $y(b_*) \in (U(\boldsymbol{x}^{t^*})\backslash U(\boldsymbol{x}^*)) \cup \{\emptyset\}$. For each $t \in \{1, \ldots, t^*\}$, we construct two sequences $\{o_{j-1/2}\}$ and $\{o_j\}$ between $\boldsymbol{x}^t$ and $\boldsymbol{x}^*$, where $j \in \{1, \ldots, |U(\boldsymbol{x}^t)| - 2\}$. Summing (13) for j from 1 to $|U(\boldsymbol{x}^t)| - 2$ and using Lemma 1, we have

$$
\begin{aligned}
g(\boldsymbol{x}^*) &\leq g(o_{|U(\boldsymbol{x}^t)|-2}) + g(\boldsymbol{x}^t) \\
&\leq g(\boldsymbol{x}^t) + \sum_{(b,i)\preceq(o_{|U(\boldsymbol{x}^t)|-2}\backslash\boldsymbol{x}^t)} [g(\boldsymbol{x}^t \sqcup (b,i)) - g(\boldsymbol{x}^t)] + g(\boldsymbol{x}^t).
\end{aligned}
\tag{18}
$$

Then applying (18) and the similar technique of (3) and (6), we can get

$$
g(\boldsymbol{x}^*) \leq 3g(\boldsymbol{x}^t) + (B - w_{\boldsymbol{x}^0})\rho_{t+1},
\tag{19}
$$

for all $t \in \{1, \ldots, t^*\}$. The detailed process of proof is shown in the Appendix. Similar to the proof of (10), using (19), we can get

$$
\frac{g((\boldsymbol{x}^{t^*} \backslash (y(b_*), j_{y(b_*)})) \sqcup (b_*, i_{b_*}))}{g(\boldsymbol{x}^*)} \geq \frac{1}{3}(1 - e^{-3}).
\tag{20}
$$

We modify inequality (8) as follows. By orthant submodularity, monotonicity and the greedy choice of $\boldsymbol{x}^\alpha$, $\boldsymbol{x}^\beta$, we have

$$
f((\boldsymbol{x}^{t^*} \backslash (y(b_*), j_{y(b_*)})) \sqcup (b_*, i_{b_*})) - f(\boldsymbol{x}^{t^*}) \leq \frac{f(\boldsymbol{x}^0)}{2}.
\tag{21}
$$

The detailed process of proof is shown in the Appendix. Combing (20) and (21), we have

$$
\begin{aligned}
f(\boldsymbol{x}^{t^*}) &= f(\boldsymbol{x}^0) + g((\boldsymbol{x}^{t^*} \backslash (y(b_*), j_{y(b_*)})) \sqcup (b_*, i_{b_*})) \\
&\quad - [f((\boldsymbol{x}^{t^*} \backslash (y(b_*), j_{y(b_*)})) \sqcup (b_*, i_{b_*})) - f(\boldsymbol{x}^{t^*})] \\
&\geq f(\boldsymbol{x}^0) + \frac{1}{3}(1 - e^{-3})g(\boldsymbol{x}^*) - \frac{f(\boldsymbol{x}^0)}{2} \\
&\geq \frac{1}{3}(1 - e^{-3})f(\boldsymbol{x}^*).
\end{aligned}
\tag{22}
$$

Hence, MK-KM' has an approximation ratio of at least $\frac{1}{3}(1 - e^{-3})$.

5 Discussion

To summarize this paper, inspired by [16] and [18], we propose a nested algorithm applicable to monotone and non-monotone k-submodular maximization with the intersection of a knapsack and a matroid constraint. For problem (1), we have a $\frac{1}{6}(1 - e^{-2})$-approximate ratio. Inspired by [12], we use a new construction method between optimal solution and current solution. For monotone k-submodular maximization with a knapsack and a matroid constraint, we achieve at least $\frac{1}{3}(1 - e^{-3})$ approximation ratio.

References

1. Bian, A.A., Buhmann, J.M., Krause, A., Tschiatschek, S.: Guarantees for greedy maximization of non-submodular functions with applications. In: Proceedings of the 34th International Conference on Machine Learning (ICML), Sydney, NSW, Australia, 2017, pp. 498–507 (2017)
2. Calinescu, G., Chekuri, C., Pál, M., Vondrák, J.: Maximizing a monotone submodular function subject to a matroid constraint. SIAM J. Comput. **40**(6), 1740–1766 (2011)
3. Ene, A., Nguyễn, H.L.: A nearly-linear time algorithm for submodular maximization with a knapsack constraint. In: Proceedings of the 46th International Colloquium on Automata, Languages and Programming (ICALP), Patras, Greece, 2019, pp. 53:1–53:12 (2019)
4. Feldman, M.: Maximization problems with submodular objective functions, Ph.D. dissertation, Computer Science Department, Technion, Haifa, Israel (2013)
5. Filmus, Y., Ward, J.: Monotone submodular maximization over a matroid via non-oblivious local search. SIAM J. Comput. **43**(2), 514–542 (2014)
6. Huang, C., Kakimura, N., Mauras, S., Yoshida, Y.: Approximability of monotone submodular function maximization under cardinality and matroid constraints in the streaming. SIAM J. Discrete Math. **36**, 355–382 (2022)
7. Huber, A., Kolmogorov, V.: Towards mininizing k-submodular functions. In: Proceedings of 2nd International Symposium on Combinatorial Optimization, pp. 451–462 (2012)
8. Iwata, S., Tanigawa, S.-I., Yoshida, Y.: Improved approximation algorithms for k-submodular function maximization. In: Proceedings of the Twenty-Seventh Annual ACM-SIAM Symposium on Discrete Algorithms (SODA), Arlington, VA, USA, 2016, pp. 404–413 (2016)
9. Liu, Q., Yu, K., Li, M., Zhou, Y.: k-Submodular Maximization with a Knapsack Constraint and p Matroid Constraints (submitted)
10. Liu, Z., Guo, L., Du, D., Xu, D., Zhang, X.: Maximization problems of balancing submodular relevance and supermodular diversity. J. Global Optim. **82**(1), 179–194 (2021). https://doi.org/10.1007/s10898-021-01063-6
11. Nemhauser, G.L., Wolsey, L.A., Fisher, M.L.: An analysis of approximations for maximizing submodular set functions-I. Math. Program. **14**(1), 265–294 (1978)
12. Nguyen, L.N., Thai, M.T.: Streaming k-submodular maximization under noise subject to size constraint. In: Proceedings of the 37th International Conference on Machine Learning (ICML), 2020, pp. 7338–7347 (2020)
13. Ohsaka, N., Yoshida, Y.: Monotone k-submodular function maximization with size constraints. Adv. Neural. Inf. Process. Syst. **28**, 694–702 (2015)
14. Oshima, H.: Improved randomized algorithm for k-submodular function maximization. SIAM J. Discret. Math. **35**(1), 1–22 (2021)
15. Sakaue, S.: On maximizing a monotone k-submodular function subject to a matroid constraint. Discret. Optim. **23**, 105–113 (2017)
16. Sarpatwar, K.K., Schieber, B., Shachnai, H.: Constrained submodular maximization via greedy local search. Oper. Res. Lett. **47**(1), 1–6 (2019)
17. Sviridenko, M.: A note on maximizing a submodular set function subject to a knapsack constraint. Oper. Res. Lett. **32**(1), 41–43 (2004)
18. Tang, Z., Wang, C., Chan, H.: On maximizing a monotone k-submodular function under a knapsack constraint. Oper. Res. Lett. **50**(1), 28–31 (2022)
19. Ward, J., Živný, S.: Maximizing k-submodular functions and beyond. ACM Trans. Algorithms **12**(4), 47:1–47:26 (2016)
20. Yoshida, Y.: Maximizing a monotone submodular function with a bounded curvature under a knapsack constraint. SIAM J. Discret. Math. **33**(3), 1452–1471 (2019)

Network Problems

Defense of Scapegoating Attack in Network Tomography

Xiaojia Xu, Yongcai Wang$^{(\boxtimes)}$, Yu Zhang, and Deying Li

School of Information, Renmin University of China, Beijing 100872, China
{xuxiaojia,ycw,2020104230,deyingli}@ruc.edu.cn

Abstract. Defending of scapegoating attack is a critical problem in network tomography. Theoretically, the ideal defending scheme is to add monitoring paths to make all the links in the network be identifiable. This requires very high monitoring cost. To overcome this problem, this paper proposes a diagnosis-based defending scheme for scapegoating attack. A scapegoating attack can be launched only when the link set manipulated by the attacker cuts the probing paths going through the scapegoat links and is not traversed by any monitoring path. This cut set is called unobserved cut set (UCS). To defense, we propose to find the UCS and add the minimum number of probing paths to traverse the UCS. A minimum set cover model is proposed to select the least number of defense links to cover the UCS, and a polynomial time algorithm is proposed. Evaluations on various network dataset show the effectiveness of the proposed strategies.

Keywords: Fault diagnosis method · Scapegoating attack · Network tomography

1 Introduction

Timely and accurately knowing the internal states of the network, such as the bandwidth, packet loss rate and link delay, is an important requirement in network management. Instead of directly measuring the elements within the network, *network tomography* which uses end-to-end path measurements to infer the internal states of the network [1,4], becomes a promising solution [21]. Network tomography deploys a set of monitors in the network [7]; measures only the end-to-end path performances between the monitors; and then infers the internal states of links and nodes by solving state recovering functions [8,10,14,16]. It avoids the issues such as high internal measurement overhead, high measurement cost of the direct measurement methods.

A critical problem in network tomography is the "identifiability" problem, which indicates whether the internal states of the network can be uniquely recovered by the end-to-end external path measurements. When identification is not

Supported by the National Natural Science Foundation of China Grant No. 61972404, 12071478.

guaranteed, methods like Pseudo-inverse and [1,12] are generally used to recover the link states. Recently, the risk of being attacked is noticed when the identifiable property is not satisfied. *Scapegoating attack (SA)* [20] refers to a kind of attack, that when an attacker manipulates a set of links to inject attacks (such as inject delay or discard packets), not only the network performances will be degraded, but also the network tomography will be misled to guilt a set of normal links as scapegoats of the attack. [2] considers to degrade the performance of the network by injecting delays to some path measurements and network tomography cannot localize the attackers. They introduce chosen-victim, maximum-damage and obfuscation scapegoating attacks. Recent work [19] proposes the conditions to successfully launch scapegoating attack. The presence of backdoor infected routers [3] and node-capture attacks [15] can be utilized to carry out above scapegoating attacks by affecting the packet delivery and the path measurement.

The scapegoating attack can cause persistent and inconspicuous performance degradation. The ideal defense scheme is to insure the probing paths of network tomography satisfying the "identifiability condition" [10], which requires *the rank of the routing matrix is equal to the number of links* [17]. However, the number of links can be very large, the identifiability condition is hard to be satisfied, unless the number of probing paths is not less than the number of links, which need very high measurement costs. Efficient methods to defend scapegoating attack without greatly increasing measurement cost is highly desired.

To address this problem, this paper propose a highly efficient fault diagnosis based scheme for defending scapegoating attacks. It doesn't require the probing paths to satisfy the identifiability condition. Instead, when network tomography detects problematic links, we propose to use very low probing cost to examine whether the problematic links are scapegoats or are real network problems. The key behind the scheme is that we investigate the necessary and sufficient condition to defense scapegoating attack and the minimum cost defending problem. The main contributions of this paper are as following.

1. We show the necessary condition to issue scapegoating attack is that the attacker manipulates all links in a cut set of the probing paths that pass through the scapegoat links and these cut set links are not observed by any other probing path except those passing through the scapegoats. Such link sets are called unobserved cut set (UCS).
2. We propose to find the UCS and add the minimum number of probing paths to traverse them, so that the condition of scapegoating attack is broken and the attacking links can be detected if any scapegoating attack exists.
3. A minimum set cover model is proposed for selecting the least number of defense links to cover all the UCS; a greedy approximation algorithm with H_K ratio is proposed to solve the defense link selection problem; and a polynomial time edge-disjoint path generation algorithm is proposed to generate defending paths to traverse these links-to-defend.
4. Extensive verifications on real network datasets show the effectiveness of the proposed defense strategies.

2   Problem Formulation

2.1   Network Tomography Model

We consider a network modeled as a weighted, undirected graph $G = (V, L, X)$. V and L are the sets of nodes and links, $|V| = n$ and $|L| = m$. Set X represents the link weights, where $x_i \in X$ is an unknown link metric that describes the link i's performance, such as latency and loss rate. We assume the measures of these link metrics are additive. This is a canonical model for representing important performance measures [8,10,11]. In the network, a subset of vertices used for injecting and extracting probing packets is defined as the set of monitors $M = \{m_i\}$. Monitor placement algorithms and probing path generation algorithms in network tomography can be referred to [8,10].

A probing path p_i for network tomography is defined as a sequence of links that starts from a source monitor s_i and ends at a destination monitor d_i. The end-to-end measurement of this probing path is denoted by y_i. $P = \{p_i\}$ is the set of probing paths and $Y = \{y_i\}$ denotes the end-to-end measurements of these paths. Routing matrix $R = \{r_{i,j}\}_{p_i \in P, l_j \in L}$ models how the probing paths traverse the links in L. $r_{i,j} = 1$ if a link $l_j \in L$ is on the path $p_i \in P$ and $r_{i,j} = 0$ otherwise. Network tomography is to solve the equation $R\hat{x} = Y$ to find a solution $\hat{x}$ that represents the estimated link metrics [8].

Note that this linear equation has a unique solution when the routing matrix R has full column rank, i.e., $\text{Rank}(R) = m$, which is the unique identification condition [17]. However, since $m \gg n$, it is generally difficult to generate enough probing paths to satisfy the identification condition. For such difficulty, network managers generally adopt the routing matrix with $\text{Rank}(R) < m$. In such case, Pseudo-inverse is generally used [2,20] to estimate $\hat{x}$ by:

$$\hat{x} = (R^T R)^{-1} R^T Y \tag{1}$$

Without loss of generality, we consider the link metric is additive, such as link delay. Then network tomography determines states of links by following method:

Definition 1 (Estimated Link States). *Given estimated link delay $\hat{x}$, a link maybe divide into three states:*

$$\psi(l_i) = \begin{cases} normal & if \ \hat{x}_i < \beta_{min} \\ uncertain & if \ \beta_{min} \leqslant \hat{x}_i \leqslant \beta_{max} \\ problematic & if \ \hat{x}_i > \beta_{max} \end{cases} \tag{2}$$

where β_{min} is a threshold to find normal links. β_{max} is a threshold to detect problematic, i.e., problematic links. When problematic links are detected, network tomography generally requires more diagnosis method to check and to fix the detailed problem.

2.2   Scapegoating Attack Model

Consider the attacker hacks a set of links, say $L_m \subset L$ is manipulated by the attackers. Attacker injects delays onto these links to affect the probing paths that pass through these links. The attackers also hope to hide their attacks by let network tomography to wrongly detect some other links as problematic. The wrongly detected problematic links are called "scapegoats". The measurement model under attack is:

$$R(x + \Delta x) = Y'$$ (3)

where $\Delta x_i > 0$ for $l_i \in L_m$ and $\Delta x_i = 0$ for $l_i \notin L_m$.

[2, 19, 20] show that scapegoating attack can be successfully launched when the probing paths don't satisfy identifiability condition. Let $L_s \subset L$ be the set of scapegoats. There should be $L_s \cap L_m = \emptyset$, since attacking links should not be discovered. The **scapegoating attack** is called successful launched if:

$$\begin{cases} \hat{x}_i \leq \beta_{min} \text{ for } l_i \in L_m \\ \hat{x}_i > \beta_{max} \text{ for some } l_i \in L_s \end{cases}$$ (4)

where $\hat{x} = \{\hat{x}_1, \hat{x}_2, \cdots \hat{x}_n\}$ is the solution of $\hat{x} = (R^T R)^{-1} R^T Y'$, which is the network tomography results under attack.

In Fig. 1, we using a simple network with five monitors to illustrate the scapegoating attack. The red markers are the monitors which are selected by MMP algorithm [8]. In order to identify the link metrics, ten probing paths are constructed among monitors, as listed in the right part of Fig. 1.

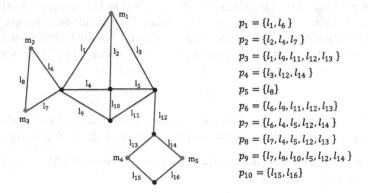

$p_1 = \{l_1, l_6\}$
$p_2 = \{l_2, l_4, l_7\}$
$p_3 = \{l_1, l_9, l_{11}, l_{12}, l_{13}\}$
$p_4 = \{l_3, l_{12}, l_{14}\}$
$p_5 = \{l_8\}$
$p_6 = \{l_6, l_9, l_{11}, l_{12}, l_{13}\}$
$p_7 = \{l_6, l_4, l_5, l_{12}, l_{14}\}$
$p_8 = \{l_7, l_4, l_5, l_{12}, l_{13}\}$
$p_9 = \{l_7, l_9, l_{10}, l_5, l_{12}, l_{14}\}$
$p_{10} = \{l_{15}, l_{16}\}$

Fig. 1. A network with five monitors selected by MMP [9] algorithm and ten probing paths.

Figure 2 shows the result of scapegoating attack. The manipulated links are l_3, l_5 and l_{11} and the scapegoating link is l_{12}. The attack misleads the network tomography to conclude that link l_{12} is a problematic link and the administrator cannot identify the attacking links by simply checking the "problematic link" l_{12}. This not only greatly degrades the network performance, but also imposes high difficulty to detect the true faults.

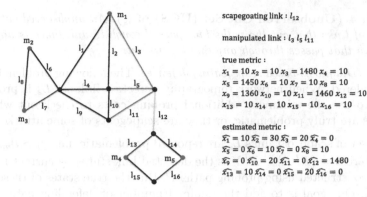

scapegoating link : l_{12}

manipulated link : $l_3\ l_5\ l_{11}$

true metric :
$x_1 = 10\ x_2 = 10\ x_3 = 1480\ x_4 = 10$
$x_5 = 1450\ x_6 = 10\ x_7 = 10\ x_8 = 10$
$x_9 = 1360\ x_{10} = 10\ x_{11} = 1460\ x_{12} = 10$
$x_{13} = 10\ x_{14} = 10\ x_{15} = 10\ x_{16} = 10$

estimated metric :
$\hat{x}_1 = 10\ \hat{x}_2 = 30\ \hat{x}_3 = 20\ \hat{x}_4 = 0$
$\hat{x}_5 = 0\ \hat{x}_6 = 10\ \hat{x}_7 = 0\ \hat{x}_8 = 10$
$\hat{x}_9 = 0\ \hat{x}_{10} = 20\ \hat{x}_{11} = 0\ \hat{x}_{12} = 1480$
$\hat{x}_{13} = 10\ \hat{x}_{14} = 0\ \hat{x}_{15} = 20\ \hat{x}_{16} = 0$

Fig. 2. The result of the scapegoating attack.

3 Defense Strategy

It is necessary to design defense strategy to avoid scapegoating attack. We firstly investigate the conditions for successful defense, and then propose an efficient, minimum cost fault diagnosis method to examine whether a detected "problematic link" is a scapegoat, so as to discover the attacking links.

3.1 What to Defend?

Note that the probing paths can be classified into three categories.

Definition 2 (Three Types of Paths). *The probing paths in P are classified as.*

- $P_m \subseteq P$ is the **manipulated path set**. Each path in it passes through at least one manipulated links.
- $P_n \subseteq P$ denotes the **normal path set**. All paths in it hasn't passed through any manipulated link.
- $P_s \subseteq P$ is the **scapegoating path set**, which contains measurement paths that pass through at least one scapegoat link.

The original delay of an attacked link is x_i and the injected delay is Δx_i. The increased delay for a path $p_i \in P_m$ is further denoted by Δt_i, where $\Delta t_i = \sum_{l_j \in p_i} \Delta x_j$. Since the injected delays only affect the manipulated path set, we can easily get the following Lemma 1.

Lemma 1. *Given P, L_s, L_m, if delays are injected onto links in L_m, there must be $\Delta t_i \geqslant 0$ for $p_i \in P_m$ and $\Delta t_i = 0$ for $p_i \in P_n$.*

Definition 3 (Cut Set of Paths). *A cut set C of a path set P is a set of links in P that for every path $p_i \in P$, the path passes through at least one link in C. In other words, when we cut all the links in C, all the paths in P will be cut.*

Definition 4 (Unobserved Cut Set (UCS) of l_s). *An unobserved cut set of l_s is a set of links that cut the paths that pass through l_s and there is no other probing path that passes through any link in this set.*

Problem 1 (Diagnosis-based scapegoating defense). The following problem is specially considered. When network tomography detects a link set L_s is problematic, how to use the minimum additional probing costs to check out whether their states are truly problematic, or they are scapegoats of some attackers.

Suppose a set of scapegoat links L_s are reported problematic, i.e., $\hat{x}_i > \beta_{max}$, for $l_i \in L_s$, where $\hat{x} = \left(R^T R\right)^{-1} R^T Y$ is the detected link states by current routing matrix R, constructed using probing paths P. But the true states of these links are normal. Our goal is to add the minimal number of defending paths P_d to construct a new routing matrix $R' = [R, R_d]^T$, where R_d is the routing matrix of P_d to recover the true states of links in L_s. By adding P_d, the recovered link states are $\hat{x}' = \left(R'^T R'\right)^{-1} R'^T Y'$. P_d is called an effective defense, if the true states of the scapegoat links are recovered. The minimal path defending problem can be stated as following:

Problem 2 (Minimal Path Defending Problem). The minimal path defending problem for scapegoating attack is to find the minimal number of additional probing paths such that:

$$
\begin{aligned}
& \min |P_d| \\
& \text{s.t.} \begin{cases} R' = [R, R_d]^T, Y' = [Y, Y_d]^T \\ \hat{x}' = \left(R'^T R'\right)^{-1} R'^T Y' \\ \hat{x}'_i \le \beta_{\max}, \forall l_i \in L_s \end{cases}
\end{aligned} \tag{5}
$$

Note that in (5), the original routing matrix R is copied, so the path constraints in original P_n and P_m are still satisfied. The key problem is how to design the defending paths.

3.2   Key Observations

For defense, although the exact locations of L_m are not known, we know that they must cover an unobserved cut set of P_s [2,19]. For a link l_i in L_s, there maybe multiple unobserved cut sets that can cut the probing paths passing through l_i. We denote the unobserved cut sets for $l_i \in L_s$ as $\{C_{i,1}, C_{i,2}, \cdots, C_{i,n_i}\}$.

Lemma 2 (Breaking an unobserved cut set for $l_i \in L_s$). *If we add a probing path p to go through any link in a cut set $C_{i,j}$ but not the link l_i, then the injected delays on the cut set $C_{i,j}$ can no longer be attributed to l_i.*

Proof. Only when all the links in a unobserved cut set of l_i are manipulated by the attacker, can the scapegoating attack to l_i be successfully launched. Therefore, if a link in the unobserved cut set $C_{i,j}$, denoted by l_k has been monitored by an added p, the added delay to this link will increase the delay of the new path p. Since p doesn't go through l_i, the increased delay cannot be attributed to l_i. We say the unobserved cut set $C_{i,j}$ is broken by the added path p.

Theorem 1 (Necessary & Sufficient Condition to Defense L_s). *Given G, P, and L_s, to discover whether each link $l_i \in L_s$ is truly problematic or not, we need to add new probing paths P_d which don't go through $l_i \in L_s$, but make each unobserved cut set of l_i have at least one link be passed by at least one path in P_d.*

Proof. Necessity: From Lemma 2, a unobserved cut set of l_i is broken if one of its link is passed through by an added path. Since every cut set may initial an attack to l_i, we need to break all cut sets of l_i to protect l_i. So l_i is protected only if all its unobserved cut sets are broken.

Sufficiency: When all the unobserved cut sets are broken, the injected delays on any cut set cannot be attributed to l_i. So try state of l_i will be recovered after adding the defending paths.

3.3 Defense Methodologies

For each link $l_i \in L_s$, there maybe many unobserved cut sets that can cut the paths going though l_i. Let $P_{l_i} \in P_s$ denote the set of paths that go through l_i. Let K denote the number of paths in P_{l_i}. Since any unobserved cut set of P_{l_i} must cover an unobserved minimal cut set (UMCS) of P_{l_i}, so in order to break all unobserved cut sets for P_{l_i}, we must break all the UMCS of P_{l_i}.

Consider there are totally K paths in the path set P_s. In order to defend all the links in L_s, all the unobserved minimal cut set of P_s need to be broken. By selecting one link from each path, an unobserved cut to the path set P_s can be formed. Using this method, all unobserved minimal cuts for P_s can be found, which is denoted by $\mathbb{C}$.

Given the unobserved minimal cut set $\mathbb{C}$, to design the minimum number of defending paths, we propose to find a link set $\mathbb{L}$ with the minimum cardinality, such that every unobserved minimal cut $C_i \in \mathbb{C}$ contains at least one link in $\mathbb{L}$. Then $\mathbb{L}$ is called a minimum cover of $\mathbb{C}$.

This problem is a typical set cover problem, which is NP-hard. A greedy algorithm is proposed to address this minimum cut set cover (MCSC) problem. Let $L_{\mathbb{C}}$ be the set of all links in $\mathbb{C}$. Let U be all uncovered cuts in $\mathbb{C}$.

In the algorithm, if a link $l \in L_{\mathbb{C}}$ appears in n unobserved cuts in U, n reveals the covering utility of the link l. In each iteration, the link with the largest covering utility is selected and is put in $\mathbb{L}$, until all unobserved cuts in $\mathbb{C}$ have been covered. gMCSC has an H_K approximation ratio [18], where K is the largest number of cut sets that share one common link in $L_{\mathbb{C}}$.

Algorithm 1: Greedy minimum set cover: $\mathbb{L}$=gMCSC($\mathbb{C}$)

   Input: $\mathbb{C}$
   Output: $\mathbb{L}$
1 Initialize $U = \mathbb{C}$, $\mathbb{L} = \emptyset$, $L_{\mathbb{C}}$= all links in $\mathbb{C}$;
2 **while** *(U is not empty)* **do**
3   select l in $L_{\mathbb{C}}$ that covers the most number of sets in U ;
4   add l to $\mathbb{L}$;
5   remove the cuts covered by l from $\mathbb{U}$ remove l from $L_{\mathbb{C}}$;
6 return $\mathbb{L}$;

Lemma 3 (Approximation ratio of gMCSC). *Let K_s be the number of unobserved cuts in $\mathbb{C}$ that have common link l. Let $K = \max_{l \in L_{\mathbb{C}}} K_s$ be the largest number of unobserved cuts that share a common link. Let $H_K = \sum_{i=1}^{K} 1/i \approx \ln K$, then the gMCSC algorithm returns $\mathbb{L}$ which has at most H_K times links than the optimal number of links to make each unobserved cut in $\mathbb{C}$ has at least one link in $\mathbb{L}$.*

3.4   Minimum Number Defending Path Generation

Since $\mathbb{L}$ covers all unobserved cuts $\mathbb{C}$, in order to break all unobserved cuts in $\mathbb{C}$, defending paths only need to be added to go through all the links in $\mathbb{L}$ and don't go through any link in L_s. We want to add the minimum number of defending paths to achieve this goal.

In network tomography, a probing path is not required to be a simple path, which can traverse the same edge more than one time. So a polynomial time algorithm is proposed to generate the minimum number of defending paths to go through all the links in $\mathbb{L}$ but no links in L_s.

The idea of the algorithm is to firstly removes the edges L_s from the graph G. Suppose the removal of L_s decomposes G into F components, denoted by $G_1, G_2, \cdots, G_F$. We show that we need to add at most F probing paths to prevent L_s from scapegoating attack.

In detail, in subgraph G_n, we select a pair of monitors and find the shortest paths P_1^* between the first monitor and one of the links l_1 in $\mathbb{L}$, P_{E+1}^* between the second monitor and another one of the links l_E in $\mathbb{L}$, and also find the shortest paths $P_2^*, \cdots, P_E^*$ between $(l_1, l_2), \cdots, (l_{E-1}, l_E)$ by using Dijkstra algorithm. The whole path $P_d^n = \{P_1^*, P_2^*, \cdots, P_E^*, P_{E+1}^*\}$ is the defending path of the subgraph G_n. Since L_s has been deleted, this path will not traverse any link in L_s. The defending paths of the F components cover all links in $\mathbb{L}$. So at most F defending paths in the F components need to be generated to cover $\mathbb{L}$ and don't traverse any link in L_S.

The detailed algorithm is given in Algorithm 2. Line 3 to line 8 generate the shortest defending path between two monitors. Line 9 to line 10 select the overall shortest defending path.

Algorithm 2: Defending path generation algorithm: $P_d = PathGen(G, M, L_s, \mathbb{L})$

Input: $G, M, L_s, \mathbb{L}$
Output: P_d
1   $G = G \setminus L_s$ and suppose it has F components ;
2   **for** $(n = 1; n \leq F; n++)$ **do**
3   select a pair of monitors m_j, m_k, $\mathbb{L} \neq \emptyset$ and suppose it has E components;
4   Find the shortest path P_1^* between (m_j, l_1);
5   **for** $(i = 1; i \leq E - 1; i++)$ **do**
6   Find the shortest path P_{i+1}^* between (l_i, l_{i+1})
7   Find the shortest path P_{E+1}^* between (l_E, m_k) ;
8      $P^* = P_1^* \cup P_2^* \cup \cdots \cup P_{E+1}^*$;
9   **if** $(P^*$ is shorter than $P_d^n)$ or $(P_d^n$ is empty) **then**
10          $P_d^n = P^*$;

11   **return** P_d ;

The obtained paths pass through all links in $\mathbb{L}$ but no link in L_s, so they satisfy the requirement to recover the true states of L_s. So at most F defending paths are generated to traverse all links in $\mathbb{L}$ but no links in L_s. The most time consuming step in Algorithm 2 is the Dijkstra algorithm, whose complexity is $O(E * n_n^2)$ where n_n is the number of nodes in G_n. Since $n_n < n$, so the complexity of Algorithm 2 is $O\left(F * E * n^2\right)$ in the worst case, where n is the number of nodes in G.

4 Performance Evaluation

4.1 Experiment Setup

We use real network topologies from the Internet Topology Zoo [6] and synthetic ER network topology, whose parameters are shown in the Table 1. Topology Zoo are real ISP network topologies which are widely used in network tomography. The ER graph [5] is a simple random graph generated by independently connecting each pair of nodes by a link with a fixed probability p. In our simulations, this probability is set to be 0.014 (Table 1). For each topology, the MMP [8]

Table 1. Parameters of topologies

Network	$L(\mathcal{G})$	$V(\mathcal{G})$	Monitors	Paths
AttMpls	56	25	5	100
Surfnet	68	50	32	845
TataNld	186	145	89	6217
ER	1215	500	61	3606

algorithm is used to select candidate monitors. After selecting the monitors, the probing paths are generated by multiple shortest paths based on Dijkstra's algorithm and Yen's algorithm [13].

4.2 Results of Defense Strategy

Figure 3 shows the average ratio of identifiable links with the increase of paths in different topologies. The evaluations on both synthetic topology and real network topologies show that when the number of paths is small, the increase of paths has a great influence on the number of identifiable links. But when the number of paths is large to a certain extent, blindly looking for new paths cannot increase the number of identifiable links. The application of our defending path generation algorithm can effectively solve this problem.

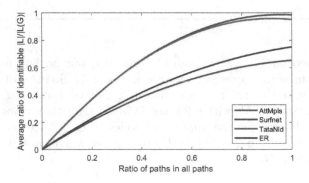

Fig. 3. Average ratio of identifiable links with the increase of paths in different topologies.

Figure 4 shows how to generate the defending paths in the AttMpls network. The first sub-figure is the network tomography of AttMpls network. The scapegoat link is link 54. There are 10 probing paths, i.e., P in the network, which are shown in different colors. Some paths are overlapped so only the top color can be seen.

To defense, we only consider the paths passing through L_s, i.e., P_s. We need to insure the minimum cover of the $\mathbb{C}$ in P_s being traversed by a defending path. Firstly, we remove the edges L_s from the graph G. We suppose the possible scapegoat, i.e., L_s is the link 54. Based on Algorithm 1, the minimum set cover of the cut set is links $\{20, 25, 29\}$. Figure 4(b) is the $\mathbb{L}$ of the given topology, each unobserved cut in $\mathbb{C}$ has at least one link in $\mathbb{L}$. According to Algorithm 1, the green highlighted links in this figure is the $\mathbb{L}$ of the given network tomography. Based on Algorithm 2, the black highlighted links in Fig. 4(c) are the shortest paths P^* between monitors and links. By using Algorithm 2, we can generate the minimum defending path in Fig. 4(d) to verify whether link 54 is a scapegoat.

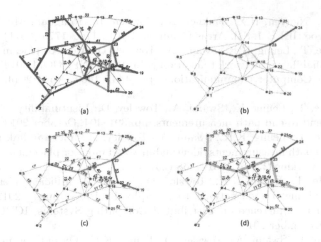

Fig. 4. An example of defending path generation

5 Conclusion

Consider the high cost of making all the links in the network be identifiable, this paper proposes a diagnosis-based defending scheme for scapegoating attack. A minimum set cover model is proposed to select the least number of defense links to cover the unobserved cut set, and a polynomial time algorithm is proposed to generate the least number of probing paths to go through the selected defense links. Theoretical analysis and simulations in the real network topology show the effectiveness of the proposed defense strategies.

In future work, how to optimize the defense strategy proposed in this paper to diagnose and defend against other types of scapegoating attacks to guarantee the security of network tomography should be further studied.

References

1. Chen, A., Cao, J., Bu, T.: Network Tomography: Identifiability and Fourier Domain Estimation, December 2007
2. Chiu, C.C., He, T.: Stealthy DGoS attack: degrading of service under the watch of network tomography. In: IEEE INFOCOM 2020 - IEEE Conference on Computer Communications, pp. 367–376. IEEE Press, Toronto, ON, Canada, July 2020
3. Constantin, L.: Attackers slip rogue, backdoored firmware onto Cisco routers — PCWorld. https://www.pcworld.com/article/2984084/attackers-install-highly-persistent-malware-implants-on-cisco-routers.html
4. Duffield, N., Presti, F.L., Paxson, V., Towsley, D.: Network loss tomography using striped unicast probes. IEEE/ACM Trans. Networking **14**(4), 697–710 (2006). Conference name: IEEE/ACM Transactions on Networking
5. Erdös, P., Rényi, A.: On the evolution of random graphs. Publ. Mah. Inst. Hung. Acad. Sci **5**, 17–60 (1960)

6. Knight, S., Nguyen, H.X., Falkner, N., Bowden, R., Roughan, M.: The internet topology zoo. IEEE J. Sel. Areas Commun. **29**(9), 1765–1775 (2011)
7. Ma, L., He, T., Leung, K.K., Swami, A., Towsley, D.: Monitor placement for maximal identifiability in network tomography. In: IEEE INFOCOM 2014 - IEEE Conference on Computer Communications, pp. 1447–1455, April 2014. ISSN: 0743-166X
8. Ma, L., He, T., Leung, K., Swami, A., Towsley, D.: Identifiability of link metrics based on end-to-end path measurements, pp. 391–404, October 2013
9. Ma, L., He, T., Leung, K.K., Swami, A., Towsley, D.: Inferring link metrics from end-to-end path measurements: identifiability and monitor placement. IEEE/ACM Trans. Networking **22**(4), 1351–1368 (2014)
10. Ma, L., He, T., Leung, K.K., Towsley, D., Swami, A.: Efficient identification of additive link metrics via network tomography. In: Proceedings - 2013 IEEE 33rd International Conference on Distributed Computing Systems, ICDCS 2013, pp. 581–590, December 2013
11. Ma, L., He, T., Swami, A., Towsley, D., Leung, K.K.: On optimal monitor placement for localizing node failures via network tomography. In: Performance Evaluation. Elsevier Science Publishers B. V. PUB568 Amsterdam, The Netherlands, The Netherlands, September 2015
12. Nguyen, H.X., Thiran, P.: The boolean solution to the congested IP link location problem: theory and practice. In: IEEE INFOCOM 2007–26th IEEE International Conference on Computer Communications, pp. 2117–2125, May 2007. ISSN: 0743-166X
13. Pepe, T., Puleri, M.: Network tomography: a novel algorithm for probing path selection. In: 2015 IEEE International Conference on Communications (ICC) (2015)
14. Qiao, Y., Jiao, J., Rao, Y., Ma, H.: Adaptive path selection for link loss inference in network tomography applications. PLOS ONE **11**(10), e0163706 (2016). Public Library of Science
15. Tague, P., Poovendran, R.: Modeling node capture attacks in wireless sensor networks. In: 2008 46th Annual Allerton Conference on Communication, Control, and Computing, pp. 1221–1224, September 2008
16. Tati, S., Silvestri, S., He, T., Porta, T.L.: Robust network tomography in the presence of failures. In: 2014 IEEE 34th International Conference on Distributed Computing Systems, pp. 481–492, June 2014. ISSN: 1063-6927
17. Tati, S., Silvestri, S., He, T., Porta, T.L.: Robust network tomography in the presence of failures. In: 2014 IEEE 34th International Conference on Distributed Computing Systems, pp. 481–492 (2014)
18. Vazirani, V.V.: Approximation Algorithms. Springer, Heidelberg (2001). https://doi.org/10.1007/978-3-662-04565-7
19. Xu, X., Wang, Y., Xu, L., Li, D.: Locate vulnerable link set to launch minimum cost scapegoating attack in network tomography. Under review (2022)
20. Zhao, S., Lu, Z., Wang, C.: When seeing isn't believing: on feasibility and detectability of scapegoating in network tomography. In: 2017 IEEE 37th International Conference on Distributed Computing Systems (ICDCS), pp. 172–182, June 2017. ISSN: 1063-6927
21. Zhao, Y., Govindan, R., Estrin, D.: Sensor network tomography: monitoring wireless sensor networks. ACM SIGCOMM Comput. Commun. Rev. **32**, 64 (2001)

Adaptive Competition-Based Diversified-Profit Maximization with Online Seed Allocation

Liman Du⬤, Wenguo Yang$^{(\boxtimes)}$ ⬤, and Suixiang Gao

School of Mathematical Sciences, University of Chinese Academy of Science,
Beijing 100049, China
duliman18@mails.ucas.edu.cn, {yangwg,sxg}@ucas.ac.cn

Abstract. The purpose of Profit Maximization (PM) problem in social media is finding some influential users as seeds to trigger large online cascading influence spread and generate profit as much as possible. Given that competitive social advertising is more common in real-world, a series of studies focus on Influence Maximization problem with competitive influence spread and propose some versions of PM problem from this perspective. However, the competition happening in the information dissemination of imperfect substitutes and the influence of potential consumers' preference have been mostly ignored. Besides, some companies may snatch seeds to limit the profits of their opponents. Motivated by the above considerations, we propose a novel Adaptive Competition-based Diversified-profit Maximization (ACDM) problem. Given the intermediate observations of each node's current state, ACDM problem aims at adaptively selecting seeds and allocating them such that the sum of profit generated by adopters for a special entity after information dissemination and social welfare with respect to adopter allocation reaches maximum. To address this problem, we design a three-steps algorithm which combines the method of online allocation and the concept of shapley value. Experimental results on three real-world data sets demonstrate the effectiveness of our proposed algorithm.

Keywords: Adaptive profit maximization · Online seed allocation · Nonsubmodularity · Competitive social advertising

1 Introduction

Due to the growing popularity of social networks and the rapid development of social advertising, Influence Maximization (IM) problem, aiming to find top-k influential nodes to influence as many nodes as possible, has become a hot research topic in the past decade. It is formulated as a discrete optimization

Supported by the National Natural Science Foundation of China under grant numbers 12071459 and 11991022 and the Fundamental Research Funds for the Central Universities under Grant Number E1E40107.

problem and is proved to be NP-hard under IC and LT model used to describe the process of information propagation of influence [11]. A greedy solution with a $1 - 1/e$ approximation guarantee for IM problem is also proposed, and its high time complexity leads to numerous subsequent researches focusing on improving greedy algorithm or designing heuristic algorithm [5,6,9,21,23]. Based on IM problem, Competitive Influence Maximization (CIM) problem is abstracted from competitive social advertising in real-world where multiple marketing campaigns compete with each other by launching comparable products over the same market and try to make the number of activated nodes reach maximum. Previous studies mainly handle this problem from two different perspectives: seed selection [2–4,13] and budget allocation [15,16,22]. A new two-phases scenario integrating the seed selection and budget allocation is proposed in [1]. Besides, Profit Maximization (PM) problem, as another extension of IM problem, is proposed from the perspective of social network hosts [7,8,20]. The purpose is changed to gain as much profit as possible and the objective function generally loses the property of submodular. An advertiser may delegate the operation of viral marketing campaigns to the social network provider, and the latter could be simultaneously paid to conduct the viral marketing campaign for many competitive companies. So, another version of PM problem proposed in [19] wants to maximize the host's profit. However, it only considers the cost to activate seeds and the objective function is still submodular.

These studies focusing on competition relationship between entities usually assume that a node can be activated by only one entity. Such assumption is likely to be unrealistic. For example, Microsoft's Surface and Apple's iPad, Coca-Cola and Pepsi, and tea and coffee, can be regarded as comparable products or imperfect substitutes. When a pair of imperfect substitutes are launched over the same market, some consumers may buy both of them and provide their review to their friends and followers. Then, the latter can make a decision according to the information they receive and their own preference. In fact, the review about different products provided by the same person is always more convincing. Under the assumption in [1,19], if the initial consumer picks iPad, his friends who prefer Microsoft cannot receive the information about Surface, to say nothing of buying it. What should be emphasized is that the probability with which a potential purchaser is persuaded to buy Surface is influenced by whether he has own iPad. What's more, the difference between potential consumers and the fact that competitive companies may snatch seeds to limit the profits of their opponents are mostly ignored. And existing works tends to assume that the information of opponent's seed are known and selects all the seeds in the beginning.

Motivated by defects mentioned above, we formulate Adaptive Competition-based Diversified-profit Maximization (ACDM) problem under Competitive Independent Cascade (CIC) model in this paper. It studies the profit related to adopters for an entity in competitive social advertising and integrates the seed selection and online allocation for competitive clients. To address ACDM problem, we divide it to two process, i.e. seed selection and allocation, and design AS and OA algorithms, respectively. Then, as a combination, the Adaptive Selection and Online Allocation (ASOA) algorithm consisting of three phases is proposed. It iteratively selects seeds and allocates them with adaptive policy. The final allocation is returned as a feasible solution.

The rest of this paper is organized as follows: The CIC model and ACDM problem is proposed in Sect. 2. In Sect. 3, the detail of ASOA algorithm is shown. And the experiments are presented in Sect. 4. We conclude in Sect. 5.

2 Problem Formulation

2.1 The CIC Model

The Com-IC model proposed in [14] captures the relationship spectrum from complementary to competitive. In this paper, we focus on the competitive relationship between only two entities and denote it as Competitive Independent Cascade (CIC) model.

There are two entities $\mathcal{A}$ and $\mathcal{B}$ which want to spread their information about promotion in a social network abstracted as a directed graph $G = (V, E)$. Each node $v \in V$ represents a user of the social network and $|V| = n$. The edge between each pair of neighbors $u, v \in V$ is denoted as $(u, v) \in E$. $p_{u,v}$ represents the probability with which information is successfully spread from u to v. During the information dissemination in G, each node selects one of {idle,accpeted,rejected} as its state for $\mathcal{A}$ and do the same thing for $\mathcal{B}$. Then, it stays in the joint state. After receiving the information, whether node v's state changes depends on its current state and a parameter set $q(v) = \{q_{\mathcal{A}|\emptyset}(v), q_{\mathcal{A}|\mathcal{B}}(v), q_{\mathcal{B}|\emptyset}(v), q_{\mathcal{B}|\mathcal{A}}(v)\}$. To reflect the competitive relationships between entities $\mathcal{A}$ and $\mathcal{B}$, we assume that $0 < q_{\mathcal{A}|\mathcal{B}}(v) < q_{\mathcal{A}|\emptyset}(v) \leq 1$ and $0 < q_{\mathcal{B}|\mathcal{A}}(v) < q_{\mathcal{B}|\emptyset}(v) \leq 1$ for each $v \in V$. For node v who does not accept $\mathcal{B}$, it transforms from $\mathcal{A}$-idle to $\mathcal{A}$-accepted with probability $q_{\mathcal{A}|\emptyset}(v)$. If v is $\mathcal{B}$-accepted and informed of $\mathcal{A}$, it becomes $\mathcal{A}$-accepted with probability $q_{\mathcal{A}|\mathcal{B}}(v)$. The meaning of $q_{\mathcal{B}|\emptyset}(v)$ and $q_{\mathcal{B}|\mathcal{A}}(v)$ are similar.

Now, we consider the CIC model as the information diffusion model. Let $S_{\mathcal{A}}, S_{\mathcal{B}} \subset V$ be two seed sets. At time $t = 0$, $v \in S_{\mathcal{A}}$ accepts $\mathcal{A}$ while $\mathcal{B}$ is accepted by $u \in S_{\mathcal{B}}$. Except for them, all the nodes initially stay in the joint state of ($\mathcal{A}$-idle, $\mathcal{B}$-idle). At each time step $t \geq 1$, for a node u becoming $\mathcal{A}$-accepted at time $t - 1$ and one of its neighbor v, information about $\mathcal{A}$ has only one chance to successfully spread from u to v with probability $p_{u,v}$. And $p_{u,v}$ is the same for both $\mathcal{A}$ and $\mathcal{B}$. If node v stays in the joint state of ($\mathcal{A}$-idle, $\mathcal{B}$-idle) and is informed about both $\mathcal{A}$ and $\mathcal{B}$ from its neighbors at the same step, tie-breaking rule is used to decide its state. It generally consists of two cases: $\mathcal{A}$ is superior to $\mathcal{B}$, that is, node always adopt $\mathcal{A}$ in competition; otherwise, $\mathcal{A}$ is inferior to $\mathcal{B}$. The process stops when there is no node can be activated. When the diffusion is terminated, each node's adoption is fixed and profit generated by adopter of $\mathcal{A}$ can be calculated.

2.2 ACDM Problem

In this part, we firstly introduce the definition of allocation and its social welfare under no-rejection condition. Given a set S_{cand} of candidates waiting to be allocated to agents, the no-rejection condition requires each candidate to choose only one agent and can not be rejected. Therefore, all the candidates should be allocated after the whole allocation process and an allocation A of S_{cand} is used to show candidates' choice. Obviously, an allocation A is a non-overlapping partition of nodes in S_{cand}. In this paper, two agents $\mathcal{A}$ and $\mathcal{B}$ are considered and $A = \{(S_{\mathcal{A}}, \mathcal{A}), (S_{\mathcal{B}}, \mathcal{B})\}$ satisfying $S_{\text{cand}} = S_{\mathcal{A}} \cup S_{\mathcal{B}}$ and $S_{\mathcal{A}} \cap S_{\mathcal{B}} = \emptyset$. Given a happiness matrix H, $h_{u,v} \in H$ denotes the happiness of candidate u when u and v choosing the same agent. $r_{u,\mathcal{A}} \in R$ denotes the appraisal of candidate u on $\mathcal{A}$ and the definition of $r_{u,\mathcal{B}}$ is similar. The social welfare is defined as $SW(A) = h_{u,v} + h_{v,u} + r_{u,\mathcal{A}} + r_{v,\mathcal{A}} + h_{x,y} + h_{y,x} + r_{x,\mathcal{B}} + r_{y,\mathcal{B}}$ for $u, v \in S_{\mathcal{A}}$ and $x, y \in S_{\mathcal{B}}$. And for each candidate v, define its utility as the sum of its happiness to all the other candidate in the same agent and its appraisal for the agent. That is to say, for $u \in S_{\mathcal{A}}$, its utility can be represented as $U_u = \sum_{v \in S_{\mathcal{A}}} h_{u,v} + r_{u,\mathcal{A}}$. We assume that all candidates arrive online in uniform randomly order. Under this assumption, when candidate i arrives, happiness value $h_{i,j}$ to all candidates j that have already arrived, as well as its appraisal to $\mathcal{A}$ and $\mathcal{B}$, are revealed. And it should be allocated immediately. The definition of weakly stable is shown below.

Definition 1 (Definition 3 in [10]). *An allocation is weakly agent stable if for any two candidates u, v choosing $\mathcal{A}$ and two candidates x, y choosing $\mathcal{B}$, switching their choices cannot increase all four candidates' utilities.*

Let $\phi_{\mathcal{A}}(v)$ represent the modified profit with respect to entity $\mathcal{A}$ generated by node v when it adopts $\mathcal{A}$. We propose some assumptions as follows.

1. For node v which does not accept $\mathcal{A}$, it can not generate profit with respect to $\mathcal{A}$ regardless of its state for $\mathcal{B}$.
2. For node v accepting both $\mathcal{A}$ and $\mathcal{B}$, it can spread the information about both $\mathcal{A}$ and $\mathcal{B}$ to its neighbors. However, v does not generate profit with respect to $\mathcal{A}$. Thus, it is ignored when calculating the total profit generated by $\mathcal{A}$-adopter.

Similar to IC model proposed in [11], the CIC model is also equivalent to a live edge graph process. Flip a coin for each edge in advance and retain it with probability $p_{u,v}$. After the process, we obtain a subgraph g of G consisting of all retained edges with randomness taken over the coin-flipping process of all edges. Such a subgraph is defined as a realization. Based on the live edge graph process, given allocation $A = \{S_{\mathcal{A}}, S_{\mathcal{B}}\}$, the profit generated by $\mathcal{A}$-adopter can be written by an expectation form. Based on the definition of realization, it can be expresses as a function $\psi : E \rightarrow \{0, 1\}$. For each $e \in E$, $\psi(e) = 1$ represent e is retained, otherwise $\psi(e) = 0$. Then, we define a partial realization as follows. Under the CIC model, for any seed sets $S_{\mathcal{A}}, S_{\mathcal{B}}$ and time step t, the status of nodes, to which length of the shortest path from nodes in $S_{\mathcal{A}} \cup S_{\mathcal{B}}$ is not bigger than t, as well as whether their in-coming edges are retained can be observed via partial realization $\psi_t \subseteq \psi$. To put it another way, for a fixed partial realization (previous observation) ψ_t and an allocation $A = \{S_{\mathcal{A}}, S_{\mathcal{B}}\}$, the status of current reachable nodes and all the edges are available. Such assumption is based on the full feedback model which is widely studied. Now, we can denote our adaptive strategy for picking seeds as a policy π. π is actually a function from a partial realization ψ_t to V, specifying which node to select at time step $t+1$ for given ψ_t.

Then, integrating seed selection and allocation, we denote diversified-profit function $DP(A)$ as the sum of social welfare and profit generated by actual $\mathcal{A}$-adopter. Hence,

$$DP(A_\psi) = \Phi_\mathcal{A}(S_\mathcal{A}|A_\psi, S_\mathcal{B}|A_\psi) + \lambda SW(A_\psi) = \sum_{v \in I_{g, A_\psi}} \phi_\mathcal{A}(v) + \lambda SW(A_\psi) \quad (1)$$

where λ is a weight of social welfare, I_{g, A_ψ} is the set of nodes which can receive the information spread from $S_\mathcal{A}$ and become ($\mathcal{A}$-accepted, $\mathcal{B}$-idle/rejected) under CIC model with allocation $A = \{S_\mathcal{A}, S_\mathcal{B}\}$ under realization ψ. Denote A_ψ^π as an allocation of seeds selected by policy π under realization ψ. Based on this notation, the expected diversified-profit of a policy π is defined as $\mathbb{E}[DP(A_\psi^\pi)]$ where the expectation is taken with respect to $p(\psi)$ which is based on a known probability distribution over realizations.

Definition 2 (ACDM problem) *Adaptive Competition-based Diversified-profit Maximization problem aims to find a weakly stable allocation A of at most K seeds with policy π^* such that the expected value of diversified-profit function is maximized under CIC-model, i.e.,*

$$A^{\pi^*} \in \arg\max \mathbb{E}[DP(A_\psi^\pi)] \quad (2)$$

In this paper, we consider the k-R (k nodes per Round) setting proposed in [18] that selects k nodes to allocate for each time round $t \in [T]$. The constraint of the number of seed nodes is divided into T equal-sized parts, i.e. $K = k \cdot T$. Take a further step of the property of IM problem, we propose a theorem as follows.

Theorem 1. *ACDM problem is NP-hard.*

3   The Algorithm

We concentrate on designing an algorithm to find a feasible solution for ACDM problem in this part. Since the diversified-profit is defined as the sum of profit influenced by seed selection and social welfare influenced by seed allocation, the ACDM problem can be divided into two sub-problems, seed selection and seed allocation. Firstly, finding a seed set S which satisfies $|S| = k$ and can maximize the profit $\phi_\mathcal{A}(S)$. Then allocate all the nodes in S such that the social welfare is maximized and the allocation is weakly stable. Based on the outcome of seed allocation, update $S_\mathcal{A}$ and $S_\mathcal{B}$. Then, based on $S_\mathcal{A}$ and $S_\mathcal{B}$, select k nodes to maximize the profit and allocate them. Repeat this process for T times.

In a nutshell, given that the result of seed selection and allocation influence each other, we propose the Adaptive Selection and Online Allocation (ASOA)

Algorithm 1. ASOA algorithm $(G = (V, E), P, Q, \phi_{\mathcal{A}}, \psi_{t-1})$

1: Initialize $S_{\mathcal{A}} = \{S_{\mathcal{A},i}\}$ with $S_{\mathcal{A},i} = \emptyset$ and $S_{\mathcal{B}} = \{S_{\mathcal{B},i}\}$ with $S_{\mathcal{B},i} = \emptyset$, for each $i \in [T]$.
2: **for** $t = 1$ to T **do**
3:     $(S, t) \leftarrow$ AS algorithm $(G = (V, E), P, k, Q, \phi_{\mathcal{A}}, \psi_{t-1})$.
4:     $(A, t) \leftarrow$ OA algorithm $(S, H, R, (A, t-1))$.
5:     $S_{\mathcal{A},t} = S_{\mathcal{A}}|(A, t)$ and $S_{\mathcal{B},t} = S_{\mathcal{B}}|(A, t)$.
6:     Obtain partial realization ψ_t.
    return $S_{\mathcal{A}}, S_{\mathcal{B}}$

---

algorithm consisting of two sub-algorithms. It can return a feasible solution for the ACDM problem and is shown as Algorithm 1.

In each round, inspired by [17], we firstly design AS algorithm whose details are shown in Algorithm 2 to find a candidate node set. Different from greedy algorithm proposed in [11], it models nodes in the social network as players in a coalitional game and captures information diffusion process as the process of coalition formation in the game. For two given current seed node sets $S_{\mathcal{A}}$ and $S_{\mathcal{B}}$, it computes a ranking list of the nodes based on the shapley value and picks the top-k nodes as candidate nodes waiting to be allocated in the next step. Subsequently, Algorithm 3 is used to allocate all the node returned by AS algorithm. Such an algorithm is based on Algorithm 2 proposed in [10] and its crucial idea is the online no-rejection bipartite matching algorithm. Regard $\Gamma = \{\mathcal{A}, \mathcal{B}\}$ as a set which is given in advance and consists of two entities. At the same time, nodes in S returned by Algorithm 2 are arriving one by one and the edges incident to each node are revealed when it arrives. In sequence of its index in S, we consider the first two nodes and consecutive allocate them to an unmatched adjacent vertex $\gamma \in \Gamma$. Then, combine each node and its choice pair as one new vertex γ and match the next two nodes with adjusted appraisal to updated vertexes.

4   Experiments Settings and Results

We conduct a series of experiments on three real social networks, Petster-Hamster-Household (PHH), Moreno-Innovation(MI) and email-Eu-core(Email). The first two can be obtained from [12] while the last one can be found in SNAP website. The number of nodes including in PHH, MI and Email are 921, 246 and 1005, respectively. Correspondingly, the number of edges are 4032, 1098 and 25571.

The propagation probability $p_{u,v}$ for each pair of neighbors (u, v) in CIC model is randomly generated from $[0, 1]$, and $q(v)$ for each $v \in V$ is fixed by the same way. The profit $\phi_{\mathcal{A}}$ of 90% nodes is set to be 1 while the rest are 0.5. In addition, the happiness matrix H and appraisal matrix R are randomly generated and satisfy that every element is confined to $[0, 1]$. T is set to be 5, and the number of seeds selected in each round varies with the seed size K which is chosen from $\{10, 20, 30, 40, 50\}$.

Algorithm 2. AS algorithm $(G = (V, E), P, k, Q, \phi_{\mathcal{A}}, \psi_{t-1})$

 1: Denote the number of repetitions and a randomly sampled set of permutations as mc and Υ, and initialize RL as an empty list.
 2: **for all** $v \in V$ **do**
 3:     Initialize each node's shapley value to 0.
 4: **for all** $\mathcal{A}_v \in \Upsilon$ **do**
 5:     **for all** nodes in $\mathcal{A}_v$ **do**
 6:         Initialize nodes' temporal shapley value to 0 and MG to a zero matrix.
 7:         **for** $k = 1$ to mc **do**
 8:             $CMG \leftarrow$ Marginal Gain $(G = (V, E), P, Q, \phi_{\mathcal{A}}, \psi_{t-1})$.
 9:             $MG = MG + CMG$.
10:         Update nodes' temporal shapley value as average of elements of MG.
11:     Update nodes' shapley value as average value of their temporal shapley value.
12: Sort the nodes in non-increasing order by their shapley value to obtain RL.
13: Initialize S as an empty set and S_{can} as an empty list.
14: **for** $i = 1$ to n **do**
15:     Denote the i-th element of RL as v.
16:     **if** $v \notin S$ and v is not adjacent to any node in S **then**
17:         Put v into S.
18:         **if** $|S| \geq k$ **then**
19:             Break.
20:     **else**
21:         Append v to S_{can}.
22: **if** $|S| < k$ **then**
23:     **for** $j = 1$ to $k - |S|$ **do**
24:         Put the j-th element of S_{can} into S.
     return (S, t)
25: **function** MARGINAL GAIN$(G = (V, E), P, Q, \phi_{\mathcal{A}}, \psi_{t-1})$
26:     Initialize $S_{\mathcal{A},temp}$ as an empty set and CMG as a n-tuple zero vector. Create two empty queues $\mathcal{A}_{can}$ and $\mathcal{B}_{can}$.
27:     **for** $j = 1$ to $|V|$ **do**
28:         Put the first j nodes in $\mathcal{A}_\pi$ into $S_{\mathcal{A},temp}$.
29:         **for** every $v \in S_{\mathcal{A},temp}$ **do**
30:             Put v into $\mathcal{A}_{can}$ and mark v as $\mathcal{A}$-accepted.
31:         **while** $\mathcal{A}_{can}$ is not empty **do**
32:             **for** every out-neighbor w of nodes in $\mathcal{A}_{can}$ **do**
33:                 Update w's joint state according to Q, ψ_{t-1} and its current state.
34:                 **if** w is $\mathcal{A}$-accepted **then**
35:                     Put w into $\mathcal{A}_{can}$.
36:         Calculate the total profit based on the first j nodes in $\mathcal{A}_\pi$. Take the difference between it and the total profit based on the first $j - 1$ nodes in $\mathcal{A}_\pi$ as the marginal gain of the j-th node in $\mathcal{A}_\pi$. Update CMG according to each node's marginal gain.
     return CMG

As this is the first work for solving ACDM problem, no direct algorithms can be compared. For comparison, we propose ASOA algorithm with one-shot policy as baseline and denote it as non-adaptive. This algorithm only conducts

---

Algorithm 3. OA algorithm $(S, H, R, (A, t-1))$

1: Denote $S = \{v_1, v_2, \ldots, v_{|S|}\}$.
2: $R' \leftarrow R$.
3: **if** $(A, t-1)$ is empty **then**
4:      **if** $R_{v_1, \mathcal{A}} > R_{v_1, \mathcal{B}}$ **then**
5:          Allocate v_1 to $\mathcal{A}$ and v_2 to $\mathcal{B}$.
6:      **else if** $R_{v_1, \mathcal{A}} < R_{v_1, \mathcal{B}}$ **then**
7:          Allocate v_1 to $\mathcal{B}$ and v_2 to $\mathcal{A}$.
8:      **else**
9:          Randomly allocate v_1 to a seed set and allocate v_2 to the other seed set.
10:     Update (A, t).
11:     **for** $i = 2, \ldots, \lceil \frac{|S|}{2} \rceil$ **do**
12:         $R'_{v_j, \mathcal{EN}} \leftarrow R_{v_j, \mathcal{EN}} + H_{v_k, j} + H_{k, v_j}$ for each k in $S_{\mathcal{EN}}$, $\mathcal{EN} = \mathcal{A}$ and $\mathcal{B}$, $j = 2 * i - 1$ and $2 * i$
13:         Update (A, t) according to the return matching of OnlineMatching on the two nodes with R'.
14:     **if** $|S| < 2 * \lceil \frac{|S|}{2} \rceil$ **then**
15:         Delete $v_{|S|}$ from (A, t).
16: **else**
17:     **for** $i = 1, \ldots, \lceil \frac{|S|}{2} \rceil$ **do**
18:         $R'_{v_j, \mathcal{EN}} \leftarrow R_{v_j, \mathcal{EN}} + H_{v_k, j} + H_{k, v_j}$ for each k in $S_{\mathcal{EN}}$, $\mathcal{EN} = \mathcal{A}$ and $\mathcal{B}$, $j = 2 * i - 1$ and $2 * i$
19:         Update (A, t) according to the return matching of OnlineMatching on the two nodes with R'.
20:     **if** $|S| < 2 * \lceil \frac{|S|}{2} \rceil$ **then**
21:         Delete $v_{|S|}$ from (A, t).
        return (A, t)
22: **function** ONLINEMATCHING
23:     $M \leftarrow \emptyset$, $L \leftarrow \emptyset$
24:     **for** each node v comes **do**
25:         $L \leftarrow L \cup \{v\}$
26:         $M^v \leftarrow$ Optimal Matching on $G[L \cup \{\mathcal{AB}\}]$.
27:         $e^v \leftarrow$ The matching edge that contains v in M^v
28:         **if** $M \cup e^v$ is a matching **then**
29:             $M \leftarrow M \cup e^v$
30:         **else**
31:             Randomly choose $\mathcal{EN}$ from $\{\mathcal{A}, \mathcal{B}\}$ and $M \leftarrow M \cup (v, \mathcal{EN})$
        return M

---

the first and second step of ASOA algorithm and the number of nodes selected in the first phase is changed to K. It picks seeds and allocates them, regardless of their realization.

Firstly, some experiments are conducted to study the influence of parameter λ. This parameter is used to weight the importance of social welfare of seeds, reflecting the evaluation of current allocation. We set $\lambda = 0.25, 0.5, 0.75$ and compare the value of DP with varying λ based on different data sets. As is shown in Fig. 1, the value of DP increases with λ. Due to the definition of DP function,

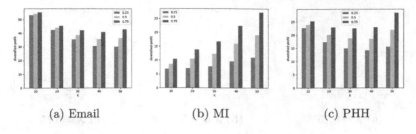

Fig. 1. The value of DP with varying of parameter λ

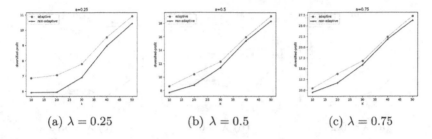

Fig. 2. The relationship between DP and seed set size on MI

this conclusion is easy to understand. Then, we compare the results obtained by two different algorithms based on three datasets and results are shown in Fig. 2, Fig. 3 and Fig. 4, respectively. According to Fig. 2 and Fig. 3, as the number of selected seeds increases, the performance of ASOA algorithm donated as adaptive are always superior to the baseline algorithm. This is because non-adaptive algorithm can not adjust its choice according to the result of allocation. However, ASOA algorithm must make a decision based on current state of each node in the social network when selecting the next k seeds. When focusing on Fig. 4, we observe that the value of DP obtained by adaptive algorithm based on PHH dataset is smaller than that calculated by non-adaptive algorithm when $K = 40$ and $\lambda = 0.5, 0.75$. In fact, it is because the difference of profit obtained by two different algorithms is smaller than that of social welfare. ASOA algorithm firstly picks nodes which can maximize the value of profit and then allocates them. Given that the value of social welfare function is related to the selected seeds, this order may influence the final result, especially when the average degree of nodes is small (such as 4.3778 in PHH). Besides, the setting of $\phi_{\mathcal{A}}$, H and R in our experiments may be another reason. In this setting, when a node v is selected, the raise of social welfare may be bigger than that of profit. Therefore, although such results seems unexpected, they show some special cases which can be avoided through more ideal setting and actually do not impact the effectiveness of ASOA algorithm. In summary, ASOA algorithm performs quite well on both small-scale and large-scale networks.

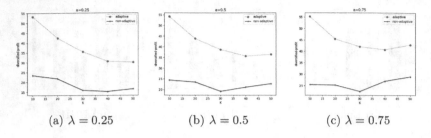

Fig. 3. The relationship between DP and seed set size on Email

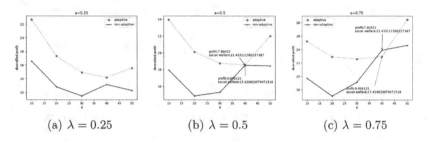

Fig. 4. The relationship between DP and seed set size on PHH

5   Conclusion

In this paper, we propose Adaptive Competition-based Diversified-profit Maximization problem integrating the process of seed selection and allocation for two competitive entities. To address such a realistic and challenging problem, Adaptive Selection and Online Allocation algorithm is designed. This algorithm consists of AS algorithm and OA algorithm focusing on seed selection and seed allocation, respectively. Combining the concept of shapley value and the method used to handle online bipartite matching problem, ASOA algorithm could obtain a better solution for ACDM problem. And we conduct experiments on real-world networks to evaluate its effectiveness. To the best of our knowledge, it is the first paper integrating seed selection and allocation in adaptive competitive profit maximization problem.

References

1. Ansari, A., Dadgar, M., Hamzeh, A., Schlötterer, J., Granitzer, M.: Competitive influence maximization: integrating budget allocation and seed selection (2019). http://arxiv.org/abs/1912.12283
2. Bharathi, S., Kempe, D., Salek, M.: Competitive influence maximization in social networks. In: Deng, X., Graham, F.C. (eds.) Internet and Network Economics, pp. 306–311. Springer, Heidelberg (2007). https://doi.org/10.1007/978-3-540-77105-0_31

3. Bozorgi, A., Samet, S., Kwisthout, J., Wareham, T.: Community-based influence maximization in social networks under a competitive linear threshold model. Knowl.-Based Syst. **134**, 149–158 (2017). https://doi.org/10.1016/j.knosys.2017.07.029

4. Carnes, T., Nagarajan, C., Wild, S.M., van Zuylen, A.: Maximizing influence in a competitive social network: a follower's perspective. In: Proceedings of the Ninth International Conference on Electronic Commerce, ICEC 2007, pp. 351–360. Association for Computing Machinery, New York, NY, USA (2007). https://doi.org/10.1145/1282100.1282167

5. Chen, W., Wang, C., Wang, Y.: Scalable influence maximization for prevalent viral marketing in large-scale social networks, pp. 1029–1038, September 2010. https://doi.org/10.1145/1835804.1835934

6. Chen, W., Wang, Y., Yang, S.: Efficient influence maximization in social networks. In: Proceedings of the 15th ACM SIGKDD International Conference on Knowledge Discovery and Data Mining, KDD 2009, pp. 199–208. Association for Computing Machinery, New York, NY, USA (2009). https://doi.org/10.1145/1557019.1557047

7. Du, L., Chen, S., Gao, S., Yang, W.: Nonsubmodular constrained profit maximization from increment perspective. J. Comb. Optim. (2021). https://doi.org/10.1007/s10878-021-00774-6

8. Du, L., Yang, W., Gao, S.: Generalized self-profit maximization in attribute networks. In: Du, D.-Z., Du, D., Wu, C., Xu, D. (eds.) COCOA 2021. LNCS, vol. 13135, pp. 333–347. Springer, Cham (2021). https://doi.org/10.1007/978-3-030-92681-6_27

9. Goyal, A., Lu, W., Lakshmanan, L.V.: CELF++: optimizing the greedy algorithm for influence maximization in social networks. In: Proceedings of the 20th International Conference Companion on World Wide Web, WWW 2011, pp. 47–48. Association for Computing Machinery, New York, NY, USA (2011). https://doi.org/10.1145/1963192.1963217

10. Huzhang, G., Huang, X., Zhang, S., Bei, X.: Online roommate allocation problem. In: Proceedings of the Twenty-Sixth International Joint Conference on Artificial Intelligence, IJCAI-2017, pp. 235–241 (2017). https://doi.org/10.24963/ijcai.2017/34

11. Kempe, D., Kleinberg, J., Tardos, E.: Maximizing the spread of influence through a social network. In: Proceedings of the ACM SIGKDD International Conference on Knowledge Discovery and Data Mining, pp. 37–146 (2003). https://doi.org/10.1145/956750.956769

12. Kunegis, J.: KONECT - the Koblenz network collection. In: Proceedings of International Conference on World Wide Web Companion, pp. 1343–1350 (2013). https://doi.org/10.1145/2487788.2488173

13. Li, H., Bhowmick, S.S., Cui, J., Gao, Y., Ma, J.: GetReal: towards realistic selection of influence maximization strategies in competitive networks. In: Proceedings of the 2015 ACM SIGMOD International Conference on Management of Data, SIGMOD 2015, pp. 1525–1537. Association for Computing Machinery, New York, NY, USA (2015). https://doi.org/10.1145/2723372.2723710

14. Lu, W., Chen, W., Lakshmanan, L.V.S.: From competition to complementarity: comparative influence diffusion and maximization. Proc. VLDB Endow. **9**(2), 60–71 (2015). https://doi.org/10.14778/2850578.2850581

15. Masucci, A., Silva, A.: Advertising competitions in social networks (2016). http://arxiv.org/abs/1608.02774

16. Masucci, A.M., Silva, A.: Strategic resource allocation for competitive influence in social networks (2014). http://arxiv.org/abs/1402.5388

17. Narayanam, R., Narahari, Y.: A Shapley value-based approach to discover influential nodes in social networks. IEEE Trans. Autom. Sci. Eng. **8**(1), 130–147 (2011). https://doi.org/10.1109/TASE.2010.2052042
18. Shi, Q., Wang, C., Ye, D., Chen, J., Feng, Y., Chen, C.: Adaptive influence blocking: minimizing the negative spread by observation-based policies. In: 2019 IEEE 35th International Conference on Data Engineering (ICDE), pp. 1502–1513 (2019). https://doi.org/10.1109/ICDE.2019.00135
19. Shi, Q., et al.: Profit maximization for competitive social advertising. Theor. Comput. Sci. **868**, 12–29 (2021). https://doi.org/10.1016/j.tcs.2021.03.036
20. Tang, J., Tang, X., Yuan, J.: Towards profit maximization for online social network providers, December 2017. https://arxiv.org/abs/1712.08963
21. Tang, Y., Shi, Y., Xiao, X.: Influence maximization in near-linear time: a Martingale approach. In: Proceedings of the 2015 ACM SIGMOD International Conference on Management of Data, SIGMOD 2015, pp. 1539–1554. Association for Computing Machinery, New York, NY, USA (2015). https://doi.org/10.1145/2723372.2723734
22. Varma, V.S., Lasaulce, S., Mounthanyvong, J., Morărescu, I.C.: Allocating marketing resources over social networks: a long-term analysis. IEEE Control Syst. Lett. **3**(4), 1002–1007 (2019). https://doi.org/10.1109/LCSYS.2019.2919959
23. Yu, H., Kim, S.K., Kim, J.: Scalable and parallelizable processing of influence maximization for large-scale social networks? In: Proceedings of the 2013 IEEE International Conference on Data Engineering (ICDE 2013), pp. 266–277. IEEE Computer Society, USA (2013). https://doi.org/10.1109/ICDE.2013.6544831

Collaborative Service Caching in Mobile Edge Nodes

Zichen Wang and Hongwei Du[✉]

School of Computer Science and Technology, Harbin Institute of Technology
(Shenzhen), Shenzhen, China
hongwei.du@ieee.org

Abstract. Recently, it has become widely accepted that moving services from original cloud servers to mobile edge nodes (MENs) can shorten the time it takes for a service to respond. In this article, we first examine the issue of service caching by numerous MENs under the assumption of market-oriented behavior, and then we propose a collaborative service caching mechanism (CSCM) to allow MENs to assist one another with services, enhancing the advantages of MENs and the service effectiveness of the overall MEC network. Then a randomized rounding (CSCM+RR) algorithm is proposed based on CSCM. Finally, we conduct experiments on a simulation platform using a real dataset and evaluate the performance of the CSCM+RR algorithm. According to the experimental findings, the CSCM+RR algorithm reduces the average delay by 31.02% to 82.90% while increasing the average profit by 5.87% to 76.78% to the baseline.

Keywords: Mobile edge computing · Service caching · Collaborative mechanism

1 Introduction

The growth of cloud computing, artificial intelligence, and communication infrastructures over the last ten years has increased the demand for high-quality, low-cost computing services. The technology used in autonomous vehicles, for instance, has a high level of quick and accurate reaction for safety drives. Virtual reality (VR) games require large volumes of data to be processed in real-time. By 2022, it is anticipated that video-related mobile data traffic would exceed 60 EB globally [1]. These applications more severely test the cloud server's computation, storage, and transmission capacities, as well as the network's overall transmission capacity.

The difficulties that network service providers (NSPs) and application service providers (ASPs) are currently facing can be solved very well thanks to the development of mobile edge computing (MEC). Mobile edge nodes (MENs) contain computing, storage, and communication capabilities and are situated closer to users than distant cloud computing facilities. Such a MEC model provides

Q. Ni and W. Wu (Eds.): AAIM 2022, LNCS 13513, pp. 195–206, 2022.
https://doi.org/10.1007/978-3-031-16081-3_17

services to users with guaranteed quality and delay and reduces the pressure on remote cloud computing centers at the same time.

In this paper, we discuss ways to increase MENs' profit by making greater use of the benefits of close proximity and quick communication. Then we propose the problem of collaborative service caching in MEC networks. Caching services must pay for both the time and bandwidth required to go to the edge from a remote cloud computing center. Making MENs exchange services can boost the use of services that are currently cached on MENs and further save service costs.

The edge cooperative caching mechanism enables MENs to provide services to each other and form coalitions, which has some challenges: (1) MENs usually belong to different NSPs that only focus on their profits. They cannot be compelled to provide services to each other, and a suitable market mechanism is needed to make them consciously form a coalition. (2) It can be difficult to create coalitions because of the size and degree of division. (3) When a coalition is formed between MENs, one MEN can buy services from another MEN and pay a certain fee. Pricing strategies are essential in these business activities, and reasonable service pricing can benefit both ASPs and NSPs.

This study investigates the issue of edge service caching by establishing a market-oriented coalition approach, comparing the cost differences between acquiring services from the edge and from a remote data center.

The main contributions of this paper are as follows.

- For the collaborative service caching problem, we elaborate a coalition mechanism to minimize the service cost of all MENs through service sharing.
- For the coalition mechanism, we propose an Integer Linear Program (ILP) solution and design a random rounding algorithm.
- We conduct simulations and implementations through an experimental platform to evaluate the performance of the proposed mechanism. The experimental results show that by establishing a service caching coalition on edge, the MEN profit is improved, and the user response delay is shortened, making the CSCM+RR algorithm outperforms the existing methods.

The rest of the paper is structured as follows. The state-of-the-art on this subject is outlined in Sect. 2. The system model and problem formulations are presented in Sect. 3. The coalition technique for the edge collaborative service caching issue is suggested in Sect. 4. The service caching issue at the edge is addressed in Sect. 5, along with a random rounding approach. The effectiveness of the proposed algorithm is assessed in Sect. 6, and the article is wrapped up in Sect. 7.

2   Related Work

The caching problem in edge computing has received extensive attention in the academic community. In existing research, people generally innovates edge caching methods to improve user experience (QoE) or reduce caching costs.

Existing researches on edge caching can be divided into two categories depending on the cached content: (1) service caching; and (2) data caching.

For service caching, the research focuses on reducing the cost of service caching or increasing the benefit. For example, Y. Liang et al. [2] studied the service entity caching problem from the utility perspective. J. Xu et al. [3] focus on minimizing the mobile user's computation delay and energy consumption. G. Zhao et al. [4] study how to efficiently offload dependent tasks to edge nodes with limited (and predetermined) service caching. G. Zeng et al. [5] focus on maximizing the total profit of service providers. Some scholars have also paid attention to the problem of cooperative service caching, for example, Z. Xu et al. [6] designed a novel coalition formation game for the problem with VM sharing and aim to minimize the total cost of all network service providers. X. Ma et al. [7] consider cooperation among edge nodes and investigate cooperative service caching.

For data caching, the research focuses on how to use the characteristics of the data to increase the request hit rate of cached content or reduce the response delay. According to the attributes of the cached content, the optimization goals of the cache scheme are different. H. Wang et al. [8] aimed to improve QoE and the user's fast joining ability and smooth viewing experience. F. Wang et al. [9] proposed an intelligent edge caching framework to minimize content access latency and traffic cost. C. Li et al. [10] proposed a collaborative caching strategy for video content in a cloud-edge collaborative environment.

Based on the above analysis, the existing research on service caching does not consider the cost of caching service behavior. The service caching needs to consume the bandwidth cost and time cost from the remote data center to the edge. If the utilization of services already cached at the edge can be increased, it should be possible to reduce the cost of services further. This paper will design the service caching mechanism from this perspective.

3 Preliminary

3.1 System Model

We considered a service caching system in a MEC network $G = (MENs \cup DC, E)$ with a set of MENs, the remote data center, and a set of communication links that connect upward to the remote data center and downward to the user terminal. The structure of the MEC network is shown in Fig. 1. Users need to offload services to MENs located at the edge in this MEC network. MENs have storage, computing, and communication capabilities. As a service provider at the edge of the network, they provide services to users and charge users a certain fee. Many coalitions are formed between MENs. When a MEN does not have the services required by users, the MEN asks other MENs in the coalition for assistance and pays a certain extra fee. When there is no user-requested service in the coalition, the MEN requests the service from the remote data center. The remote data center will provide the service and return the calculation results to the user via the MEN. Each MEN has limited computing and caching resources to provide services to users.

Users pay MENs for the amount of computation resource required for service, regardless of whether MENs are busy or need to ask the DC for help. Charging a service fee at a fixed rate can attract users better. Thus, MENs can only increase its profits by reducing service costs to ensure user service quality without increasing service fees.

In this paper, the user is represented by $u \in U$ the MEN is represented by $n \in N$ and the service is represented by $l \in L$ With each MEN as the center, a coalition $a_n \in A$ is formed, and the MEN can cooperate with the MENs in its coalition to provide services.

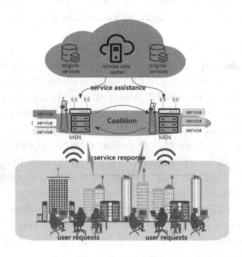

Fig. 1. System model

3.2   Cache Services at Mobile Edge Nodes

We consider caching a portion of services deployed in the DC to the MEC network to reduce the request latency of users, which is called service caching.

This paper proposes to form a service cache coalition among MENs, which can increase the utilization of services on MENs and reduce the service cost of MENs. After deploying some services to the MEC network, a user sends a service request to MEN a. If MEN a can provide services for the user, it can directly serve service calculations and return the calculation results to the user. When MEN a does not have the service requested by the user, MEN a can request the service from other MENs in its coalition. Assuming that the MEN b in the coalition can provide the service, the MEN b provides the service for the user and sends the calculation result back to the user through MEN a. Due to the close distance between MENs in the Coalition, it can save a lot of time to ask for service assistance from MENs in the coalition instead of going to DC to obtain services. On the other hand, a collaborative service mechanism can increase the utilization of services on MENs. Through reasonable service pricing, increased service utilization can translate into additional profits for MENs.

3.3   The Cost of Serving at the Edge and the Remote Data Center

Base Cost: For MENs, it needs to cache the service on the infrastructure and perform computations after receiving user requests. Since MENs' infrastructure has limited cache space and limited computing resources, the provision of computing services by MENs means that computing costs and storage costs are needed. At the same time, MENs need to consume a specific network bandwidth for transmitting data, so MENs also need to spend the bandwidth cost.

Denote by $c_{n,l}^c$ the cost of using a unit of computing resource in MEN n by service l. And denote by $c_{n,l}^b$ the cost of using a unit of bandwidth resource in MEN n by service l. Let $c_{n,l}$ be the usage cost of caching a service l in MEN n, then,

$$c_{n,l} = c_{n,l}^c \cdot C_l + c_{n,l}^b \cdot B_l \tag{1}$$

where C_l is the amount of computing resource units required to calculate once service l and B_l is the amount of bandwidth resource units required for service l.

Collaboration Cost: When a MEN cannot provide services, it needs to request service assistance from other MENs within its coalition. MENs only pay attention to their own interests. In order to get help from other MENs, the MEN needs to pay extra "assistance" fees in addition to the cost of base calculation and bandwidth. If there are no MENs in the coalition that can help, the MEN needs to send an assistance request to the DC. Since the DC is far away, the MEN needs to pay more service fees to get the help of the DC. These fees are part of the MEN service cost.

3.4   Pricing Strategy of the Edge and the Remote Data Center

MENs need to charge a certain fee for providing services to users, and the fee is priced according to the resources required by the service. The cost of a user requesting a service l from MEN n is $P_{n,l}$, then,

$$P_{n,l} = T_l \cdot p_n \tag{2}$$

where T_l is the duration of serving the request of service l, and p_n is the price that MEN n provides the service per unit time. When a MEN requests services from other MENs in its coalition, the cost of computation and extra cost of assistance fees apply. Since the pricing of fees paid by users is not changed by which MEN provides services, that is, the income is fixed, the service pricing of MENs providing services to other MENs is particularly important. Since each MEN only pays attention to their own income, both the MEN requesting the service and the MEN providing the service expect to obtain part of the income. We assume that the cost of MEN n requesting a service l from MEN m is,

$$P_{n,l}^m = T_l \cdot p_m \cdot \alpha \qquad (3)$$

where α is the coalition cost parameter. When MENs request services from the DC, they need to pay more service fees to the DC to obtain the assistance of the DC. Since the DC is too far away from MENs, and the distances between different MENs and the DC are quite different, in order to facilitate the calculation, we assume that the fee of services by the DC is the service fee paid by users, namely

$$P_{n,l}^{DC} = P_{n,l} \qquad (4)$$

3.5   The Utility of Serving at the Edge and the Remote Data Center

We believe that the service transactions between MENs and between MENs and users make MENs profitable. The service fees obtained by MENs for providing services to users are generally higher than the cost of providing services. Therefore, MENs sell services to users to obtain profits. When a MEN provides services to other MENs, it ought to charge a part of the extra assistance fee, so the MEN who provide service assistance can also get certain profits. The MEN who ask for assistance still has a part of the profits after paying the service cost and assistance fee. When MENs request services from the DC, the service fee is the service fee paid by users, so the profits of MENs are 0.

Let $V_{n,l}^n$ be profits obtained by MEN n for providing a service l.

$$V_{n,l}^n = P_{n,l} - c_{n,l} \qquad (5)$$

Let $V_{n,l}^m$ be profits obtained by MEN n requesting MEN m to provide a service l.

$$V_{n,l}^m = P_{n,l} - P_{n,l}^m \qquad (6)$$

Therefore, we have

$$V_{n,l} = \begin{cases} V_{n,l}^n, & \text{MEN provide service by itself} \\ V_{n,l}^m, & \text{other MEN in the coalition assists to provide service} \\ 0, & \text{DC assists to provide service} \end{cases} \qquad (7)$$

3.6   The Response Delay of the Edge and the Remote Data Center

The response delay of a user requesting service is mainly composed of the queuing delay waiting to be served on the MEN, the computing delay of the service, and the transmission delay of transmitting data in the network. MEN can place multiple service tasks in multiple threads for parallel computing. When all threads are occupied, each thread will have a queue. When a new service request arrives, it only needs to select the thread with the shortest queue to wait for the service. After the calculation is completed, the MEN sends the calculation result back to the user. Since the MEN is very close to the user, this period of delay is very short. We assume that is a fixed value, which is represented by d_{base}. We use $d_{n,l}$ to represent the user response delay when MEN n provides service l,

$$d_{n,l} = (\sum_{o \in q_n} C_o + C_l) \cdot t_c + d_{base} \tag{8}$$

where q_n represents the task set in the queue with the shortest queue on MEN n, and t_c is the unit task processing time. When the MEN n requests the MEN m in the coalition to provide services, a transmission delay from the MEN m to the MEN n will be added. We use $d_{n,l}^m$ to represent the user response delay when MEN n requests service l from MEN m, that is,

$$d_{n,l}^m = (\sum_{o \in q_m} C_o + C_l) \cdot t_c + B_l \cdot t_b \cdot D_{mn} + d_{base} \tag{9}$$

where t_b is the required transmission time per unit bandwidth resource, and D_{mn} is the distance between m and n. When MEN n requests DC to provide services, the transmission delay from MEN n to DC will be much longer than between MENs. For the convenience of calculation, we assume that the transmission delay from MEN to DC is a fixed value, using $d_{n,l}^{DC}$ to represent.

4   Coalition Mechanism for the Collaborative Service Caching Problem with Service Sharing

4.1   Coalition Mechanism Among MENs

We set up a MENs collaboration coalition in the edge network so that MENs in the Coalition can performed collaborative caching, increase the utilization of the service cached on MENs, reduce the user response delay, and increase the profits of MENs.

The basic idea of the algorithm is to form a stable optimal cache deployment strategy through repeated iterations, including service deployment and service response, which is called Collaborative Service Caching Mechanism (CSCM). The detailed mechanism is given in Algorithm 1.

After a MEN receives a service request from a user, the MEN calculates the expected profits of caching the service. If the expected profits are higher than the minimum expected profits among the services currently pre-cached by the MEN, the MEN adds the service to the pre-cache list. Cache services after all service requests from all MENs have been considered.

After all of the services are cached on MENs as much as possible, we consider the mechanism by which MENs respond to user service requests.

4.2   Mechanism Analysis

After MENs are formed into a coalition, each MEN can provide services to users who are in the service scope of other MENs in the coalition, which is equivalent to

---

Algorithm 1. CSCM

Input: A set of service requests. Each request contains the requested MEN and the
service, as well as the start time and end time of the request
Output: A set of caching schemes, where each MEN caches as many services as possible to generate greater profit
1: **while** there is a service request that is not pre-cached **do**
2:     A user sends a request of service l to his nearest MEN n;
3:     MEN n considers adding the service l to the pre-cache list to maximize expected
profit without violating its capacity of itself;
4: **end while**
5: Cache the services in the pre-cache list on MENs;
6: **while** there is a service request that is not responded **do**
7:     MEN n checks whether there is a cache of the service;
8:     **if** MEN n cached service l **then**
9:         MEN n puts the service l in the queue with the shortest queuing time;
10:    **else**
11:        MEN n asks other MENs in its collaborative coalition;
12:        **if** there is a MEN in the coalition who can provide the service l **then**
13:            MEN n selects the MEN m with the lowest service price;
14:            MEN m puts the service in the queue with the shortest queuing time;
15:        **else**
16:            MEN n asks the DC for service l;
17:        **end if**
18:    **end if**
19: **end while**

---

expanding the service scope of each MEN. Assuming that there are N MENs in
each coalition on average, there are u users who sent service requests within each
MEN's scope. The repetition ratio of the requested service is Δ. The repetition
ratio of all requests is α, the average number of services that each MEN can
store is m, and the repetition ratio of services stored between MENs is β, then
forming a MEN coalition can make the average hit rate increase $\frac{m(\Delta-\alpha\beta)}{u\Delta\alpha\beta}$. It
can be inferred that when the condition $(\Delta > \alpha\beta)$ is satisfied, forming coalitions
will improve the average hit rate of MENs.

On the other hand, when a MEN cannot provide services, it can request the
assistance of other MENs in the coalition, which is equivalent to improving the
storage capacity of each MEN. For MENs, the number of services that can be
provided at a lower cost is as $(N\beta - 1)$ times as original.

5   Approximation Algorithm for the Service Caching Problem

5.1   Problem Formulation

Let x_{nl} be a binary variable that indicates whether service l is cached in MEN
n. The problem can be formulated as follows.

$$max \sum_{n\in N} \sum_{l\in L} x_{nl} \cdot V_{n,l} \tag{10}$$

subject to

$$\sum_{l \in L} x_{nl} \leq Cap^c \tag{11}$$

$$V_{n,l} \geq 0 \tag{12}$$

$$x_{nl} \in \{0, 1\} \tag{13}$$

where constraint (11) says that the number of services cached per MEN cannot exceed the capacity of the MEN, and Cap^c is the capacity of the MEN. Constraint (12) ensures that the profit of a service provided by MEN is non-negative.

5.2   Randomized Algorithm

Since the objective function (10) is an Integer Linear Programming (ILP) problem, it is NP-hard. We consider relaxing the constraints on the original problem to find a feasible approximate solution in a limited time. First, we relax constraint (13) as,

$$0 < x_{nl} < 1 \tag{14}$$

Then, the ILP problem is relaxed into an LP problem with the objective function as (10) and constraints as (11), (12), and (14).

Finally, the relaxed objective function is solved with a randomized rounding algorithm. The random rounding algorithm obtains a feasible solution in each iteration and finally obtains a feasible global solution. The detailed algorithm is given in Algorithm 2, which is called Randomized Rounding (RR).

Algorithm 2. RR

Input: A set of service requests. Each request contains the requested MEN and the service, as well as the start time and end time of the request

Output: A set of caching schemes, where each MEN caches as many services as possible to generate greater profit

 Relax constraint (13) of ILP into constraint (14) and obtain an LP;

2: **while** there is a service request that is not pre-cached **do**

   Assign a requested service a cached probability of x_{nl} which satisfies constraint (14);

4:   **if** x_{nl} is a feasible solution **then**

      Put service l into the pre-cache list;

6:   **end if**

   end while

8: Cache the services in the pre-cache list on MENs;

---

6   Simulation Results and Discussions

6.1   Experiment Settings

We used a real communication dataset in our experiments. The dataset [11–13], provided by Shanghai Telecom, contains more than 7.2 million records of accessing the Internet through 3,233 base stations from 9,481 mobile phones for six months.

We compared the algorithm proposed in this paper with three algorithms: (1) A distributed coalition formation (**coalition**) [6]: different service providers share computing resources on the same cloudlet (2) A coalition formation with temporal VM sharing (**CoalitionVMS**) [6]: different service providers share VMs in different time periods on the same cloudlet (3) **Random**: MEN randomly selects cached services.

6.2   Performance Evaluation

The metrics we evaluate performance are the average response delay of users and the average profit of all MENs over the same time period.

We first evaluate the impact of the number of service types on the algorithm. The service types are 500, 1000, 2000, 4000, and 8000, respectively. The MEN's capacity was 100. The experimental results are shown in Fig. 2. As shown in Fig. 2, the increase in the number of service types will reduce the performance of all algorithms The performance of algorithms that collaborative caching among MENs changes fast than algorithms that only cooperate among service providers. The algorithm combining CSCM and RR (CSCM+RR) has the best performance, while the coalition algorithm has the worst.

Then, we evaluate the impact of the MEN's cache capacity on the algorithm's performance. The cache capacities of MENs are 100, 200, 300, 400, and 500, respectively. The service type is 1000. The experimental results are shown in Fig. 3. It can be seen from Fig. 3 that the CSCM + RR algorithm has the best performance under different MEN's cache capacities, and the performance improves as the MEN's cache capacity increases. It is worth noting that when the MEN's cache capacity exceeds 200, the average profit of coalitionVMS exceeds that of the algorithm combining CSCM and random(CSCM + random). When the MEN cache capacity further exceeds 400, the average latency of coalition-VMS is also lower than the CSCM + random algorithm.

Overall, the experimental results show that the CSCM + RR algorithm increases the average profit by 5.87% to 76.78% and reduces the average delay by 31.02% to 82.90% compared with coalitionVMS, which is the best performing in the baseline.

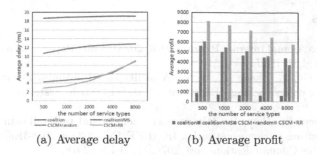

(a) Average delay   (b) Average profit

Fig. 2. The impact of the number of service types on algorithms.

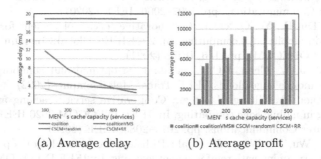

(a) Average delay   (b) Average profit

Fig. 3. The impact of the MEN's cache capacity on algorithms.

7 Conclusion

In this paper, we investigated the problem of service caching for collaboration among MENs. We proposed to form a collaborative caching coalition among MENs and proposed a mechanism CSCM for cooperative service caching so that MENs can provide service assistance to each other. In addition, we also implement this mechanism nicely using the randomized rounding algorithm. Finally, we conduct simulation experiments using a real communication dataset to evaluate the performance of the proposed algorithm. The experimental results show that the proposed algorithm has good performance in comparing various dimensions, which is better than the existing algorithms. The future work of this research includes: (1) proposing an algorithm to handle the transfer of user interest; and (2) considering a method for dynamic replacement of services on MEN in continuous time.

Acknowledgement. This work is supported by National Natural Science Foundation of China (No. 62172124), the Shenzhen Basic Research Program (Project No. JCYJ20190806143011274) and the Shenzhen Colleges and Universities Stable Support Program No. GXWD20201230155427003-20200822080602001.

References

1. Cisco: Cisco Annual Internet Report (2018–2023) White Paper, white paper (2020). http://goo.gl/l77HAJ
2. Liang, Y., Ge, J., Zhang, S., et al.: A utility-based optimization framework for edge service entity caching. IEEE Trans. Parallel Distrib. Syst. **30**(11), 2384–2395 (2019)
3. Xu, J., Chen, L., Zhou, P.: Joint service caching and task offloading for mobile edge computing in dense networks. In: IEEE INFOCOM 2018-IEEE Conference on Computer Communications, pp. 207–215. IEEE (2018)
4. Zhao, G., Xu, H., Zhao, Y., et al.: Offloading dependent tasks in mobile edge computing with service caching. In: IEEE INFOCOM 2020-IEEE Conference on Computer Communications, pp. 1997–2006. IEEE (2020)
5. Zeng, G., Du, H., Ye, Q., et al.: Collaborative service placement for maximizing the profit in mobile edge computing. In: 2021 IEEE Global Communications Conference (GLOBECOM), pp. 1–6. IEEE (2021)
6. Xu, Z., Zhou, L., Chau, S.C.K., et al.: Near-optimal and collaborative service caching in mobile edge clouds. IEEE Trans. Mob. Comput. (2022)
7. Ma, X., Zhou, A., Zhang, S., et al.: Cooperative service caching and workload scheduling in mobile edge computing. In: IEEE INFOCOM 2020-IEEE Conference on Computer Communications, pp. 2076–2085. IEEE (2020)
8. Wang, H., Wu, K., Wang, J,. et al.: Rldish: edge-assisted QoE optimization of HTTP live streaming with reinforcement learning. In: IEEE INFOCOM 2020-IEEE Conference on Computer Communications, pp. 706–715. IEEE (2020)
9. Wang, F., Wang, F., Liu, J., et al.: Intelligent video caching at network edge: a multi-agent deep reinforcement learning approach. In: IEEE INFOCOM 2020-IEEE Conference on Computer Communications, pp. 2499–2508. IEEE (2020)
10. Li, C., Zhang, Y., Song, M., et al.: An optimized content caching strategy for video stream in edge-cloud environment. J. Netw. Comput. Appl. **191**, 103158 (2021)
11. Li, Y., Zhou, A., Ma, X., et al.: Profit-aware edge server placement. IEEE Internet Things J. **9**(1), 55–67 (2022)
12. Guo, Y., Wang, S., Zhou, A., et al.: User allocation-aware edge cloud placement in mobile edge computing. Softw. Pract. Exp. **50**(5), 489–502 (2020)
13. Wang, S., Guo, Y., Zhang, N., et al.: Delay-aware microservice coordination in mobile edge computing: a reinforcement learning approach. IEEE Trans. Mob. Comput. **20**(3), 939–953 (2021)

A Decentralized Auction Framework with Privacy Protection in Mobile Crowdsourcing

Jianxiong Guo[1,2], Qiufen Ni[3], and Xingjian Ding[4(✉)]

[1] Advanced Institute of Natural Sciences,
Beijing Normal University, Zhuhai 519087, China
[2] Guangdong Key Lab of AI and Multi-Modal Data Processing,
BNU-HKBU United International College, Zhuhai 519087, China
jianxiongguo@bnu.edu.cn
[3] School of Computers, Guangdong University of Technology,
Guangzhou 510006, China
niqiufen@gdut.edu.cn
[4] Faculty of Information Technology, Beijing University of Technology,
Beijing 100124, China
dxj@bjut.edu.cn

Abstract. With the rapid popularization of mobile devices, the mobile crowdsourcing has become a hot topic in order to make full use of the resources of mobile devices. To achieve this goal, it is necessary to design an excellent incentive mechanism to encourage more mobile users to actively undertake crowdsourcing tasks, so as to achieve maximization of certain economic indicators. However, most of the reported incentive mechanisms in the existing literature adopt a centralized platform, which collects the bidding information from workers and task requesters. There is a risk of privacy exposure. In this paper, we design a decentralized auction framework where mobile workers are sellers and task requesters are buyers. This requires each participant to make its own local and independent decision, thereby avoiding centralized processing of task allocation and pricing. Both of them aim to maximize their utilities under the budget constraint. We theoretically prove that our proposed framework is individual rational, budget balanced, truthful, and computationally efficient, and then we conduct a group of numerical simulations to demonstrate its correctness and effectiveness.

Keywords: Decentralization · Incentive mechanism · Auction theory · Utility maximization · Truthfulness

---

This work was supported in part by the Start-up Fund from Beijing Normal University under grant 310432104, the Start-up Fund from BNU-HKBU United International College under grant UICR0700018-22, the Guangdong Basic and Applied Basic Research Foundation under grant 2021A1515110321 and 2022A1515010611, and Guangzhou Basic and Applied Basic Research Foundation under grant 202201010676.

1   Introduction

During the last ten years, mobile devices have been getting stronger and stronger by installing multiple sensors, microcomputers, and communication equipments. It forms a new pattern, mobile crowdsourcing (MC), which has attracted wide attention from academia and industry because of its great commercial value. The MC refers to the use of computing resources or sensory abilities of a group of mobile users to accomplish different types of tasks. A lot of different applications based on MC are gradually being industrialized, such as traffic monitoring, healthcare, and electric vehicle charging. As for a crowdsourcing platform, it wants as many mobile users as possible to participate in the crowdsourcing task. However, mobile users are reluctant to do so, because it not only consumes energy and time, but also risks exposing privacy. Therefore, while ensuring security, it is a core issue to promote users' participation.

In the existing works, a great deal of research has been done to promote users' participation in crowdsourcing by designing incentive mechanisms [2,7]. Auction theory [11,16] is commonly used in incentive mechanisms for its advantage in dealing with the interaction between buyers and sellers. In MC, it can be exploited to determine the hired mobile users and the fees paid to them, in which mobile workers are sellers and task requesters are buyers. Nevertheless, there are several severe issues in the existing auction mechanisms [8,13,16,18]. First, the complexity and generalization of both tasks and workers are not fully studied. (1) A task is published by a requester, and the task can be undertaken by many workers, but the total remuneration paid by the requester for the task is limited; (2) A worker can take on several different tasks at the same time, but the resource and energy investment it puts into each tasks are different; (3) For each worker, its total investment to tasks is limited; (4) For the same investment, different workers bear the cost of it differently. Second, in the existing auction design, they always assume that there is a fair and right-minded platform/auctioneer to collect the bidding information from buyers and sellers, and then determine the auction results. Such a centralized management pattern is not conducive to managing the joining and leaving of requesters and workers in a dynamic environment, but also threatens the users' privacy, thus reducing the incentive effect. Taking these two issues into consideration, it is a challenge to incentivize both requesters and workers that adapts to complexity and privacy protection simultaneously.

Based on the above careful observation, we propose a decentralized auction framework (DAF), and take DAF as an incentive mechanism to achieve the assignment and pricing between tasks and workers. Here, we consider the flexibility of real crowdsourcing scenarios as much as possible, and generalize it to a concise mathematical expression, which mainly covering the following points: (1) Budget constraints: the total payment paid to workers for each task has been constrained by the requester and the total investment in the task has been constrained by the worker; (2) For each worker, it can offer a arbitrary investment with its own bid according to its cost; (3) Our proposed DAF can be carried out in a decentralized manner, in which each requester can determine the winning workers to complete its task with the corresponding payment and each worker can determine whether it is willing to undertake the tasks locally and independently. The decentralized mechanism can be further integrated with emerging technologies (such as distributed computing, edge computing, and blockchain). At the same time, our DAF satisfies several design rationales of auction theory: individual rationality, budget balance, truthfulness, and computational efficiency. This will avoid price manipulation and guarantees a fair and competitive market environment. To our best knowledge, this is the first time to put forward a decentralized and truthful auction mechanism like us to address the MC problem. For convenience, we will use buyers and sellers interchangeably with task requesters and mobile users (workers) in the rest of this paper. Finally, we conduct intensive simulations to evaluate the performance of our proposed algorithms, whose results verify the correctness and effectiveness of our theoretical analysis.

2 Crowdsourcing Model and Problem Formulation

2.1 Crowdsourcing Model

In this paper, we consider a decentralized mobile crowdsourcing application where each task requester initiates its request to the mobile users. In general, we denote by $T = \{t_1, \cdots, t_j, \cdots, t_n\}$ the set of tasks submitted by requesters and $W = \{w_1, \cdots, w_i, \cdots, w_m\}$ the set of mobile users who are willing to act as workers. For each task $t_j \in T$, it needs to recruit a subset of workers in W to complete this task together, but the total payment paid to them cannot be larger than its budget B_j. For each worker $w_i \in W$, it invests resource and energy to complete the tasks, thus the total hardware/software investment to the tasks it undertakes cannot exceed R_i. In this crowdsourcing scenario, workers have high mobility. Here, we use a function $\delta : T, W \to \mathbb{R}_+$ to measure the matching degree between tasks and works, where a smaller delta means a higher degree of matching. For example, $\delta(t_j, w_i)$ can be defined as the distance between task t_j and worker w_i. Let d_j be the maximum tolerance of t_j, in other words, only those worker w_k with $\delta(t_j, w_k) \leq d_j$ can be assigned to complete task t_j.

To characterize the trading between task requesters and mobile users, we use the auction model where task requesters are buyers, who want to attract several workers to contribute to their tasks; and workers are sellers, who want to exploit their limited resources to make a profit through undertaking takes. However, in the traditional auction model, both buyers and sellers should submit all their bidding information to a third-party auctioneer, and then the auctioneer determines the assignment and pricing between buyers and sellers. It is at the risk of exposing private information such as buyer's budget and seller's investment. To overcome this drawback, we design a decentralized auction framework that does not need to submit bidding information to an auctioneer, but requires the participants to make decisions by themselves locally and independently. Thereby it greatly reduces the risk of privacy leakage.

2.2   Formulation of Decentralized Auction

We begin to talk about our decentralized auction framework (DAF) for mobile crowdsourcing, which is executed round by round. In each round, it can be decomposed into the following five steps:

- Each requester (buyer) broadcasts its crowdsourcing task t_j to the mobile users in the neighborhood.
- After receiving the task set $T = \{t_1, \cdots, t_n\}$, each worker (seller) w_i give its bidding information $\{b_{ij}, r_{ij}\}_{t_j \in T}$, where the b_{ij} is the total bid to provide r_{ij} resource investment to task t_j. If worker w_i are not willing to trade with task t_l, it sets $\{b_{il}, r_{il}\} = \emptyset$.
- After receiving all workers' bidding information $\{b_{ij}, r_{ij}\}_{w_i \in W}$, each requester has to choose a subset of workers it wants to trade as candidates because of its limited budget and corresponding payment to them.
- Up to now, each worker should know which requesters it can cooperate with to complete the crowdsourcing task, and then it chooses a subset of tasks from feasible requesters because of its limited resource and give confirmation to the selected requesters.
- Finally, each requester can cooperate with the winning workers to complete its task and finish the transaction in this round.

In each round, the auction are executed between requesters and workers in a local manner, which will determine two matrix: assignment matrix $X = \{x_{ij} : w_i \in W, t_j \in T\}$ and pricing matrix $P = \{p_{ij} : w_i \in W, t_j \in T\}$. Here, we have $x_{ij} = 1$ if w_i is a winner of t_j, else $x_{ij} = 0$. The p_{ij} is the payment paid to worker w_i because of participating in task t_j. The bid b_{ij} given by seller may not be its truthful cost, thus we denote by c_{ij} the truthful cost of worker w_i to task t_j. There is a limited budget for each requester and limited resource for each worker. Based on that, we can define the utilities of sellers and buyers. For each seller $w_i \in W$, its utility $\bar{U}_i$ can be defined as follows:

$$\bar{U}_i = \sum_{t_j \in T} \bar{u}_{ij} = \sum_{t_j \in T} x_{ij} \cdot (p_{ij} - c_{ij}) \text{ s.t. } \sum_{t_j \in T} x_{ij} \cdot r_{ij} \leq R_i, \qquad (1)$$

where each seller aims to maximize its economic return by undertaking crowd-sourcing tasks and the total resource investment cannot exceed its maximum load. For each buyer $t_j \in T$, its utility $\hat{U}_j$ can be defined as follows:

$$\hat{U}_j = \sum_{w_i \in W} \hat{u}_{ij} = \sum_{w_i \in W} x_{ij} \cdot (r_{ij} - p_{ij}) \text{ s.t. } \sum_{w_i \in W} x_{ij} \cdot p_{ij} \leq B_j, \qquad (2)$$

where each buyer aims to maximize the total investment from workers and the total payment to workers cannot exceed requester's budget.

3   Design Rationales and Algorithms

Based on the above problem defined above, a reasonable algorithm design for DAF between requesters and workers should satisfy the following properties.

- Individual rationality: In a reserve auction, it must ensure that the utility of each seller must be positive (payment is larger than cost).
- Budget balance: In our case, it implies that the total payment to sellers cannot exceed buyer's budget and each buyer is profitable by requesting a crowdsourcing task.
- Truthfulness: In a reserve auction, it is better to guarantee that no seller can get a higher utility by giving an untruthful bidding information, including total bid and investment. Thus, truthful bidding is seller's dominant strategy.
- Computational efficiency: The auction results (assignment and payment) can be obtained in polynomial time.

To the truthfulness, we assume the resource investment r_{ij} given by each worker w_i are truthful and cannot be falsified since this is monitored by other equipments. We only need to consider b_{ij} when analyzing the truthfulness. When it is truthful, the DAF prevents it from being manipulated deliberately and there is no seller having the motivation to untruthfully bid since they can get the best utility by truthful bidding. Thus, the truthfulness simplifies the strategic decisions for sellers and make sure a fair market environment, which plays an important role in mechanism design. Here, we regard the worker $w_i \in W$ as a truthful bid if $b_{ij} = c_{ij}, \forall t_j \in T$, otherwise regard it as an untruthful bid if $b_{il} \neq c_{il}, \exists t_l \in T$.

The detailed process of DAF is shown in Algorithm 1, which consists of two stages, winning seller candidate determination stage and assignment & pricing stage. In the first stage, after receiving all workers' bidding information $\{b_{ij}, r_{ij}\}_{w_i \in W}$, each requester $t_j \in T$ selects a subset of workers which are qualified to undertake its crowdsourcing task. Obviously, it is inevitable that $p_{ij} = b_{ij}$ to maximize $\hat{U}_j$. The utility $\hat{U}_j$ then becomes $\hat{U}_j = \sum_{w_i \in W} \hat{u}_{ij} = \sum_{w_i \in W} x_{ij} \cdot (r_{ij} - b_{ij})$ s.t. $\sum_{w_i \in W} x_{ij} \cdot b_{ij} \leq B_j$. For the task t_j, the qualified worker set is denoted by W_j, that is

$$W_j = \{w_i \in W : b_{ij} \leq B_j, \delta(t_j, w_i) \leq d_j\}, \tag{3}$$

where the worker's bid should be less than the total budget and the matching degree with t_j should be within the maximum tolerance. Then, we sort the W_j according to their investment per bid, denoted by $W_j = \{w_1, w_2, \cdots, w_{|W_j|}\}$ such that $r_{1j}/b_{1j} \geq r_{2j}/b_{2j} \geq \cdots \geq r_{|W_j|j}/b_{|W_j|j}$. From this sorted W_j, we select the maximum L with $L \leq |W_j|$ such that $r_{Lj} - b_{Lj} \geq 0$ and $\sum_{q=1}^{L} b_{qj} \leq B_j$. The first L workers with the highest investment per bid in W_j is denoted by $W_j^L = \{w_1, w_2, \cdots, w_L\}$. We call W_j^L as winning seller candidate set, and we have $|W_j^L| \leq |W_j|$ definitely.

Next, in the second stage, we have to determine the payment p_{ij} for each worker $w_i \in W_j^L$. For each worker $w_i \in W_j^L$, we define a sorted list $W_{j:-i} = W_j \backslash \{w_i\} = \left\{ w_{i_1}, w_{i_2}, \cdots, w_{i_{|W_j|-1}} \right\}$. From this sorted $W_{j:-i}$, we select the maximum L' with $L' \leq |W_j| - 1$ such that $r_{L'j} - b_{L'j} \geq 0$ and $\sum_{q=1}^{L'-1} b_{i_qj} + b_{ij} \leq B_j$. The first L' workers with the highest investment per bid in $W_{j:-i}$ is denoted by

---

Algorithm 1: Decentralized Auction Framework (DAF)

---

1   // **Action of each requester (buyer)** $t_j \in T$:

2   $W_j = \{w_i \in W : b_{ij} \leq B_j, \delta(t_j, w_i) \leq d_j\}$;

3   $p_{ij} = 0$ for each $w_i \in W$;

4   **if** $W_j = \emptyset$ **then**

5   No worker can be used to undertake its crowdsourcing task;

6   **return**;

7   Sort $W_j = \{w_1, \cdots, w_{|W_j|}\}$ where $r_{1j}/b_{1j} \geq \cdots \geq r_{|W_j|j}/b_{|W_j|j}$;

8   $L = \max\{L : w_L \in W_j, r_{Lj} - b_{Lj} \geq 0, \sum_{q=1}^{L} b_{qj} \leq B_j\}$;

9   $W_j^L = \{w_1, w_2, \cdots, w_L\}$;

10   **foreach** $w_i \in W_j^L$ **do**

11   Sort $W_{j:-i} = W_j \backslash \{w_i\} = \{w_{i_1}, w_{i_2}, \cdots, w_{i_{|W_j|-1}}\}$;

12     $L' = \max\{L' : w_{L'} \in W_{j:-i}, r_{L'j} - b_{L'j} \geq 0, \sum_{q=1}^{L'-1} b_{i_q j} + b_{ij} \leq B_j\}$;

13     $W_{j:-i}^{L'} = \{w_{i_1}, w_{i_2}, \cdots, w_{i_{L'}}\}$;

14   **if** $\sum_{q=1}^{L'} b_{i_q j} + b_{ij} \geq B_j$ **then**

15       $p_{ij} = r_{ij} \cdot b_{i_{L'} j}/r_{i_{L'} j}$;

16   **else**

17       $p_{ij} = \max\{r_{ij} \cdot b_{i_{L'} j}/r_{i_{L'} j}, r_{ij}\}$

18   **return**;

19   // **Action of each worker (seller)** $w_i \in W$:

20   $T_i = \{t_j \in T : w_i \in W_j^L\}$;

21   Sort $T_i = \{t_1, \cdots, t_{|T_i|}\}$ where $(p_{i1} - c_{i1})/r_{i1} \geq \cdots \geq (p_{i|T_i|} - c_{i|T_i|})/r_{i|T_i|}$;

22   $x_{ij} = 0$ for each $t_j \in T$; $R = R_i$;

23   **foreach** $t_j \in T_i$ **do**

24   **if** $R - r_{ij} \geq 0$ **then**

25       $x_{ij} = 1$; $R = R - r_{ij}$;

26   **return**;

$W_{j:-i}^{L'} = \{w_{i_1}, w_{i_2}, \cdots, w_{i_{L'}}\}$. Totally, there is a sorted list $W_{j:-i}^{L'}$ associated with each worker $w_i \in W_j^L$. For each position i_k in $W_{j:-i}^{L'}$, we can get the maximum bid $b'_{ij(k)}$ that the worker w_i could bid in order to replace w_{i_k} at the position i_k. To achieve this goal, we have $r_{ij}/b'_{ij(k)} \geq r_{i_k j}/b_{i_k j}$. Thus, we define a $b'_{ij(k)}$ for each worker $w_{i_k} \in W_{j:-i}^{L'}$ as

$$b'_{ij(k)} = r_{ij} \cdot b_{i_k j}/r_{i_k j}. \tag{4}$$

Similarly, we can deal with other workers in $W_{j:-i}^{L'}$ in the same way. Based on Eq. (4), we can get that $b'_{ij(1)} \leq b'_{ij(2)} \leq \cdots \leq b'_{ij(L')}$ where each $b'_{ij(k)}$ is the worker w_i's maximum bid to replace the worker w_{i_k} in $W_{j:-i}^{L'}$. If the worker w_i replaces w_{i_k} in $W_{j:-i}^{L'}$, we can see that

$$W_{j:-i}^{L'} \cup \{w_i\} = \{w_{i_1}, w_{i_2}, \cdots, w_{i_{k-1}}, w_i, w_{i_k}, \cdots, w_{i_{L'}}\}, \tag{5}$$

here the worker w_{i_k} will be moved after w_i. Then, there are two cases we need to discuss as follows.

- $\sum_{q=1}^{L'} b_{i_q j} + b_{ij} > B_j$: In this case, if the worker w_i want to win, it must replace at least one worker in $W_{j:-i}^{L'}$. Based on $\sum_{q=1}^{L'-1} b_{i_q j} + b_{ij} \leq B_j$, when the worker w_i replaces any w_{i_k} in $W_{j:-i}^{L'}$, we have $\sum_{q=1}^{k-1} b_{i_q j} + b_{ij} \leq B_j$, which will not exceed the requester's budget. If the worker w_i can replace one of the workers in $W_{j:-i}^{L'}$, it will be selected as a winning candidate by task t_j. Thus, the payment to work w_i is

$$p_{ij} = \max \left\{ b'_{ij(1)}, b'_{ij(2)}, \cdots, b'_{ij(L')} \right\} = b'_{ij(L')}. \tag{6}$$

- Otherwise: In this case, we have $\sum_{q=1}^{L'} b_{i_q j} + b_{ij} \leq B_j$, which indicates that $W_{j:-i}^{L'} = |W_j| - 1$ or $r_{L'+1,j} - b_{L'+1,j} < 0$. In addition to replace one of the workers in $W_{j:-i}^{L'}$, the worker w_i can be placed after all the workers in $W_{j:-i}^{L'}$. The worker w_i will be selected as a winning candidate when satisfying $r_{ij} - b_{ij} \geq 0$. Thus, the payment to worker w_i is

$$p_{ij} = \max \left\{ b'_{ij(L')}, r_{ij} \right\} \tag{7}$$

After the above winning seller candidate determination stage at requester side, each worker $w_i \in W$ may be selected as a candidate by multiple requesters. Thus, it needs to choose the tasks that can achieve as much utility as possible. For each worker $w_i \in W$, we denoted by $T_i = \{t_j \in T : w_i \in W_j^L\}$ the task set that the worker w_i can choose. Since the total resource investment of each worker cannot exceed its maximum load, we sort the T_i according to their utility per investment, denoted by $T_i = \{t_1, t_2, \cdots, t_{|T_i|}\}$ such that $(p_{i1} - c_{i1})/r_{i1} \geq (p_{i2} - c_{i2})/r_{i2} \geq \cdots \geq (p_{i|T_i|} - c_{i|T_i|})/r_{i|T_i|}$. From this sorted list T_i, we traverse every task from left to right, and the worker will undertake the current task if have enough space in investment.

4   Theoretical Analysis

In this section, we begin to analyze whether our proposed DAF can meet the requirement of design rationales, including individual rationality, budget balance, truthfulness, and computational efficiency, respectively.

Lemma 1. *The DAF is individually rational to seller.*

Proof. Given any worker $w_i \in W_j^L$ for any task $t_j \in T$, we denote by $w_{i_i j}$ the replacement of worker w_i that is placed in i-th position in the $W_{j:-i}$. When the winning worker w_i joins in this sorted list, the $w_{i_i j}$ cannot be placed in i-th position by now. When $i \leq L'$, we have $r_{ij}/b_{ij} \geq r_{i_i j}/b_{i_i j}$, it implies that $b_{ij} \leq r_{ij} \cdot b_{i_i j}/r_{i_i j} \leq p_{ij}$. When $i > L'$, we have $b_{ij} \leq r_{ij} \leq p_{ij}$ because it is a winner. Thus, the utility of worker w_i is always non-negative. ∎

Lemma 2. *The DAF is budget balanced.*

Proof. For any requester $t_j \in T$, its total payment to sellers $\sum_{w_i \in W} x_{ij} \cdot p_{ij} \leq B_j$ has been relaxed to $\sum_{w_i \in W} x_{ij} \cdot b_{ij} \leq B_j$ so as to guarantee the truthfulness. Since the payment to a buyer is larger than its bid, the total payment to sellers cannot be ensured to be less than the requester's budget. However, according to the auction process given by DAF, not all the workers in W_j^L undertake this task, and then total payment will not exceed the budget with a high probability.

Suppose each worker $w_i \in W_j^L$ undertake task t_j, the buyer's utility $\hat{U}_j = \sum_{w_i \in W_j^L} (r_{ij} - p_{ij})$. Thus, it is sufficient to show that $r_{ij} \geq p_{ij}$ for each $w_i \in W_j^L$. For each worker $w_i \in W_j^L$, we can consider two cases as before. When $\sum_{q=1}^{L'} b_{i_q j} + b_{ij} > B_j$, we have $p_{ij} = \max \left\{ b'_{ij(1)}, b'_{ij(2)}, \cdots, b'_{ij(L')} \right\} = b'_{ij(L')}$. Based on that, we have

$$p_{ij} = b'_{ij(L')} = r_{ij} \cdot b_{i_{L'} j}/r_{i_{L'} j} \leq r_{ij}, \tag{8}$$

because $r_{i_{L'} j} - b_{i_{L'} j} \geq 0$. Otherwise, we have $p_{ij} = \max \left\{ b'_{ij(L')}, r_{ij} \right\} \leq r_{ij}$. Thus, the utility of requester t_j is non-negative, and the DAF is budget balanced. ∎

According to Myerson theory [11], in a reserve auction, it is truthful if and only if: (1) The bid of each seller is monotonic. For each worker w_i, if it wins the task t_j with bid b_{ij}, then it must win this task with any bid that is smaller than b_{ij}. (2) The payment to each seller is critical. For each worker w_i with bid b_{ij} to task t_j, the payment p_{ij} is the maximum bid it can win. ∎

Lemma 3. *The DAF is truthful.*

Proof. First, we need to show that the bid of each seller is monotonic. Let us consider a worker $w_i \in W_j^L$ with its bid b_{ij} to task $t_j \in T$. When it gives a bid $b_{ij}^* \leq b_{ij}$, this worker w_i will not be moved backward in the sorted list W_j since we have $r_{ij}/b_{ij}^* > r_{ij}/b_{ij}$ definitely. Therefore, if the worker w_i is a winner in the winning seller candidate determination stage with bid b_{ij}, then it will be a winner as well with bid b_{ij}^*.

Second, we need to show that the payment to each seller is critical. For each worker $w_i \in W_j^L$, we can consider two cases as before. When $\sum_{q=1}^{L'} b_{i_q j} + b_{ij} > B_j$, the worker w_i must replace at least one worker in $W_{j:-i}^{L'}$. If it gives a bid $b_{ij}^* > p_{ij} = b'_{ij(L')}$, we have $r_{ij}/b_{ij}^* < r_{i_{L'} j}/b_{i_{L'} j}$. The worker w_i cannot replace any worker in $W_{j:-i}^{L'}$ because of buyer's budget constraint, and then it cannot be a winner. Otherwise, suppose the worker w_i cannot replace any workers in $W_{j:-i}^{L'}$, it still wins this auction if $r_{ij} \geq b_{ij}$ because the corresponding requester has enough budget. At this time, if it gives a bid $b_{ij}^* > p_{ij} = r_{ij}$, it cannot be added into the W_j^L and then fails. Thus, the DAF is truthful. ∎

Lemma 4. *The DAF is computationally efficient.*

Proof. Let us deal with requester side and worker side respectively. Here, we denoted by $|T| = n$ and $|W| = m$ as before. For each requester $t_j \in T$, it takes $O(m)$ time to check all the workers and determine the qualified worker set W_j, and then takes $O(m \log m)$ time to sort. We can see $|W_j^L| \leq |W_j| \leq W$. To determine the payment of all sellers in W_j^L, it takes at most m loop, and consume at most $O(m)$ time in each loop. Thus, the total time complexity for each requester to determine the winning seller candidates and their payment is $O(m^2) + O(m \log m) = O(m^2)$. For each worker $w_i \in W$, it takes $O(n)$ to check all the requesters and determine the possible task set W_i and then takes $O(n \log n)$ time to sort. We can see $T_i \leq T$, thereby it consume at most $O(n)$ to choose the tasks. Thus, the total time complexity for each worker to choose the tasks it want to undertake is $O(n \log n) + O(n) = O(n \log n)$. Finally, the DAF is computationally efficient. ∎

Theorem 1. *The DAF given by Algorithm 1 is an effective decentralized auction mechanism, which can satisfy the above four design rationales: Individual rationality, budget balance, truthfulness, and computational efficiency.*

Proof. It can be proven by combining Lemma 1 to Lemma 4. ∎

5   Numerical Simulations

In this section, we implement our DAF in a pre-defined virtual crowdsourcing application, which is located in a area with 1000×1000 m². There are n tasks and m workers distributed uniformly in this area. The matching degree $\delta(t_j, w_i)$ between task t_j and worker w_i is defined as their distance, $\delta(t_j, w_i) = \sqrt{(x_i - x_j)^2 + (y_i - y_j)^2}$. For each task t_j, its tolerance d_j is randomly sampled from $[200, 400]$ and its budget B_j is randomly sampled from $[30, 50]$. For each worker w_i, it is a critical setting that how do determine the relationship between cost c_{ij} and r_{ij}. Intuitively, c_{ij} will be larger if t_j is far away from w_i. Thus, we give $c_{ij} = r_{ij} \cdot (c_i' + \sigma(t_j, w_i)/(1000\sqrt{2}))$, where c_i' is its unit cost per investment. Here, its investment r_{ij} is randomly sampled from $[2, 8]$, its unit cost c_i' is randomly sampled from $[0.5, 0.7]$, and maximum load is sampled from $[15, 25]$. By default, we assume that $n = 20$ and $m \geq n$.

Figure 1 represents the utilities of tasks and workers in different settings. Shown as Fig. 1 (a), the utility of each task is positive, and we can use it to check whether our algorithm satisfies the budget balance. Shown as Fig. 1 (b), we test the changing trend in average utilities of tasks or workers with the increase of the number of workers. First, as the number of workers increases, the average

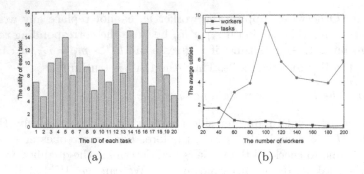

(a)   (b)

Fig. 1. The utilities of tasks and workers in different settings.

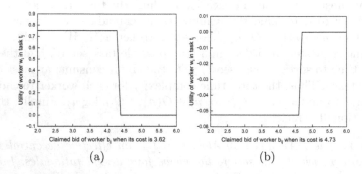

(a)   (b)

Fig. 2. The utility of a representative worker. (a) w_i's utility at t_j when winning with truthful bid; (b) w_i's utility at t_j when losing with truthful bid.

utility of workers will decrease since their competition is heating up when the total tasks is certain. However, the average utility of tasks will first increase and then decrease. This shows that workers are not enough at first, and then gradually reach saturation. Increasing the number of workers after saturation does not significantly increase the utility of tasks. Figure 2 represents the utility of a representative worker to verify the truthfulness of sellers. Shown as Fig. 2 (a), when worker w_i give a truthful bid $b_{ij} = c_{ij} = 3.62$, it will win the auction and get utility $\bar{u}_{ij} = 0.75$. If it reduces its bid, its utility will keep constant. If it increases its bid, its utility will change to zero when being larger than the critical price. Shown as Fig. 2 (b), when worker w_i give a truthful bid $b_{ij} = c_{ij} = 4.73$, it will lose the auction and get utility $\bar{u}_{ij} = 0$. If it reduces its bid and win the auction with its untruthful bid, its utility will be negative. If it increases its bid, it will lose as well. Thus, our mechanism achieve the design rationales.

6   Related Works

In this section, we summarize some important literatures about mechanism design in MC problem. The auction, as a technique of game theory, has been commonly used to deal with users' strategic behaviors in a variety of network-based applications, not only in MC [8,13,16,18], but also in cloud/edge computing [5,9], spectrum trading [19,20], and so on. Tong *et al.* [12] summarized the main problems in spatial crowdsourcing, including task assignment, quality control, incentive mechanism, and privacy protection. The existing auction-based incentive mechanisms can be divided into several categories according to their objectives. The first is to maximize the social welfare [4,15]. Wang *et al.* [15] proposed a truthful two-stage auction algorithm with location privacy-preserving for MC systems. Gao *et al.* [4] designed a reverse-auction-based mechanism with a nearly minimum social cost. The second is to maximize the total utility of the platform [6,17]. Zhang *et al.* [17] proposed two optimization models to maximize the user cardinality and sensing utility function in mobile crowdsensing applications. Guo *et al.* [6] designed a combinatorial double auction mechanism to maximize the revenue of edge computing platform.

In fact, all these works considered specific and restricted scenarios in MC applications, and they all depended on a third-party auctioneer and thus neglect the privacy protection. Then, there were some researchers introducing differential privacy [1,10] to prevent the adversary from inferring participants' sensitive information. However, such a method increases the difficulty of algorithm design and affects the efficiency of algorithm operation. Thus, we adopt a decentralized approach to design our auction mechanism, which effectively reduces the sharing of some sensitive information, such as budget and investment. The only works similar to us were in [3,14], which gave us a distributed auction for MC, but their problems and algorithms are totally different from us.

7 Conclusion

In this paper, we design and implement a decentralized auction framework (DAF) to effectively achieve task assignment and pricing between task requesters and mobile users in generalized MC applications. Different from previous works, our DAF is decentralized while ensuring the design rationales, where each participant can make decisions locally, thus avoiding sharing some confidential information and increasing security. Theoretical analysis and simulation results validate that both buyers and sellers can optimize their own utilities and guarantee the truthfulness in a decentralized manner.

References

1. Chen, Z., Ni, T., Zhong, H., Zhang, S., Cui, J.: Differentially private double spectrum auction with approximate social welfare maximization. IEEE Trans. Inf. Forensics Secur. **14**(11), 2805–2818 (2019)
2. Ding, X., Guo, J., Li, D., Wu, W.: An incentive mechanism for building a secure blockchain-based internet of things. IEEE Trans. Netw. Sci. Eng. **8**(1), 477–487 (2020)

3. Duan, Z., Li, W., Cai, Z.: Distributed auctions for task assignment and scheduling in mobile crowdsensing systems. In: 2017 IEEE 37th International Conference on Distributed Computing Systems, pp. 635–644. IEEE (2017)
4. Gao, G., Xiao, M., Wu, J., Huang, L., Hu, C.: Truthful incentive mechanism for nondeterministic crowdsensing with vehicles. IEEE Trans. Mob. Comput. **17**(12), 2982–2997 (2018)
5. Guo, D., Gu, S., Xie, J., Luo, L., Luo, X., Chen, Y.: A mobile-assisted edge computing framework for emerging IoT applications. ACM Trans. Sens. Netw. **17**(4), 1–24 (2021)
6. Guo, J., Ding, X., Jia, W.: Combinatorial resources auction in decentralized edge-thing systems using blockchain and differential privacy. arXiv preprint arXiv:2108.05567 (2021)
7. Guo, J., Ding, X., Wu, W.: A blockchain-enabled ecosystem for distributed electricity trading in smart city. IEEE Internet Things J. **8**(3), 2040–2050 (2020)
8. Guo, J., Ding, X., Wu, W.: Reliable traffic monitoring mechanisms based on blockchain in vehicular networks. IEEE Trans. Reliab. (2021). https://doi.org/10.1109/TR.2020.3046556
9. Jiao, Y., Wang, P., Niyato, D., Suankaewmanee, K.: Auction mechanisms in cloud/fog computing resource allocation for public blockchain networks. IEEE Trans. Parallel Distrib. Syst. **30**(9), 1975–1989 (2019)
10. McSherry, F., Talwar, K.: Mechanism design via differential privacy. In: 48th Annual IEEE Symposium on Foundations of Computer Science (FOCS 2007), pp. 94–103. IEEE (2007)
11. Nisan, N., Roughgarden, T., Tardos, É., Vazirani, V.V.: Algorithmic Game Theory. Cambridge University Press, Cambridge (2007)
12. Tong, Y., Zhou, Z., Zeng, Y., Chen, L., Shahabi, C.: Spatial crowdsourcing: a survey. VLDB J. **29**(1), 217–250 (2019). https://doi.org/10.1007/s00778-019-00568-7
13. Wang, J., Tang, J., Yang, D., Wang, E., Xue, G.: Quality-aware and fine-grained incentive mechanisms for mobile crowdsensing. In: 2016 IEEE 36th International Conference on Distributed Computing Systems, pp. 354–363. IEEE (2016)
14. Wang, X., Tushar, W., Yuen, C., Zhang, X.: Promoting users' participation in mobile crowdsourcing: a distributed truthful incentive mechanism (DTIM) approach. IEEE Trans. Veh. Technol. **69**(5), 5570–5582 (2020)
15. Wang, Y., Cai, Z., Tong, X., Gao, Y., Yin, G.: Truthful incentive mechanism with location privacy-preserving for mobile crowdsourcing systems. Comput. Netw. **135**, 32–43 (2018)
16. Yang, D., Xue, G., Fang, X., Tang, J.: Incentive mechanisms for crowdsensing: crowdsourcing with smartphones. IEEE/ACM Trans. Netw. **24**(3), 1732–1744 (2015)
17. Zhang, X., Jiang, L., Wang, X.: Incentive mechanisms for mobile crowdsensing with heterogeneous sensing costs. IEEE Trans. Veh. Technol. **68**(4), 3992–4002 (2019)
18. Zhou, R., Li, Z., Wu, C.: A truthful online mechanism for location-aware tasks in mobile crowd sensing. IEEE Trans. Mob. Comput. **17**(8), 1737–1749 (2017)
19. Zhu, K., et al.: Privacy-aware double auction with time-dependent valuation for blockchain-based dynamic spectrum sharing in IoT systems. IEEE Internet Things J. (2022). https://doi.org/10.1109/JIOT.2022.3165819
20. Zhu, R., Liu, H., Liu, L., Liu, X., Hu, W., Yuan, B.: A blockchain-based two-stage secure spectrum intelligent sensing and sharing auction mechanism. IEEE Trans. Industr. Inf. **18**(4), 2773–2783 (2021)

Profit Maximization for Multiple Products in Community-Based Social Networks

Qiufen Ni[1] and Jianxiong Guo[2,3](✉)

[1] School of Computers, Guangdong University of Technology,
Guangzhou 510006, China
niqiufen@gdut.edu.cn
[2] Advanced Institute of Natural Sciences, Beijing Normal University,
Zhuhai 519087, China
[3] Guangdong Key Lab of AI and Multi-Modal Data Processing,
BNU-HKBU United International College, Zhuhai 519087, China
jianxiongguo@bnu.edu.cn

Abstract. In this paper, we studies the profit maximization problem for multiple kinds of products in social networks. It is formulated as a Profit Maximization Problem for Multiple Products (PMPMP), which aims at selecting a set of seed users within the total budget B such that the total profit for k kinds of products is maximized. We introduce the community structure and assume that different kinds of products are adopted by different groups of people, and different product information spread in different communities under the IC information propagation model. We prove that the objective function satisfies the k-submodularity, and then use the multilinear extension to relax the objective function. A continuous greedy algorithm is put forward for the relaxed function, which can obtain an $\frac{1}{2}$ approximation performance guarantee, respectively. The experimental results on two real world social network datasets show the effectiveness of the proposed continuous greedy algorithm.

Keywords: Social network · Profit maximization · k-submodular · Multilinear extension

1 Introduction

Social networks, such as WeChat, Facebook, Twitter, are embedded in our daily lives and are important platforms for people to communicate and for business to advertise. Companies make use of the word-of mouth effect of social networks to promote their products, this application in social networks is called viral marketing. The information spread process in viral marketing is formulated as Influence Maximization (IM) by Kempe *et al.* in [1], which aims at selecting a set of users as seeds to maximize the expected number of users who are influenced by seeds. In [1], they propose two classic information propagation models: Independent Cascade (IC) model and Linear Threshold (LT) model, both of them simulate

© The Author(s), under exclusive license to Springer Nature Switzerland AG 2022
Q. Ni and W. Wu (Eds.): AAIM 2022, LNCS 13513, pp. 219–230, 2022.
https://doi.org/10.1007/978-3-031-16081-3_19

how influence spreads from a initial seed set in the social networks. They also formulate the influence maximization problem as a monotone submodular function, and present a greedy algorithm to solve it, the returned solution can get an $1 - 1/e$ approximation performance guarantee.

Most of the existing works focus on the influence maximization related problem with a single information diffusion, i.e., there is only one product information spreading in the social network. While in our real life, multiple products may propagate in the social network at the same time. Given k kinds of products, a budget B, assume that different products have different activation costs and different profits can be obtained when the product is purchased, how to allocate the budget to seeds for k kinds of products such that the total profit is maximized? We study the profit maximization problem for multiple products in this paper, and assume that each person can purchase only one product. We aim to allocate discounts to k sets of seed users for k kinds of products. As we all know, social networks have the characteristics of community structures [2]. Nodes in the same community are closely connected, whereas nodes in different communities are sparsely connected. What's more, influence spreads quickly within communities, but much more slowly across different communities [3]. We introduce the community structure of a social network, and let k sets of seeds are selected from k different communities, respectively. The objective function that maximizing the total profit of k kinds of products can be formulated as a k-submodular function. Approximation algorithms with theoretical guarantee are proposed in our work. We summarize the main contributions in this paper as follows:

- Community structure is introduced, and we consider that k kinds of product information spread in k different communities.
- We formulate the profit maximization for multiple products problem as a k-submodular function with knapsack constraint problem.
- We relax the k-submodular objective function with multilinear extension technique. A continuous greedy algorithm with constant approximation ratio is proposed.
- Intensive experiments are conducted to test the performance of the proposed algorithm, which shows the superiority of our algorithms.

The rest of the paper is arranged as follows. Section 2 is the related work. Section 3 is the network model and problem formulation. In Sect. 4, the k-submodularity, relaxation for objective function, continuous greedy algorithm, the theoretical results are proposed. Comprehensive experiments are conducted on two real-world social network datasets in Sect. 5. Section 6 is the conclusion of this work.

2 Related Work

In this paper, we consider the profit maximization problem with multiple products in the network. We summarize the related studies on our work as follows.

Influence Maximization with Multiple Information: Most existing research consider the IM related problem with a single information cascade in the social network. In recent years, some studies on multiple products information propagation in the network have been put forward. Bharathi *et al.* [4] firstly propose the competitive influence maximization problem, which is solved with the method of game theory. Wu *et al.* [5] study the Influence Blocking Maximization (IBM) problem under two competitive IC diffusion model, they devise Maximum Influence Arborescence based heuristic algorithms to solve the proposed problem. Wu *et al.* [6] study Multiple Influence Maximization (MIM) problem that multiple information diffuse simultaneously in a network. The objective for this problem is to maximize the accumulative influence of all the information within k seed budget. The greedy algorithm is presented to solve the MIM problem with an $1/3$ approximation ratio.

Profit Maximization: Profit maximization is a problem that solves how to devise optimal strategies to allocate the limited budget such that the total profit of the product is maximized, it is a transformation of the influence maximization. Han *et al.* [7] propose a discount allocation strategy to maximize the revenue of one product. The objective function is proved to be non-monotone and non-submodular. Then a "surrogate optimization" algorithm and two randomized algorithms with constant approximation ratio are put forward to solve the problem. Zhang *et al.* [8] investigate the Profit Maximization with Multiple Adoptions (PM^2A) problem. They design two approximation algorithms to maximize the total profit of multiple products by selecting a set of seeds. Zhang *et al.* [9] put forward a multiple product IC (MPIC) model for the viral marketing of multiple products. Each user can purchase more than one product at the same time. The objective function is to maximize the total profit of multiple products. They propose a series of algorithms with approximation performance guarantee and a heuristic algorithm which has less running time.

k-Submodular: Huber *et al.* [10] generalize the submodular function as k-submodular function. Different from the submodular function where the input only has one set, the input of k-submodular function is k disjoint sets. The k submodular function becomes the submodular function for $k = 1$ and bisubmodular function for $k = 2$, respectively [11]. Ohsaka *et al.* [12] propose greedy algorithms for maximizing a monotone k-submodular function under the total size constraint and individual size constraint, respectively, both of which obtain constant approximation factor. Ward and Živný [13] study the maximization of unconstrained non-monotone and k-submodular function. The algorithm they proposed can obtain an approximation ratio of $\max\{\frac{1}{3}, \frac{1}{1+a}\}$ where $a = max\{1, \sqrt{\frac{k-1}{4}}\}$. They also propose a greedy algorithm for the maximization problem of unconstrained monotone k-submodular function, which has an $\frac{1}{2}$ approximation performance guarantee.

3 Network Model and Problem Formulation

3.1 The Network Model

A social network is represented as a directed graph $G(V, E)$, where each node $v \in V$ represents a user, and each edge $(u, v) \in E$ is the follow relationship of user u and v. $N^-(v)$ and $N^+(v)$ are used to denote the incoming neighbor set and outgoing neighbor set of a node v, respectively. The information propagation process of the IC model and LT model is formulated in literature [1].

We adopt the IC model for our problem in this paper. Then we introduce the definition of IC model.

Definition 1 (IC model). *Nodes in social networks are either active or inactive state, and the initial state of each node is inactive. Each edge $e = (u, v) \in E$ is associated with an influence probability $p_{uv} \in (0, 1]$. When node u is activated at time t, he has one chance to activate his each inactive outgoing neighbor $v \in N^+(u)$ with probability p_{uv}. The influence spread process terminates if there are no new nodes can be activated in the current round.*

3.2 Problem Formulation

Give a social network $G = (V, E)$, we consider the community structure characteristic of social networks. We use the algorithm proposed in literature [14] to partition the network into k disjoint communities $C = (C_1, C_2, \cdots, C_k)$. Assume that marketers wants to promote k kinds of products in the social network. k kinds of products information propagate under the IC model at the same time. We aim to choose k seed sets $\boldsymbol{S} = \{S_1, S_2, \cdots, S_k\}$ from k different communities respectively and provide discounts to them. In our problem, we consider that the influence rarely crosses different communities, i.e., the influence of product i only spreads within community i. Let $\sigma(S_i|C_i)$ be the expected influence spread within community i for seed set S_i, i.e., the expected number of users who adopt product i in community i. Let $f(S_i|C_i)$ be the total profit that obtained by purchasing product i. Moreover, $\sigma(\boldsymbol{S}|G)$ and $f(\boldsymbol{S}|G)$ are the expected number of influenced people and the total profit obtained by adopting k kinds of products, respectively.

The profit maximization problem for the multiple products marketing at the same time in the community-based social network can be formulated as follows:

Problem 1 (Profit Maximization Problem for Multiple Products (PM PMP)). *Given a social network graph $G = (V, E, C)$ with community structure $C = \{C_1, C_2, \cdots, C_k\}$, where $V = \bigcup_{i=1}^k C_i$, k kinds of products, the IC model, the cost c_i that activating a node to purchase product i, the profit p_i that a node can gain when he adopts product i, the total activation cost B for seed set $\boldsymbol{S}$, the expected influence spread $\sigma(S_i|C_i)$ for seed set S_i. Our target is to select an optimal seed set $\boldsymbol{S} = \{S_1, S_2, \cdots, S_k\}$ where S_i is the seed set from community C_i such that the total profit $f(\boldsymbol{S}|G)$ is maximized, i.e.,*

$$S^* = \arg \max_{S \in (k+1)^V} f(S|G)$$

$$s.t. \ \sum_{i=1}^{k} c_i |S_i| \leq B \tag{1}$$

From the definition of PMPMP, we can know that $f(S|G) = \sum_{i=1}^{k} p_i \sigma(S_i|C_i)$. Kempe et al. [1] proved that the influence maximization problem under IC and LT model are both NP-hard. When there is only one product, our PMPMP is equivalent to the traditional influence maximization problem under the IC model. Therefore, the PMPMP is also a NP-hard problem.

4 Solution for PMPMP

In this section, we solve the PMPMP. Firstly, we analyze the properties of the objective function for PMPMP.

4.1 Properties of Profit Maximization Function f

Firstly, we introduce the monotonicity and submodularity of a set function. A function $h: 2^X \to \mathbb{R}$ is monotone if it satisfies $h(C) \leq h(D)$ when $C \subseteq D \subseteq X$ and submodular if $h(C \cup \{v\}) - h(C) \geq h(D \cup \{v\}) - h(D)$ when $C \subseteq D \subseteq X$ and $v \notin D$.

Then, we introduce the k-submodularity. Let X be a finite non-empty set, and let $(k+1)^X := \{(U_1, \cdots, U_k)|U_i \subseteq X, \forall i \in \{1, 2, \cdots, k\}, U_i \cap U_j = \emptyset, \forall i \neq j\}$ be the family of k disjoint sets. A function $h: (k+1)^X \to \mathbb{R}$ is k-submodular if for any $U = \{U_1, \cdots, U_k\}$ and $W = \{W_1, \cdots, W_k\}$ in $(k+1)^X$, we can get

$$h(U) + h(W) \geq h(U \sqcup W) + h(U \sqcap W), \tag{2}$$

where

$$U \sqcap W := (U_1 \cap W_1, \cdots, U_k \cap W_k),$$

$$U \sqcup W := (U_1 \cup U_1 \backslash (\bigcup_{i \neq 1} U_i \cup W_i), \cdots, U_k \cup W_k \backslash (\bigcup_{i \neq k} U_i \cup W_i)).$$

When $k = 1$, the definition of k-submodular in Eq. 2 agrees with the definition of submodular function, which shows that k-submodular function is a generalization of submodular function.

A k-submodular function indicates that it satisfies the properties of orthant submodularity and pairwise monotonicity. Then, we introduce the marginal gain in a k-submodular function for better understanding these two properties. Let $\preceq$ denotes the partial order on $(k+1)^X$, for $U = (U_1, \cdots, U_k)$ and $W = (W_1, \cdots, W_k)$, if $U_i \subseteq W_i$ for every $i \in [k]$, $U \preceq W$ holds. When an element $e(e \notin \bigcup_{l \in [k]} U_l$ and $i \in [k])$ is added to the i-th set of U, the marginal gain of h can be denoted as

$$\Delta_{e,i} h(U) = h(U_1, \cdots, U_{i-1}, U_i \cup \{e\}, U_{i+1}, \cdots, U_k) - h(U_1, \cdots, U_k).$$

When the marginal gain $\Delta_{e,i}h(U) \geq 0$, the function h is monotonicity. It is *pairwise monotonicity*, if

$$\Delta_{e,i}h(U) + \Delta_{e,j}h(U) \geq 0$$

for $j \in [k]$ and $i \neq j$. And it is *orthant submodularity* if the marginal gain satisfies:

$$\Delta_{e,i}h(U) \geq \Delta_{e,i}h(W)$$

for any $U, W \in (k+1)^X$ with $U \preceq W$, $e \notin \bigcup_{l \in [k]} W_l$, and $i \in [k]$.

Theorem 1 [15]. *A function h: $(k+1)^X \to \mathbb{R}$ is a k-submodular function if and only if h satisfies orthant submodularity and pair monotonicity.*

The profit maximization function f for PMPMP is clearly monotone since more cost can use to select more seeds, and more seeds will have more influences in the network. Then, we explore the submodularity of the objective function f in Theorem 2 as follows.

Theorem 2. *The profit function f for the PMPMP under the IC model is k-submodular.*

Proof. We omit the proof as the limitation of conference pages.

4.2   Relaxation of Profit Function f

The objective function f for PMPMP is a k-submodular function with knapsack constraint. We are inspired by [16] to devise a continuous greedy algorithm to solve it, and get an efficient approximation performance guarantee at the same time. We introduce the definition of multilinear extension at first, which is a good tool to relax a submodular set function.

Definition 2 (Multilinear Extension). *Let h be a monotone submodular set function h: $2^X \to \mathbb{R}^+$. The Multilinear Extension of h is the function H: $x \in [0,1]^X \to \mathbb{R}$, and it is defined as:*

$$H(x) = \mathbb{E}_{T \sim x}[h(T)] = \sum_{T \subseteq X} h(T) \prod_{i \in T} x_i \prod_{i \in X \setminus T} (1 - x_i).$$

The multilinear extension can be explained in terms of probability. Given $x \in [0,1]^X$, let T be a random subset of X, where each element $i \in X$ is independently included to 1 with probability x_i and not included with probability $1 - x_i$. Based on the definition of multilinear extension above, we define the multilinear extension for a k-submodular function.

Definition 3 (Multilinear Extension for a k-submodular function). *Let h be a k-submodular function $h\colon (k+1)^X \to \mathbb{R}^+$. The Multilinear Extension of h is the function $H\colon P \to \mathbb{R}$, and it is defined as:*

$$H(\boldsymbol{x}) = \sum_{T_1 \uplus \cdots \uplus T_k = T \subseteq X} h(T_1, T_2, \cdots, T_k) \Big(\prod_{j \in [k]} \prod_{i \in T_j} x_{i,j} \Big) \prod_{i \in X \setminus T} \Big(1 - \sum_{j=1}^{k} x_{i,j} \Big).$$

where the polytope $P = \{ \boldsymbol{x} \in [0,1]^{n \times k} : \sum_{j=1}^{k} x_{i,j} \leq 1, \forall i \in [n] \}$ is the domain of function H, $n = |X|$, $\uplus$ denotes disjoint union, $\boldsymbol{x} = (x_{1,1}, \cdots, x_{1,k}, \cdots, x_{n,1}, \cdots, x_{n,k}) \in [0,1]^{n \times k}$. The definition of multilinear extension for a k-submodular function also can be interpreted from the perspective of probability, each element $i \in [n]$ is selected independently with probability $\sum_{j=1}^{k} x_{i,j}$, and element i is allocated to set T_j exclusively obeying a categorical distribution $Pr[i \in T_j] = x_{i,j} / \sum_{j=1}^{k} x_{i,j}$.

Corollary 1 [17]. *For a k-submodular function h, its multilinear extension function H satisfies the following properties:*

1. *H is concave along any direction $d \geq 0$.*
2. *When all the $x_{i',j}$ are fixed where $i' \neq i$ and $j \in [k]$, $\frac{\partial H}{\partial x_{i,j}}$ is a constant.*
3. *Let $z_{i,j}$ be a one-hot vector, where the (i,j)-th element is equal to 1. Then for any $i_1, i_2 \in [n]$, H is convex along any direction $z_{i_1,j_1} - z_{i_2,j_2}$ such that $i_1 \neq i_2$ and $j_1, j_2 \in [k]$.*
4. *Given $\boldsymbol{x} \leq \boldsymbol{y}$, then $\frac{\partial H(\boldsymbol{x})}{\partial x_{i,j}} \geq \frac{\partial H(\boldsymbol{y})}{\partial x_{i,j}}$, where $i \in [n]$, $j \in [k]$.*

We introduce a decision vector $\boldsymbol{x} \in [0,1]^{n \times k}$, where $n = |V|$ is the number of nodes in the given social network graph, k is the number of products. $x_{i,j} \in \boldsymbol{x}$ denotes that node $i \in \{1, 2, \cdots, n\}$ is selected for spreading the information of product $j \in \{1, 2, \cdots, k\}$. Then we relax our objective function for PMPMP with the multilinear extension as follows:

$$\max_{\boldsymbol{x} \in P \cap P_c} F(\boldsymbol{x})$$

$$\text{s.t. } P_c = \{ \boldsymbol{x} \in [0,1]^{n \times k} : \sum_{i=1}^{n} c_i \sum_{j=1}^{k} x_{i,j} \leq B \}, \tag{3}$$

where $F(\boldsymbol{x})$ is the multilinear extension of profit function $f(S|G)$, $P = \{ \boldsymbol{x} \in [0,1]^{n \times k} : \sum_{j=1}^{k} x_{i,j} \leq 1, \forall i \in [n] \}$, $P_c = \{ \boldsymbol{x} \in [0,1]^{n \times k} : \sum_{i=1}^{n} c_i \sum_{j=1}^{k} x_{i,j} \leq B \}$ is the knapsack constraint of the individual seed activation cost c_i and the total seed cost budget B.

Algorithm 1. Continuous Greedy Algorithm (CGA)

Input: Graph G; Number of product k; Cost c_i; Budget B.
Output: $x \in [0,1]^{n \times k}$.
1: Initialize $x(0) \leftarrow 0$, $t \leftarrow 0$, timestep Δt;
2: **while** $x \in P \cap P_c$ **do**
3:    Calculate $(i(t), j(t)) = arg \max \{ \frac{\partial_{i,j} F(x(t))}{c_i} : x(t) + \Delta(t) \cdot z_{i,j} \}$.
4:

$$w_{i,j} = \begin{cases} 1, & if(i,j) = (i(t), j(t)). \\ 0, & otherwise. \end{cases}$$

5:    **for** $i \in [n]$ and $j \in [k]$ **do**
6:        $x_{i,j}(t+1) \leftarrow x_{i,j}(t) + \Delta t w_{i,j}(t)$.
7:    **end for**
8:    $t \leftarrow t + 1$.
9: **end while**
10: **return** $x(T)$.

4.2.1 Continuous Greedy Algorithm

As our objective function is a k-submodular function and the vector x moves in direction constrained by $P \cap P_c$, we can know that $F(x)$ is concave from the property 1 in Corollary 1. For a concave function, we can make advantage of the feature of its partial derivative to solve the problem.

In Algorithm 1, we set w to be one-hot vector which takes 1 only on the node and product pair (i,j) with the largest gradient $\frac{\partial_{i,j} F(x)}{c_i}$ in time t. $z_{i,j}$ is the changing direction of the gradient $\frac{\partial_{i,j} F(x(t))}{c_i}$ and $\sum_j z_{i,j} = 1$ for any $i \in [n]$. $w_{i,j} = 1$ indicates that when node i is selected as seed for product j at time t, it is the direction with the largest rate of change of tangent to curve of $F(x)$. If t increases continuously, we have to calculate the integral of the objective function, which will be very hard. So we set the timestep as Δt, then t increases discretely by Δt at each step, which simplifies the calculation.

After we get the fractional vector solution $x(T)$ returned by Algorithm 1, we need to convert it to the integer solution by randomized rounding technique: selecting node i to spread the influence of product j with probability $x_{i,j}(T)$ independently and making sure that each node can propagate at most one product's influence, i.e., $x_{i,j} = 1$ with probability $x_{i,j}(T)$, while $x_{i,j} = 0$ with probability $1 - x_{i,j}(T)$, where $\sum_{j \in [k]} x_{i,j} \leq 1$ for any $i \in [n]$.

4.2.2 Performance Analysis for Continuous Greedy Algorithm

Theorem 3. *Let $F(x)$ be the multilinear extension of a monotone k-submodular function f for PMPMP. $x(T)$ returned by Algorithm 1 satisfies: $F(x(T)) \geq \frac{1}{2} OPT$.*

Proof. We omit the proof as the limitation of conference pages.

Then we have to do the randomized rounding for the solution at the second stage of Algorithm. It is known that the relationship between the continuous greedy solution $F(\boldsymbol{x}(T))$ and the randomized rounding solution $F_R(\boldsymbol{x}(T))$ is $F_R(\boldsymbol{x}(T)) \geq F(\boldsymbol{x}(T))$. Then we get the final result $F_R(\boldsymbol{x}(T)) \geq \frac{1}{2}OPT$.

5   Experiments

In this section, we compare the efficiency and effectiveness of the proposed CGA with other algorithms.

5.1   Experimental Setup

All the experiments are done on two different datasets. Dataset 1 is called NetScience which it is a co-authorship network. The edges represent the co-authorship that scientists publish papers in the field of network science. It is a undirected graph and has 400 nodes and 1010 edges, and the average degree is 5. Dataset 2 is called Wikivote which is from a Wikipedia voting set. The edges represents the relationship of who votes to whom. It is a directed graph and has 914 nodes and 2914 edges. Both of the datasets are from [18].

Influence Model: The information propagation model is IC model in this paper. The influence spread probability on each edge (u_i, v_i) is set as $p_{uv}^i = 1/|N^-(v_i)|$, which is widely used in prior works [19]. The number of Monte Carlo simulation which is used to compute the maximum influence marginal gain for each node is set as 200.

As mentioned before, a social graph can be partitioned into k communities. For any node in each community C_i, it has a cost c_i and a profit p_i. In this experiment, we default the number of community by $k = \{2, 3, 4\}$, as well as the cost c_i and profit p_i in each community C_i is uniformly sampled from $[0.8, 1.2]$. The timestep Δt is set as 0.05. Then, we compare our GCA algorithm with some commonly used baselines, which can be summarized as follows:

- CGA: Continuous Greedy Algorithm, which is shown as Algorithm 1.
- k-Greedy-TS [12]: It selects the node v in C_i such that maximizing the gain per cost $\Delta_{v, i \in [k]} f(S|G)/c_i$ iteratively until using all budget up.
- MaxDegree-1: It selects the node v in C_i with the largest outdegree per cost $N^+(v)/c_i$ iteratively until using all budget up.
- MaxDegree-2: It selects the node v with the largest outdegree $N^+(v)$ iteratively until using all budget up.
- Random: It selects a node in the graph randomly until using all budget up.

5.2   Experimental Results

In Fig. 1, the results are collected from dataset 1. We compare the proposed algorithms and with four baseline algorithms. The results in Fig. 1 (a) shows how the total profit changes when the budget increases from 0 to 50 and the number of product types is $k = 2$. We can see that when the budget increases, the total profit increases for all the five different algorithms, this is because that more budget can be used to choose more seeds, as the number of seeds selected increases, it increases the influence of the products, i.e., the number of users purchasing the products increases, then this leads to the total profits increases. We can also observe that the total profit obtained by CGA is greater than the four baseline algorithms, which show the efficiency of the proposed algorithm. When the number of products increases to $k = 3$, we can see the changes from the Fig. 1 (b). The results in Fig. 1 (b) have the same variation trends as it shows in Fig. 1 (a), therefore, the conclusion found in Fig. 1 (a) can also be verified in Fig. 1 (b). Comparing the result in Fig. 1 (b) with that in Fig. 1 (a), we can get that when the budget is the same, the total profit in Fig. 1 (a) is greater than that in Fig. 1 (b) for the same algorithm, this is because that when k increases, the number of communities increases, as we know that the influence is difficult to cross between different communities, therefore, this will reduce the total influence of k kinds of products in the social network, and the total profits will also decrease. Similarly, when k goes up to 4 in Fig. 1 (c), the total profit goes down compared with that in Fig. 1 (b) for the same algorithm, but not as much as when k increases from 2 to 3.

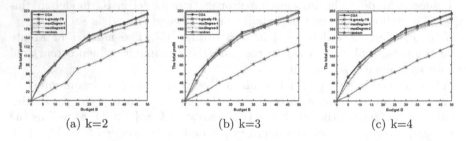

(a) k=2   (b) k=3   (c) k=4

Fig. 1. Performance comparison achieved by different algorithms with different k under dataset 1

We also conduct experiments on a larger dataset 2. The effectiveness of the proposed algorithms can be further verified by observing the conclusion in Fig. 2. Comparing the total profit in Fig. 1 with that in Fig. 2, we can have that the profit in Fig. 2 is larger than that in Fig. 1 with the same budget and same algorithm, which is because the result in Fig. 2 is conducted based on a larger dataset 2. Based on the results in dataset 2, we can see that the performance gap between our algorithms and MaxDegree is larger than that in dataset 1, which further validates the effectiveness of our proposed algorithms in large and directed social networks.

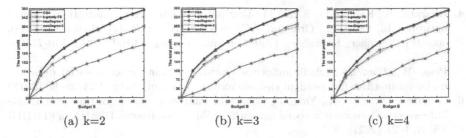

(a) k=2   (b) k=3   (c) k=4

Fig. 2. Performance comparison achieved by different algorithms with different k under dataset 2

6 Conclusion

We study the profit maximization problem for k kinds of products in social networks in this paper. We assume that one seed user can only spread one kind of product's influence. Then we introduce the community structure of a social network to the proposed problem, and constrain that each community is used for promoting only one kind of product. So the goal of the proposed problem is to select k subsets of users as seeds from k different communities such that the overall profit for k kinds of products is maximized. We prove that the objective function satisfies k-submodularity. Then the multilinear extension is introduced to relax the objective function, and a Continuous Greedy Algorithm is devised to solve the profit maximization problem, which can obtain an $\frac{1}{2}$ approximation ratio. Comprehensive experiments are conducted on two real world datasets in social networks, and the experimental results verify the correctness as well as the effectiveness of the proposed algorithm.

Acknowledgment. This work is supported in part by the Guangdong Basic and Applied Basic Research Foundation under Grant No. 2021A1515110321 and No. 2022A1515010611 and in part by Guangzhou Basic and Applied Basic Research Foundation under Grant No. 202201010676.

References

1. Kempe, D., Kleinberg, J., Tardos, É.: Maximizing the spread of influence through a social network. In: Proceedings of the Ninth ACM SIGKDD International Conference on Knowledge Discovery and Data Mining, pp. 137–146. ACM (2003)
2. Ni, Q., Guo, J., Huang, C., Weili, W.: Community-based rumor blocking maximization in social networks: algorithms and analysis. Theoret. Comput. Sci. **840**, 257–269 (2020)
3. Fan, L., Lu, Z., Wu, W., Thuraisingham, B., Ma, H., Bi, Y.: Least cost rumor blocking in social networks. In: 2013 IEEE 33rd International Conference on Distributed Computing Systems, pp. 540–549 (2013)

4. Bharathi, S., Kempe, D., Salek, M.: Competitive influence maximization in social networks. In: Deng, X., Graham, F.C. (eds.) WINE 2007. LNCS, vol. 4858, pp. 306–311. Springer, Heidelberg (2007). https://doi.org/10.1007/978-3-540-77105-0_31
5. Peng, W., Pan, L.: Scalable influence blocking maximization in social networks under competitive independent cascade models. Comput. Netw. **123**, 38–50 (2017)
6. Guanhao, W., Gao, X., Yan, G., Chen, G.: Parallel greedy algorithm to multiple influence maximization in social network. ACM Trans. Knowl. Disc. Data (TKDD) **15**(3), 1–21 (2021)
7. Han, K., Xu, C., Gui, F., Tang, S., Huang, H., Luo, J.: Discount allocation for revenue maximization in online social networks. In: Proceedings of the Eighteenth ACM International Symposium on Mobile Ad Hoc Networking and Computing, pp. 121–130 (2018)
8. Zhang, H., Zhang, H., Kuhnle, A., Thai, M.T.: Profit maximization for multiple products in online social networks, pp. 1–9 (2016)
9. Zhang, Y., Yang, X., Gao, S., Yang, W.: Budgeted profit maximization under the multiple products independent cascade model. IEEE Access **7**, 20040–20049 (2019)
10. Huber, A., Kolmogorov, V.: Towards minimizing k-submodular functions. In: Mahjoub, A.R., Markakis, V., Milis, I., Paschos, V.T. (eds.) ISCO 2012. LNCS, vol. 7422, pp. 451–462. Springer, Heidelberg (2012). https://doi.org/10.1007/978-3-642-32147-4_40
11. Singh,, A., Guillory, A., Bilmes, J.: On bisubmodular maximization. In: Artificial Intelligence and Statistics, pp. 1055–1063 (2012)
12. Ohsaka, N., Yoshida, Y.: Monotone k-submodular function maximization with size constraints. In: Advances in Neural Information Processing Systems, pp. 694–702 (2015)
13. Ward, J., Živný, S.: Maximizing k-submodular functions and beyond. ACM Trans. Algorithms (TALG) **12**(4), 47 (2016)
14. Ni, Q., Guo, J., Weili, W., Wang, H., Jigang, W.: Continuous influence-based community partition for social networks. IEEE Trans. Netw. Sci. Eng. **9**(3), 1187–1197 (2021)
15. Ward, J., Zivny, S.: Maximizing k-submodular functions and beyond. ACM Trans. Algorithms **12**(4), 1–26 (2016)
16. Vondrák, J.: Optimal approximation for the submodular welfare problem in the value oracle model. In: Proceedings of the Fortieth Annual ACM Symposium on Theory of Computing, pp. 67–74 (2008)
17. Wang, B., Zhou, H.: Multilinear extension of k-submodular functions. arXiv preprint arXiv:2107.07103 (2021)
18. Rossi, R., Ahmed, N.: The network data repository with interactive graph analytics and visualization. In: Twenty-Ninth AAAI Conference on Artificial Intelligence, pp. 4292–4293 (2015)
19. Ni, Q., Guo, J., Wu, W., Wang, H.: Influence-based community partition with sandwich method for social networks. IEEE Trans. Comput. Soc. Syst. (2022). https://doi.org/10.1109/TCSS.2022.3148411

MCM: A Robust Map Matching Method by Tracking Multiple Road Candidates

Wanting Li[ID], Yongcai Wang(✉)[ID], Deying Li[ID], and Xiaojia Xu[ID]

School of Information, Renmin University of China, Beijing 100872,
People's Republic of China
ycw@ruc.edu.cn

Abstract. Map matching is to track the positions of vehicles on the road network based on the positions provided by GPS (Global Positioning System). Balancing localization accuracy and computation efficiency is a key problem in online map matching, for which, existing methods HMM and MHT mainly use Markov assumption which drops early unused data. Although the roads to explore can be remarkably reduced by the Markov assumption, miss-of-match and matching breaks may occur if the GPS data is highly noisy. To address these problems, this paper presents Multiple Candidate Matching (MCM) to improve the robustness of map matching by using historical trajectory data. MCM tracks multiple road candidates in the map matching process while limiting the number of road candidates by excluding the routes whose likelihood are below a threshold. Numerical experiments in large-scale data sets show that MCM is very promising in terms of accuracy, computational efficiency, and robustness. Mismatching problems caused by Markov assumption can be resolved effectively when compared with state-of-the-art online map matching methods.

Keywords: Map matching · Multiple candidate · Road continuity · Online

1 Introduction

Tracking vehicles' exact locations on the road network is a critical problem for vehicle navigation and various location-based services. Due to the measurement noise of the GPS equipment, GPS reported positions might deviate from the real road. The GPS noises may be caused by different reasons, such as when the vehicles are under bridges or in tunnels [6]. Other errors may occur due to multi-path satellite signals [2,3,16] that arrive at a receiver via a non-direct path, such as being reflected off high buildings in built-up city areas.

In order to locate vehicles accurately, researchers proposed to match the trajectory of a vehicle with the road network. By using the continuity constraint of the vehicle's motion and the continuity characteristics of the roads,

This work was supported in part by the National Natural Science Foundation of China Grant No. 61972404, 12071478, Public Computing Cloud, Renmin University of China.

the vehicle localization problem becomes to find the most likely road that can be matched with the GPS trajectory, which is called *map matching* problem [1]. Map matching can be classified into online map matching [8,9,12–14] and offline map matching [5,7,11,15,18–20]. The former estimates the current road segment the vehicle is on immediately after a GPS data is collected. The latter recovers the travelled roads by offline processing the collected whole GPS trajectory.

Online map matching requires both matching accuracy and efficiency. For efficiency purpose, Markov assumption is widely adopted, which assumes the road-matching state at time t depends only on the states at time $t-1$, which is not related to the states and observations of early times. Based on the Markov assumption, two categories of methods, i.e., *Hidden Markov Model (HMM)* based methods [8,9,12–14] and *Multiple Hypothesis Techniques (MHT)* [10,13,17] are mainly proposed in literature for online map matching.

HMM utilizes the state transition probabilities learned from the road topology, in which a latent state sequence indicating the route is inferred based on the observed sequence of the GPS trajectory. Two methods are mainly used to calculate the hidden Markov chain in real time. The first method uses the approximate algorithm to greedily calculate the current optimal result [12,14]. The approximation algorithm only retains one current optimal path in each step, which can not guarantee that the predicted state sequence as a whole is the most likely state sequence, because the predicted state sequence may have parts that do not actually occur. The other method uses the sliding window Viterbi algorithm to calculate the optimal path in a period of time [8,9,13]. However, when using Viterbi or approximate algorithm for sequence matching, the matching results before time $t-1$ are not traced back. One mismatch at t may lead to cascaded errors in the later associations, which may cause errors, such as matching break [4] shown in Fig. 1. The green trajectory is the estimated trajectory using online HMM method. Because the side road ends at that point, a matching break happens. In addition, the HMM-based method needs to train the probability model in advance. The sliding window-based method has a certain delay.

To address the above problems of HMM, MHT [10,13,17] generates a variety of road hypotheses at time t through utilizing historical transportation information to evaluate the probabilities of choosing the subsequent road. This method mainly uses Bayesian filtering. The goal is to obtain the probability distribution of the state quantity at time $t-1$ when the prior probability is known and to estimate the posterior probability distribution of the state quantity at time t when the observation and transition probability matrix at time t are known. This method does not need to train the model in advance, and it is a limited state candidate without delay. However, the Markov hypothesis is also used without considering the influence of historical states. This method will still be affected by noises. In complex environments with large GPS errors, it still leads to subsequent matching errors due to a wrong early matching.

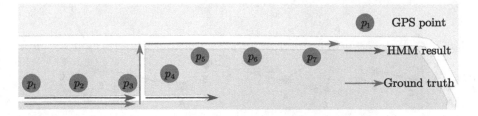

Fig. 1. Matching break. The matching break is a common problem in map-matching, which is mainly caused by trajectory outliers. Because of Markov assumption, HMM algorithm matches the wrong side road with a higher probability at point p_4, but there is no way to correct it, and a break occurs.

Historical trajectory data has great value and should be fully utilized. This paper relaxes the Markov assumption but still designs a highly efficient map matching algorithm, i.e., *Multiple route Candidate Matching (MCM)*. MCM is essentially to find the longest common sub-sequence between the GPS trajectory and the potential routes generated from the road network. MCM considers the topological continuity between the trajectory points and the continuity among the route segments. These continuity constraints are embedded into the similarity model. MCM has three key points to balance accuracy and efficiency.

(1) The likelihood of multiple route candidates is tracked by a dynamic programming process using a similarity matrix.
(2) A "last" label is used in each row of the similarity table to record the historical matching point.
(3) The most unlikely routes are autonomously excluded to control the number of "alive" candidates so that the computation is efficient.

MCM shows strong fault tolerance. Even if the matching at t is wrong, the trajectory points can be corrected to the correct route by subsequent matching. The validity of MCM is investigated theoretically and experimentally, regarding both the matching accuracy and matching efficiency. Experiments on one widely used map matching dataset show that the proposed MCM method provides the highest mapping accuracy compared with state-of-the-art online map matching algorithms. MCM is also easy to be carried out without the pain of training the probability model. We will outsource the codes and provide online demos for MCM for potential use by society.

2   Problem Model

2.1   Point-to-Road Similarity

There are many ways to define the similarity between a GPS point and a road segment. In our case, we use point-to-line distance with a threshold to define the point-to-road similarity. It truncates very low similarity cases for efficiency.

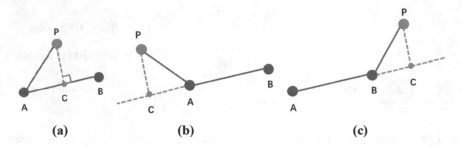

Fig. 2. AB is the segment; P is the point and C is P's projection on AB. (a) C is on AB; (b) C is not on AB and is closer to A; (c) C is not on AB and is closer to B

Definition 1 (Point-to-Road Similarity). *The point-to-road similarity* $\mathbf{S}(p_i, e_j)$ *between a point* $p_i \in \mathbf{T}$ *and a road* $e_j \in \mathbf{G}$ *is defined by Eq. (1).*

$$\mathbf{S}(p_i, e_j) = \begin{cases} \varepsilon - \|p_i - e_j\|_2, \text{ if } \|p_i - e_j\|_2 < \varepsilon \\ 0, \ otherwise \end{cases} \tag{1}$$

ε here is a threshold to eliminate the similarity calculation if the point is too far away from the road segment, which is helpful to reduce the amount of subsequent calculation. $\|p_i - e_j\|_2$ is the distance from the point p_i to the road segment e_j, which is defined by *the shortest distance from the point to the line segment* as shown in Fig. 2.

Let's assume the line segment is AB and the point is P. The projection from P to AB is denoted by C. If C is on AB, $\|p_P - e_{AB}\|_2 = \|PC\|$. If C is not on AB and C is closer to A, then $\|p_P - e_{AB}\|_2 = \|PA\|$; If C is not on AB and C is closer to B, then $\|p_P - e_{AB}\|_2 = \|PB\|$;

2.2 Trajectory-to-Route Similarity

We then consider to evaluate the similarity between a GPS trajectory $\mathbf{T}$ and a route R on G. Suppose $\mathbf{T}$ is composed by a set of successively measured GPS points, i.e., $\mathbf{T} = \{p_1, p_2, \cdots, p_n\}$. Suppose R is composed by a set of sequentially connected edge segments, i.e., $R = \{e_1, e_2, \cdots, e_m\}$.

Definition 2 (Trajectory-to-route Similarity). *Given* $\mathbf{T} = \{\mathbf{p_1}, \mathbf{p_2}, \cdots, \mathbf{p_n}\}$ *and* $R = \{e_1, e_2, \cdots, e_m\}$*, the trajectory to route similarity* $M(\mathbf{T}, R)$ *is defined as:*

$$M(\mathbf{T}, R) = \sum_{i=1}^{n} S(p_i, e_{nearest}(p_i)) \tag{2}$$

where $e_{nearest}(p_i)$ *is the route on* R *which has the minimum distance to* p_i *as shown in Fig. 3.*

Then, let's denote R^* the route on G, which matches best with $\mathbf{T}$. Then the goal is to find the route with the best similarity score with $\mathbf{T}$.

$$R^* = arg \max_R M\left(\mathbf{T}, R\right) \tag{3}$$

where R is any route that can be generated from G.

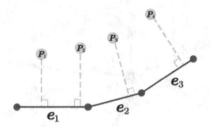

Fig. 3. Trajectory-to-route similarity by trajectory p_1, p_2, p_3, p_4 to route $e_1 \rightarrow e_2 \rightarrow e_3$

3 Multiple Candidate Matching

Enumerating all the possible routes on G is computational complexity explosive. MCM generates routes in a controlled way. Instead of training the transitional model or using additional road level or travel speed information, MCM uses only the trajectory and the roads' continuity information.

MCM proposes a method by matching the trajectory to the map via generating multiple candidate routes. It tracks the matching probabilities of multiple routes and outputs the best matching result. MCM is mainly divided into two steps. Firstly, the candidate routes are generated based on the current "alive" matches on the matching table and the continuity constraint of the routes on the route graph; Secondly, the trajectory-to-road similarities for the multiple route candidates are evaluated using a dynamic programming model. In this step, the matching similarities to all potential routes will be evaluated; the best match is output as the current result, and some unlikely routes will be pruned for keeping computation efficiency.

MCM proposes an *score matrix* $\mathbf{M}$ to explore the likelihoods of all potential routes that may match with the GPS trajectory. The values in the matrix $\mathbf{M}$ represent the trajectory-to-road similarities of the candidate routes. The rows of the matrix represent the edge segments in the road network, and the columns of the matrix represent the GPS points on the trajectory ordered by the collecting time. The map matching process is indeed to update the score matrix $\mathbf{M}$. Because the trajectory points are collected in order, at each time t, when a new point p_t is obtained, we need to fill a new column, i.e., the tth column of $\mathbf{M}$.

The summation of all the point-to-road similarity values of a trajectory sequence as trajectory-to-route similarity. We call each value in the score matrix as trajectory-to-route similarity value. Figure 5 shows an example of a score matrix where $n = m = 7$. Map matching uses this sparse score matrix. The concepts used in MCM and the steps to fill the score matrix are as follows.

3.1   Route Candidates

For finding the route candidates on G that may match with $\mathbf{T}$, we use roads'
continuity information. Based on the neighbor edges we got in the previous step,
If there is a path that conforms to the road topology in the last matching pairs,
it indicates that one of the route candidates can be continued.

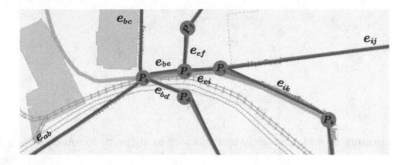

Fig. 4. Preceding and successive road set.

Definition 3 (Alive Routes, i.e., Route Candidates). *Alive routes in* $\mathbf{M}$
records the potential candidates of routes that may match with the trajectory $\mathbf{T}$.
Each route candidate is composed by a sequence of connected edges.

Suppose at time $t-1$, there are K alive route candidates in $\mathbf{M}$. For each
route candidate, we record only the last edge of each route to represent the
route. This is because when a route's similarity score is obtained at time t, we
can trace back the whole route from the last edge of that route in $\mathbf{M}$.

Definition 4 (Last Matching Pair). *The last edge on each route is saved as*
a last matching pair $last_k = (e, p)$, *where* e *is an edge index and* p *is a point*
index. It means on the kth route, the last matching point p on $\mathbf{T}$ *matches with*
the edge e on G. It also means that the entry (e, p) in $\mathbf{M}$ *is the endpoint of the*
kth alive route.

At the time t, we assume the total number of alive routes is K, and these
K alive routes' *last matching pairs* are saved in a queue data structure $QLast$.
The following functions are defined to return the edge index and point index in
the kth route's last matching pair.

Definition 5 (The $last(\cdot)$ Function). *Suppose $l_k = (e, p)$ is the last matching*
pair of the kth alive route, the function $e(l_k) = e$ returns the edge index saved in
l_k and $p(l_k) = p$ returns the point index in l_k.

Then route generation considers the route continuity information on map G.

Definition 6 (The $near(\cdot)$ **Function).** *A near() function is designed to restrict MCM to generate only reasonable routes based on the road network topology.* $near(p_i, e_j) = Pre(e_j) \cup Suc(e_j) \cup e_j$ *where* $Pre(e_j)$ *is the preceding road set of road* e_j *and* $Suc(e_j)$ *is successive road set of road* e_j *in the road network. As shown in Fig. 4,* e_j *is the road segment* e_{be}, *the start point* $e_j.\mathcal{S}$ *is point* p_b *and the end point* $e_j.\mathcal{E}$ *is point* p_e. *So* $Pre(e_{be}) = \{e_{ab}, e_{bc}, e_{bd}\}$ *and* $Suc(e_{be}) = \{e_{ef}, e_{ei}\}$.

3.2   Updating the Sparse Score Matrix

We use the dynamic programming method to calculate the score matrix, that is, the summation of all the similarity values of a trajectory sequence.

At time t, when a new GPS point p_t is obtained, we check all the alive routes' last edges, i.e., all the $l_k \in QLast$. $\mathbf{M}(e_i, p_t)$ is filled by one of the following two cases:

(1) For each l_k we find $near(e(l_k))$, i.e., all connected edges of the last edge. Then we calculate the similarity scores $\mathbf{S}(e_i, p_t)$ for every $e_i \in near(e(l_k))$. If $\mathbf{S}(e_i, p_t) > 0$, the score of e_i obtained from the kth alive route, denoted by $\mathbf{M}_k(e_i, p_t)$ is calculated by:

$$\begin{aligned} &\text{if } \mathbf{S}(e_i, p_t) > 0 \& e_i \in near(e(l_k)), \\ &\mathbf{M}_k(e_i, p_t) = \mathbf{M}(e(l_k), p(l_k)) + \mathbf{S}(e_i, p_t) \end{aligned} \tag{4}$$

Then all the K alive routes will be processed to calculate Eq. (4). The updated score of $\mathbf{M}(e_i, p_t)$, i.e., the score at the ith row and tth column in $\mathbf{M}$ is filled by the highest score calculated from all the K route candidates.

$$\mathbf{M}(e_i, p_t) = \max_{k=1:K} \mathbf{M}_k(e_i, p_t) \tag{5}$$

(2) If an e_i is not in the near edge set of any route's last edge, but $\mathbf{S}(e_i, p_t) > 0$, a new route candidate will be generated. Its matching score is filled as:

$$\mathbf{M}(e_i, p_t) = \mathbf{S}(e_i, p_t), \text{if } \mathbf{S}(e_i, p_t) > 0 \& \forall l_k, e_i \notin near(l_k) \tag{6}$$

The overall equation to fill the matching score at $\mathbf{M}(e_i, p_t)$ is therefore given in

$$\mathbf{M}(e_i, p_t) = \begin{cases} \max\limits_{k=1:K} \{\mathbf{M}(e(l_k), p(l_k)) + \mathbf{S}(e_i, p_t)\}, & \text{if } \mathbf{S}(e_i, p_t) > 0 \& e_i \in near(e(l_k)) \\ \mathbf{S}(e_i, p_t), & \text{if } \mathbf{S}(e_i, p_t) > 0 \& e_i \notin near(e(l_k)), \forall l_k \end{cases} \tag{7}$$

3.3   MCM for Online Map Matching

In online map matching, the problem is to find the associated roads for the trajectory up to time t. MCM outputs the candidate matching roads with the best matching score up to time t. So it find the maximum $\mathbf{M}(e_i, p_t), i \in [1, m]$ at time t in case $\mathbf{S}(e_i, p_t) > 0$. Then (e_i, p_t) is the last matching pair of the best-matched route, and the route can be traced back using this last edge.

So the overall routine in MCM for online matching is as described below. The pseudocode for MCM is given in Algorithm 1.

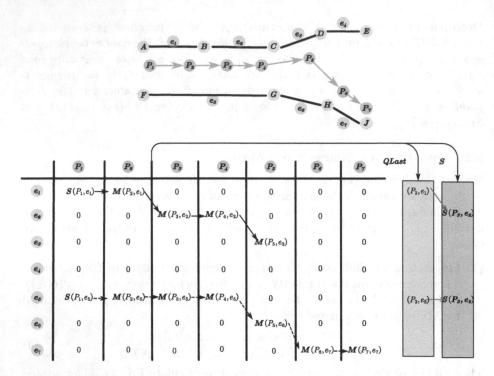

Fig. 5. The figure shows a sparse score matrix. All the route candidates are shown in the figure. We calculate neighbors for each node (only cells within radius in ε in each column). The last alive matching pairs are store in queue $Qlast$ and the point-to-road similarity matrix **S**. The dotted line is the trajectory which has maximum score.

(1) Initialize sparse score matrix with zero and the last matching pair as empty. (Line 1–2)
(2) Find neighbor roads within radius in ε for the current view of the vehicle. (Line 3–7)
(3) Find the route candidates based on the last matching pairs and the neighboring roads. (Line 10)
(4) For each route candidate, calculate the similarity score with point-to-road similarity values, and update the sparse score matrix. (Line 11 & 14)
(5) Finally, find the best matching road at t and update the last matching pairs. (Line 18)

An example is shown in Fig. 5. The process of matching by MCM from P_2 to P_3 is shown in Fig. 5.

After completing the matching at time t_2, we get the alive candidate routes as $e_1 \rightarrow e_1$ and $e_5 \rightarrow e_5$, and the two last matching pairs are (e_1, p_2) and (e_5, p_2). At time t_3, we first find neighbor roads $\{e_2, e_5\}$. Then we match the route candidates based on the last matching pairs and the neighboring roads. In the route $e_1 \rightarrow e_1$,

(e_2, p_3) satisfies the $near(\cdot)$ function. So we get $\mathbf{M}(p_3, e_2) = \mathbf{M}(p_2, e_1) + \mathbf{S}(p_3, e_2)$ by Eq. (7). The other alive routes obtained in the same way.

At time t_6, only one alive route can be found and the last matching pair is updated to be (e_5, p_5) to (e_5, p_6). Finally we can get the score matrix M which stores the alive candidate routes as shown in Fig. 5.

Algorithm 1. MCM algorithm

Input: Graph $G = \{V, E\}$ and trajectory T
Parameter: Threshold ε
Output: matches=Avector containing matching indices

1: global M = initialize(len(G), len(T))// score matrix.
2: global QLast = initialize(1,len(G))// record last alive matching node.
3: **for** $j = 1$ to n **do**
4:     **if** $\max_{i \in [1,m]} \mathbf{S}(i, j) > 0$ **then**
5:         $\mathbf{M}(i, :) = \mathbf{S}(i, :)$; $Qlast(i) = j$;Update $near(p_i, e_j)$;break;
6:     **end if**
7: **end for**
8: **for** j to n **do**
9:     **for** $i = 1$ to m **do**
10:         **if** $\mathbf{S}(i, j) > 0 \&\& Qlast(k) > 0$ **then**
11:             $\mathbf{M}(i, j) = \max_{k \in [1,m]} \{\mathbf{M}(k, j - 1) + \mathbf{S}(i, j)\}$; $Qlast(i) = j$;$Qlast(k) = 0$;Update $near(p_i, e_j)$;
12:         **else**
13:             **if** $\mathbf{S}(i, j) > 0$ **then**
14:                 $\mathbf{M}(i, j) = \mathbf{S}(i, j)$; $Qlast(i) = j$;Update $near(p_i, e_j)$;// Begin again.
15:             **end if**
16:         **end if**
17:     **end for**
18:     matches(j)= $\max_{i \in [1,m]} M(e_i, p_j)$;
19: **end for**

For online map matching, we only need to output the road corresponding to the maximum M value at the current time, that is, select the maximum value of each column.

4 Experiment

The Washington Dataset was presented by Newson et al. [15] and is one of the most widely used benchmark data sets for testing online map-matching algorithms. It contains GPS data from a drive around Seattle, WA, USA using SiRF Star III GPS chipset with WAAS (Wide Area Augmentation System) enabled. The journey was sampled 1 Hz and contains just over two hours of driving in both challenging inner-city environments and the outer suburbs. The total route was 80 km long with 7531 data samples containing latitude and longitude pairs.

Effects of Parameter Selection. Figure 6(a) shows the relationship between the map matching efficiency and the parameter selection in online matching. The point-to-road similarity threshold varies from 5 to 40. We can see that the running time increases with ε. We can also see (Table 1) that even the maximum running time is much less than 1 s (on a laptop with Intel i7-8550U CPU), so it can well support online map matching, which generally takes GPS data 1 Hz. Figure 6(b) shows the accuracy of MCM for various sampling intervals and various threshold parameters. The horizontal axis represents the similarity threshold, the bar colors represent the sampling intervals, and the vertical axis represents the accuracy. We can see larger similarity threshold can help to improve the matching accuracy.

Table 1. Online run-time of our proposed method with Washington dataset when $\varepsilon = 30$

Average run-time	Maximum run-time	Minimum run-time
0.0482 s	0.0731 s	0.0057 s

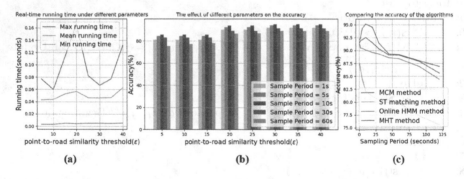

| (a) | (b) | (c) |

Fig. 6. (a) Efficiency under different ε for Washington dataset. (b) Different ε and different GPS sampling rate v.s. matching accuracy (% of points matched to correct road segment) for Washington dataset. (c) MCM vs HMM based method for Washington dataset.

Accuracy Comparison with Other Methods. In this paper, we hope to use road information as little as possible to make the MCM method be suitable for different cases. So for performance comparison, we choose ST matching [12], Online HMM [8] and MHT [17] as the comparative methods.

Figure 6(c) illustrates that the matching accuracy is the highest when the sampling rate 10 Hz. The term 'accuracy' in this context refers to the percentage of GPS points that have been matched to the correct segment of the road network. When the sampling rate decreases, the matching accuracy will decrease because the previous GPS points provide less information about the current GPS point. When the sampling rate is too high, unnecessary detour [4] will appear, and the accuracy of matching will also be reduced. So we need to balance various factors and choose the best matching sampling rate. In the Washington data set, 10 Hz is the best choice, and we choose $\varepsilon = 30$.

From the matching accuracy comparison in Fig. 6(c), we can see that the matching accuracy of MCM is 95%, which is better than all other three methods with a high sampling rate. With a low sampling rate, all online methods get less accuracy, under 90%. But our MCM method still performs better than the Online-HMM method and ST matching method. MHT is most close to MCM, and they are always better than the other two methods. MCM also uses less information than MHT, which is therefore highly suitable for online map matching (Fig. 7).

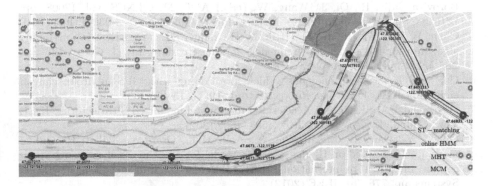

Fig. 7. Result for Washington dataset. Here is an overpass scene, we can see that our method can match the side road well, while other methods will match wrong.

5   Conclusion and Future Work

In this paper, a new method, i.e., MCM for online map matching is proposed. MCM tracks multiple alive route candidates while controlling the scale of candidates according to the continuity of the road by excluding unnecessary matching candidates. MCM does not need to set the noise distribution function and the transition probability. It needs to set a threshold to specify the maximum acceptable offset. MCM works well without the pain of preliminary data processing work while providing better robustness and accuracy. In future work, we can also modify the road continuity function, such as introducing semantic information to better distinguish overpass sections, and conduct road navigation while conducting map matching.

References

1. Bernstein, D., Kornhauser, A., et al.: An introduction to map matching for personal navigation assistants (1996)
2. Blanco-Delgado, N., Nunes, F.D.: Multipath estimation in multicorrelator GNSS receivers using the maximum likelihood principle. IEEE Trans. Aerosp. Electron. Syst. **48**(4), 3222–3233 (2012)
3. Chaggara, R., Macabiau, C., Chatre, E.: Using GPS multicorrelator receivers for multipath parameters estimation. In: Proceedings of the 15th International Technical Meeting of the Satellite Division of The Institute of Navigation (ION GPS 2002), pp. 477–492 (2002)
4. Chao, P., Xu, Y., Hua, W., Zhou, X.: A survey on map-matching algorithms. In: Borovica-Gajic, R., Qi, J., Wang, W. (eds.) ADC 2020. LNCS, vol. 12008, pp. 121–133. Springer, Cham (2020). https://doi.org/10.1007/978-3-030-39469-1_10
5. Chen, B.Y., Yuan, H., Li, Q., Lam, W.H., Shaw, S.L., Yan, K.: Map-matching algorithm for large-scale low-frequency floating car data. Int. J. Geogr. Inf. Sci. **28**(1), 22–38 (2014)
6. Cui, Y., Ge, S.S.: Autonomous vehicle positioning with GPS in urban canyon environments. IEEE Trans. Robot. Autom. **19**(1), 15–25 (2003)
7. Dogramadzi, M., Khan, A.: Accelerated map matching for GPS trajectories. IEEE Trans. Intell. Transp. Syst. **23**(5), 4593–4602 (2021)
8. Goh, C.Y., Dauwels, J., Mitrovic, N., Asif, M.T., Oran, A., Jaillet, P.: Online map-matching based on Hidden Markov Model for real-time traffic sensing applications. In: 2012 15th International IEEE Conference on Intelligent Transportation Systems, pp. 776–781. IEEE (2012)
9. Jagadeesh, G.R., Srikanthan, T.: Online map-matching of noisy and sparse location data with hidden Markov and route choice models. IEEE Trans. Intell. Transp. Syst. **18**(9), 2423–2434 (2017)
10. Li, G., Lou, L., Zheng, P., et al.: Route restoration method for sparse taxi GPS trajectory based on Bayesian network. Tehnički vjesnik **28**(2), 668–677 (2021)
11. Li, Y., Huang, Q., Kerber, M., Zhang, L., Guibas, L.: Large-scale joint map matching of GPS traces. In: Proceedings of the 21st ACM SIGSPATIAL International Conference on Advances in Geographic Information Systems, pp. 214–223 (2013)
12. Lou, Y., Zhang, C., Zheng, Y., Xie, X., Wang, W., Huang, Y.: Map-matching for low-sampling-rate GPS trajectories. In: Proceedings of the 17th ACM SIGSPATIAL International Conference on Advances in Geographic Information Systems, pp. 352–361 (2009)
13. Luo, L., Hou, X., Cai, W., Guo, B.: Incremental route inference from low-sampling GPS data: an opportunistic approach to online map matching. Inf. Sci. **512**, 1407–1423 (2020)
14. Mohamed, R., Aly, H., Youssef, M.: Accurate real-time map matching for challenging environments. IEEE Trans. Intell. Transp. Syst. **18**(4), 847–857 (2016)
15. Newson, P., Krumm, J.: Hidden Markov map matching through noise and sparseness. In: Proceedings of the 17th ACM SIGSPATIAL International Conference on Advances in Geographic Information Systems, pp. 336–343 (2009)
16. Spangenberg, M., Giremus, A., Poire, P., Tourneret, J.Y.: Multipath estimation in the global positioning system for multicorrelator receivers. In: 2007 IEEE International Conference on Acoustics, Speech and Signal Processing-ICASSP'07, vol. 3, pp. III-1277. IEEE (2007)

17. Taguchi, S., Koide, S., Yoshimura, T.: Online map matching with route prediction. IEEE Trans. Intell. Transp. Syst. **20**(1), 338–347 (2018)
18. Wei, H., Wang, Y., Forman, G., Zhu, Y., Guan, H.: Fast Viterbi map matching with tunable weight functions. In: Proceedings of the 20th International Conference on Advances in Geographic Information Systems, pp. 613–616 (2012)
19. Yuan, J., Zheng, Y., Zhang, C., Xie, X., Sun, G.Z.: An interactive-voting based map matching algorithm. In: 2010 Eleventh International Conference on Mobile Data Management, pp. 43–52. IEEE (2010)
20. Zeng, Z., Zhang, T., Li, Q., Wu, Z., Zou, H., Gao, C.: Curvedness feature constrained map matching for low-frequency probe vehicle data. Int. J. Geogr. Inf. Sci. **30**(4), 660–690 (2016)

Pilot Pattern Design with Branch and Bound in PSA-OFDM System

Shuchen Wang[ORCID], Suixiang Gao[✉], and Wenguo Yang[ORCID]

School of Mathematical Sciences, University of Chinese Academy of Sciences,
Beijing 100049, China
wangshuchen19@mails.ucas.ac.cn, {sxgao,yangwg}@ucas.ac.cn

Abstract. Pilot symbol assisted (PSA) channel estimation is an important means to improve the communication quality of orthogonal frequency division multiplexing (OFDM) systems. The insertion position of pilot in frequency domain and time domain of OFDM symbol is called pilot pattern. Appropriate pilot pattern design can greatly reduce the channel estimation error and enhance the communication quality. In this paper, the branch and bound (BnB) method is adopted to design the pilot pattern BnB-PP for the first time. Specifically, the result of the linear least mean square error (LMMSE) method is taken as the target value of channel estimation in PSA-OFDM systems. For branching, pilot positions are randomly selected one by one in the form of binary tree. For boundary, the correction term is subtracted from the result to replace it after the node is filled randomly. The results show that BnB-PP is better than the common pilot pattern. The average MSE under all signal to noise ratio (SNR) of channel estimation for 32 and 64 pilots in 1344 data signals is reduced by 80.68% and 3.88% respectively compared with lattice-type pilot pattern.

Keywords: Pilot pattern · Branch and bound · PSA-OFDM · Channel estimation · LMMSE

1 Introduction

OFDM modulation realizes the parallel transmission of serial data through frequency division multiplexing, which makes it have the ability to resist multipath fading [24]. In 3G and 4G, OFDM is gradually mature [28], and it is still one of the most important technologies in 5G wireless communication network [2]. The wireless channel will fade in both frequency domain [16] and time domain [5], and there is randomness, which affects the communication quality. Therefore, it is necessary to estimate the channel state information (CSI). Channel estimation can get the channel impulse response, so as to improve the performance of OFDM system [4].

© The Author(s), under exclusive license to Springer Nature Switzerland AG 2022
Q. Ni and W. Wu (Eds.): AAIM 2022, LNCS 13513, pp. 244–254, 2022.
https://doi.org/10.1007/978-3-031-16081-3_21

Channel estimation methods can be divided into blind, semi-blind and unblinded estimation. Blind and semi-blind channel estimation, such as subspace decomposition [15], can effectively improve the system capacity, but it is difficult to deal with fast fading wireless channels due to its poor flexibility. And PSA-OFDM is the most commonly communication system for non blind estimation. The pilot signal is the known specific signal data, which is placed at the specific position of the transmission symbol data of the PSA-OFDM system [3,8]. The channel condition is estimated by comparing the difference between the pilot signal at the receiving end and the transmitting end. Least Squares (LS) [9] can estimate the channel impulse response of the pilot position, and then it needs to estimate other positions by interpolation [1]. MMSE method is based on LS [25]. MMSE can obtain the impulse response of the whole channel with higher accuracy by incorporating the influence of noise into the calculation, but it needs to obtain or estimate some other characteristics of the channel in advance [23]. LMMSE is an improved version of MMSE with lower computational complexity [29].

The three most important problems in PSA are where to insert pilots [6], what data to insert as pilots [11], and channel estimation methods [10,13,21]. In this paper we only focus on the first problem, that is, pilot pattern design. There are three common pilot pattern in Fig. 1. These are mainly designed by experience and proved feasible in practice [27]. Intuitively, block-type pilot pattern are more suitable for channels with large frequency selective fading [19], while comb-type pilot pattern are for large time selective fading [7]. And the lattice-type pilot pattern has better stability [30]. Some researches have tried some advanced methods to design non-uniform pilot pattern, such as deep neural network [22].

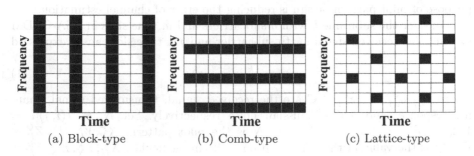

(a) Block-type   (b) Comb-type   (c) Lattice-type

Fig. 1. Common pilot pattern. The whole is PSA-OFDM symbol, in which the black squares are the position of the pilot signal, and the others are the position of the data signal.

Most of the pilot pattern are designed according to the channel characteristics, while the pilot pattern design in this paper is based on data-driven, which is more in line with the complex channel in the actual situation. It is essentially an NP-hard combinatorial optimization problem. And it can be modeled as $0-1$

mixed integer linear programming (MILP), but due to the poor structure and large scale, it is still difficult to solve by solver of MILP directly [12,20]. BnB is a classical algorithm for solving integer programming problems [14,18,26]. We adopt the framework of BnB with random binary trees branches. It is difficult to obtain the boundary on each node, so this paper presents an expected boundary method with correction term. The simulation results confirm the superiority of BnB-PP.

The main contributions of this paper are summarized as follows:

Method: BnB framework is adopted to solve the pilot pattern design problem for the first time.

Bound: A heuristic calculation method of boundary in BnB is given.

Accuracy and Generalization: The results show that BnB-PP is better than the common pilot pattern under most SNR, especially when the number of pilots is small.

The rest of this paper is outlined as follows: Sect. 2 provides the principle of channel estimation in PSA-OFDM system, while Sect. 3 discusses the method of designing pilot pattern with BnB. Section 4 gives the simulation results with discussion. The conclusion and future works have been provided in Sect. 5.

2   Channel Estimation in PSA-OFDM System

This section introduces channel estimation methods in PSA-OFDM system. The purpose of pilot pattern design is reducing the error of channel estimation.

Consider the subframe k with n subcarriers and m time slots in PSA-OFDM system. The received signal in ith subcarrier and jth time slot can be expressed as:

$$y_{i,j}^k = h_{i,j}^k x_{i,j}^k + z_{i,j}^k \qquad (1)$$

where $x_{i,j}^k, y_{i,j}^k, h_{i,j}^k, z_{i,j}^k \in \mathbb{C}$ are the received signal, transmitted signal, channel response, and white Gaussian noise respectively. Set $N = \{(i,j)|i = 1,2,\cdots,n, j = 1,2,\cdots,m\}$. $P \subseteq N$ is the pilot pattern. $X_P^k, Y_P^k, H_P^k, Z_P^k \in \mathbb{C}^{|P|\times 1}$ is the value in the pilot position while in particular, $X_N^k, Y_N^k, H_N^k, Z_N^k$ is the value at all symbols. Channel estimation in PSA-OFDM system is estimating the value of H_N^k by X_P^k and Y_N^k and pilot pattern design is constructing P.

LS estimates the channel at the pilot position by minimizing (2) where $\| \cdot \|_2$ is ℓ_2 norm and *diag* is vector diagonalization. The result of the optimization is $\hat{H}_P^k = diag(X_P^k)^{-1}Y_P^k$. In order to obtain other channel response, the most commonly method is interpolation, which is based on the fact that the channel changes are continuous and small between adjacent symbols in both time domain and frequency domain.

$$\hat{H}_P^k = \arg\min_H \|Y_P^k - diag(X_P^k)H\|_2^2 \qquad (2)$$

MMSE method obtains the channel estimation $\hat{H}(k, P)$ of the whole subframe k by multiplying $\hat{H}_P^k$ by a coefficient matrix $\hat{W}_P^k \in \mathbb{C}^{|N| \times |P|}$. That is $\hat{H}(k, P) = \hat{W}_P^k \hat{H}_P^k$. $\hat{W}_P^k$ is determined by minimizing the estimation error:

$$\hat{W}_P^k = \arg\min_W \|H_N^k - W\hat{H}_P^k\|_2^2 \qquad (3)$$

The result of $\hat{W}_P^k$ by solving optimization problem (3) is

$$\hat{W}_P^k = R_{NP}(R_{PP} + \frac{\sigma_z^2}{\sigma_x^2}I)^{-1} \qquad (4)$$

where $I \in \mathbb{R}^{|P| \times |P|}$ is the identity matrix, $R_{PP} = \mathbb{E}\{H_P H_P^\dagger\} \in \mathbb{C}^{|P| \times |P|}$ is the auto-correlation matrix of the pilot and $R_{NP} = \mathbb{E}\{H_N H_P^\dagger\} \in \mathbb{C}^{|N| \times |P|}$ is the cross-correlation matrix between the pilot and all symbols. σ_x^2 and σ_z^2 are the variances of the transmitted signal and the channel noise, respectively.

In fact, R_{NP} and R_{PP} are unknown while sometimes the correlation between position (i_1, j_1) and (i_2, j_2) can be approximated as follows:

$$r_{(i_1,j_1),(i_2,j_2)} = \frac{J_0(2\pi f_{max}T(j_1 - j_2))}{1 + j2\pi\tau_{max}\Delta f(i_1 - i_2)} \qquad (5)$$

where J_0 is the first kind of zero order Bessel function, f_{max} is the maximum Doppler frequency, T is the symbol block time, τ_{max} is the maximum delay spread, and Δf is the subcarrier spacing.

In LMMSE, $\frac{\sigma_z^2}{\sigma_x^2}$ is replaced by expectation $\frac{\beta}{SNR}$. β is a channel modulation type parameter. In this method, $\hat{W}_P^k$ is uniquely determined by the modulation mode and SNR, which saves a lot of calculation time of matrix multiplication and inversion in practical application. Therefore, the channel estimation methods in Sects. 3 and 4 are LMMSE.

3 Pilot Pattern Design with Branch and Bound

Let's first describe the pilot pattern design problem. Set $S_P = diag\{s_{(i,j)}^P\} \in \{0,1\}^{|N| \times |N|}$ and $s_{(i,j)}^P = 1$ when $(i, j) \in P$. Then the result of LMMSE for subframe k with pilot pattern P can be expressed as:

$$\hat{H}(k, P) = R_{NN}[S_P(R_{NN} + \frac{\beta}{SNR}I)S_P]^{-1}\hat{H}_P^k = W_{LMMSE}(S_P)\hat{H}_P^k \qquad (6)$$

The inverse operation here is the pseudo inverse operation.

The pilot pattern is designed to determine no more than p pilot insertion positions to make the channel estimation results on all subframes more accurate:

$$Obj(P) = \frac{1}{K} \sum_{k}^{K} \|H_N^k - \hat{H}(k, P)\|_2^2 \tag{7}$$

$$= \frac{1}{K} \sum_{k}^{K} \|H_N^k - W_{LMMSE}(S_P)\hat{H}_P^k\|_2^2 \tag{8}$$

$$\hat{P} = \underset{P, \, |P| \leq p}{\arg \min} \, Obj(P) \tag{9}$$

where K is the number of PSA-OFDM symbols used to design pilot pattern.

In order to solve (9), BnB is adopted. The diagram of BnB is shown in Fig. 2.

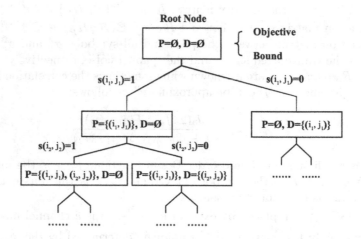

Fig. 2. Branch and bound diagram.

3.1   Branch

Branching repeatedly divides the solution space into smaller and smaller subsets. Each node determines not only a feasible solution P_{Node}, but also a set of positions that are discarded D_{Node}. In this work, we make full use of binary tree branching and random strategy. When branching each node, randomly select a position from $U_{Node} = N - (P_{Node} \cup D_{Node})$ to generate two new nodes based on whether to select the location as the pilot location. At the leaf node, $P_{Node} = p$ or $N = (P_{Node} \cup D_{Node})$.

For the node branch priority, the worst boundary strategy and the optimal target value strategy are adopted alternately.

3.2 Bound

Boundary is an important index of nodes, which refers to the lower boundary here. On the one hand, after each branch, no further branch will be made for any child node whose boundary exceeds the known feasible solution value. In this way, many nodes can be ignored, thus narrowing the search scope. On the other hand, the boundary is also an important basis to determine the node branch priority.

However, the boundary cannot be estimated accurately, so we give a heuristic method. Take $P_r \subseteq U_{Node}$ at random such that $|P_{Node} \cup P_r| = p$, which means that the pilot number is supplemented to p at random in the alternative set. The boundary B_{Node} is defined as follows:

$$B_{Node} = Obj(P_{Node} \cup P_r) - \alpha \cdot Obj(P_{Norm}) \tag{10}$$

$$\alpha = \gamma(1 - \frac{|P_{Node}|}{p})(1 - \frac{|P_{Node}| + |D_{Node}|}{|N|}) \tag{11}$$

where P_{Norm} is a traditional standard pilot pattern and γ is the correction rate parameter. Instead of directly taking the target value corresponding to the randomly supplemented pilot mode as the boundary, we subtract a correction term. The meaning of the correction term is the gap between the random boundary and the real boundary. The smaller the number of pilot positions selected and determined not to be selected, the more inaccurate the random term and the larger the correction term. Here, we default the number of pilot is positive correlation with the accuracy of channel estimation. And according to this calculation, the B_{Node} of the child node may be smaller than that of the parent node, so sometimes max operation is performed on the boundary of the node and its parent node.

The whole process can be expressed as the following algorithm.

The total number of leaf nodes is $\sum_{i=0}^{p} \binom{n \times m}{i}$. For each different P on the node, $W_{LMMSE}(S_P)$ needs to be calculated. Since there are many 0 in S_P, the calculation of the pseudo inverse in $W_{LMMSE}(S_P)$ is actually equivalent to the calculation of the pseudo inverse of a $|P| \times |P|$ matrix. This still requires a lot of computation and generally, the optimal solution cannot be obtained by this algorithm. However, due to the consideration of node computing priority and pruning operation, BnB is still an efficient solution finding strategy in this problem and has achieved good results in practice.

4 Simulation Results

In this section, we get BnB-PP with different pilot numbers on the simulation data, and compare the channel estimation results of the pilot pattern we designed with the common pilot pattern under different SNR.

In the simulation, Vienna 5G Link Level Simulator [17] is introduced to simulate wireless signal transmission under 5G NR standard. Only one antenna is set at both transmitter and receiver. The subcarrier interval is 60 kHz. Each

---

Algorithm . Branch and Bound

Require: n, m, p, ε

Ensure: $BestNode$

$\quad N = \{(i,j)|i = 1, 2, \cdots, n, j = 1, 2, \cdots, m\}$

$\quad Obj = Root.obj = +\infty, Root.Bound = -\infty,$

$\quad Root.P = Root.D = \emptyset$

$\quad Node = \{Root\}, BsetNode = Root$

$\quad$ **while** $Node \neq \emptyset$ and $Obj > \varepsilon$ **do**

$\qquad node = \underset{node \in Node}{\arg\max} \ Priority(Node)$

$\qquad Node = Node - \{node\}$

$\qquad$ **if** $|node.P| < p$ and $|node.P| + |node.D| < |N|$ **then**

$\qquad\quad nodeLeft = nodeRight = node$

$\qquad\quad$ Random select $(i,j) \in N - node.P - node.D$

$\qquad\quad nodeLeft.P = nodeLeft.P \cup \{(i,j)\}$

$\qquad\quad nodeRight.D = nodeRight.D \cup \{(i,j)\}$

$\qquad\quad$ update $nodeLeft.Bound$ and $nodeRight.Bound$

$\qquad\quad$ **if** $nodeLeft.Bound < Obj$ **then**

$\qquad\qquad Node = Node \cup \{nodeLeft\}$

$\qquad\quad$ **end if**

$\qquad\quad$ **if** $nodeRight.Bound < Obj$ **then**

$\qquad\qquad Node = Node \cup \{nodeRight\}$

$\qquad\quad$ **end if**

$\qquad\quad nodeLeft.obj = \frac{1}{K} \sum_k^K \|H_N^k - \hat{H}(k, nodeLeft.P)\|_2^2$

$\qquad\quad$ **if** $nodeLeft.obj < Obj$ **then**

$\qquad\qquad Obj = nodeLeft.Obj, BestNode = nodeLeft$

$\qquad\quad$ **end if**

$\qquad$ **end if**

$\quad$ **end while**

---

frame consists of $m = 56$ time slot and $n = 24$ subcarriers. Vehicle-A (VehA) wireless channel model is adopted, and the center frequency is 2.1 GHz. The modulation mode is 64 Quadrature Amplitude Modulation (16QAM) and $\beta = 2.6854$ at this time. The speed of user equipment (UE) is 20 m/s.

This paper mainly studies the case when the number of pilots is 32 and 64. As shown in Fig. 3, the pilot pattern is designed by BnB under the data of $SNR = 36$. Here, just take a relatively small data scale $K = 5$, so that the result can achieve a certain generalization performance. Set correction factor $\gamma = 1$. It can be seen that the pilot pattern are uneven and relatively scattered, which is consistent with the design idea of the conventional pilot pattern. Taking the result when the number of pilots is 64 as an example. From the frequency domain, there is only one subcarrier frequency without pilot. Most of the subcarriers are equipped with 2 or 3 pilots. From the time domain, there must be pilots on every three consecutive time slots, and the number of pilots in each time slot does not exceed 4. And there are few cases where two pilots are adjacent.

Figure 4 shows the impact of different pilot numbers on channel estimation when $SNR = 36$, and compares BnB-PP with lattice-type pilot pattern.

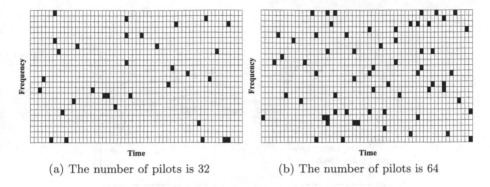

(a) The number of pilots is 32   (b) The number of pilots is 64

Fig. 3. Pilot pattern obtained by branch and bound

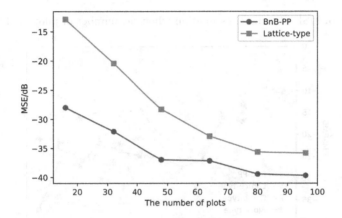

Fig. 4. MSE of channel estimation with different pilot numbers.

The channel estimation method here is LMMSE and for visualization, MSE is converted to dB. The effect is better than lattice type under each pilot number, and the smaller the pilot number, the better the effect. A smaller number of pilots means a larger channel capacity, and OFDM symbols can transmit more information, which fully demonstrates the value of this method.

Figure 5 and Fig. 6 respectively show the comparison of MSE estimated by the channel under different SNR when the number of pilots is 32 and 64. Due to the frequency selective fading of this channel, the comb pilot mode also shows good performance. It can be seen that when the pilot number is 32 and the SNR is greater than 10, BnB-PP is better than other pilot modes. And in other cases, the performance of BnB-PP is stable. Table 1 shows the average decline rate of MSE under all SNR. For example, the average MSE under all SNR of channel estimation for 32 and 64 pilots is reduced by 80.68% and 3.88% respectively compared with lattice-type pilot pattern.

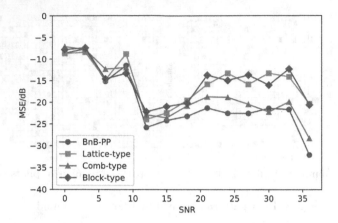

Fig. 5. MSE of channel estimation when the number of pilots is 32.

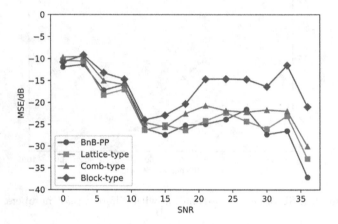

Fig. 6. MSE of channel estimation when the number of pilots is 64.

Table 1. Average decline rate of MSE under all SNR

The number of pilots	Block-type	Comb-type	Lattice-type
32	45.75%	17.92%	80.68%
64	60.34%	34.00%	3.88%

5   Conclusion

In this work, we propose a pilot pattern design scheme for BnB based LMMSE channel estimation in PSA-OFDM systems. And a large number of random strategies are adopted in branching and boundary calculation. The simulation results show that the pilot pattern BnB-PP we designed is better than the common pilot pattern under most SNR, especially when the number of pilots is small, which is of great significance to improve the system capacity and communication quality of PSA-OFDM system. However, this method is not fully integrated with LMMSE. The computing speed on each node is slow, and there may be a lot of computational redundancy. In the future work, we will try to reduce the computing time by matrix operation. And we will consider MIMO system and analyze the impact of different channel models on pilot pattern design.

References

1. Adegbite, S., Stewart, B., McMeekin, S.: Least squares interpolation methods for LTE system channel estimation over extended ITU channels. Int. J. Inf. Electron. Eng. **3**(4), 414 (2013)
2. Andrews, J.G., et al.: What will 5G be? IEEE J. Sel. Areas Commun. **32**(6), 1065–1082 (2014)
3. Coleri, S., Ergen, M., Puri, A., Bahai, A.: Channel estimation techniques based on pilot arrangement in OFDM systems. IEEE Trans. Broadcast. **48**(3), 223–229 (2002)
4. Colieri, S., Ergen, M., Puri, A., Bahai, A.: A study of channel estimation in OFDM systems. In: Proceedings IEEE 56th Vehicular Technology Conference, vol. 2, pp. 894–898. IEEE (2002)
5. Fu, A., Modiano, E., Tsitsiklis, J.: Optimal energy allocation for delay-constrained data transmission over a time-varying channel. In: IEEE INFOCOM 2003. Twenty-Second Annual Joint Conference of the IEEE Computer and Communications Societies (IEEE Cat. No. 03Ch37428), vol. 2, pp. 1095–1105. IEEE (2003)
6. He, S., Zhang, Q., Qin, J.: Pilot pattern design for two-dimensional OFDM modulations in time-varying frequency-selective fading channels. IEEE Trans. Wireless Commun. **21**, 1335–1346 (2021)
7. Hsieh, M.H., Wei, C.H.: Channel estimation for OFDM systems based on comb-type pilot arrangement in frequency selective fading channels. IEEE Trans. Consum. Electron. **44**(1), 217–225 (1998)
8. Kewen, L., et al.: Research of MMSE and LS channel estimation in OFDM systems. In: The 2nd International Conference on Information Science and Engineering, pp. 2308–2311. IEEE (2010)
9. Lin, J.C.: Least-squares channel estimation for mobile OFDM communication on time-varying frequency-selective fading channels. IEEE Trans. Veh. Technol. **57**(6), 3538–3550 (2008)
10. Liu, Y., Tan, Z., Hu, H., Cimini, L.J., Li, G.Y.: Channel estimation for OFDM. IEEE Commun. Surv. Tutor. **16**(4), 1891–1908 (2014)
11. Ma, J., Xue, E., Dong, X.: New pilot signal design on compressive sensing based random access for machine type communication. In: 2020 IEEE 8th International Conference on Information, Communication and Networks (ICICN), pp. 69–73. IEEE (2020)
12. Mallach, S.: Improved mixed-integer programming models for the multiprocessor scheduling problem with communication delays. J. Comb. Optim. **36**(3), 871–895 (2017). https://doi.org/10.1007/s10878-017-0199-9
13. Morelli, M., Mengali, U.: A comparison of pilot-aided channel estimation methods for OFDM systems. IEEE Trans. Signal Process. **49**(12), 3065–3073 (2001)
14. Morrison, D.R., Jacobson, S.H., Sauppe, J.J., Sewell, E.C.: Branch-and-bound algorithms: a survey of recent advances in searching, branching, and pruning. Discret. Optim. **19**, 79–102 (2016)

15. Muquet, B., De Courville, M., Duhamel, P.: Subspace-based blind and semi-blind channel estimation for OFDM systems. IEEE Trans. Signal Process. **50**(7), 1699–1712 (2002)
16. Panta, K.R., Armstrong, J.: Effects of clipping on the error performance of OFDM in frequency selective fading channels. IEEE Trans. Wireless Commun. **3**(2), 668–671 (2004)
17. Pratschner, S., et al.: Versatile mobile communications simulation: the Vienna 5G Link Level Simulator. EURASIP J. Wirel. Commun. Netw. **2018**(1), 1–17 (2018). https://doi.org/10.1186/s13638-018-1239-6
18. Quadri, D., Soutif, E., Tolla, P.: Exact solution method to solve large scale integer quadratic multidimensional knapsack problems. J. Comb. Optim. **17**(2), 157–167 (2009)
19. Shi, L., Guo, B., Zhao, L.: Block-type pilot channel estimation for OFDM systems under frequency selective fading channels (2009)
20. So, J., Kim, D., Lee, Y., Sung, Y.: Pilot signal design for massive MIMO systems: a received signal-to-noise-ratio-based approach. IEEE Signal Process. Lett. **22**(5), 549–553 (2014)
21. Soltani, M., Pourahmadi, V., Mirzaei, A., Sheikhzadeh, H.: Deep learning-based channel estimation. IEEE Commun. Lett. **23**(4), 652–655 (2019)
22. Soltani, M., Pourahmadi, V., Sheikhzadeh, H.: Pilot pattern design for deep learning-based channel estimation in OFDM systems. IEEE Wireless Commun. Lett. **9**(12), 2173–2176 (2020)
23. Soman, A.M., Nakkeeran, R., Shinu, M.J.: Pilot based MMSE channel estimation for spatial modulated OFDM systems. Int. J. Electron. Telecommun. **67**(4), 685–691 (2021)
24. Stuber, G.L., Barry, J.R., Mclaughlin, S.W., Li, Y., Ingram, M.A., Pratt, T.G.: Broadband MIMO-OFDM wireless communications. Proc. IEEE **92**(2), 271–294 (2004)
25. Sutar, M.B., Patil, V.S.: LS and MMSE estimation with different fading channels for OFDM system. In: 2017 International conference of Electronics, Communication and Aerospace Technology (ICECA), vol. 1, pp. 740–745 (2017). https://doi.org/10.1109/ICECA.2017.8203641
26. Tian, Y., Li, K., Yang, W., Li, Z.: A new effective branch-and-bound algorithm to the high order MIMO detection problem. J. Comb. Optim. **33**(4), 1395–1410 (2017)
27. Tong, L., Sadler, B.M., Dong, M.: Pilot-assisted wireless transmissions: general model, design criteria, and signal processing. IEEE Signal Process. Mag. **21**(6), 12–25 (2004)
28. Wang, X.: OFDM and its application to 4G. In: 14th Annual International Conference on Wireless and Optical Communications, WOCC 2005, p. 69. IEEE (2005)
29. Wu, H.: LMMSE channel estimation in OFDM systems: a vector quantization approach. IEEE Commun. Lett. **25**(6), 1994–1998 (2021)
30. Zhang, L., et al.: Lattice pilot aided DMT transmission for optical interconnects achieving 5.820 bits/HZ per lane (2019)

AoI Minimizing of Wireless Rechargeable Sensor Network Based on Trajectory Optimization of Laser-Charged UAV

Chuanwen Luo[1,2], Yunan Hou[1,2], Yi Hong[1,2(✉)], Zhibo Chen[1,2], Ning Liu[1,2], and Deying Li[3]

[1] School of Information Science and Technology, Beijing Forestry University, Beijing 100083, China
{chuanwenluo,hongyi,zhibo}@bjfu.edu.cn
[2] Engineering Research Center for Forestry-Oriented Intelligent Information Processing of National Forestry and Grassland Administration, Beijing 100083, China
[3] School of Information, Renmin University of China, Beijing 100872, People's Republic of China
deyingli@ruc.edu.cn

Abstract. This paper investigates a new UAV-assisted Wireless Rechargeable Sensor Network (WRSNs) based on wireless powered technologies, where the Unmanned Aerial Vehicle (UAV) can not only be used as wireless aerial mobile base station for gathering data from sensors but also be used as mobile charger to replenish energy for sensors, and it also can be charged energy by Laser Beam Directors (LBDs) to overcome the limitation of on-board energy shortage. In such a network, we study the average Age of Information Optimization (AoIO) problem whose objective is to minimize the average AoI of data collected from sensors by UAV. We first prove that the AoIO problem is NP-hard. Since the average AoI depends on UAV's flight trajectory, the hovering time required for data collection, energy power transfer and laser charging, we propose an approximation algorithm to solve the AoIO problem by jointly optimizing these factors. Afterwards, we conduct extensive simulation experiments to validate effectiveness of the proposed algorithm.

Keywords: WRSN · Laser-charged UAV · Age of Information

1 Introduction

Wireless Sensor Networks (WSNs) have wide applications in human production and life. Data gathering is a fundamental issue in WSNs since they are typically data centric. Specially, in some real-time status information updating

---

This work was supported in part by the Fundamental Research Funds for the Central Universities under grants (2021ZY88), also supported by the National Natural Science Foundation of China under Grant (62002022, 32071775).

applications, such as earthquake monitoring and forest fire protection, the data generated by sensors are required to delivered to the destination as quickly as possible for further data processing and analysis. Many research use the metric of Age of Information (AoI) to characterize the freshness of sensory data [1,2]. The AoI depicts the time difference between the time of data arriving destination and the time of data collection beginning for any sensor [3].

However, due to the battery capacity of sensors, the energy replenishing issues of sensors are becoming one of the most critical issues in the applications of WSNs. With the development of the wireless energy transfer technologies, such as Frequency Signal (RF) [4], the sensors can be charged through equipping wireless energy receiving antenna in WSNs. Such networks are called Wireless Rechargeable Sensor Networks (WRSNs). In WRSNs, the sensors can complete data transmission task through wireless received energy from mobile chargers.

In WSNs, Unmanned Aerial Vehicle (UAV) can be used as mobile station for collecting data from sensors [5] and used as mobile Frequency Signal (RF) energy source to replenish energy for sensors [6] due to their high maneuverability, good speed, flexibility, and increasing carrying capacity. However, since the UAVs are powered by batteries, they suffer from a lack of energy during performing data collection and wireless energy transfer tasks in the network. To solve the energy limitation problem of UAV, the Laser Beam Directors (LBDs) deployed at ground are used for providing energy by emitting wireless laser beams [7].

Inspired by the novel wireless energy transfer and laser charging technologies, joint the benefit of UAV as mobile station for data gathering, in this paper, we study the average Age of Information Optimization (AoIO) problem in UAV-assisted WRSNs, whose objective aims at minimizing the average AoI of data collected from sensors such that the UAV and sensors have enough power to complete tasks. The contribution of this paper can be summarized as below.

(1) We propose a new laser-charged UAV-assisted WRSN structure, in which UAV can not only be used as wireless mobile collector for gathering data from sensors but also be used as mobile charger to replenish energy for sensors, and it can also be replenished energy by LBDs to work for long periods of time.
(2) We first prove that the AoIO is NP-hard. Then we propose an approximation to solve the AoIO problem, which is to design the optimal trajectory of UAV and hovering times during data gathering, wireless energy transfer and laser charing to minimize the average AoI of data collected from sensors.
(3) We conduct extensive simulations to illustrate the effectiveness of the proposed algorithm for the AoIO problem.

The rest of the paper is organized as follows. In Sect. 2, we briefly introduce the literatures related with the investigated problem. In Sect. 3, we introduce the model and definition for the AoIO problem. In Sect. 4, we propose an approximation algorithm to solve the AoIO problem. In Sect. 5, we present the simulation results to verify the validity of the proposed algorithm. In Sect. 6, the paper is summarized.

2   Related Works

In this section, we briefly review the literatures related with the AoIO problem.

In [3], Hu et al. investigated the average AoI minimizing problem based on UAV's trajectory and time allocation for energy harvesting and data collection in UAV-assisted wireless powered WSN, where the UAV was used as both mobile data collector and charger for sensors without considering the limitation of the energy. In [8], Benmad et al. investigated the data collection problem in UAV-assisted WSNs powered by harvested energy, whose objective is to minimize the mission total time of the UAV while serving all sensors by ensuring that each sensor receives its required energy and transfers its sensed data since the limited energy of its battery. In [9], Zhang et al. studied the average AoI problem by optimizing the UAV trajectory in energy recharging UAV-based WSNs, where the UAV is used as a mobile data collector to gather data from sensors and it can be replenished energy by ground chargers through wireless energy transfer. In [10], Lahmeri et al. proposed a Internet of Things (IoTs) structure which consists of UAVs, IoT devices and LBDs, where the UAVs collect data from IoT devices and LBDs provide energy for UAVs. Zhao et al. [11] proposed an optimization network framework for joint optimizing the source, UAV and LBD's transmitting powers along with the UAV's trajectory in a laser-charged UAV-enabled mobile relaying system to maximize the weighted sum of the energy efficiency, where the UAV is charged by LBD through emitting laser beams.

Based on the above literatures, this paper proposes a new network framework which consists a UAV, LBDs and sensors, where the UAV is used as mobile collector for gathering data from sensors and also used to be charger to replenishing energy for sensors, and it is charged by ground LBDs. In such a network, we investigate the average AoI minimization problem by optimizing jointly UAV's trajectory and hovering times required for wireless energy transfer, laser charging and data collection.

3   Model and Problem Definition

3.1   Network Model

In this paper, we consider a laser charged UAV-assisted WRSN that consists of a UAV f equipped with half-duplex hybrid access point, n ground sensors represented as $S = \{s_1, s_2, \cdots, s_n\}$, m ground LBDs denoted as $D = \{b_1, b_2, \cdots, b_m\}$, a data center $s_0 = \{x_0^s, y_0^s\}$, as shown in Fig. 1. In the network, sensors are randomly deployed in a monitoring area $\mathcal{A}$ to detect environment and each $s_i \in S$ with E_i energy and V_i data is located at $s_i = (x_i^s, y_i^s)$. The UAV can collect data from sensors and also transfer energy power to sensors when it hovers. Assume that the UAV stores initial energy E and flies at a fixed altitude H with a constant speed v. LBDs that have the same laser transmitting power P_L are distributed in the monitoring area $\mathcal{A}$ with uniform distribution, and each $b_j \in D$ is located at $b_j = (x_j^b, y_j^b)$.

Since the flight altitude of UAV is H, the UAV completes the charging and data collection tasks or receiving laser beam energy only when it hovers at H altitude right above of sensors and LBDs. For any $s_i \in S$ (or $b_j \in D$), let s'_i (or b'_j) be the point at H altitude right above s_i (or b'_j), and $S' = \{s'_1, s'_2, \cdots, s'_n\}$ and $D' = \{b'_1, b'_2, \cdots, b'_m\}$. For any pair of points u and t in the three dimensional space, let $d(u,t)$ denote the Euler distance between them, for example, $d(s'_i, s'_k)$ represents the Euler distance between s'_i and s'_k. For simplicity, let $d(u,t,w)$ denotes the sum of $d(u,t)$ and $d(t,w)$.

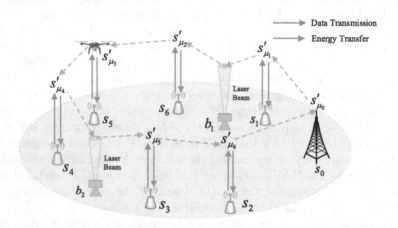

Fig. 1. An illustrative model of laser-charged UAV-assisted WRSN, where UAV as mobile collector and charger flies in trajectory $s'_{\mu_0} \rightarrow s'_{\mu_1}, \cdots, \rightarrow s'_{\mu_6} \rightarrow s'_{\mu_0}$ and is replenished energy by LBD.

3.2   Data Transmission Model from Sensor to UAV

In this paper, we adopt the LoS ground-to-air channel model between sensor and UAV with the path loss exponent $2 \leq \alpha < 4$ used by [12]. Therefore, the data transmission rate from s_i to UAV can be expressed as

$$R_i^f = W \log_2(1 + \frac{\beta_0 P_i}{\sigma^2 d^\alpha(s_i, f)}), \tag{1}$$

where W represents the channel bandwidth, P_i denotes the data transmission power of s_i, β_0 denotes the channel power at the reference distance $d_0 = 1\,\text{m}$, and σ^2 is the Gaussian noise power at the UAV.

3.3   Energy Harvesting Model from UAV to Sensor

In the wireless power transfer stage, the UAV keeps transmitting RF signals to the sensor with a fixed power P_u. In this paper, we use the RF energy transfer model proposed in [13]. Therefore, the received power at s_i from UAV can be described by

$$P_i^r = \frac{G_t G_r \eta}{L_p} \left(\frac{\lambda}{4\pi(d(s_i, f) + \beta)} \right)^2 P_u, \tag{2}$$

where β is a parameter to adjust the Friis equation for the short distance transmission, λ denotes the average wavelength, G_t represents the transmit gain parameter, G_r denotes the received gain parameter, L_p is the polarization loss.

3.4 Laser Charging Model from LBD to UAV

In this paper, the LBDs as laser beams transmitters charge for the UAV. We adopt the theoretical framework of power transfer in [14] from LBDs to UAV, and the received power of UAV from b_j is expressed as

$$P_j^f = P_L \cdot \eta_{le} \cdot e^{-\delta \cdot d(b_j, f)}, \tag{3}$$

where δ is the laser attenuation coefficient, $d(b_j, f)$ represents the distance between b_j and f, η_{le} denotes the laser to electricity conversion efficiency. The value of δ can be depict as $\frac{\ell}{\mu}(\frac{\partial}{\chi})^{-\varsigma}$, where ℓ and χ are two constants, μ is the visibility of environment, ∂ denotes the wavelength, ς represents the size distribution of the scattering particles.

3.5 Problem Formulation

AoI Model. Suppose that the UAV completes the energy transfer and data collection tasks for n sensors one by one based on obtained flight trajectory of UAV, i.e. $s'_{\mu_0} \rightarrow s'_{\mu_1} \rightarrow s'_{\mu_2} \cdots \rightarrow s'_{\mu_n} \rightarrow s'_{\mu_0}$, where $s'_{\mu_0} = s'_0$ and s'_{μ_i} denotes the i-th sensor on the flight trajectory. We denote the total flight time of UAV for completing charging and data collection tasks of network as T. Let $t^0_{\mu_i}$ be the time of UAV reaching s_{μ_i}. According to the definition of AoI [15], we can obtain that the AoI of data collected from s_{μ_i} can be expressed as

$$\Delta_{\mu_i} = T - t^0_{\mu_i}. \tag{4}$$

The average AoI of all data collected from sensors is defined

$$\bar{\Delta} = \frac{1}{n} \sum_{i=1}^{n} \Delta_{\mu_i}. \tag{5}$$

Mathematical Problem Model. For any i from 0 to n, we use $t^c_{\mu_i}$ and $t^h_{\mu_i}$ to denote the hovering time of UAV for data collection and energy transfer for s_{μ_i}, respectively. Let $E^0_{\mu_i}$ be the remaining energy of UAV when it is at the time of leaving s'_{μ_i}, where $E^0_{\mu_0} = E$. We use $t^b_{\mu_i}$ and $t^b_{\mu_{i,i+1}}$ to denote the charging time of UAV hovering at b'_{μ_i} and $b'_{\mu_{i,i+1}}$, respectively, where we can obtain b'_{μ_i} and $b'_{\mu_{i,i+1}}$ by $d(s'_{\mu_i}, b'_{\mu_i}) = \min\{d(s'_{\mu_i}, b'_j) | b'_j \in D'\}$ and $d(s'_{\mu_i}, b'_{\mu_{i,i+1}}, s'_{\mu_{i+1}}) = \min\{d(s'_{\mu_i}, b'_j, s'_{\mu_{i+1}}) | b'_j \in D'\}$. For simplicity, we let $\mu_{n+1} = \mu_0$.

We define the binary variables x_{μ_i} and $x_{\mu_{i,i+1}}$ as below.

$$x_{\mu_i} = \begin{cases} 1 & \text{the UAV is charged by } b_{\mu_i} \\ 0 & \text{otherwise} \end{cases} \tag{6}$$

$$x_{\mu_{i,i+1}} = \begin{cases} 1 & \text{the UAV is charged by } b_{\mu_{i,i+1}} \\ 0 & \text{otherwise} \end{cases} \tag{7}$$

Our goal is to design the optimal trajectory of UAV and also allocate the time for data collection, energy transfer and harvesting energy in order to minimize $\bar{\Delta}$ of the network. The optimization problem which is called the average <u>A</u>ge of <u>I</u>nformation <u>O</u>ptimization (AoIO) can be mathematically expressed as

$$min \ \bar{\Delta} \tag{8}$$

s.t.

$$t_{\mu_i}^c R_{\mu_i}^f \geq V_{\mu_i} \tag{9}$$

$$E_{\mu_i} + t_{\mu_i}^h P_{\mu_i}^r \geq t_{\mu_i}^c P_{\mu_i} + E_\theta \tag{10}$$

$$E \geq \max\{\frac{d(s'_{\mu_i}, b'_{\mu_i}) + d(s'_{\mu_i}, s'_{\mu_{i+1}})}{v} + t_{\mu_i}^c + t_{\mu_i}^h | 1 \leq i \leq n\} \tag{11}$$

$$\min\{E, E_{\mu_i}^0 + t_{u_{i,i+1}}^b P_{\mu_{i,i+1}}^f - \frac{d(s'_{\mu_i}, b_{\mu_{i,i+1}}) + d(b_{\mu_{i,i+1}}, s'_{\mu_{i+1}})}{v}\} \tag{12}$$
$$\geq \frac{d(s'_{\mu_{i+1}}, b'_{\mu_{i+1}})}{v} + t_{\mu_{i+1}}^c + t_{\mu_{i+1}}^h$$

$$t_{\mu_i}^0 = \sum_{k=1}^i (\frac{d(s'_{\mu_{k-1}}, s'_{\mu_k})}{v} \cdot x_{\mu_{k-1,k}} + (\frac{d(s'_{\mu_{k-1}}, b'_{\mu_{k-1,k}}, s'_{\mu_k})}{v} + t_{\mu_{k-1,k}}^b)$$
$$\cdot x_{\mu_{k-1,k}} + (\frac{2d(s'_{\mu_{k-1}}, b'_{\mu_{k-1}})}{v} + t_{\mu_{k-1}}^b) \cdot x_{\mu_{k-1}} + t_{\mu_{k-1}}^c + t_{\mu_{k-1}}^h) \tag{13}$$

$$E_{\mu_i}^0 = t_{\mu_i}^0 - \sum_{k=1}^i ((t_{\mu_k}^b - t_{\mu_k}^b P_{\mu_k}^f) \cdot x_{\mu_k} + (t_{\mu_{k-1,k}}^b - t_{\mu_{k-1,k}}^b P_{\mu_{k-1,k}}^f) \cdot x_{\mu_{k-1,k}})$$
$$+ (\frac{2d(s'_{\mu_i}, b'_{\mu_i})}{v} + t_{\mu_i}^b) \cdot x_{\mu_i} + t_{\mu_i}^c + t_{\mu_i}^h \tag{14}$$

$$T = t_{\mu_{n+1}}^0 + \kappa, t_{\mu_0}^0 = \kappa \tag{15}$$

$$x_{\mu_i}, x_{\mu_{i,i+1}} \in \{0,1\}, 0 \leq i \leq n \tag{16}$$

Constraint (9) is to make sure s_{μ_i} successfully transmits V_{μ_i} data to the UAV within $t^c_{\mu_i}$ time. Constraint (10) ensures that the sum of the initial energy and charging energy of s_{μ_i} is greater than or equal to the sum of threshold and the energy consumption for data transmission. Constraint (11) guarantees that the AoIO problem has feasible solution. Constraint (12) ensures that after replenishing energy at $b_{\mu_{i,i+1}}$, the UAV can complete charging and data collection tasks for $s'_{\mu_{i+1}}$ and arrive to $b'_{\mu_{i+1}}$. The Eq. (13) is to compute the time $t^0_{\mu_i}$ which contains the flying time on the trajectory and all hovering times for data collection, energy transfer and harvesting energy before arriving s'_{μ_i}. The Eq. (14) is to compute the remaining energy of UAV when is leaving s'_{μ_i}. The Eq. (15) is to illustrate that T is the time of UAV returning s_{μ_0}.

Theorem 1. *The AoIO problem is NP-hard.*

Proof. If we set E is a very large value, then the problem can be reduced to the P_1 problem in [3] since the UAV does not need to replenished energy by LBDs. Since the P_1 problem is proved NP-hard and it is a special case of the problem, the problem is also NP-hard.

4   Algorithm for the AoIO Problem

In this section, we propose an approximation algorithm to solve the AoIO problem, which is called AoIOA. The algorithm aims at finding an optimal trajectory of UAV such that the average AoI of the network is minimized. The algorithm consists of the following four steps.

In the first step, we use the 1.5-approximation algorithm for the TSP problem to compute the flight trajectory of UAV for $S' \cup \{s'_0\}$ [16], and the order visited by UAV is denoted as $s'_{\mu_0}, s'_{\mu_1}, s'_{\mu_2}, \cdots, s'_{\mu_n}$, where $s'_{\mu_0} = s'_0$.

In the second step, for any i from 0 to n, we compute the data transmission rate $R^f_{\mu_i}$ from s_{μ_i} to UAV as $R^f_{\mu_i} = W \log_2(1 + \frac{\beta_0 P_{\mu_i}}{\sigma^2 H^\alpha})$. Then we compute the data collection time $t^c_{\mu_i} = \frac{V_i}{R^f_{\mu_i}}$ and the energy wireless transfer time $t^h_{\mu_i} = \frac{\max\{E_\theta + t^c_{\mu_i} P_{\mu_i} - E_{\mu_i}, 0\}}{P^r_{\mu_i}}$ of UAV at s'_i. Afterwards, we select b'_{μ_i} which is closest to s'_{μ_i} from D', i.e. $d(s'_{\mu_i}, b'_{\mu_i}) = \min\{d(s'_{\mu_i}, b'_j)|b'_j \in D'\}$.

In the third step, for any i $(0 \leq i \leq n)$, we let parameters $t^0_{\mu_i}$ and $T^\ell_{\mu_i}$ denote the time of UAV arriving s'_{μ_i} and leaving s'_{μ_i}, respectively. Then we set $t^0_{\mu_0} = \kappa$ and $T^\ell_{\mu_0} = \kappa$, where κ is the vertical flight time from s_{μ_0} to s'_{μ_0}. After that, for any i from 0 to $n-1$, we set $E^{i,i+1}_c = \frac{d(s'_{\mu_i}, s'_{\mu_{i+1}})}{v} + \frac{d(s'_{\mu_{i+1}}, b'_{\mu_{i+1}})}{v} + t^c_{\mu_{i+1}} + t^h_{\mu_{i+1}}$, which is to judge whether the remanning energy of UAV can support UAV to fly from s'_{μ_i} to $s'_{\mu_{i+1}}$ and complete data collection and energy transfer tasks of $s'_{\mu_{i+1}}$ meanwhile can arrive the nearest LBD for replenishing energy. Then we select the flight trajectory of UAV as the following three cases by comparing $E^{i,i+1}_c$ with the remanning energy E_r of UAV.

Algorithm 1: AoIOA

Data: $D = \{b_1, b_2, \cdots, b_m\}$, $S = \{s_1, s_2, \cdots, s_n\}$, V_i and E_i for each
$s_i \in S$, v, H, E, $E_r = E$;

Result:

1 Using the 1.5-approximation algorithm for the TSP problem to compute
the flight trajectory of UAV for $S' \cup \{s'_0\}$ [16], and the order visited by
UAV is denoted as $s'_{\mu_0}, s'_{\mu_1}, s'_{\mu_2}, \cdots, s'_{\mu_n}$, where $s'_{\mu_0} = s'_0$;

2 **for** *i from 1 to n* **do**

3 $\quad R^f_{\mu_i} = W \log_2(1 + \frac{\beta_0 P_{\mu_i}}{\sigma^2 H^\alpha})$, $t^c_{\mu_i} = \frac{V_i}{R^f_{\mu_i}}$, $t^h_{\mu_i} = \frac{\max\{E_\theta + t^c_{\mu_i} P_{\mu_i} - E_{\mu_i}, 0\}}{P^r_{\mu_i}}$;

4 $\quad d(s'_{\mu_i}, b'_{\mu_i}) = \min\{d(s'_{\mu_i}, b'_j) | b'_j \in D'\}$;

5 **end**

6 $t^0_{\mu_0} = \kappa$, $T^\ell_{\mu_0} = \kappa$;

7 **for** *i from 0 to n − 1* **do**

8 $\quad E^{i,i+1}_c = \frac{d(s'_{\mu_i}, s'_{\mu_{i+1}}) + d(s'_{\mu_{i+1}}, b'_{\mu_{i+1}})}{v} + t^c_{\mu_{i+1}} + t^h_{\mu_{i+1}}$;

9 $\quad$ **if** $E^{i,i+1}_c \leq E_r$ **then**

10 $\quad\quad t^0_{\mu_{i+1}} = T^\ell_{\mu_i} + \frac{d(s'_{\mu_i}, s'_{\mu_{i+1}})}{v}$, $T^\ell_{\mu_{i+1}} = t^0_{\mu_{i+1}} + t^c_{\mu_{i+1}} + t^h_{\mu_{i+1}}$;

11 $\quad\quad E_r = E_r - E^{i,i+1}_c + \frac{d(s'_{\mu_{i+1}}, b'_{\mu_{i+1}})}{v}$;

12 $\quad$ **else**

13 $\quad\quad$ **if** $E^{i,i+1}_c - t^c_{\mu_{i+1}} - t^h_{\mu_{i+1}} \leq E_r < E^{i,i+1}_c$ **then**

14 $\quad\quad\quad P^f_{\mu_{i+1}} = P_L \cdot \eta_{le} \cdot e^{-\delta H}$, $t^0_{\mu_{i+1}} = T^\ell_{\mu_i} + \frac{d(s_{\mu_i}, s_{\mu_{i+1}})}{v}$;

15 $\quad\quad\quad E_r = E - E^{i,i+1}_c + E_r - \frac{d(s'_{\mu_{i+1}}, b'_{\mu_{i+1}})}{v}$, $t^b_{\mu_{i+1}} = \frac{E}{P^f_{\mu_{i+1}}}$;

16 $\quad\quad\quad T^\ell_{\mu_{i+1}} = t^0_{\mu_{i+1}} + t^c_{\mu_{i+1}} + t^h_{\mu_{i+1}} + t^b_{\mu_{i+1}} + \frac{2d(s'_{\mu_{i+1}}, b'_{\mu_{i+1}})}{v}$;

17 $\quad\quad$ **else**

18 $\quad\quad\quad$ **for** *j from 1 to m* **do**

19 $\quad\quad\quad\quad$ **if** $d(s'_{\mu_i}, b'_j) \leq v \cdot E_r$ **then**

20 $\quad\quad\quad\quad\quad Q_{\mu_i} = Q_{\mu_i} \cup \{d(s'_{\mu_i}, b'_j, s'_{\mu_{i+1}})\}$;

21 $\quad\quad\quad\quad$ **end**

22 $\quad\quad\quad$ **end**

23 $\quad\quad\quad d(s'_{\mu_i}, b'_{\mu_{i,i+1}}, s'_{\mu_{i+1}}) = \min\{d(s'_{\mu_i}, b'_j, s'_{\mu_{i+1}}) | d(s'_{\mu_i}, b'_j, s'_{\mu_{i+1}}) \in Q_{\mu_i}\}$;

24 $\quad\quad\quad P^f_{\mu_{i,i+1}} = P_L \cdot \eta_{le} \cdot e^{-\delta H}$, $t^b_{\mu_{i,i+1}} = \frac{E - (E_r - \frac{d(s'_{\mu_i}, b'_{\mu_{i,i+1}})}{v})}{P^f_{\mu_{i,i+1}}}$;

25 $\quad\quad\quad t^0_{\mu_{i+1}} = T^\ell_{\mu_i} + \frac{d(s_{\mu_i}, b'_{\mu_{i,i+1}}) + d(b'_{\mu_{i,i+1}}, s_{\mu_{i+1}})}{v} + t^b_{\mu_{i,i+1}}$;

26 $\quad\quad\quad T^\ell_{\mu_{i+1}} = t^0_{\mu_{i+1}} + t^c_{\mu_{i+1}} + t^h_{\mu_{i+1}}$;

27 $\quad\quad$ **end**

28 $\quad$ **end**

29 **end**

30 $T = T^\ell_{\mu_n} + \frac{d(s'_{\mu_n}, s'_{\mu_0})}{v} + \kappa$, $\bar{\Delta} = \frac{1}{n} \sum_{i=1}^{n}(T - t^0_{\mu_i})$;

---

(1) $E_c^{i,i+1} \leq E_r$. The UAV with remanning energy E_r not only can arrive $s'_{\mu_{i+1}}$ for completing data collection and energy transfer tasks but also can arrive the nearest point $b'_{\mu_{i+1}}$ for replenishing energy. Then we compute $t_{\mu_{i+1}}^0 = T_{\mu_i}^{\ell} + \frac{d(s'_{\mu_i}, s'_{\mu_{i+1}})}{v}$ and $T_{\mu_{i+1}}^{\ell} = t_{\mu_{i+1}}^0 + t_{\mu_{i+1}}^c + t_{\mu_{i+1}}^h$. Afterwards, we update the remaining time of UAV as $E_r = E_r - E_c^{i,i+1} + \frac{d(s'_{\mu_{i+1}}, b'_{\mu_{i+1}})}{v}$.

(2) $E_c^{i,i+1} - t_{\mu_{i+1}}^c - t_{\mu_{i+1}}^h \leq E_r < E_c^{i,i+1}$. The UAV with remanning energy E_r can only arrive $s'_{\mu_{i+1}}$ to complete a part of energy transfer and data collection tasks. Therefore, the UAV needs to fly to $b'_{\mu_{i+1}}$ for replenishing energy and returns to $s'_{\mu_{i+1}}$ for completing the remaining tasks. Firstly, we compute the energy transfer power from $b'_{\mu_{i+1}}$ to UAV as $P_{\mu_i, i+1}^f = P_L \cdot \eta_{le} \cdot e^{-\delta H}$ and compute the charging time at $b'_{\mu_{i+1}}$ as $t_{\mu_{i+1}}^b = \frac{E}{P_{\mu_{i+1}}^f}$. Secondly, we compute the leaving time at $s'_{\mu_{i+1}}$ as $T_{\mu_{i+1}}^{\ell} = t_{\mu_{i+1}}^0 + t_{\mu_{i+1}}^c + t_{\mu_{i+1}}^h + t_{\mu_{i+1}}^b + \frac{2d(s'_{\mu_{i+1}}, b'_{\mu_{i+1}})}{v}$. Finally, we update the remaining energy at the $T_{\mu_{i+1}}^{\ell}$ time as $E_r = E - E_c^{i,i+1} + E_r - \frac{d(s'_{\mu_{i+1}}, b'_{\mu_{i+1}})}{v}$.

(3) $E_r \leq E_c^{i,i+1} - t_{\mu_{i+1}}^c - t_{\mu_{i+1}}^h$. Firstly, for any j from 1 to m, we use the condition $d(s'_{\mu_i}, b'_j) \leq v \cdot E_r$ to judge whether the remanning energy of UAV can support UAV to arrive b'_j. If it is, then we add $d(s'_{\mu_i}, b'_j, s'_{\mu_{i+1}})$ to Q_{μ_i}. Then we let $d(s'_{\mu_i}, b'_{\mu_i, i+1}, s'_{\mu_{i+1}}) = \min\{d(s'_{\mu_i}, b'_j, s'_{\mu_{i+1}}) | d(s'_{\mu_i}, b'_j, s'_{\mu_{i+1}}) \in Q_{\mu_i}\}$, where $b'_{\mu_i, i+1}$ is an intermediate node connecting s'_{μ_i} and $s'_{\mu_{i+1}}$ to replenish energy of UAV. Afterwards, we compute the hovering time of UAV for charging at $b'_{\mu_{i+1}}$ as $t_{\mu_i, i+1}^b = \frac{E - (E_r - \frac{d(s'_{\mu_i}, b'_{\mu_i, i+1})}{v})}{P_{\mu_i, i+1}^f}$. Finally, we compute $t_{\mu_{i+1}}^0 = T_{\mu_i}^{\ell} + \frac{d(s'_{\mu_i}, b'_{\mu_i, i+1}) + d(b'_{\mu_i, i+1}, s'_{\mu_{i+1}})}{v} + t_{\mu_i, i+1}^b$ and $T_{\mu_{i+1}}^{\ell} = t_{\mu_{i+1}}^0 + t_{\mu_{i+1}}^c + t_{\mu_{i+1}}^h$.

Consequently, we can obtain the total time consumption of UAV is $T = T_{\mu_n}^{\ell} + \frac{d(s'_{\mu_n}, s'_{\mu_0})}{v} + \kappa$ and the average AoI of data is $\bar{\Delta} = \frac{1}{n} \sum_{i=1}^{n} (T - t_{\mu_i}^0)$. The pseudo-code of the algorithm is given in Algorithm 1.

5   Performance Evaluation

In this section, we give the performance analysis of the AoIOA algorithm through a large number of experiments using MATLAB and Java programming. In the simulations, sensors are randomly deployed in the 2000 m * 2000 m square detecting area and each result is the average of 100 runs.

To prove the effectiveness of the AoIOA algorithm, we first propose an Enumerating Optimal Algorithm (EOA) to compute an optimal solution for a given laser-charged UAV-assisted WRSN that consists of one base station, one UAV, 3 LBDs and 15 sensors. Then we compare AoIOA with EOA for any sensor in the network when we set $E = 10000$ s, $P_i = 20$ W and $V_i = 500$ Mb for any $s_i \in S$, $P_u = 200$ W, $P_b = 300$ W, $\eta_{le} = 0.15$, $v = 10\,\text{m/s}$, $\frac{\beta_0}{\sigma^2} = 80$ dB, $H = 60$ m, and $W = 1\,\text{MB/s}$. Figure 2(a) gives the AoI $\triangle_{\mu_i}$ of s_{μ_i} achieved by the two algorithms for any $1 \le i \le 15$. The results show that the AoI of each sensor obtained by our AoIOA is close to the optimal EOA algorithm. According to Fig. 2(b), we can find that the gap of the average AoI between AoIOA and EOA is small and it maintains changeless, which proves the validity of the AoIOA algorithm.

In the following, we evaluate the impact of the different parameter settings on the average AoI of the network.

Firstly, we measure the average AoI of the network when we set $n = 60$, $m = 4$, $E = 10000$ s, $P_i = 20$ W and $V_i = 500$ MB for each $s_i \in S$, $\eta_{le} = 0.15$, $\frac{\beta_0}{\sigma^2} = 80$ dB, $v = 10$ m/s, $P_u = 200\,\text{W}$, $P_L = 300\,\text{W}$, $H = 50\,\text{m}$, 60 m, 70 m, 80 m, 90 m and change W from 1 MB to 3.5 MB. As shown in Fig. 3(a), we can find that the average AoI of the network decreases as W increases. This is because when W increases, the hovering time of UAV for collecting data from sensors decreases, which leads to the descend of AoI for each sensor. And we also observe that the average AoI decreases with the increasing of H since the data transmission rate from sensors to UAV grows as H decreases.

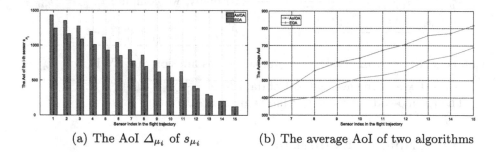

(a) The AoI $\triangle_{\mu_i}$ of s_{μ_i} (b) The average AoI of two algorithms

Fig. 2. The comparison result between AoIOA and EOA.

Secondly, we evaluate the impact of n on the average AoI as we set $E = 10000$ s, $P_i = 20$ W and $V_i = 500$ MB for each $s_i \in S$, $\eta_{le} = 0.15$, $\frac{\beta_0}{\sigma^2} = 80$ dB, $v = 10$ m/s, $H = 60\,\text{m}$, $P_u = 200\,\text{W}$, $P_L = 300\,\text{W}$, $m = 2$, 3, 4, 5, 6, and vary n from 50 to 110 increased by 10. As shown in Fig. 3(b), we can find that the average AoI grows as n increases since the time arriving at s_0 increases with the increasing of n. We also observe that the average AoI decreases as m grows. This is because that the flying distance from sensors to LBDs decreases as the deployment density of LBDs increases.

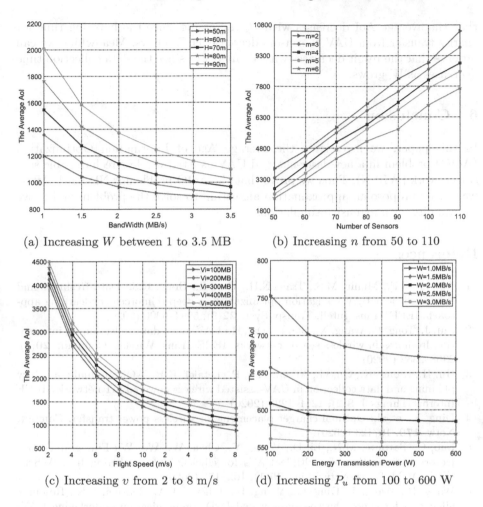

Fig. 3. The performance of AoIOA under different configurations.

Thirdly, we illustrate the impact of flying speed v on the average AoI when we set $n = 60$, $E = 10000$ s, $m = 4$, $P_i = 20$ W for each $s_i \in S$, $\eta_{le} = 0.15$, $\frac{\beta_0}{\sigma^2} = 80$ dB, $v = 10$ m/s, $H = 60$ m, $P_u = 200$ W, $P_L = 300$ W, $V_i = 100$ MB, 200 MB, 300 MB, 400 MB, 500 MB, and change v from 2 m/s to 8 m/s. As shown in Fig. 3(c), we can find that the average AoI decreases with v increasing since the decrease of flying time give rise to the decrease of AoI for each sensor. We also observe that the average AoI decreases as V_i diminishes since the hovering time for collecting time decrease as V_i reduces.

Finally, we measure the average AoI of the network when we set $n = 60$, $E = 10000$ s, $m = 4$, $P_i = 20$ W for each $s_i \in S$, $\eta_{le} = 0.15$, $\frac{\beta_0}{\sigma^2} = 80$ dB, $v = 10$ m/s, $H = 60$ m, $P_u = 200$ W, $P_L = 300$ W, $W = 1$ MB, 1.5 MB, 2 MB, 2.5 MB, 3 MB, and change P_u from 100 W to 600 W. As shown in Fig. 3(d), we can find that the average AoI decreases with P_u increasing since the hovering time for energy transfer from UAV to sensors decrease as P_u grows. Meanwhile, we can observe that the result decreases with W increases since the data collection time decreases as W grows.

6 Conclusion

In this paper, we investigate the average Age of Information Optimization (AoIO) problem in a new laser-charged UAV-assisted wireless rechargeable sensor network (WRSN). Then we prove that the AoIO problem is NP-hard. Afterwards, we propose an approximation algorithm to solve the problem and prove the validity with extensive simulations.

References

1. Abedin, S.F., Munir, M.S., Tran, N.H., Han, Z., Hong, C.S.: Data freshness and energy-efficient UAV navigation optimization: a deep reinforcement learning approach. IEEE Trans. Intell. Transp. Syst. **22**(9), 5994–6006 (2020)
2. Liu, J., Tong, P., Wang, X., Bai, B., Dai, H.: UAV-aided data collection for information freshness in wireless sensor networks. IEEE Trans. Wireless Commun. **20**(4), 2368–2382 (2020)
3. Hu, H., Xiong, K., Qu, G., Ni, Q., Fan, P., Letaief, K.B.: AOI-minimal trajectory planning and data collection in UAV-assisted wireless powered IoT networks. IEEE Internet Things J. **8**(2), 1211–1223 (2020)
4. Ding, X., et al.: Optimal charger placement for wireless power transfer. Comput. Netw. **170**, 107123 (2020)
5. Luo, C., Satpute, M.N., Li, D., Wang, Y., Chen, W., Wu, W.: Fine-grained trajectory optimization of multiple UAVs for efficient data gathering from WSNs. IEEE/ACM Trans. Networking **29**(1), 162–175 (2020)
6. Wang, H., Wang, J., Ding, G., Wang, L., Tsiftsis, T.A., Sharma, P.K.: Resource allocation for energy harvesting-powered D2D communication underlaying UAV-assisted networks. IEEE Trans. Green Commun. Networking **2**(1), 14–24 (2017)
7. Liu, W., Zhang, L., Ansari, N.: Laser charging enabled DBS placement for downlink communications. IEEE Trans. Netw. Sci. Eng. **8**(4), 3009–3018 (2021)
8. Benmad, I., Driouch, E., Kardouchi, M.: Data collection in UAV-assisted wireless sensor networks powered by harvested energy. In: 2021 IEEE 32nd Annual International Symposium on Personal, Indoor and Mobile Radio Communications (PIMRC), pp. 1351–1356. IEEE (2021)
9. Zhang, C., Liu, J., Xie, L., He, X.: Age-optimal data gathering and energy recharging of UAV in wireless sensor networks. In: 2021 3rd International Conference on Advanced Information Science and System (AISS 2021), pp. 1–6 (2021)
10. Lahmeri, M.-A., Kishk, M.A., Alouini., M.-S.: Charging techniques for UAV-assisted data collection: is laser power beaming the answer? IEEE Commun. Mag. **60**(5), 50–56 (2022)
11. Zhao, M.-M., Shi, Q., Zhao, M.-J.: Efficiency maximization for UAV-enabled mobile relaying systems with laser charging. IEEE Trans. Wireless Commun. **19**(5), 3257–3272 (2020)

12. Zeng, Y., Zhang, R.: Energy-efficient UAV communication with trajectory optimization. IEEE Trans. Wireless Commun. **16**(6), 3747–3760 (2017)
13. He, S., Chen, J., Jiang, F., Yau, D.K.Y., Xing, G., Sun, Y.: Energy provisioning in wireless rechargeable sensor networks. IEEE Trans. Mobile Comput. **12**(10), 1931–1942 (2012)
14. Zhang, Q., Fang, W., Liu, Q., Jun, W., Xia, P., Yang, L.: Distributed laser charging: a wireless power transfer approach. IEEE Internet Things J. **5**(5), 3853–3864 (2018)
15. Kaul, S., Yates, R., Gruteser, M.: Real-time status: How often should one update? In: 2012 Proceedings IEEE INFOCOM, pp. 2731–2735. IEEE (2012)
16. Du, D., Ko, K.-I., Hu, X., et al.: Design and Analysis of Approximation Algorithms, vol. 62. Springer, New York (2012). https://doi.org/10.1007/978-1-4614-1701-9

Energy-Constrained Geometric Coverage Problem

Huan Lan[(✉)]

School of Mathematics and Statistics, Yunnan University,
Kunming 650504, People's Republic of China
lanhuan0714@163.com

Abstract. Wireless sensor networks have many applications in real life. We are given m sensors and n users on the plane. The coverage of each sensor s is a disc area, whose radius $r(s)$ and energy $p(s)$ satisfy that $p(s) = r(s)^\alpha$, where $\alpha \geq 1$ is the attenuation factor. In this paper, we study the energy-constrained geometric coverage problem, which is to find an energy allocation scheme such that the total energy does not exceed a given bound P, and the total profit of the covered points is maximized. We propose a greedy algorithm whose approximation ratio is $1 - \frac{1}{\sqrt{e}}$.

Keywords: Geometric covering · Energy-constrained · Greedy algorithm · Approximation ratio

1 Introduction

Wireless sensor networks utilize sensors at fixed locations to serve users in a target area. Assuming that the sensor and the user are in the same plane. The service area of each sensor s is a disc $D(s, r(s))$ with a radius $r(s)$, which depends on the energy provided by the sensor. The relationship between the sensor energy and the radius is $p(s) = r(s)^\alpha$, where $\alpha \geq 1$ is the attenuation factor. A user u is covered by sensor s if the location of u is in the service disc $D(s, r(s))$. The minimum energy coverage problem [1] is to find a minimum energy distribution scheme to ensure that all users on the plane are covered by the sensor network. In fact, the minimum energy coverage problem is NP-hard when $\alpha > 1$ [1,2].

However, in most real environments, the wireless sensor network cannot cover all users due to energy constraints [4,12–14]. Trade-offs between energy and performance of sensor are important [11]. The energy-constrained geometric coverage problem aims to utilize limited energy to maximize coverage profits, which is defined as follows. Given a user set U, a sensor set S on the plane and an energy budget P, the profit of user u is w_u for any $u \in U$. The goal is to find an energy allocation scheme such that the total energy does not exceed P and the total profit of covered users is maximized.

The energy-constrained geometric coverage problem is a special case of the submodular cost constrained submodular function maximization problem, which is to find a subset S' of a given collection S of sets such that the submodular profit of the union of S' is maximized and the submodular cost of S' is no more than a given upper bound P, where an element in S' can be seen as the set of users covered by a sensor in S'. Zhang et al. [19] proposed a bi-criterion generalized greedy algorithm which can achieve an approximation ratio of $1/2(1-1/e)$. Iyer et al.[8] proposed a bi-criterion algorithm which achieves an approximation ratio of $1/k_g(1 - (\frac{K_f - K_g}{K_f})^{k_f})$, where $K_f = max\{|\ x\ |: f(x) < b\}$, $k_f = min\{|\ x\ |: f(x) < b \quad and \quad \forall j \notin x : f(x \cup j) > b\}$. If the cost function is a modular function, Wolsey et al. [17] proved that the modified greedy algorithm can achieve an approximation ratio of $(1 - 1/e^\beta)$, where β is the unique root of the equation $e^x = 2 - x$. Khuller et al. [9] proposed an algorithm which achieves an approximation ratio of $1/2(1-1/e)$. Recently, Tang et al. [16] proved that the algorithm in [9] can achieve an approximation ratio of $(1 - 1/\sqrt{e})$. Sviridenko [15] proposed an algorithm by enumerating three elements which achieves an approximation ratio of $(1 - 1/e)$. Most recently, Ariel et al. [10] proposed an algorithm by enumerating only two elements which achieves an approximation ratio of $(1 - 1/e)$. Yaroslavtsev et al. [18] proposed the $Greed^+$ algorithm, which adds a single element to each iteration of the greedy algorithm and achieves an approximation ratio of $1/2$. Feldman et al. [6] proposed an improved algorithm which enumerates an element in combination with $Greed^+$ and achieves an approximation ratio of 0.6174.

The energy-constrained geometric coverage problem is a special case of the maximum coverage problem under grouped budgets, which is to find a subset in each group such that the number of elements in each group does not violate the constraint and the total cost does not violate the constraint, and the coverage profit is maximized. Chekuri et al. [3] proposed the greedy algorithms for the cardinality constraint and the cost constraints and achieves an approximation ratio of $1/(1+\alpha)$ and $1/[6(1+\alpha)]$, respectively, where α is the approximate ratio for calling the sub-algorithm.

For the cost grouping constraint, Farbstein et al. [5] proposed an approximate algorithm and achieves an approximation ratio of $\alpha/(3 + 2\alpha)$. Recently, Guo et al. [7] used linear programming and rounding techniques to obtain a pseudopolynomial-time approximation algorithm and achieves an approximation ratio of $(1 - 1/e)$. They extended the algorithm to the maximum coverage problem of cost constraints and cardinality grouping constraint.

In this paper, we study the energy-constrained geometric coverage problem, which is given a user set U, a sensor set S on the plane and an energy budget P, the profit of user u is w_u for any $u \in U$. The goal is to find an energy allocation scheme such that the total energy does not exceed P and the total profit of covered users is maximized. We proposed an greedy algorithm for the energy-constrained geometric coverage problem and prove that the approximate ratio of the greedy algorithm is $1 - \frac{1}{\sqrt{e}}$.

The main structure of this paper is as follows: in Sect. 2, we introduce the fundamentals of energy-constrained geometric coverage problem. In Sect. 3, we propose a greedy algorithm. In Sect. 4, we make a summary.

2 Preliminaries

We are given a ground set $U = \{u_1, u_2, \ldots, u_n\}$ and a set function $f : 2^U \to \mathbb{R}$. A set function f is monotonic if $\forall X \subseteq Y, f(X) \leq f(Y)$. A set function f is submodular if $\forall X \subseteq Y \subseteq U, u \notin Y$,

$$f(X \cup \{u\}) - f(X) \geq f(Y \cup \{u\}) - f(Y).$$

We define $f(u \mid X) = f(X \cup \{u\}) - f(X)$.

The energy-constrained geometric coverage problem is described as follows:

Definition 1. Given a ground set of users $U = \{u_1, u_2, \ldots, u_n\}$, each user has a corresponding weight w_u for any $u \in U$. Given a set of sensors $S = \{s_1, s_2, \ldots, s_m\}$, each sensor $s \in S$ can form a disc $D(s, r(s))$ by adjusting the energy $p(s)$. The relationship between energy and radius is $p(s) = r(s)^\alpha, \alpha \geq 1$. Given $P \in \mathbb{R}^+$, the energy-constrained geometric coverage problem is to find an energy allocation scheme such that the total energy does not exceed P and the total profit of covered users is maximized.

For any instance of the energy-constrained geometric coverage problem, the distance between the sensor and the user is at most n. So a sensor s_i has at most n optional discs whose radius is $r_{i1} \leq r_{i2} \leq \cdots \leq r_{in}$. Let the optional disc set of the sensor s_i be $D_i = \{D(s_i, r_{i1}), \ldots, D(s_i, r_{in})\}$ and $\mathcal{D} = \{D_1, \ldots, D_m\}$. The user set covered by $D(s_i, r_{ij})$ is $cov(D(s_i, r_{ij}))$. The energy-constrained geometric coverage problem is that each sensor selects a disc from n optional discs while satisfying the energy constraint and the total profit of covered users is maximized. The mathematical programming is as follows:

$$\max \quad \sum_{k=1}^{n} w_u \cdot z_k$$

$$s.t. \quad \sum_{\{D(s_i, r_{ij}) | u_k \in cov(D(s_i, r_{ij}))\}} x_{ij} \geq z_k \quad k = 1, 2, \ldots, n$$
$$\sum_{i=1}^{m} \sum_{j=0}^{n} r_{ij}^\alpha x_{ij} \leq P$$
$$\sum_{j=1}^{n} x_{ij} \leq 1 \qquad\qquad\qquad i = 1, 2, \ldots, m$$
$$x_{ij} \in \{0, 1\} \qquad\qquad\qquad j = 0, 1, \ldots, n; i = 1, 2, \ldots, m$$
$$z_k \in \{0, 1\} \qquad\qquad\qquad k = 1, 2, \ldots, n$$

Among them, $x_{ij} = 1$ indicates that the sensor s_i selects the disc $D(s_i, r_{ij})$, r_{ij}^α is the sensor s_i selects the disc $D(s_i, r_{ij})$ required energy. The first condition indicates whether u_k is covered. The second condition indicates that the energy constraint and the third condition indicates that each sensor can only select one disc.

3 Greedy Algorithm

In this section, we introduces the greedy algorithm for the energy-constrained geometric coverage problem. Each iteration of the greedy algorithm is to select the disc with the largest profit per unit of energy. Until all the discs are considered. Compare the selected disc set with the maximum profit disc and choose the better one as the algorithm output.

In the energy-constrained geometric coverage problem, the discs of the same sensor have a containment relationship, $D(s_i, r_{i1}) \subseteq D(s_i, r_{i2}) \subseteq \cdots \subseteq D(s_i, r_{in})$. Let

$$c(D^*) = \sum_{i=1}^{m} \max_{D^* \cap D_i} \{r_{ij}^{\alpha}\}$$

be sum of energy of the largest disc of the sensor. Let

$$f(D^*) = \sum_{u \in cov(D^*)} w_u$$

be sum of the sensor coverage benefits.

For the energy-constrained geometric coverage problem, the cost function $c(\cdot)$ and the benefit function $f(\cdot)$ are monotone submodular function. In the following, we use their properties and sensor can only choose one disc to prove the lower bound of the approximation ratio of the greedy algorithm.

Algorithm 1:

Input: Objective function f, cost function $c(\cdot)$, budget P

Output: $\arg\max\limits_{D^* \in \{D_g, \{d_s\}\}} f(D^*)$

1 Initially, let $D_g = \emptyset$ and $D' = \mathcal{D}$.
2 **while** $D' \neq \emptyset$ **do**
3   **for** $i = 1, 2, \ldots, m$ **do**
4          $d_i \in \arg\max_{d \in Disk_i} \frac{f(D_g \cup d) - f(D_g)}{c(D_g \cup d) - c(D_g)}$
5      $d' \in \arg\max\limits_{i} \frac{f(D_g \cup d_i) - f(D_g)}{c(D_g \cup d_i) - c(D_g)}$
6   **if** $c(D_g \cup d') \leq P$ *and* $f(D_g \cup d') - f(D_g) > 0$ **then**
7          $D_g := D_g \cup d'$
8      $D' = D' \backslash \{d'\}$
9 $d_s \in \arg\max\limits_{d \in \mathcal{D}, c(d) \leq P} f(d)$
10 $D_g := \{\arg\max\limits_{D_g \cap Disk_i} c(d) \mid i = 1, 2, \ldots, m\}$

Let d_i be the i-th disc added to the greedy solution D_g and $D_i = \{d_1, d_2, \ldots, d_i\}$, $0 \leq i \leq \mid D_g \mid$ as the intermediate solution when the d_i is added. Let A_i is the set of discs discarded due to budget constraints until d_i is selected. Let o and o' be the first and second disc in the OPT but not added to the D_g due to the budget constraints. Let Q be the intermediate solutions of the greedy algorithm when o be chosen.

Lemma 1. For the intermediate solution $\bar{D}$ of the greedy algorithm after a certain iteration, $\bar{A}$ is the corresponding discarded disc set and $k = |\bar{D}|$. Given any disc assignment T, if $T \cap \bar{A} = \emptyset$, then

$$f(\bar{D}) \geq (1 - e^{\frac{-c(\bar{D})}{c(T)}})f(T).$$

Proof. If $f(\bar{D}) \geq f(T)$, the lemma has been proven. We consider $f(\bar{D}) \leq f(T)$. Since f is a monotone sub-modular function, we have:

$$f(T) \leq f(D_i) + \sum_{d \in T \backslash D_i} f(d \mid D_i) = f(D_i) + \sum_{d \in T \backslash D_i} (c(d)\frac{f(d \mid D_i)}{c(d)}).$$

According to $c(\cdot)$ is also a monotone submodular function and the greediness, for any $i \leq k - 1$ we have

$$\frac{f(d \mid D_i)}{c(d)} \leq \frac{f(d \mid D_i)}{c(d \mid D_i)} \leq \frac{f(d_{i+1} \mid D_i)}{c(d_{i+1} \mid D_i)}.$$

Since $T \cap \bar{A} = \emptyset$ and sensors can only be arranged on a disc, we have

$$f(T) \leq f(D_i) + \frac{f(d_{i+1} \mid D_i)}{c(d_{i+1} \mid D_i)} \sum_{d \in T \backslash D_i} c(d) \leq f(D_i) + \frac{f(d_{i+1} \mid D_i)}{c(d_{i+1} \mid D_i)}c(T).$$

Then

$$f(T) - f(D_{i+1}) \leq (1 - \frac{c(d_{i+1} \mid D_i)}{c(T)})(f(T) - f(D_i)) \leq e^{-\frac{c(d_{i+1} \mid D_i)}{c(T)}}(f(T) - f(D_i)).$$

The first inequality holds because the rearrangement inequalities. The second inequality holds because the inequalities $1 - x \leq e^{-x}$.

Repeating the above inequality, we have

$$\begin{aligned}
f(T) - f(\bar{D}) &= f(T) - f(D_k) \\
&\leq e^{-\frac{c(d_k \mid D_{k-1})}{c(T)}}(f(T) - f(D_{k-1})) \\
&\leq e^{-\frac{c(d_k \mid D_{k-1})}{c(T)}}e^{-\frac{c(d_{k-1} \mid D_{k-2})}{c(T)}}(f(T) - f(D_{k-2})) \\
&= e^{-\frac{c(D_k - c(D_{k-2}))}{c(T)}}(f(T) - f(D_{k-2})) \\
&\cdots \\
&\leq e^{-\frac{c(D_k) - c(D_0)}{c(T)}}(f(T) - f(D_0)) \\
&= e^{-\frac{c(\bar{D})}{c(T)}}f(T)
\end{aligned}$$

From Lemma 1, the intermediate solution D_g of the greedy algorithm satisfies the following relation.

Lemma 2. The intermediate solution D_g of the greedy algorithm satisfies

$$f(D_g) \geq (1 - e^{\frac{-c(Q)}{P}})f(OPT).$$

Proof. By Lemma 1, let $\bar{D} = Q$ and $OPT \cap \bar{A} = \emptyset$, let $T = OPT$, then we have

$$f(Q) \geq (1 - e^{\frac{-c(Q)}{OPT}})f(OPT).$$

Since $Q \subseteq D_g$ and $c(OPT) \leq P$ then

$$f(D_g) \geq (1 - e^{\frac{-c(Q)}{P}})f(OPT).$$

Lemma 3. If $i \leq j - 1$ then

$$\frac{f(d_{j+1} \mid D_i)}{c(d_{j+1} \mid D_i)} \geq \frac{f(d_{j+1} \mid D_j)}{c(d_{j+1} \mid D_j)}.$$

Proof. The following will prove from three cases.

If the generated intermediate solutions $D_j \setminus D_i$ do not add a disc to the sensor where d_{j+1} is located, then $c(d_{j+1} \mid D_i) = c(d_{j+1} \mid D_j)$. The following holds:

$$\frac{f(d_{j+1} \mid D_i)}{c(d_{j+1} \mid D_i)} \geq \frac{f(d_{j+1} \mid D_j)}{c(d_{j+1} \mid D_j)}.$$

If $D_{j-1} \setminus D_i$ does not have the disc of the sensor where d_{j+1} is located and d_j and d_{j+1} on the same sensor, then:

$$\frac{f(d_{j+1} \mid D_i)}{c(d_{j+1} \mid D_i)} \geq \frac{f(d_{j+1} \mid D_{j-1})}{c(d_{j+1} \mid D_{j-1})}.$$

The inequality above has been proved.

$$\frac{f(d_j \mid D_{j-1})}{c(d_j \mid D_{j-1})} \geq \frac{f(d_{j+1} \mid D_{j-1})}{c(d_{j+1} \mid D_{j-1})}.$$

The inequality due to greediness. At the same time, $f(d_{j+1} \mid D_{j-1}) = f(d_j \mid D_{j-1}) + f(d_{j+1} \mid D_j)$. Then there is the following inequality:

$$\frac{f(d_{j+1} \mid D_i)}{c(d_{j+1} \mid D_i)} \geq \frac{f(d_{j+1} \mid D_{j-1})}{c(d_{j+1} \mid D_{j-1})} = \frac{c(d_j \mid D_{j-1})\frac{f(d_j \mid D_{j-1})}{c(d_j \mid D_{j-1})} + c(d_{j+1} \mid D_j)\frac{f(d_{j+1} \mid D_j)}{c(d_{j+1} \mid D_j)}}{c(d_{j+1} \mid D_{j-1})}.$$

Then

$$\frac{f(d_{j+1} \mid D_i)}{c(d_{j+1} \mid D_i)} \geq \frac{f(d_{j+1} \mid D_j)}{c(d_{j+1} \mid D_j)}.$$

From D_i to D_j, if k discs are added to the sensor where d_{j+1} is located, the above two cases and the greediness can be used repeatedly. The Lemma 3 has been proven. Another inequality of D_g will be proved below.

Lemma 4. Given an arbitrary disc assignment T, the intermediate solution D_g of the greedy algorithm satisfies

$$f(D_g) \geq (1 - \frac{c(T)}{P})f(T).$$

Proof. If the $T \subseteq D_g$, the lemma has been proven. We consider $T \setminus D_g \neq \emptyset$. Let $D_l = \{d_1, \ldots, d_l\}$ be intermediate solution of the greedy algorithm when the first disc from T is be selected but not added to D_g due to the budget constraint. $T' = T \setminus D_l$. We have

$$\frac{f(d_1 \mid D_0)}{c(d_1 \mid D_0)} \geq \frac{f(d_2 \mid D_0)}{c(d_2 \mid D_0)} \geq \frac{f(d_2 \mid D_1)}{c(d_2 \mid D_1)} \geq \cdots \geq \frac{f(d_l \mid D_{l-1})}{c(d_l \mid D_{l-1})}$$

$$\geq \max_{d \in T'} \frac{f(d \mid D_l)}{c(d \mid D_l)} \geq \frac{f(T' \mid D_l)}{c(T' \mid D_l)}.$$

The first inequality holds because of greediness and the second inequality holds because of Lemma 3. Repeat the above inequality. The last inequality holds because $f(T' \mid D_l) \leq \sum_{d \in T'} f(d \mid D_l)$ and $c(T' \mid D_l) = \sum_{d \in T'} c(d \mid D_l)$. Then

$$\frac{f(T' \mid D_l)}{c(T' \mid D_l)} \leq \frac{\sum_{d \in T'} f(d \mid D_l)}{\sum_{d \in T'} c(d \mid D_l)} \leq \frac{\sum_{d \in T'} c(d \mid D_l)\frac{f(d \mid D_l)}{c(d \mid D_l)}}{\sum_{d \in T'} c(d \mid D_l)} \leq \max_{d \in T'} \frac{f(d \mid D_l)}{c(d \mid D_l)}.$$

Because

$$f(T) \leq f(D_l) + f(T' \setminus D_l)$$

and

$$f(D_l) = \sum_{i=1}^{l} f(d_i \mid D_{i-1}) = \sum_{i=1}^{l} c(d_i \mid D_{i-1})\frac{f(d_i \mid D_{i-1})}{c(d_i \mid D_{i-1})}$$

$$\geq \sum_{i=1}^{l} c(d_i \mid D_{i-1})\frac{f(T' \mid D_l)}{c(T' \mid D_l)} \geq \sum_{i=1}^{l} c(d_i \mid D_{i-1})\frac{f(T' \mid D_l)}{c(T')}$$

$$= c(D_l)\frac{f(T' \mid D_l)}{c(T')}.$$

By the definition of T' we have $c(D_l) + c(T') > P$. Then

$$f(T) \leq f(D_l) + f(T' \setminus D_l) \leq f(D_l) + \frac{c(T')}{c(D_l)}f(D_l) < (1 + \frac{c(T')}{P - c(T')})f(D_l).$$

We use Lemma 4 to prove another relation between the intermediate solution D_g of the greedy algorithm and OPT.

Lemma 5. Let $OPT' = OPT \setminus (Q \cup \{o\})$ then

$$f(D_g) \geq f(Q) + (1 - \frac{c(Q)}{P - c(Q)})f(OPT' \mid Q).$$

Proof. We know that $f(\cdot)$ is a monotone submodular function, then $f(S \mid Q)$ is also a monotone submodular function about S. We have

$$f(D_g) = f(Q) + f((D_g \setminus Q) \mid Q) \geq f(Q) + (1 - \frac{c(OPT')}{P - c(Q)})f(OPT' \mid Q).$$

We also have $c(Q) + c(o) > P$ and $c(o) + c(OPT') \leq c(OPT) \leq P$ then $c(OPT') < c(Q)$. We have

$$f(D_g) \geq f(Q) + (1 - \frac{c(Q)}{P - c(Q)})f(OPT' \mid Q).$$

Theorem 1. For energy-constrained geometric coverage problem, the greedy algorithm can obtain an approximate lower bound of $1 - \frac{1}{\sqrt{e}}$.

Proof. Let α^* be the minimum of the following mathematical programming problem about α, x_1, x_2, x_3.

$$
\begin{aligned}
\min \quad & \alpha \\
s.t. \quad & \alpha \geq x_1 \\
& \alpha \geq 1 - x_1 - x_2 \\
& \alpha \geq x_1 + (1 - \frac{x_3}{1 - x_3})x_2. \\
& x_1 \geq 1 - e^{-x_3} \\
& \alpha, x_1, x_2, x_3 \in [0, 1]
\end{aligned}
$$

where $\alpha = \frac{f(D^*)}{f(OPT)}$, $x_1 = \frac{f(Q)}{f(OPT)}$, $x_2 = \frac{f(OPT' \mid Q)}{f(OPT)}$, $x_3 = \frac{c(Q)}{P}$.

The first condition is due to $f(D^*) \geq f(Q)$. The second condition is due to $f(Q) + f(o \mid Q) + f(OPT' \mid Q) \geq f(Q) + f(\{o\} \cup OPT' \mid Q) = f(OPT)$ and $f(D^*) \geq f(d_s) \geq f(o \mid Q)$. The third condition is because of Lemma 5. The fourth condition is because of Lemma 2.

Lemma 6 [16]. $\alpha^* \geq 1 - \frac{1}{\sqrt{e}}$.

4   Discussion

In this paper, we study the energy-constrained geometric coverage problem. For this problem we propose a greedy algorithm and achieves an approximation ratio of $1 - \frac{1}{\sqrt{e}}$.

Uncovered users are not considered in the energy-constrained geometric coverage problem. Giving corresponding penalties to uncovered users and maximizing the benefit function is a future research direction.

References

1. Alt, H., Arkin, E.M., Brönnimann, H., et al.: Minimum-cost coverage of point sets by disks. In: Amenta, N., Cheong, O. (eds.) Symposium on Computational Geometry, pp. 449–458. ACM, New York (2006)
2. Bilò, V., Caragiannis, I., Kaklamanis, C., Kanellopoulos, P.: Geometric clustering to minimize the sum of cluster sizes. In: Brodal, G.S., Leonardi, S. (eds.) ESA 2005. LNCS, vol. 3669, pp. 460–471. Springer, Heidelberg (2005). https://doi.org/10.1007/11561071_42
3. Chekuri, C., Kumar, A.: Maximum coverage problem with group budget constraints and applications. In: Jansen, K., Khanna, S., Rolim, J.D.P., Ron, D. (eds.) APPROX/RANDOM 2004. LNCS, vol. 3122, pp. 72–83. Springer, Heidelberg (2004). https://doi.org/10.1007/978-3-540-27821-4_7
4. Dai, H., Deng, B., Li, W., et al.: A note on the minimum power partial cover problem on the plane. J. Comb. Optim. (2022). https://doi.org/10.1007/s10878-022-00869-8
5. Farbstein, B., Levin, A.: Maximum coverage problem with group budget constraints. Journal of Combinatorial Optimization **34**, 725-735 (2017). https://doi.org/10.1007/s10878-016-0102-0
6. Feldman, M., Nutov, Z., Shoham, E.: Practical budgeted submodular maximization. arXiv:2007.04937 (2020)
7. Guo, L., Li, M., Xu, D.: Approximation algorithms for maximum coverage with group budget constraints. In: Gao, X., Du, H., Han, M. (eds.) COCOA 2017. LNCS, vol. 10628, pp. 362–376. Springer, Cham (2017). https://doi.org/10.1007/978-3-319-71147-8_25
8. Iyer, R., Bilmes, J.: Submodular optimization subject to submodular cover and submodular knapsack constraints. In: Koyejo, S., Agarwal, A. (eds.) Twenty-Seventh Conference on Neural Information Processing Systems, Lake Tahoe, Nevada, USA, vol. 2, pp. 2436–2444 (2013)
9. Khuller, S., Moss, A., Naor, J.: The budgeted maximum coverage problem. Inf. Process. Lett. **70**(1), 39–45 (1999)
10. Kulik, A., Schwartz, R., Shachnai, H.: A refined analysis of submodular greedy. Oper. Res. Lett. **49**, 507–514 (2021)
11. Li, W., Liu, X., Cai, X., Zhang, X.: Approximation algorithm for the energy-aware profit maximizing problem in heterogeneous computing systems. J. Parallel Distrib. Comput. **124**, 70–77 (2019)
12. Liu, X., Li, W., Dai, H.: Approximation algorithms for the minimum power cover problem with submodular/linear penalties. Theoret. Comput. Sci. **923**, 256–270 (2022)
13. Liu, X., Li, W., Xie, R.: A primal-dual approximation algorithm for the k-prize-collecting minimum power cover problem. Optim. Lett. (2021).https://doi.org/10.1007/s11590-021-01831-z
14. Liu, X., Li, W., Yang, J.: A primal-dual approximation algorithm for the k-prize-collecting minimum vertex cover problem with submodular penalties. Front. Comput. Sci. (2022). https://doi.org/10.1007/s11704-022-1665-9
15. Sviridenko, M.: A note on maximizing a submodular set function subject to a knapsack constraint. Oper. Res. Lett. **32**(1), 41–43 (2004)
16. Tang, J., Tang, X., Lim, A., Han, K., Li, C., Yuan, J.: Revisiting modified greedy algorithm for monotone submodular maximization with a knapsack constraint. In: Proceedings of the ACM on Measurement and Analysis of Computing Systems, vol. 5, no. 1, pp. 1–22 (2021)

17. Wolsey, L.: Maximising real-valued submodular functions: primal and dual heuristics for location problems. Math. Oper. Res. **7**(3), 410–425 (1982)
18. Yaroslavtsev, G., Zhou, S., Avdiukhin, D.: "Bring Your Own Greedy"+Max: near-optimal 1/2-approximations for submodular knapsack. In: Chiappa, S., Calandra, R. (eds.) International Conference on Artificial Intelligence and Statistics, Palermo, Sicily, Italy, pp. 3263–3274 (2020)
19. Zhang, H., Vorobeychik, Y.: Submodular optimization with routing constraints. In: Schuurmans, D, Wellman, M.P. (eds.) Proceedings of the 30th AAAI Conference on Artificial Intelligence, Lake Tahoe, Nevada, pp. 819–826 (2016)

Incremental SDN Deployment to Achieve Load Balance in ISP Networks

Yunlong Cheng[1(✉)], Hao Zhou[1], Xiaofeng Gao[1], Jiaqi Zheng[2], and Guihai Chen[1]

[1] MoE Key Lab of Artificial Intelligence, Department of Computer Science and Engineering, Shanghai Jiao Tong University, Shanghai, China
{aweftr,h-zhou}@sjtu.edu.cn, {gao-xf,gchen}@cs.sjtu.edu.cn
[2] Department of Computer Science and Technology, Nanjing University, Nanjing, China
jzheng@nju.edu.cn

Abstract. Software defined network (SDN) decouples control planes from data planes and integrates them into a logically centralized controller. With capture of the global view, the controller can dynamically and timely reply to the changes of network states. However, replacing the entire traditional networks, e.g., Internet Service Provider (ISP) networks, with SDNs is difficult and computationally expensive. Hence, incremental deployment of partial SDN devices has received much attention. In this paper, we consider the k-LB problem, i.e., upgrading at most k legacy switches to SDN switches to achieve load balance. We claim that k-LB problem is NP-hard and there is no polynomial time $(N + M)^{1-\epsilon}$-approximation algorithm for any constant $\epsilon > 0$ unless $\mathbf{P} = \mathbf{NP}$, where N (M) is the number of switches (links) in the network. Given these negative results, we propose an effective greedy algorithm and claim that it reaches an approximation guarantee of $\frac{c_{avg}}{c_{min}} M$, where c_{avg} (c_{min}) is the average (minimum) link capacity. Large-scale simulations on real ISP networks show that our greedy algorithm achieves near optimal performance and decreases the maximum link utilization by 30% on average compared with the state of the art.

Keywords: SDN · Incremental deployment · Load balance · Approximation algorithm

1 Introduction

Software defined network (SDN) is a new network architecture [3,11], which decouples control planes from data planes and integrates them into a logically centralized controller. With the global view of network states, the controller can make flexible management and configuration to data planes. Unlike traditional networks, SDNs can dynamically and timely adjust network routes to solve many unexpected problems, e.g., link failures or traffic variations, which provide great advantages in traffic engineering.

© The Author(s), under exclusive license to Springer Nature Switzerland AG 2022
Q. Ni and W. Wu (Eds.): AAIM 2022, LNCS 13513, pp. 278–290, 2022.
https://doi.org/10.1007/978-3-031-16081-3_24

Traditional routing protocols, e.g., Open Shortest Path First (OSPF) protocol, aim to forward packets based on the shortest path, which may cause unbalanced traffic distribution. Equal Cost Multi-Path (ECMP) strategy makes attempts to relieve this situation by partitioning flows equally into multiple shortest paths, which still cannot balance the traffic for the whole networks. On the contrary, in SDNs, instead of multiple shortest paths with equal partition, the controller can determine the routing path of each flow and compute the splitting ratio at each switch to achieve load balance. Because of the great advantages, many enterprises hope to upgrade the entire traditional networks by SDNs. However, it is impractical for many reasons. Firstly, traditional networks are required to provide uninterrupted services but replacing the entire network causes a temporary break, which results in a huge monetary loss for network operators. Secondly, upgrading all the legacy switches or routers is expensive especially in large-scale networks with many devices, e.g., Internet Service Provider (ISP) networks. Therefore, incremental deployment of SDN devices to replace partial networks is a reasonable alternative when making a tradeoff between the cost and the benefits. In addition, our experiments show that upgrading only 10% legacy switches to SDN switches can achieve nearly equal performance compared with upgrading all the switches.

Incremental deployment of SDN devices has received much attention in recent years. Some previous works [9,12,15] believed that programmable traffic, i.e., the traffic which traverses at least one SDN switch, can increase flexibility in performing traffic engineering. In order to obtain more programmable traffic, M. Huang et al. [9] would like to upgrade k nodes with more incident links and fewer traffic workload on these links. Different from it, K. Poularakis et al. [12] made attempts to upgrade the entire ISP networks during several phases. Due to the hardness of the direct incremental deployment to maximize the network throughput, H. Xu et al. [15] divided this problem into two sub-problems, i.e., programmable traffic maximization and route selection. However, as shown by our experiments in real ISP networks, the algorithm to maximize so-called programable traffic does not behave well for load balance. Another work [8] believed that the node with higher degree can contribute to greater advantages in performing load balance. Unfortunately, it focused on heuristics algorithms and could not provide performance guarantee. Based on the above, an effective algorithm and an in-depth analysis are necessary for the incremental deployment problem with load balance.

In this paper, we consider the problem of upgrading at most a given number (e.g. k) of legacy nodes to SDN nodes to achieve load balance with the limited budget, which is defined as k-LB problem. We formulate it as an optimization problem and conduct the theoretical analysis for computational complexity and inapproximability. In addition, we propose a greedy algorithm to solve k-LB problem and show the effectiveness of this greedy algorithm both theoretically and practically. More specially, our contributions can be summarized as follows:

1. We define the k-LB problem into an optimization problem, and we formulate it as a complex mixed interger program, which cannot be directly solved as the scale of networks increases.
2. We analysis the complexity and inapproximability for k-LB problem.

3. We design an effective greedy algorithm, which has an approximation guarantee of $\frac{c_{avg}}{c_{min}}M$.
4. We perform numerical experiments in real ISP networks to evaluate the proposed algorithm and show the excellent and stable performance of our algorithm.

2   Network Model and Problem Formulation

2.1   Network Model and Problem Definition

A network is a directed graph $G = (V, E, c, w)$, where V is a set of the legacy
routers, E is a set of the links connecting two nodes, $c(\cdot)$ and $w(\cdot)$ are capacity
and weight functions of links, respectively. In this paper, we assume that all the
legacy nodes (routers) run the traditional IP routing protocol, like OSPF, i.e.,
the legacy node forwards packets based on the shortest path. Further, we use
$SP(s, d)$ to denote one of the shortest paths from source s to destination d. We
want to upgrade at most k legacy nodes to SDN nodes due to the budget limits,
where we assume that SDN node can split flows and determine the splitting
ratios. In some cases there   may be some constraints when splitting a flow,
e.g., unsplittable flows [7,10]. We ignore these constraints because the problem
remains difficult and hard to find approximations when removing the constraints,
and we can get the optimal traffic distribution as a reference for the deployment
of SDN nodes in the removed version. In addition, our analysis and algorithm
can also be easily extended to address these constraints. In all, given the traffic
demand matrix T, we aim to achieve load balance when upgrading at most k
legacy nodes. A common variant of load balance is to minimize the maximum
link utilization. Formally, we present the k-LB problem in Definition 1. More
notations are shown in Table 1.

Definition 1 (k-LB Problem). *Given a directed graph $G = (V, E, c, w)$, a
traffic demand matrix T and an integer k, the k-LB problem is to minimize the
maximum link utilization when upgrading at most k legacy nodes to SDN nodes
with flexible flow splitting capacities.*

2.2   Problem Formulation

Based on the above model and definition, we formulate the k-LB problem as a
mixed integer program (1). Our objective function is to minimize the maximum
link utilization λ when given Constraint (2)–(7). Constraint (2) ensures that for
each intermediate node v, the sum of flows routed to destination d from source
v plus flows entering into node v from other sources to destination d is equal
to flows exiting from node v destined for d. Notice that if $x_v = 1$, node v is an
SDN node and can split the flows into multiple sub-flows; otherwise, node v is
a legacy node and the flows can only be forwarded to the next hop of node v
according to the shortest path $SP(v, d)$.

Table 1. Key notations

Notation	Description
V	The set of switch nodes
E	The set of links
N	The number of nodes, $|V| = N$
M	The number of links, $|E| = M$
$w(u,v)$	The weight of link (u,v)
$c(u,v)$	The capacity of link (u,v)
G	The directed network graph $G = (V, E, w, c)$
T	The traffic demand matrix
$SP(s,d)$	One of the shortest paths from s to d
$NH_p(v)$	The next hop of v in path p
$f_d(u,v)$	The sum of flows routed to destination d on the link (u,v)
k	The maximum number of legacy nodes to upgrade
λ	The maximum link utilization
x_v	The binary variable, where node v is upgraded if $x_v = 1$

$$T_{v,d} + \sum_{(u,v)\in E} f_d(u,v) = (1 - x_v)\, f_d(v, NH_{SP(v,d)}(v))$$

$$+ x_v \sum_{(v,u)\in E} f_d(v,u), \quad \forall v \in V, d \in V \backslash \{v\}, \tag{2}$$

Constraint (3) ensures that for each destination d, all flows must enter into their sink node d completely.

$$\sum_{(u,d)\in E} f_d(u,d) = \sum_{(d,u)\in E} f_d(d,u) + \sum_{v\in V} T_{v,d}, \quad \forall d \in V, \tag{3}$$

Constraint (4) guarantees that for each link (u,v), the utilization does not exceed λ. Constraint (5) means that we can only upgrade at most k legacy nodes to SDN nodes.

$$\sum_{d\in V} f_d(u,v) \leq \lambda\, c(u,v), \quad \forall (u,v) \in E, \tag{4}$$

$$\sum_{v\in V} x_v \leq k, \quad \forall v \in V, \tag{5}$$

Therefore, the whole mixed integer program can be formulated as follows:

$$\min \quad \lambda, \tag{1}$$

$$\text{s.t.} \quad \text{Constraint } (2), (3), (4), (5)$$

$$f_d(u,v) \geq 0, \quad \forall d \in V, (u,v) \in E, \tag{6}$$

$$x_v \in \{0,1\}, \quad \forall v \in V, \tag{7}$$

3　Complexity and Inapproximability Analysis

3.1　Hardness of k-LB Problem

We prove that k-LB problem is NP-hard by a reduction from the set cover problem [4] in Theorem 1.

Definition 2 (Set Cover Problem). *Given an universe $\mathcal{U} = \{u_1, u_2, \ldots, u_m\}$, a family $\mathcal{S} = \{\mathcal{S}_1, \mathcal{S}_2, \ldots, \mathcal{S}_n\}$ where $\mathcal{S}_i$ is a subset of $\mathcal{U}$ and an integer k, the set cover problem is to determine whether there is a set cover with size no more than k, where a cover is defined as a subfamily $\mathcal{C} \subseteq \mathcal{S}$ of sets whose union is $\mathcal{U}$.*

Theorem 1. *k-LB problem is NP-hard.*

Proof. Given a set cover problem instance, we construct a polynomial time reduction to a special instance of k-LB problem. As shown in Fig. 1a, for each element $u_j \in \mathcal{U}$, we construct an edge $e_j = (v_j, d_j)$; for each subset $\mathcal{S}_i \in \mathcal{S}$, we construct a node s_i. In the set cover problem instance, if $u_j \in \mathcal{S}_i$, we add two edges (s_i, v_j) and (s_i, d_j). Now we consider the weight of each edge. We set $w(s_i, v_j) = w(v_j, d_j) = 0$ (dotted red lines) and $w(s_i, d_j) = 1$ (solid black lines). Therefore, the shortest path from s_i to d_j is path $(s_i \to v_j \to d_j)$ as Fig. 1a (dotted red lines). Further, there is a flow demand of one unit size from s_i to d_j. At the beginning, all the flows are along the shortest path (dotted red lines) since there is no SDN node. For the capacity, we set $c(v_j, d_j) = f(j)$, where $f(j)$ is the indegree of node v_j, and the capacity of other edges is set as 2 as shown by Fig. 1b. Thus, each edge e_j ($\forall 1 \le j \le m$) has the maximum link utilization $\lambda = 1$. The constructed k-LB problem instance is to upgrade at most k legacy node to SDN node to minimize the maximum link utilization λ in the network which is given by Fig. 1. Notice that upgrading node v_j or d_j is unhelpful since their outdegree is no more than one. In addition, if node s_i is upgraded, the flow from s_i to d_j can be along the arbitrary path, e.g., path $(s_i \to d_j)$, which decreases the utilization of edge e_j but increases the utilization of edge (s_i, d_j) to only $1/2$. After the upgrade of at most k legacy nodes, let λ^* be the solution for this k-LB instance. Thus, $\lambda^* < 1$ means that the flow on each edge e_j ($\forall 1 \le j \le m$) has been decreased. In other words, each element $u_j \in \mathcal{U}$ (the corresponding edge e_j) has been covered by at most k subsets $\mathcal{S}_i \in \mathcal{S}$ (the corresponding legacy node s_i). However, $\lambda^* = 1$ means that there is at least one edge e_{j^*} whose utilization is still equal to 1. In other words, element u_{j^*} cannot be covered and thus there is no set cover whose size does not exceed k.

3.2　Inapproximability Analysis

ρ-inapproximability. Similarly, we present this inapproximability result by a reduction from the set cover problem in Lemma 1.

Lemma 1. *For any fixed constant $\rho \ge 1$, there is a gap-introducing reduction that transforms a set cover problem instance $\mathbf{S}(\mathcal{U}, \mathcal{S}, k)$ to a k-LB problem instance $\mathbf{K}(G, T, k + |\mathcal{U}|)$ such that*

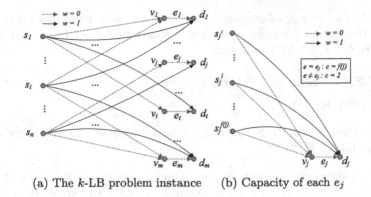

(a) The k-LB problem instance   (b) Capacity of each e_j

Fig. 1. Reduction from set cover to k-LB. (Color figure online)

– if $\mathbf{S}(\mathcal{U}, \mathcal{S}, k)$ is satisfied, $\mathrm{OPT}(\mathbf{K}) < \dfrac{1}{\rho}$;

– if not, $\mathrm{OPT}(\mathbf{K}) = 1$.

Theorem 2. *There is no polynomial time ρ-approximation algorithm for k-LB problem assuming* $\mathbf{P} \neq \mathbf{NP}$, *where ρ is a constant defined in Lemma 1.*

Proof. Suppose that on the contrary, there is a polynomial time ρ-approximation algorithm A for k-LB problem and denote the solution by $h_A(\mathbf{K})$, which satisfies

$$\frac{h_A(\mathbf{K})}{\rho} \leq \mathrm{OPT}(\mathbf{K}) \leq h_A(\mathbf{K})$$

Consider the constructed k-LB problem instance in Lemma 1, where $\mathrm{OPT}(\mathbf{K})$ has only two cases, i.e., 1 or $< 1/\rho$. We have $h_A(\mathbf{K}) \leq 1$ due to $\lambda = 1$ even though we do not upgrade any node. On the one hand, $h_A(\mathbf{K}) = 1$ induces that

$$\mathrm{OPT}(\mathbf{K}) \geq \frac{h_A(\mathbf{K})}{\rho} = \frac{1}{\rho}$$

Based on Lemma 1, we further get $\mathrm{OPT}(\mathbf{K}) = 1$ and $\mathbf{S}(\mathcal{U}, \mathcal{S}, k)$ cannot be satisfied. On the other hand, $h_A(\mathbf{K}) < 1$ means

$$\mathrm{OPT}(\mathbf{K}) \leq h_A(\mathbf{K}) < 1$$

when $\mathbf{S}(\mathcal{U}, \mathcal{S}, k)$ can be satisfied, we have $\mathrm{OPT}(\mathbf{K}) < 1/\rho$. Therefore, we can solve set cover problem according to $h_A(\mathbf{K})$. It contradicts with our assumption of $\mathbf{P} \neq \mathbf{NP}$.

$(N+M)^{1-\epsilon}$**-inapproximability.** In fact, we cannot even give an $(N+M)^{1-\epsilon}$-approximation algorithm to solve k-LB problem for any constant $\epsilon > 0$. Similarly, we construct a gap-introducing reduction from set cover problem as shown by Lemma 2.

Lemma 2. *For any fixed constant $\epsilon > 0$, there is a gap-introducing reduction that transforms a set cover problem instance $\mathbf{S}(\mathcal{U}, \mathcal{S}, k)$ to a k-LB problem instance $\mathbf{K}(G_x, T, k + |\mathcal{U}|)$ such that*

- *if $\mathbf{S}(\mathcal{U}, \mathcal{S}, k)$ is satisfied, $\mathrm{OPT}(\mathbf{K}) < \dfrac{1}{(N+M)^{1-\epsilon}}$;*
- *if not, $\mathrm{OPT}(\mathbf{K}) = 1$.*

where x is a function of $(\mathcal{U}, \mathcal{S}, k)$, N is the number of nodes and M is the number of edges in $\mathbf{K}(G_x, T, k + |\mathcal{U}|)$.

Based on the above gap-introducing reduction, we can further show the $(N + M)^{1-\epsilon}$-inapproximability for k-LB problem. Its proof is somewhat similar to that of Theorem 2. where we just need to replace ρ with $(N + M)^{1-\epsilon}$.

Theorem 3. *There is no polynomial time $(N+M)^{1-\epsilon}$-approximation algorithm for k-LB problem assuming $\mathbf{P} \neq \mathbf{NP}$, where ϵ is a constant defined in Lemma 2.*

4   Algorithm Design and Analysis

Given these negative results, we try to solve k-LB problem using a greedy algorithm in this subsection. Furthermore, the greedy algorithm can reach a good approximation bound.

Before further discussion, we first make attempts to minimize the maximum link utilization when given the upgraded node set C, which is a subroutine in the greedy algorithm. We denote this problem instance by $\mathbf{K}(G, T, C)$ and formulate it as Program (8). If node v is a SDN node, i.e., $v \in C$, it can split flows to multiple sub-flows as shown by Constraint (8a). Otherwise, node v forwards them to next hop based on the shortest path as Constraint (8b). Other constraints are similar to Program (1). Since the upgraded node set C is determined, Program (8) is a linear program which can be solved in polynomial time by using the standard solver, e.g., Gurobi [2].

$$\min \quad \lambda$$

$$\text{s.t.} \quad T_{v,d} + \sum_{(u,v)\in E} f_d(u,v) = \sum_{(v,u)\in E} f_d(v,u), \quad \forall v \in C, d \in V\setminus\{v\} \tag{8a}$$

$$T_{v,d} + \sum_{(u,v)\in E} f_d(u,v) = f_d(v, NH_{SP(v,d)}(v)), \quad \forall v \in V\setminus C, d \in V\setminus\{v\} \tag{8b}$$

$$\sum_{(u,d)\in E} f_d(u,d) = \sum_{(d,u)\in E} f_d(d,u) + \sum_{v\in V} T_{v,d}, \; \forall d \in V \tag{8c}$$

$$\sum_{d\in V} f_d(u,v) \leq \lambda \, c(u,v), \; \forall (u,v) \in E \tag{8d}$$

$$f_d(u,v) \geq 0, \; \forall d \in V, (u,v) \in E \tag{8e}$$

Now we present the details of the proposed algorithm in Algorithm 1. Let C be the set of nodes which have been determined to upgraded. First, set C is initialized with $\emptyset$ (line 1). Next, we add nodes into C one by one (line 2), where the selected node has to satisfy two conditions. Firstly, the upgraded node can minimize the maximize link utilization when given the current set C (line 3), and let λ^* be the minimum value (line 4). Secondly, the upgraded node can minimize the number of edges whose utilization is equal to λ^* (line 5). Finally, we randomly select one from the nodes which satisfy the above two conditions (line 6) and add it into C (line 7).

Algorithm 1. A Greedy Algorithm for k-LB Problem

Input: $G = (V, E, c, w), T, k$.
Output: The upgraded node set C.
1: $C \leftarrow \emptyset$.
2: **for** $i = 1$ **to** k **do**
3:    $C_1 \leftarrow \arg\min_{v \in V} \mathbf{K}(G, T, C \cup \{v\})$.
4:    $\lambda^* \leftarrow \min_{v \in V} \mathbf{K}(G, T, C \cup \{v\})$.
5:    $C_2 \leftarrow \arg\min_{v \in C_1} |\{e : u(e) = \lambda^*\}|$.
6:    select v randomly from C_2.
7:    $C \leftarrow C \cup \{v\}$.
8: **end for**
9: **return** C.

Now let us begin the formal analysis for Algorithm 1.

Theorem 4. *Algorithm 1 achieves an approximation guarantee of $\frac{c_{avg}}{c_{min}} M$, where c_{avg} (c_{min}) is the average (minimum) capacity of edges and M is the number of edges.*

Proof. We consider the optimal solution and the worst case, respectively. First, consider the flow distribution in the optimal solution. Since all the flows reach their destinations through certain edges, the sum of all the demands is smaller than the sum of flows in all the edges, i.e.,

$$\sum_{u,v \in V} T_{u,v} \leq \sum_{(u,v) \in E} f(u, v) \qquad (9)$$

$$\leq \sum_{(u,v) \in E} c(u, v) \text{OPT}(\mathbf{K}) \qquad (10)$$

Inequality (10) is true since $\text{OPT}(\mathbf{K})$ is the maximum link utilization, which further induces that

$$\text{OPT}(\mathbf{K}) \geq \frac{\sum_{u,v \in V} T_{u,v}}{\sum_{(u,v) \in E} c(u, v)}$$

In addition, the worst case is that all flows go through the same edge with the minimum capacity, denoted by WST($\mathbf{K}$), i.e.,

$$\text{WST}(\mathbf{K}) = \frac{\sum_{u,v \in V} T_{u,v}}{c_{min}}$$

where c_{min} is the minimum link capacity. We denote the solution of Algorithm 1 by $G(\mathbf{K})$ and further get

$$\frac{G(\mathbf{K})}{\text{OPT}(\mathbf{K})} \leq \frac{\text{WST}(\mathbf{K})}{\text{OPT}(\mathbf{K})}$$

$$\leq \frac{\sum_{u,v \in V} T_{u,v}}{c_{min}} \frac{\sum_{(u,v) \in E} c(u,v)}{\sum_{u,v \in V} T_{u,v}}$$

$$= \frac{\sum_{(u,v) \in E} c(u,v)}{c_{min}} = \frac{c_{avg}}{c_{min}} M$$

where c_{avg} is the average link capacity.

5   Numerical Evaluation

In this section, we evaluate the performance of our greedy algorithm, using real-world and synthetic ISP topologies and traffic matrices. In addition, we also test several other algorithms with the same dataset as a comparison.

5.1   Experimental Setup

We conduct the whole evaluation based on four real-world ISP topologies and one synthetic topology. The real-world ISP topologies are large-scale ISP networks in US and Europe, measured with Rocketfuel in [14]. The synthetic one was generated using the Delaunay's Triangulation Algorithm introduced in IGen [13]. The whole dataset can be downloaded in [1] and also used in [6] and [5].

In the original network, we define all the nodes in the network as legacy nodes which transfer flows based on the OSPF routing protocol. Then we choose a certain percentage of legacy nodes to upgrade according to different algorithms.

The six compared situations are showed as follows:

1. **Legacy Network:** This is the original network where all the nodes follow the OSPF routing protocol.
2. **DEG** [8]: An heuristic algorithm which upgrades the nodes with the highest degree (the number of incoming and outgoing adjacent links) in the whole topology.
3. **VOL** [8]: An heuristic algorithm which upgrades the nodes with the highest traffic volume traversing in the whole topology.
4. **MODG** [12]: An algorithm upgrades nodes to maximize the programmable traffic, which is defined as the total volume of flows passing through one or more SDN nodes. In each step it chooses the node which can bring the largest increment on programmable traffic.

5. **GREEDY:** The greedy algorithm in Algorithm 1.
6. **Optimal:**   The optimal solution of SDN nodes selection. We use Gurobi
   7.52 [2] to solve the mixed integer program (MIP) presented in Sect. 2.

When given the upgraded nodes which are computed by the above algorithms, we use a linear program to get the optimal traffic distribution and compare the maximum link utilizations.

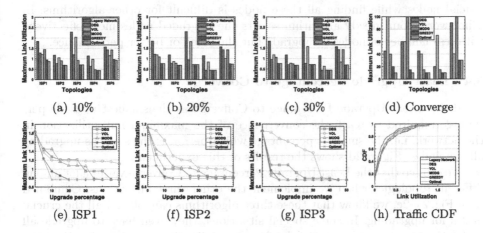

| (a) 10% | (b) 20% | (c) 30% | (d) Converge |

| (e) ISP1 | (f) ISP2 | (g) ISP3 | (h) Traffic CDF |

Fig. 2. Performance comparison.

5.2   Maximum Link Utilization

Figure 2a–2c show the maximum link utilizations computed by different algorithms on different topologies when upgrade percentage is 10%, 20% and 30%.
Notice that the utilization larger than one means congestion.

Overall, we find that all the node-upgrading algorithms work effectively compared to the legacy network, while our greedy algorithm (GREEDY) is more
effective and is very close to the optimal solution. On the contrast, the other
three algorithms work unsteadily and the results vary a lot among different
topologies. With enlarging the upgrade percentage from 10% to 30%, we find
that the maximum link utilizations become smaller and the performance of these
algorithms becomes closer to that of the optimal solution in different degrees.
For ISP2, ISP3 and ISP4, the maximum link utilization of the optimal solution
show no improvement. In fact, it results from the convergence of the maximum
link utilization, which will be further discussed in Subsect. 5.4.

5.3   Upgrade Percentage

Upgrade percentage is the percentage of legacy nodes we choose to upgrade.
Intuitively, the larger the upgrade percentage, the smaller the maximum link

utilization. Figure 2a–2c show the maximum link utilizations computed by different algorithms with different upgrade percentages under topologies ISP1, ISP2 and ISP3, where other topologies have similar results. Here we can find that in the early stages of upgrading, all the algorithms decrease the utilization significantly while the greedy algorithm decreases fastest and is very close to the optimal solution. However, other algorithms seem to stagnate or improve very slowly as upgrade percentage gradually increases while the greedy algorithm convergences very fast. This means that all the algorithms can find some (or major) crucial nodes while finding all these nodes is difficult for other algorithms. In other words, our greedy algorithm selects the upgraded nodes more effectively, with less number of nodes to upgrade but the same or better performance.

5.4   Least Upgrade Percentage to Converge

We define Least Upgrade Percentage to Converge as the smallest upgrade percentage required to reach the convergence of the maximum link utilization in the network, i.e. the smallest percentage which performs the same as upgrading all the nodes. Figure 2d shows the results of this experiment.

We can see that the Least Upgrade Percentage to Converge varies a lot among different topologies when we implement the other three algorithms. Combined with Fig. 2a–2c, we know that these three algorithms cannot find all the crucial nodes for upgrading. In some special situations, they even have to upgrade all nodes in order to get the optimal solution. Obviously, in such a situation, the node-upgrade algorithm loses its meaning. As for our greedy algorithm, we find that the average Least Upgrade Percentage is 17%, varying from 10% to 30%, which is quite close to the optimal solution. This result verifies the conclusion that our greedy algorithm can upgrade the node more effectively.

5.5   Cumulative Distribution Function (CDF)

To observe the details of traffic distribution in the whole network, we conduct this experiment. We plot the cumulative distribution function (CDF) curve on link utilization with topology ISP1 and upgrade percentage 10%. Figure 2h shows that the major part of the link utilizations are less than 0.5. The link utilizations of the legacy network (in the top left corner of Fig. 2h) distribute disproportionately, many of which aggregate on smaller values and larger values.

After the upgrade, the link utilizations present a more concentrated (or balanced) distribution, which means that more links have a utilization of medium value, neither too small nor too large. Comparing all the CDF curves, we claim that our greedy algorithm performs better than other algorithms and is quite close to the optimal solution (in the bottom right corner of Fig. 2h). In addition, we can find that the curve has a jump when CDF reaches 1 especially for GREEDY and Optimal. The reason is that the number of links reaching the maximum link utilization is large for these two situations. This fact also proves that our greedy algorithm can produce a more balanced utilization distribution.

6   Conclusion

In this paper, we studied the incremental SDN deployment to achieve load balance, named k-LB problem. We conducted the theoretical analysis for computational complexity and inapproximability of this problem. Given these negative results, we proposed an effective greedy algorithm to solve k-LB problem. Large-scale simulations in the real ISP networks showed that our algorithm achieved near optimal performance and decreased the maximum link utilization by 30% on average compared to the state of the art.

Acknowledgments. This work was supported by the National Key R&D Program of China [2020YFB1707900]; the National Natural Science Foundation of China [61872238, 61972254], Shanghai Municipal Science and Technology Major Project [2021SHZDZX0102], and the Huawei Cloud [TC20201127009].

References

1. Declarative and expressive forwarding optimizer. https://sites.uclouvain.be/defo/
2. GUROBI. https://www.gurobi.com/
3. Casado, M., Freedman, M.J., Pettit, J., Luo, J., McKeown, N., Shenker, S.: Ethane: taking Control of the Enterprise. In: ACM International Conference on Applications, Technologies, Architectures, and Protocols for Computer Communications (SIGCOMM), pp. 1–12 (2007)
4. Garey, M.R., Johnson, D.S.: Computers and Intractability: A Guide to the Theory of NP-Completeness. W. H. Freeman (1979)
5. Hartert, R., Schaus, P., Vissicchio, S., Bonaventure, O.: Solving segment routing problems with hybrid constraint programming techniques. In: Pesant, G. (ed.) CP 2015. LNCS, vol. 9255, pp. 592–608. Springer, Cham (2015). https://doi.org/10.1007/978-3-319-23219-5_41
6. Hartert, R., et al.: A declarative and expressive approach to control forwarding paths in carrier-grade networks. In: ACM International Conference on Applications, Technologies, Architectures, and Protocols for Computer Communications (SIGCOMM), vol. 45, no. 5, pp. 15–28 (2015)
7. Hartman, T., Hassidim, A., Kaplan, H., Raz, D., Segalov, M.: How to split a flow? In: IEEE International Conference on Computer Communications (INFOCOM), pp. 828–836 (2012)
8. Hong, D.K., Ma, Y., Banerjee, S., Mao, Z.M.: Incremental deployment of SDN in hybrid enterprise and ISP networks. In: Symposium on SDN Research (SOSR), pp. 1–7 (2016)
9. Huang, M., Liang, W.: Incremental SDN-enabled switch deployment for hybrid software-defined networks. In: IEEE International Conference on Computer Communication and Networks (ICCCN), pp. 1–6 (2017)
10. Kleinberg, J.M.: Single-source unsplittable flow. In: IEEE Symposium on Foundations of Computer Science (FOCS), pp. 68–77 (1996)
11. McKeown, N., et al.: OpenFlow: enabling innovation in campus networks. SIGCOMM Comput. Commun. Rev. **38**(2), 69–74 (2008)
12. Poularakis, K., Iosifidis, G., Smaragdakis, G., Tassiulas, L.: One step at a time: optimizing SDN upgrades in ISP networks. In: IEEE Conference on Computer Communications (INFOCOM), pp. 1–9 (2017)

13. Quoitin, B., den Schrieck, V.V., François, P., Bonaventure, O.: IGen: generation of router-level internet topologies through network design heuristics. In: IEEE International Teletraffic Congress (ITC), pp. 1–8 (2009)

14. Spring, N.T., Mahajan, R., Wetherall, D., Anderson, T.E.: Measuring ISP topologies with rocketfuel. IEEE/ACM Trans. Netw. (TON) **12**(1), 2–16 (2004)

15. Xu, H., Li, X., Huang, L., Deng, H., Huang, H., Wang, H.: Incremental deployment and throughput maximization routing for a hybrid SDN. IEEE/ACM Trans. Netw. (TON) **25**(3), 1861–1875 (2017)

Graph Theory

Polynomial Time Algorithm
for k-vertex-edge Dominating Problem
in Interval Graphs

Peng Li and Aifa Wang[✉]

Chongqing University of Technology, 69 Hongguang Road, Chongqing, China
{lipengcqut,wangaf}@cqut.edu.cn

Abstract. Let G be an interval graph with n vertices and m edges. For any positive integer k and any subset S of $E(G)$, we design an $O(n|S|+m)$ time algorithm to find a minimum k-vertex-edge dominating set of G with respect to S. This shows that the vertex-edge domination problem and the double vertex-edge domination problem can be solved in linear time. Furthermore, the k-vertex-edge domination problem can be solved in $O(nm)$ time algorithm in interval graphs.

Keywords: Vertex-edge domination · Double vertex-edge domination · k-vertex-edge domination · Polynomial time algorithm · Interval graphs

1 Introduction

In this paper, all graphs are assumed to be finite, simple, undirected, and loopless. Let G be any simple graph with vertex set $V = V(G)$ and edge set $E = E(G)$. For any $v \in V(G)$, the *open neighborhood* $N_G(v)$ is the set $\{u \in V(G) : uv \in E(G)\}$ and the *closed neighborhood* of v is the set $N_G[v] = N_G(v) \cup \{v\}$. A sutset S of V in G is called a *dominating set* if every vertex of G is either in S or adjacent to a vertex of S. The *domination number* $\gamma(G)$ is the minimum cardinality of a dominating set in G. In recent years, the domination and its variations have attracted considerable attention and have been widely studied, see [5,6].

For any graph G, a map $\mathcal{I}$ that assigns to each vertex $x \in V(G)$ a nonempty closed interval $\mathcal{I}(x) = [\ell_\mathcal{I}(x), r_\mathcal{I}(x)]$ is called an *interval representation* of G provided $xy \in E(G)$ if and only if $x \neq y$ and $\mathcal{I}(x) \cap \mathcal{I}(y) \neq \emptyset$ for all $x, y \in V(G)$. If all the endpoints of $\{\mathcal{I}(x) : x \in V(G)\}$ are distinct, then we say $\mathcal{I}$ is *distinguishing*. Without loss of generality, we assume that all interval representations are distinguishing in this paper. If $r_\mathcal{I} - \ell_\mathcal{I}$ takes a constant value, we refer to the interval representation $\mathcal{I}$ as a *unit interval representation*. A graph is an *interval graph* if and only if it has an interval representation and a graph is a *unit interval graph* if and only if it has a unit interval representation. In fact, interval graphs arise naturally and frequently in modeling real-life situations, especially those involving time dependencies or other restrictions that are linear

Q. Ni and W. Wu (Eds.): AAIM 2022, LNCS 13513, pp. 293–302, 2022.
https://doi.org/10.1007/978-3-031-16081-3_25

in nature. Since 1985, the study of interval graphs has displayed intrinsic beauty and interest which attracts much attention of mathematicians simply due to its elegance [3, 4, 11–13, 18, 19].

Let G be a simple graph and k be some positive integer. For any $vw \in E(G)$ and $u \in V(G)$, we say vw is *vertex-dominated* by u if $u \in N_G[v] \cup N_G[w]$. A set $D \subseteq V(G)$ is a *vertex-edge dominating set* (resp. *double vertex-edge dominating set, k-vertex-edge dominating set*), or *VEDS* (resp. *DVEDS, k-VEDS*) for short, if each edge of $E(G)$ is vertex-dominated by one (resp. two, k) vertex of D. Let S be any subset of $E(G)$. A set $D \subseteq V(G)$ is a k-VEDS of G with respect to S if each edge of S is vertex-dominated by at least k vertices of D.

The vertex-edge dominating number $\gamma_{ve}(G)$ (resp. double vertex-edge dominating number $\gamma_{dve}(G)$, k-vertex-edge dominating number $\gamma_{kve}(G)$) of G is the minimum cardinality of a vertex-edge dominating set (resp. double vertex-edge dominating set, k-vertex-edge dominating set) of G. A set S is called an *independent vertex-edge dominating set* if S is both an independent set and a vertex-edge dominating set. The *independent vertex-edge domination number* of a graph G is the minimum cardinality of an independent vertex-edge dominating set and is denoted by $i_{ve}(G)$.

The *vertex-edge domination problem*, namely MIN-VEDS, (resp. *double vertex-edge domination problem*, namely MIN-DVEDS, *k-vertex-edge domination problem*, namely MIN-k-VEDS, *independent vertex-edge domination problem*) is to find a minimum VEDS (resp. minimum DVEDS, minimum k-VEDS, minimum independent vertex-edge dominating set) of G. In 1986, Peters [15] introduced the vertex-edge domination problem in his PhD thesis. In 2007, Lewis [15] introduced some new parameters related to it and obtained some lower bounds on $\gamma_{ve}(G)$ for different graph classes like connected graphs, k-regular graphs, cubic graphs etc. In addition, Lewis proved that the vertex-edge domination problem is NP-hard for bipartite, chordal, planar and circle graphs, and the independent vertex-edge domination problem is NP-hard even when restricted to a bipartite and chordal graph. Also, approximation algorithms and approximation hardness results are obtained in [15]. In 2010, Lewis et al. [10] characterized the trees with equal dominating and vertex-edge dominating number. In 2014, Krishnakumari et al. [8] got both upper and lower bounds on $\gamma_{ve}(G)$ of a tree. In 2016, Boutrig et al. [1] found some upper bounds on $\gamma_{ve}(G)$ and $i_{ve}(G)$ and some relationship between $\gamma_{ve}(G)$ and other domination parameters have been proved. In 2019, Zylinski [20] showed that for every connected graph G with $|V(G)| \geq 6$, it holds $\gamma_{ve}(G) \leq \frac{n}{3}$. Boutrig and Chellali [2] in 2018 and Krishnakumari et al. [7] in 2017 also studied other variations of vertex-edge dominations. On the other hand, in 2007, Lewis [15] proposed a linear time algorithm for vertex-edge domination problem for trees. Recently, Paul and Ranjan [14] constructed an example for which the algorithm proposed by Lewis fails. They proposed linear time algorithms for vertex-edge domination and independent vertex-edge domination problem in block graphs, and a linear time algorithm for weighted vertex-edge domination problem in trees. They also proved that the MIN-VEDS problem is NP-hard for undirected path graphs. Furthermore, they characterized the trees with equal vertex-edge dominating and independent vertex-edge dominating numbers.

In this paper, we study the MIN-VEDS and MIN-k-VEDS problems in interval graphs. Given some interval graph G, some positive integer k and any $S \subseteq E(G)$, we proposed an $O(|V(G)||S| + |E(G)|)$ time algorithm to find a minimum k-VEDS of G with respect to S, hence the MIN-VEDS and MIN-k-VEDS problems in interval graphs can be solved in polynomial time.

2 Preliminaries and Notation

For any two integers i and j with $i \leq j$, we write $[i, j]$ for the set of integers k such that $i \leq k \leq j$. Let S be any subset of $V(G)$. We write $N_G(S)$ for $(\cup_{v \in S} N_G(v)) \setminus S$ and write $N_G[S]$ for $N_G(S) \cup S = \cup_{v \in S} N_G[v]$. Let $G[S]$ denote the subgraph of G induced by S. Reference to G will be omitted when the context makes it obvious. For simplicity, we often write $G[V(G) - S]$ as $G - S$ and write $G - S$ as $G - v$ when S is a singleton set $\{v\}$. Let V be a n-vertex set and $\sigma = (\sigma_1, \ldots, \sigma_n)$ be an ordering of V. For any $1 \leq i \leq j \leq n$, denote $(\sigma_i, \ldots, \sigma_j)$ by $\sigma[i, j]$ and denote $(\sigma_1, \ldots, \sigma_i)$ by $\sigma[i]$. Let G be a graph and $\sigma = (\sigma_1, \ldots, \sigma_n)$ be an ordering of $V(G)$. We call σ an I-ordering if $v_i v_k \in E(G)$ implies $v_i v_j \in E(G)$ for all $1 \leq i < j < k \leq n$. Take any $v \in V(G)$. We denote $\max\{i : \sigma_i \in N_G[v]\}$ by $r_\sigma(v)$ and denote $\min\{i : \sigma_i \in N_G[v]\}$ by $\ell_\sigma(v)$.

Lemma 2.1. *[16, 17] A graph is an interval graph if and only if there is an I-ordering of its vertex set.*

The rest of this paper is organized as follows. We begin in Sect. 3 with the MIN-k-VEDS algorithm for interval graphs as well as the necessary terminologies. Two examples are given to describe this algorithm in Sect. 3. Next we prove the the correctness of this algorithm and explains how to implement it in Sect. 4. We give some interesting problems related to k-vertex edge domination in Sect. 5.

3 The MIN-k-VEDS Algorithm for Interval Graphs

Let G be an interval graph and $\mathcal{I}$ be an interval representation of G. For any edge ab of $E(G)$, denote $N_G[a] \cup N_G[b]$ by $N_G[ab]$. Take any positive integer q. Let T be a subset of $V(G)$. Suppose that $T = \{u_1, \ldots, u_s\}$, where $r_{\mathcal{I}}(u_1) > \cdots > r_{\mathcal{I}}(u_s)$ and $q \leq s$. Denote the set $\{u_1, \ldots, u_q\}$ by $M_{G,\mathcal{I}}^q(T)$. Let uv and ab be two edges of $E(G)$. Suppose $r_{\mathcal{I}}(u) < r_{\mathcal{I}}(v)$ and $r_{\mathcal{I}}(a) < r_{\mathcal{I}}(b)$. If $r_{\mathcal{I}}(v) < r_{\mathcal{I}}(b)$, or $v = b$ and $r_{\mathcal{I}}(u) < r_{\mathcal{I}}(a)$, then we say $uv <_{\mathcal{I}} ab$. Next is the algorithm to find a minimum k-VEDS of G with respect to a subset $S \subseteq E(G)$.

MIN-κVED(G, S)
1   ▷ **Input** an interval graph G, a subset S of $E(G)$
2   ▷ **Output** a minimum k-VEDS D of G with respect to S
3   If there is some $ab \in S$ with $|N_G[ab]| < k$, then return "there
4   is no k-VEDS of G with respect to S", exit.
5   Else, construct an interval representation $\mathcal{I}$ of G.
6   Suppose $S = \{u_1w_1, \ldots, u_pw_p\}$ and $u_1w_1 <_{\mathcal{I}} \cdots <_{\mathcal{I}} u_pw_p$.
7     $D \leftarrow \emptyset$;
8   **for** $i \leftarrow 1$ **to** p
9   **do if** $|D \cap (N_G[u_iw_i])| < k$
10   **then** Let $q = k - |D \cap (N_G[u_iw_i])|$
11   Do $D \leftarrow D \cup M_{G,\mathcal{I}}^q((N_G[u_iw_i]) \setminus D)$.
12   output D.
13 ▷ **exit**

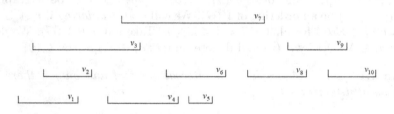

Fig. 1. An interval graph and its interval representation.

Example 3.1. *Let $G, \mathcal{I}$ be the interval graph and interval representation as depicted in Fig. 1. Let $S = \{v_1v_2, v_2v_3, v_3v_4, v_4v_6, v_3v_7, v_8v_9, v_9v_{10}\}$. We want to find a minimum double vertex-dominating set of G with respect to S. Consider the algorithm MIN-2VED(G, S).*

Iteration 0, $D = \emptyset$ and i is being increased to 1;

Iteration 1, $i = 1$, $N_G[v_1v_2] = \{v_3, v_2, v_1\}$, $|D \cap (N_G[v_1v_2])| = 0 < 2$. At this iteration, $q = 2 - |D \cap (N_G[v_1v_2])| = 2$, $M_{G,\mathcal{I}}^q((N_G[v_1v_2]) \setminus D) = \{v_3, v_2\}$, then we do $D \leftarrow D \cup M_{G,\mathcal{I}}^q((N_G[v_1v_2]) \setminus D)$, now we see $D = \emptyset \cup \{v_3, v_2\} = \{v_3, v_2\}$ and i is being increased to 2;

Iteration 2, $i = 2$, we see $D = \{v_3, v_2\}$, $N_G[v_2v_3] = \{v_7, v_4, v_3, v_2, v_1\}$ and $|N_G[v_2v_3] \cap D| = 2$. At this iteration, i is being increased to 3;

Iteration 3, $i = 3$, it holds $D = \{v_3, v_2\}$, $N_G[v_3v_4] = \{v_7, v_6, v_4, v_3, v_2\}$ and $|N_G[v_3v_4] \cap D| = 2$, We do nothing and i is being increased to 4;

Iteration 4, $i = 4$, we find $D = \{v_3, v_2\}$, $N_G[v_4v_6] = \{v_7, v_6, v_5, v_4, v_3\}$ and $|N_G[v_4v_6] \cap D| = 1 < 2$. $q = 2 - |D \cap (N_G[v_4v_6])| = 1$, $M_{G,\mathcal{I}}^q((N_G[v_4v_6]) \setminus D) = \{v_7\}$, then we do $D \leftarrow D \cup M_{G,\mathcal{I}}^q((N_G[v_4v_6]) \setminus D)$, we get $D = \{v_3, v_2\} \cup \{v_7\} = \{v_7, v_3, v_2\}$ and i is being increased to 5;

Iteration 5, $i = 5$, $D = \{v_7, v_3, v_2\}$, $N_G[v_3v_7] = \{v_8, v_7, v_6, v_5, v_4, v_3, v_2, v_1\}$ and $|N_G[v_3v_7] \cap D| = 3 > 2$. At this iteration, we do nothing and i is being increased to 6;

Iteration 6, $i = 6$, we have $D = \{v_7, v_3, v_2\}$, $N_G[v_8v_9] = \{v_{10}, v_9, v_8, v_7\}$ and $|N_G[v_8v_9] \cap D| = 1 < 2$. $q = 2 - |D \cap (N_G[v_8v_9])| = 1$, $M_{G,\mathcal{I}}^q((N_G[v_8v_9]) \setminus D) = \{v_{10}\}$, then we do $D \leftarrow D \cup M_{G,\mathcal{I}}^q((N_G[v_8v_9]) \setminus D)$, we obtain $D = \{v_7, v_3, v_2\} \cup \{v_{10}\} = \{v_{10}, v_7, v_3, v_2\}$ and i is being increased to 7;

Iteration 7, $i = 7$, $D = \{v_{10}, v_7, v_3, v_2\}$, $N_G[v_9v_{10}] = \{v_{10}, v_9, v_8\}$ and $|N_G[v_9v_{10}] \cap D| = 1 < 2$. $q = 2 - |D \cap (N_G[v_9v_{10}])| = 1$, $M_{G,\mathcal{I}}^q((N_G[v_9v_{10}]) \setminus D) = \{v_9\}$, then we do $D \leftarrow D \cup M_{G,\mathcal{I}}^q((N_G[v_9v_{10}]) \setminus D)$, we obtain $D = \{v_{10}, v_7, v_3, v_2\} \cup \{v_9\} = \{v_{10}, v_9, v_7, v_3, v_2\}$. Output D, exit. It is easy to check that $D = \{v_{10}, v_9, v_7, v_3, v_2\}$ is minimum double vertex-dominating set of G with respect to a subset S.

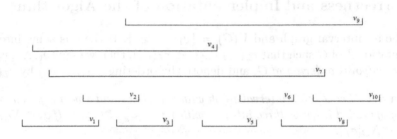

Fig. 2. An interval graph and its interval representation.

Example 3.2. *Let $G, \mathcal{I}$ be the interval graph and interval representation as depicted in Fig. 2. Let $S = \{v_1v_2, v_2v_3, v_4v_5, v_5v_6, v_2v_7, v_6v_7, v_7v_8, v_8v_9, v_8v_{10}\}$. In order to find a minimum 4-vertex-dominating set of G with respect to S, we consider the algorithm MIN-4VED(G, S).*

Iteration 0, $D = \emptyset$ and i is being increased to 1;

Iteration 1, $i = 1$, $N_G[v_1v_2] = \{v_9, v_7, v_4, v_3, v_2, v_1\}$, $|D \cap (N_G[v_1v_2])| = 0 < 4$. At this step, $q = 4 - |D \cap (N_G[v_1v_2])| = 4$, $M_{G,\mathcal{I}}^q((N_G[v_1v_2]) \setminus D) = \{v_9, v_7, v_4, v_3\}$, then we do $D \leftarrow D \cup M_{G,\mathcal{I}}^q((N_G[v_1v_2]) \setminus D)$, we see $D = \emptyset \cup \{v_9, v_7, v_4, v_3\} = \{v_9, v_7, v_4, v_3\}$ and i is being increased to 2;

Iteration 2, $i = 2$, $D = \{v_9, v_7, v_4, v_3\}$, $N_G[v_2v_3] = \{v_9, v_7, v_4, v_3, v_2, v_1\}$ and $|N_G[v_2v_3] \cap D| = 4$. We do nothing and i is being increased to 3;

Iteration 3, $i = 3$, $D = \{v_9, v_7, v_4, v_3\}$, $N_G[v_4v_5] = \{v_9, v_7, v_6, v_5, v_4, v_3, v_2, v_1\}$ and $|N_G[v_4v_5] \cap D| = 4$. At this iteration, we do nothing and i is being increased to 4;

Iteration 4, $i = 4$, $D = \{v_9, v_7, v_4, v_3\}$, $N_G[v_5v_6] = \{v_9, v_8, v_7, v_6, v_5, v_4\}$ and $|N_G[v_4v_6] \cap D| = 3 < 4$. $q = 4 - |D \cap (N_G[v_4v_6])| = 1$, $M_{G,\mathcal{I}}^q((N_G[v_5v_6]) \setminus D) = \{v_8\}$, then we do $D \leftarrow D \cup M_{G,\mathcal{I}}^q((N_G[v_5v_6] \setminus D)$, we get $D = \{v_9, v_7, v_4, v_3\} \cup \{v_8\} = \{v_9, v_8, v_7, v_4, v_3\}$ and i is being increased to 5;

Iteration 5, $i = 5$, $D = \{v_9, v_8, v_7, v_4, v_3\}$, $N_G[v_2v_7] = \{v_{10}, v_9, v_8, v_7, v_6, v_5, v_4, v_3, v_2, v_1\}$ and $|N_G[v_3v_7] \cap D| = 5 > 4$, i is being increased to 6;

Iteration 6, $i = 6$, $D = \{v_9, v_8, v_7, v_4, v_3\}$, $N_G[v_6v_7] = \{v_{10}, v_9, v_8, v_7, v_6, v_5,$ $v_4, v_3, v_2, v_1\}$ and $|N_G[v_6v_7] \cap D| = 5 > 4$, i is being increased to 7;

Iteration 7, $i = 7$, $D = \{v_9, v_8, v_7, v_4, v_3\}$, $N_G[v_7v_8] = \{v_{10}, v_9, v_8, v_7, v_6, v_5,$ $v_4, v_3, v_2, v_1\}$ and $|N_G[v_7v_8] \cap D| = 5 > 4$, i is being increased to 8;

Iteration 8, $i = 8$, $D = \{v_9, v_8, v_7, v_4, v_3\}$, $N_G[v_8v_9] = \{v_{10}, v_9, v_8, v_7, v_6, v_5,$ $v_4, v_3, v_2\}$ and $|N_G[v_8v_9] \cap D| = 5 > 4$, i is being increased to 9;

Iteration 9, $i = 9$, $D = \{v_9, v_8, v_7, v_4, v_3\}$, $N_G[v_8v_{10}] = \{v_{10}, v_9, v_8, v_7, v_6\}$ and $|N_G[v_8v_{10}] \cap D| = 3 < 4$. $q = 4 - |D \cap (N_G[v_9v_{10}])| = 1$, $M_{G,\mathcal{I}}^q((N_G[v_8v_{10}]) \setminus D) = \{v_{10}\}$, then we do $D \leftarrow D \cup M_{G,\mathcal{I}}^q((N_G[v_8v_{10}]) \setminus D)$, we obtain $D = \{v_9, v_8, v_7, v_4, v_3\} \cup \{v_{10}\} = \{v_{10}, v_9, v_8, v_7, v_4, v_3\}$ and output D, end. We can check that $D = \{v_{10}, v_9, v_8, v_7, v_4, v_3\}$ is a minimum 4-vertex-dominating set of G with respect to a subset S.

4   Correctness and Implementation of the Algorithm

Let G be an interval graph and $V(G) = \{v_1, \ldots, v_n\}$. If there is some interval representation $\mathcal{I}$ of G such that $r_{\mathcal{I}}(v_1) < \cdots < r_{\mathcal{I}}(v_n)$, then we say $(v_1, \ldots, v_n)$ is an *right endpoint ordering* of G, and denote the ordering $(v_1, \ldots, v_n)$ by $r_{\mathcal{I}}(G)$.

Lemma 4.1. *Let G be an interval graph with n vertices and σ be a right endpoint ordering of G. Let i, j, k be three integers with $1 \le i \le j \le k \le n$. If $\sigma_i \in N_G[\sigma_k]$, then $\sigma_j \in N_G[\sigma_k]$.*

Proof. Let $\mathcal{I}$ be an interval representation of G and $\sigma = r_{\mathcal{I}}(G)$. By the definition of right endpoint ordering, we have $r_{\mathcal{I}}(\sigma_1) < \cdots < r_{\mathcal{I}}(\sigma_n)$. Because $i \le k$ and $\sigma_i \in N_G[\sigma_k]$, we obtain $\ell_{\mathcal{I}}(\sigma_k) < r_{\mathcal{I}}(\sigma_i) \le r_{\mathcal{I}}(\sigma_k)$. As we see $i \le j \le k$, it holds $r_{\mathcal{I}}(\sigma_i) \le r_{\mathcal{I}}(\sigma_j) \le r_{\mathcal{I}}(\sigma_k)$, hence $\ell_{\mathcal{I}}(\sigma_k) < r_{\mathcal{I}}(\sigma_j) \le r_{\mathcal{I}}(\sigma_k)$, which implies that $\sigma_j \in N_G[\sigma_k]$, finishing the proof. □

Lemma 4.2. *Let G be an interval graph with n vertices and σ be a right endpoint ordering of G. Take any $i \in [n]$. Then for each $j \in [\ell_\sigma(i), i]$, it holds $\sigma_j \in N_G[\sigma_i]$. In addition, for each $k \in [i, r_\sigma(i)]$, it holds $\sigma_j \in N_G[\sigma_{r_\sigma(i)}]$.*

Proof. By the definitions of $\ell_\sigma(i)$ and $r_\sigma(i)$, it holds that $\sigma_{\ell_\sigma(i)} \in N_G[\sigma_i]$ and $\sigma_{r_\sigma(i)} \in N_G[\sigma_i]$. Then the lemma follows from Lemma 4.1. □

Lemma 4.3. *Let G be an interval graph with n vertices and $\mathcal{I}$ be an interval representation of G. Let $S = \{u_1w_1, \ldots, u_pw_p\}$ be a subset of $E(G)$ with $u_1w_1 <_{\mathcal{I}} \cdots <_{\mathcal{I}} u_pw_p$ and $r_{\mathcal{I}}(u_i) < r_{\mathcal{I}}(w_i)$ holds for each $j \in [p]$. Take any positive integer k and any integer $i \in [p]$. Suppose that D is some k-vertex-edge dominating set of G with respect to $S_{i-1} = \{u_1w_1, \ldots, u_{i-1}w_{i-1}\}$, and there is some minimum k-VEDS D' of G with respect to S such that $D \subseteq D'$. Assume that $|D \cap (N_G[u_iw_i])| < k$. Let $q = k - |D \cap (N_G[u_iw_i])|$, $W = M_{G,\mathcal{I}}^q((N_G[u_iw_i]) \setminus D)$ and $\tilde{D} = D \cup W$. Then $\tilde{D}$ is a k-vertex-edge dominating set of G with respect to $S_i = \{u_1w_1, \ldots, u_iw_i\}$. Furthermore, there is some minimum k-VEDS D'' of G with respect to S such that $\tilde{D} \subseteq D''$.*

Proof. By the facts that D is a k-vertex-edge dominating set of G with respect to $S_{i-1}, q = k - |D \cap (N_G[u_i w_i])|$ and $\tilde{D} = D \cup W = D \cup M_{G,\mathcal{I}}^q((N_G[u_i w_i]) \setminus D)$, we see $\tilde{D}$ is a k-vertex-edge dominating set of G with respect to $S_i = \{u_1 w_1, \ldots, u_i w_i\}$. Next we turn to prove that there is some minimum k-VEDS D'' of G with respect to S such that $\tilde{D} \subseteq D''$.

Since $|D \cap (N_G[u_i w_i])| < k$ and D' is a minimum k-VEDS of G with respect to S such that $D \subseteq D'$, it holds $|(D' \setminus D) \cap (N_G[u_i w_i])| \geq k - |D \cap (N_G[u_i w_i])| = q$. Let $(N_G[u_i w_i]) \setminus D = \{\sigma_1, \ldots, \sigma_t\}$ with $r_{\mathcal{I}}(\sigma_1) > \cdots > r_{\mathcal{I}}(\sigma_t)$. Let $(D' \setminus D) \cap (N_G[u_i w_i]) = \{\sigma_{j_1}, \ldots, \sigma_{j_q}\}$ while $q \leq s \leq t$ and $1 \leq j_1 < \cdots < j_s \leq t$. Let $T = \{\sigma_{j_1}, \ldots, \sigma_{j_q}\}$. Notice that $W = M_{G,\mathcal{I}}^q((N_G[u_i w_i]) \setminus D) = \{\sigma_1, \ldots, \sigma_q\}$.

Consider $D'' = (D' \setminus T) \cup W$. It is not hard to check that $|D''| = |D'|$ since $W \cap (D' \setminus T) = \emptyset$ and $|W| = |T| = q$. To finish the proof, we just need to show that D'' is a k-VEDS of G with respect to S. If this were not true, then there is some edge $u_{j'} w_{j'} \in S$ such that $|D'' \cap (N_G[u_{j'} w_{j'}])| < k$. Note that $r_{\mathcal{I}}(w_{j'}) > r_{\mathcal{I}}(u_{j'})$. Since D' is a k-VEDS of G with respect to S, there is some vertex σ_{j_h} of $D' \setminus D'' = T \setminus W$ which satisfies that $\sigma_{j_h} \in N_G[u_{j'} w_{j'}]$ and $\sigma_h \notin N_G[u_{j'} w_{j'}]$. Notice that $j_h > h$, hence $r_{\mathcal{I}}(\sigma_h) > r_{\mathcal{I}}(\sigma_{j_h})$. Because $\sigma_{j_h} \in N_G[u_{j'} w_{j'}]$ and $\sigma_h \notin N_G[u_{j'} w_{j'}]$, we obtain $r_{\mathcal{I}}(w_{j'}) < \ell_{\mathcal{I}}(\sigma_h)$. Recall that $\sigma_h \in N_G[u_i w_i]$ and $r_{\mathcal{I}}(u_i) < r_{\mathcal{I}}(w_i)$, hence $\ell_{\mathcal{I}}(\sigma_h) < r_{\mathcal{I}}(w_i)$, which implies that $r_{\mathcal{I}}(w_{j'}) < r_{\mathcal{I}}(w_i)$ and $u_{j'} w_{j'} <_{\mathcal{I}} u_i w_i$. We further deduce that $j' \leq i - 1$ and $u_{j'} w_{j'} \in S_{i-1}$ from the fact $u_{j'} w_{j'} <_{\mathcal{I}} u_i w_i$. But we know D is a k-vertex-edge dominating set of G with respect to $S_{i-1} = \{u_1 w_1, \ldots, u_{i-1} w_{i-1}\}$, so it must holds $|D \cap (N_G[u_{j'} w_{j'}])| \geq k$, contradicting with the fact that $D \subseteq D''$ and $|D'' \cap (N_G[u_{j'} w_{j'}])| < k$.

Since $|D''| = |D'|$ and D' is a minimum k-VEDS of G with respect to S, we find that D'' is also a minimum k-VEDS of G with respect to S, completing the proof. $\qquad\square$

Theorem 4.4. *Let k be any positive integer, G be a n-vertex interval graph and S be a subset of $E(G)$. Assume that $|N_G[ab]| \geq k$ holds for each edge $uv \in S$. Then the output of MIN-kVED(G, S), say D, is a minimum k-VEDS of G with respect to S.*

Proof. Let $V(G) = \{v_1, \ldots, v_n\}$ and $\mathcal{I}$ be an interval representation of G with $r_{\mathcal{I}}(v_1) < \cdots < r_{\mathcal{I}}(v_n)$. Suppose $S = \{u_1 w_1, \ldots, u_p w_p\}$ and $u_1 w_1 <_{\mathcal{I}} \cdots <_{\mathcal{I}} u_p w_p$. For each $i \in [p]$, denote the set D right after iteration i is completed by D_i and denote the set $\{u_1 w_1, \ldots, u_i w_i\}$ by S_i. Let $D_0 = S_0 = \emptyset$.

We want to prove that for each $i \in [p]$, D_i is a k-vertex-edge dominating set of G with respect to $S_i = \{u_1 w_1, \ldots, u_i w_i\}$ and there is some minimum k-VEDS D_i' of G with respect to S such that $D_i \subseteq D_i'$. We shall proceed by induction on i.

When $i = 1$, we see $D_0 = S_0 = \emptyset$, hence D_0 is a k-vertex-edge dominating set of G with respect to S_0. Notice that for any minimum k-VEDS D' of G with respect to S, $D_0 = \emptyset \subseteq D'$. Note that $|D_0 \cap (N_G[u_1 w_1])| = 0 < k$. Let $q = k - |D_0 \cap (N_G[u_1 w_1])|$, $W = M_{G,\mathcal{I}}^q((N_G[u_1 w_1]) \setminus D_0)$ and $\tilde{D}_0 = D_0 \cup W$. By the rule of algorithm MIN-kVED(G, S), we see $D_1 = \tilde{D}_0 = D_0 \cup W$. By

Lemma 4.3, we get D_1 is a k-vertex-edge dominating set of G with respect to $S_1 = \{u_1 w_1\}$ and there is some minimum k-VEDS D'_1 of G with respect to S such that $D_1 \subseteq D'_1$.

Suppose $i > 1$ and the statement holds for each smaller i. If $|D_{i-1} \cap (N_G[u_i w_i])| \geq k$, then by the rule of algorithm MIN-kVED(G, S), we get $D_i = D_{i-1}$, hence $D_i = D_{i-1}$ is also a k-vertex-edge dominating set of G with respect to S_i, and $D'_i = D'_{i-1}$ is also a minimum k-VEDS of G with respect to S such that $D_i \subseteq D'_i$.

Else if $|D_{i-1} \cap (N_G[u_i w_i])| < k$. Let $q = k - |D_{i-1} \cap (N_G[u_i w_i])|$, $W = M^q_{G,\mathcal{I}}((N_G[u_i w_i]) \setminus D_{i-1})$ and $\tilde{D}_{i-1} = D_{i-1} \cup W$. By the rule of algorithm MIN-kVED(G, S), we see $D_i = \tilde{D}_{i-1} = D_{i-1} \cup W$. By Lemma 4.3, we get D_i is also a k-vertex-edge dominating set of G with respect to S_i and there is some minimum k-VEDS D'_i of G with respect to S such that $D_i \subseteq D'_i$.

Now, we have reach the fact that the output of MIN-kVED(G, S), say $D = D_p$, is a k-VEDS of G with respect to S and there is some minimum k-VEDS D'_p of G with respect to S such that $D = D_p \subseteq D'_p$. Since D is already a k-VEDS of G with respect to S, it must holds that $D = D_p = D'_p$, hence D itself is a minimum k-VEDS D'_p of G with respect to S, finishing the proof. □

Theorem 4.5. Let G be any interval graph with n vertices and m edges. For each positive integer k and any subset S of $E(G)$, the algorithm MIN-kVED(G, S) can be implemented in $O(n|S| + m)$ time.

Proof. Suppose $|S| = p$. To determine whether there is some $ab \in S$ with $|N_G[ab]| < k$ or not takes us $O(kp)$ time. Construct an interval representation $\mathcal{I}$ of G needs $O(n + m)$ time. We need $O(p^2)$ time to order the edges of S such that $S = \{u_1 w_1, \ldots, u_p w_p\}$ and $u_1 w_1 <_\mathcal{I} \cdots <_\mathcal{I} u_p w_p$.

For i from 1 to p, at each iteration, computing $|D \cap (N_G[u_i w_i])|$ and $q = k - |D \cap (N_G[u_i w_i])|$ can be done in $O(n)$ time. Do $D \leftarrow D \cup M^q_{G,\mathcal{I}}((N_G[u_i w_i]) \setminus D)$ can also be done in $O(n)$ time. So the algorithm MIN-kVED(G, S) can be implemented in $O(np + m) = O(n|S| + m)$ time. □

Corollary 4.6. *Let G be any interval graph with n vertices and m edges. The MIN-VEDS and MIN-DVEDS problems can be solved in linear time. In addition, for any positive integer k, the MIN-k-VEDS problem can be solved in $O(nm)$ time in interval graphs.*

5   Conclusion

In this paper, we study the MIN-k-VEDS problem for the class of interval graphs. Let G be an interval graph with n vertices and m edges. For any positive integer k and any subset S of $E(G)$, we design an $O(n|S| + m)$ time algorithm to find a minimum k-vertex-edge dominating set of G with respect to S, this shows that the MIN-VEDS and MIN-DVEDS problems can be solved in linear time. Furthermore, the MIN-k-VEDS problem can be solved in $O(nm)$ time in interval graphs. There are still many interesting problems about k-vertex-edge domination, we list as follows:

(1) Polynomial time algorithms to solve the k-vertex-edge dominating problem in weighed unit interval graphs;
(2) Polynomial time algorithms to solve the k-vertex-edge dominating problem in weighed interval graphs;
(3) Polynomial time algorithms to solve the k-vertex-edge dominating problem in (weighed) strongly chordal graphs, permutation graphs, cocomparability graphs, or other graph classes.

Acknowledgement. We thank the referees and editors for their constructive input. This work was supported by the National Natural Science Foundation of China (11701059), the Natural Science Foundation of Chongqing (cstc2019jcyj-msxmX0156, cstc2020jcyj-msxmX0272, cstc2021jcyj-msxmX0436), the Youth project of science and technology research program of Chongqing Education Commission of China(KJQN202 001130, KJQN202001107, KJQN202101130).

References

1. Żyliński, P.: Vertex-edge domination in graphs. Aequationes Math. **93**(4), 735–742 (2018). https://doi.org/10.1007/s00010-018-0609-9
2. Boutrig, R., Chellali, M.: Total vertex-edge domination. Int. J. Comput. Math. **95**(9), 1820–1828 (2018)
3. Fishburn, P.C.: Interval Orders and Interval Graphs: A Study of Partially Ordered Sets, John Wiley & Sons Inc., (1985)
4. Golumbic, M.C.: Algorithmic Graph Theory and Perfect Graphs, 2nd ed., Annals of Discrete Mathematics, 57, Elsevier, Amsterdam, The Netherlands (2004)
5. Haynes, T.W., Hedetniemi, S.T., Slater, P.J.: Domination in Graphs: Advanced Topics. Marcel Dekker Inc, New York (1998)
6. Haynes, T.W., Hedetniemi, S.T., Slater, P.J.: Fundamentals of Domination in Graphs. Marcel Dekker Inc, New York (1998)
7. Krishnakumari, B., Chellali, M., Venkatakrishnan, Y.B.: Double vertex-edge domination. Discrete Math Algorithms Appl. **09**(04), 1750045 (2017)
8. Krishnakumari, B., Venkatakrishnan, Y.B., Krzywkowski, M.: Bounds on the vertex-edge domination number of a tree. C R Math **352**(5), 363–366 (2014)
9. Lewis, J.R.: Vertex-edge and edge-vertex parameters in graphs, Ph.D. thesis, Clemson, SC, USA, (2007)
10. Lewis, J.R., Hedetniemi, S.T., Haynes, T.W., Fricke, G.H.: Vertex-edge domination. Util Math **81**, 193–213 (2010)
11. Li, P., Wu, Y.: Spanning connectedness and Hamiltonian thickness of graphs and interval graphs. Discrete Math. Theor. Comput. Sci. **16**(2), 125–210 (2015)
12. Li, P., Wu, Y.: A linear time algorithm for the 1-fixed-endpoint path cover problem on interval graphs. SIAM J. Discret. Math. **31**(1), 210–239 (2017)
13. Möhring, R.H.: Algorithmic aspects of comparability graphs and interval graphs. In: Rival, I. (ed.) Graphs and Orders, pp. 41–101. D. Reidel, Boston (1985)
14. Paul, S., Ranjan, K.: Results on vertex-edge and independent vertex-edge domination. J. Comb. Optim. **4**, 1–28 (2021). https://doi.org/10.1007/s10878-021-00832-z
15. Peters, J.K.W.: Theoretical and algorithmic results on domination and connectivity (NordhausCGaddum, Gallai type results, maxCmin relationships, linear time, seriesCparallel), Ph.D. thesis, Clemson, SC, USA (1986)

16. Ramalingam, G., Rangan, C.P.: A uniform approach to domination problems on interval graphs. Inf. Process. Lett. **27**, 271–274 (1988)
17. Raychaudhuri, A.: On powers of interval and unit interval graphs. Congr. Numer. **59**, 235–242 (1987)
18. Shang, J., Li, P., Shi, Y.: The longest cycle problem is polynomial on interval graphs. Theoret. Comput. Sci. **859**, 37–47 (2021)
19. Trotter, W.T.: New perspectives on interval orders and interval graphs. In: Bailey, R.A. (ed.) London Mathematical Society Lecture Note Series 241, pp. 237–286. Cambridge University Press, Cambridge (1997)
20. Żyliński, P.: Vertex-edge domination in graphs. Aequationes Math. **93**(4), 735–742 (2018). https://doi.org/10.1007/s00010-018-0609-9

Cyclically Orderable Generalized Petersen Graphs

Xiaofeng Gu[1]([⊠]) [iD] and William Zhang[2]

[1] Department of Computing and Mathematics, University of West Georgia,
Carrollton, GA 30118, USA
xgu@westga.edu
[2] Harker High School, San Jose, CA 95129, USA

Abstract. A cyclic base ordering of a connected graph G is a cyclic ordering of $E(G)$ such that every cyclically consecutive $|V(G)| - 1$ edges induce a spanning tree of G. The density of G is defined to be $d(G) = |E(G)|/(|V(G)| - 1)$; and G is uniformly dense if $d(H) \leq d(G)$ for every connected subgraph H of G. It was conjectured by Kajitani, Ueno and Miyano that G has a cyclic base ordering if and only if G is uniformly dense. In this paper, we study cyclic base ordering of generalized Petersen graphs to support this conjecture.

Keywords: Cyclic ordering · Spanning tree · Generalized Petersen graph

1 Introduction

Let $G = (V, E)$ be a connected graph. A **cyclic base ordering** or for short **CBO** of G is a cyclic ordering of $E(G)$ such that every cyclically consecutive $|V(G)|-1$ edges induce a spanning tree of G. Equivalently, a cyclic base ordering is a bijection $\mathcal{O} : E(G) \longrightarrow \{1, 2, \ldots, |E(G)|\}$ such that for each $i = 1, 2, \ldots, |E(G)|$, $\{\mathcal{O}^{-1}(k) : k = i, i+1, \ldots, i + |V(G)| - 2\}$ induces a spanning tree of G, where the labelling k is equivalent to $k - |E(G)|$ if $k > |E(G)|$. If G has a CBO, then we say G is **cyclically orderable**. Clearly, cycles and trees are cyclically orderable.

Following the terminology of Catlin et al. [1,2], we define the **density** $d(G)$ of a connected graph G to be $d(G) = \frac{|E(G)|}{|V(G)|-1}$. A connected graph G is **uniformly dense** if $d(H) \leq d(G)$ for every connected subgraph H of G.

Cyclic base ordering is closely related to uniformly dense graphs. Kajitani, Ueno and Miyano [6] conjectured that they are actually equivalent. We may point out that the original conjecture in [6] was posed for matroids. We study this conjecture for graphs only and thus describe it in graph terminology below.

Conjecture 1 (Kajitani, Ueno and Miyano [6]). A connected graph G is cyclically orderable if and only if G is uniformly dense.

© The Author(s), under exclusive license to Springer Nature Switzerland AG 2022
Q. Ni and W. Wu (Eds.): AAIM 2022, LNCS 13513, pp. 303–315, 2022.
https://doi.org/10.1007/978-3-031-16081-3_26

The necessity was confirmed in [6], however, the sufficiency is still unsolved. Several families of uniformly dense graphs have been verified to be cyclically orderable in [4–6], including any uniformly dense simple connected graph with at most 5 vertices, any graph consisting of exactly 2 edge-disjoint spanning trees, complete graphs, complete bipartite graphs, maximum 2-degenerate graphs, and many others.

The **generalized Petersen graph** was introduced by Coxeter [3], and here is a notation by Watkins [8]. For $k < n/2$, let $G(n,k)$ denote the graph with vertex set $\{v_1, v_2, \ldots, v_n, u_1, u_2, \ldots, u_n\}$ and edge set $\{v_i v_{i+1}, u_i v_i, u_i u_{i+k} : i = 1, 2, \ldots, n\}$, where the subscripts are equivalent modulo n, that is, if a subscript j of a vertex is greater than n, then the subscript is actually $j - n$. The vertices $v_1, v_2, \ldots, v_n$ induce the outer rim and the vertices $u_1, u_2, \ldots, u_n$ induce the inner rim. Among all generalized Petersen graphs, $G(n,1)$ and $G(n,2)$ are two most important graph families. In particular, $G(n,1)$ has an n-gonal prism as its skeleton, which is also called the **prism graph**, and equivalently, it can be constructed as the Cartesian product of the cycle C_n and a single edge K_2. The graph $G(n,2)$ has a similar structure to the Petersen graph and actually $G(5,2)$ is exactly the Petersen graph.

In this paper, we discover cyclic base ordering of generalized Petersen graphs $G(n,1)$ and $G(n,2)$. Our results also show that $G(n,1)$ and $G(n,2)$ are uniformly dense. Given an edge ordering of a graph G, every cyclically consecutive $|V(G)| - 1$ edges is called a **progression**. To verify an edge ordering is a CBO, it suffices to show that any progression of this edge ordering induces a spanning tree.

2 Cyclic Base Ordering of $G(n, 1)$

Theorem 1. *For $n \geq 3$, $G(n,1)$ is cyclically orderable.*

Define 3 sets of edges, a, b, c. For every integer i such that $1 \leq i \leq n$, define a_i to be the edge $v_i v_{i+1}$, b_i to be the edge $v_i u_i$, and c_i to be the edge $u_i u_{i+1}$. We can call them a-edges, b-edges and c-edges, respectively. When the subscript $k > n$ or $k \leq 0$, u_k, v_k, a_k, b_k, c_k are the same as u_r, v_r, a_r, b_r, c_r respectively, where r is the (unique) integer such that $1 \leq r \leq n$ and $r \equiv k \pmod{n}$. We prove Theorem 1 by considering three cases: $n = 3m$, $n = 3m + 1$ or $n = 3m + 2$ for some positive integer m.

2.1 The Case $n = 3m$

Proof. Define an edge ordering $\mathcal{O}$ of $G(n,1)$ by

$$\mathcal{O} = (a_1, c_{2m+1}, b_{m+2}, a_2, c_{2m+2}, b_{m+3}, \ldots, a_{3m}, c_{5m}, b_{4m+1}).$$

The subscripts are equivalent modulo n, that is, if a subscript k of an edge is greater than $3m$, then the subscript is actually $k - 3m$. For example, c_{5m} is actually c_{2m}, and b_{4m+1} is b_{m+1}. Here is an example for $G(6, 1)$ in Fig. 1.

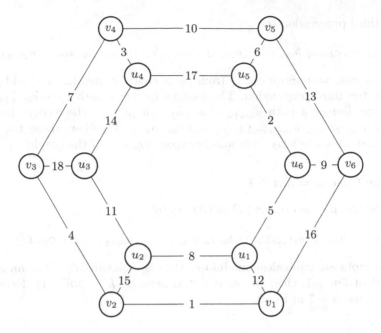

Fig. 1. CBO of $G(6,1)$

We will show that $\mathcal{O}$ is a CBO of $G(n,1)$. By symmetry, any progression starting with an a-edge has the same structure, and similarly for any progression starting with a b-edge or a c-edge. Thus it suffices to verify the first 3 progressions. Recall that $n = 3m$ and $G(n,1)$ has $2n = 6m$ vertices. Thus any progression consists of exactly $2n - 1 = 6m - 1$ edges.

The first progression is

$$a_1, c_{2m+1}, b_{m+2}, a_2, c_{2m+2}, b_{m+3}, \ldots, a_{2m-1}, c_{4m-1}, b_{3m}, a_{2m}, c_{4m}.$$

The a-edges $a_1, a_2, \ldots, a_{2m}$ in this progression induce a path $v_1 v_2 \cdots v_{2m} v_{2m+1}$. The c-edges are $c_{2m+1}, c_{2m+2}, \ldots, c_{3m}, c_1, \ldots, c_m$, which induce a path $u_{2m+1} \cdots u_{3m} u_1 \cdots u_m u_{m+1}$. The b-edges $b_{m+2}, b_{m+3}, \ldots, b_{3m}$ induces a matching $u_{m+2} v_{m+2}, u_{m+3} v_{m+3}, \ldots, u_{3m} v_{3m}$. It is not hard to see that all vertices are connected by this progression without any cycle, and thus the first progression induces a spanning tree.

The second progression is

$$c_{2m+1}, b_{m+2}, a_2, c_{2m+2}, b_{m+3}, \ldots, a_{2m-1}, c_{4m-1}, b_{3m}, a_{2m}, c_{4m}, b_{3m+1}.$$

In other words, we remove a_1 from the first progression and add b_{3m+1} (i.e. b_1) to obtain the second progression. The a-edges induce a path $v_2 \cdots v_{2m} v_{2m+1}$. The c-edges induce a path $u_{2m+1} \cdots u_{3m} u_1 \cdots u_m u_{m+1}$. The b-edges induces a matching $u_{m+2} v_{m+2}, u_{m+3} v_{m+3}, \ldots, u_{3m} v_{3m}, u_1 v_1$. Together, these two paths and the matching form a spanning tree of the graph.

The third progression is

$$b_{m+2}, a_2, c_{2m+2}, b_{m+3}, \ldots, a_{2m-1}, c_{4m-1}, b_{3m}, a_{2m}, c_{4m}, b_{3m+1}, a_{2m+1}.$$

In other words, we remove c_{2m+1} from the second progression and add a_{2m+1} to obtain the third progression. The a-edges induce a path $v_2 \cdots v_{2m+1} v_{2m+2}$. The c-edges induce a path $u_{2m+2} \cdots u_{3m} u_1 \cdots u_m u_{m+1}$. The b-edges induce a matching $u_{m+2} v_{m+2}, u_{m+3} v_{m+3}, \ldots, u_{3m} v_{3m}, u_1 v_1$. Together, these two paths and the matching similarly form another spanning tree of the graph.

2.2   The Case $n = 3m + 1$

Proof. Define an edge ordering $\mathcal{O}$ of $G(n, 1)$ by

$$\mathcal{O} = (a_1, b_{m+2}, c_{2m+3}, a_2, b_{m+3}, c_{2m+4}, \ldots, a_{3m+1}, b_{m+1}, c_{2m+2}).$$

The subscripts are equivalent modulo n, that is, if a subscript k of an edge is greater than $3m + 1$, then the subscript is actually $k - (3m + 1)$. Here is an example when $n = 7$ in Fig. 2.

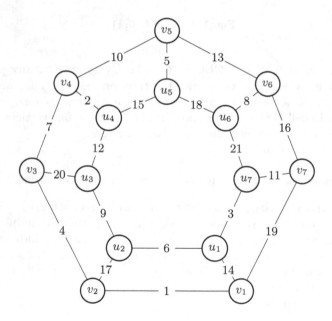

Fig. 2. CBO of $G(7, 1)$

We will show that $\mathcal{O}$ is a CBO of $G(n, 1)$. Similar to the above, all the progressions starting with a-edges have the same structure, as do those starting with b-edges and c-edges. There are $2n = 6m + 2$ vertices and so each progression contains exactly $6m + 1$ edges.

The first progression is

$$a_1, b_{m+2}, c_{2m+3}, a_2, b_{m+3}, c_{2m+4}, \ldots, a_{2m}, b_{3m+1}, c_{4m+2}, a_{2m+1}.$$

The a-edges $a_1, a_2, \ldots, a_{2m+1}$ in this progression induce a path $v_1 v_2 \cdots v_{2m+1}$ v_{2m+2}, and the b-edges $b_{m+2}, b_{m+3}, \ldots, b_{3m+1}$ induce a matching $u_{m+2} v_{m+2}$, $u_{m+3} v_{m+3}$, $\ldots$, $u_{3m+1} v_{3m+1}$. The c-edges are $c_{2m+3}, c_{2m+4}, \ldots, c_{3m+1}, c_1, \ldots$, c_{m+1}, which induce a path $u_{2m+3} u_{2m+4} \cdots u_{3m+1} u_1 \cdots u_{m+1} u_{m+2}$. Much similar to the above, all the vertices are connected without any cycles, thus the progression induces a spanning tree.

The second progression is

$$b_{m+2}, c_{2m+3}, a_2, b_{m+3}, c_{2m+4}, a_3, \ldots, a_{2m}, b_{3m+1}, c_{4m+2}, a_{2m+1}, b_{3m+2}.$$

In other words, we remove a_1 from the first progression and add b_{3m+2} (i.e. b_1) to obtain the second progression. In this progression, the a-edges induce a path $v_2 v_3 \cdots v_{2m+1} v_{2m+2}$, the b-edges induce a matching $u_{m+2} v_{m+2}$, $u_{m+3} v_{m+3}$, $\ldots$, $u_{3m+1} v_{3m+1}$, $u_1 v_1$ and the c-edges induce a path $u_{2m+3} u_{2m+4} \cdots u_{3m+1} u_1 \cdots$ $u_{m+1} u_{m+2}$. Together, these two paths and the matching form a spanning tree of the graph.

The third progression is

$$c_{2m+3}, a_2, b_{m+3}, c_{2m+4}, a_3, \ldots, a_{2m}, b_{3m+1}, c_{4m+2}, u_{2m+1}, b_{3m+2}, c_{4m+3}.$$

In other words, we remove b_{m+2} from the second progression and add c_{4m+3} (i.e. c_{m+2}) to obtain the third progression. In this progression, the a-edges induce a path $v_2 v_3 \cdots v_{2m+1} v_{2m+2}$, the b-edges induce a matching $u_{m+3} v_{m+3}$, $u_{m+4} v_{m+4}$, $\ldots$, $u_{3m+1} v_{3m+1}$, $u_1 v_1$ and the c-edges induce a path $u_{2m+3} u_{2m+4} \cdots u_{3m+1} u_1 \cdots u_{m+2} u_{m+3}$. Together, the two paths and matching form a spanning tree of the graph once again.

2.3   The Case $n = 3m + 2$

Proof. This case was originally proved in the summer project [7] mentored by the first author, but did not publish anywhere. For completeness, we include a proof here. Define an edge ordering $\mathcal{O}$ of $G(n, 1)$ by

$$\mathcal{O} = (a_1, b_3, c_3, a_4, b_6, c_6, a_7, b_9, c_9, \ldots, a_{3n-2}, b_{3n}, c_{3n}),$$

where the subscripts are equivalent modulo n. For example, b_{3n} is actually b_n. This ordering can be considered as a list of n ordered triples $(a_{3i+1}, b_{3i+3}, c_{3i+3})$ for $i = 0, 1, \ldots, n - 1$. Here is an example for $G(8, 1)$ in Fig. 3.

We will show that $\mathcal{O}$ is a CBO of $G(n, 1)$. By symmetry, it suffices to verify the first three progressions. Notice that $n = 3m + 2$ and $G(n, 1)$ has $2n = 6m + 4$ vertices. Each progression has $2n - 1 = 6m + 3$ edges.

The first progression is

$$a_1, b_3, c_3, a_4, b_6, c_6, a_7, b_9, c_9, \ldots, a_{6m+1}, b_{6m+3}, c_{6m+3}.$$

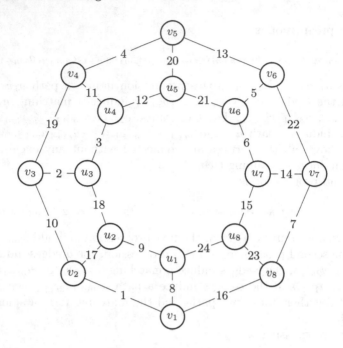

Fig. 3. CBO of $G(8,1)$

All the a-edges in this progression are $a_1, a_4, \ldots, a_{3m-2}, a_{3m+1}, a_{3m+4}, \ldots,$ a_{6m+1}. If the subscript k is greater than $3m + 2$, then the subscript is actually $k - (3m + 2)$. Thus the a-edges here actually are $a_1, a_4, \ldots, a_{3m-2},$ $a_{3m+1}, a_2, a_5, \ldots, a_{3m-1}$. In other words, the a-edges induce $m + 1$ short paths $v_1 v_2 v_3, v_4 v_5 v_6, \ldots, v_{3m-2} v_{3m-1} v_{3m}$ and $v_{3m+1} v_{3m+2}$.

All the c-edges in this progression are $c_3, c_6, \ldots, c_{3m}, c_{3m+3}, \ldots, c_{6m+3}$. If the subscript k is greater than $3m+2$, then the subscript is actually $k-(3m+2)$. Thus the c-edges in this progression actually are $c_3, c_6, \ldots, c_{3m}, c_1, c_4, c_7, \ldots, c_{3m+1}$. In other words, the c-edges also induce $m + 1$ short paths, they are $u_1 u_2, u_3 u_4 u_5,$ $u_6 u_7 u_8, \ldots, u_{3m} u_{3m+1} u_{3m+2}$.

All the b-edges in this progression are $b_3, b_6, \ldots, b_{3m}, b_{3m+3}, \ldots, b_{6m+3}$. If the subscript k is greater than $3m+2$, then the subscript is actually $k-(3m+2)$. Thus the b-edges in this progression actually are $b_3, b_6, \ldots, b_{3m}, b_1, b_4, b_7, \ldots, b_{3m+1}$. It is not hard to see that each b_{3i} connects the path $v_{3i-2} v_{3i-1} v_{3i}$ and the path $u_{3i} u_{3i+1} u_{3i+2}$ for $i = 1, 2, \ldots, m$. Each b_{3i+1} connects the path $v_{3i+1} v_{3i+2} v_{3i+3}$ and the path $u_{3i} u_{3i+1} u_{3i+2}$ for $i = 1, 2, \ldots, m - 1$, and b_1 connects the path $v_1 v_2 v_3$ and the edge $u_1 u_2$, while b_{3m+1} connects the path $v_{3m+1} v_{3m+2}$ and the path $u_{3m} u_{3m+1} u_{3m+2}$. All vertices are connected by this progression and the progression contains exactly $2n - 1 = |V| - 1$ edges, thus it induces a spanning tree.

The second progression is

$$b_3, c_3, a_4, b_6, c_6, a_7, b_9, c_9, \ldots, a_{6m+1}, b_{6m+3}, c_{6m+3}, a_{6m+4}.$$

Notice that a_{6m+4} is actually a_{3m+2}. It is like to remove a_1 from the first progression and add a_{3m+2}. Removing a_1 will disconnect v_1 from other v_k's but adding a_{3m+2} will connect v_1 back to v_{3m+2}. Thus this progression still induces a spanning tree.

The third progression is

$$c_3, a_4, b_6, c_6, a_7, b_9, c_9, \ldots, a_{6m+1}, b_{6m+3}, c_{6m+3}, a_{6m+4}, b_{6m+6}.$$

Notice that b_{6m+6} is actually b_2. Similarly, it is like to remove b_3 from the second progression and add b_2. It is not hard to verify this progression still induces a spanning tree.

Therefore $\mathcal{O}$ is a CBO of $G(n, 1)$.

3   Cyclic Base Ordering of $G(n, 2)$

Theorem 2. *For $n \geq 5$, $G(n, 2)$ is cyclically orderable.*

Similarly, define 3 sets of edges, a, b, c. For every integer i such that $1 \leq i \leq n$, define a_i to be the edge $v_i v_{i+1}$, b_i as the edge $v_i u_i$, and c_i as the edge $u_i u_{i+2}$. We can call them a-edges, b-edges and c-edges, respectively. When the subscript $k > n$ or $k \leq 0$, u_k, v_k, a_k, b_k, c_k are the same as u_r, v_r, a_r, b_r, c_r respectively, where r is the (unique) integer such that $1 \leq r \leq n$ and $r \equiv k \pmod{n}$. When

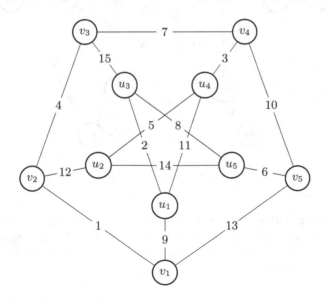

Fig. 4. CBO of the Petersen graph $G(5, 2)$

$n = 5$, $G(n, 2)$ is the Petersen graph and a CBO is shown in Fig. 4. Thus we focus on $n \geq 6$ and it suffices to consider three cases: $n = 3m$, $n = 3m + 1$ or $n = 3m + 2$ for some integer $m \geq 2$.

3.1   The Case $n = 3m$ for $m \geq 2$

Proof. Define an edge ordering $\mathcal{O}$ of $G(n, 2)$ by

$$\mathcal{O} = (a_1, c_{2m}, b_{m+2}, a_2, c_{2m+1}, b_{m+3}, \ldots, a_{3m}, c_{5m-1}, b_{4m+1}).$$

The subscripts are equivalent modulo n, that is, if a subscript k of an edge is greater than $3m$, then the subscript is actually $k - 3m$. To show that $\mathcal{O}$ is a CBO, we only consider even n. The case of odd n is quite similar and will be omitted. Let n be even, and so m is also even. An example of $G(6, 2)$ is shown in Fig. 5.

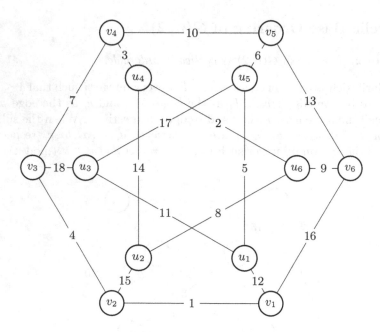

Fig. 5. CBO of $G(6, 2)$

As the edge ordering $\mathcal{O}$ is symmetric, it suffices to verify each of the first three progressions induces a spanning tree. Since $n = 3m$ and the graph has $2n = 6m$ vertices, any progression consists of exactly $6m - 1$ edges.

The first progression is

$$a_1, c_{2m}, b_{m+2}, a_2, c_{2m+1}, b_{m+3}, \ldots, a_{2m-1}, c_{4m-2}, b_{3m}, a_{2m}, c_{4m-1}.$$

The a-edges in this progression induce a path $v_1v_2 \cdots v_{2m+1}$, while the b-edges induce a matching $u_{m+2}v_{m+2}, u_{m+3}v_{m+3}, \ldots, u_{3m}v_{3m}$. The c-edges are $c_{2m}, c_{2m+1}, \ldots, c_{3m}, c_1, \ldots, c_{m-1}$, which induce two paths $u_{2m}u_{2m+2} \cdots u_{3m}u_2u_4 \cdots u_m$ and $u_{2m+1}u_{2m+3} \cdots u_{3m-1}u_1u_3 \cdots u_{m+1}$. It is not hard to see that all vertices have been connected by this progression. Since it has exactly $|V| - 1 = 2n - 1 = 6m - 1$ edges, it is a spanning tree.

The second progression is

$$c_{2m}, b_{m+2}, a_2, c_{2m+1}, b_{m+3}, \ldots, a_{2m-1}, c_{4m-2}, b_{3m}, a_{2m}, c_{4m-1}, b_{3m+1}.$$

In other words, it can be obtained from the first progression by deleting the edge $a_1 = v_1v_2$ and adding the edge $b_{3m+1} = b_1 = u_1v_1$. Since the first progression induce a spanning tree, the deletion of a_1 creates an isolated vertex v_1 however the addition of b_1 will connect v_1 back. Thus the second progression also induces a spanning tree.

The third progression is

$$b_{m+2}, a_2, c_{2m+1}, b_{m+3}, \ldots, a_{2m-1}, c_{4m-2}, b_{3m}, a_{2m}, c_{4m-1}, b_{3m+1}, a_{2m+1}.$$

In other words, it can be obtained from the second progression by deleting the edge $c_{2m} = u_{2m}u_{2m+2}$ and adding the edge $a_{2m+1} = v_{2m+1}v_{2m+2}$. Since the second progression induce a spanning tree, the deletion of c_{2m} will disconnect this spanning tree into two components. It is not hard to see that v_{2m+1} and v_{2m+2} are in different components. Clearly the addition of $a_{2m+1} = v_{2m+1}v_{2m+2}$ will connect the two components, and thus the third progression also induces a spanning tree.

3.2 The Case $n = 3m + 1$ for $m \geq 2$

Proof. Define an edge ordering $\mathcal{O}$ by

$$\mathcal{O} = (a_1, b_{m+2}, c_{2m+3}, a_2, b_{m+3}, c_{2m+4}, \ldots, a_{3m+1}, b_{4m+2}, c_{5m+3}).$$

The subscripts are equivalent modulo n, that is, if a subscript k of an edge is greater than $3m + 1$, then the subscript is actually $k - (3m + 1)$. To show that $\mathcal{O}$ is a CBO, we only consider odd n. The case of even n is similar and will be omitted. Let n be odd, and so m is even. An example of $G(7, 2)$ is shown in Fig. 6.

By symmetry, it suffices to verify each of the first three progressions induces a spanning tree. The first progression is

$$a_1, b_{m+2}, c_{2m+3}, a_2, b_{m+3}, c_{2m+4}, \ldots, a_{2m}, b_{3m+1}, c_{4m+2}, a_{2m+1}.$$

The a-edges in this progression induce a path $v_1v_2 \cdots v_{2m+1}v_{2m+2}$, while the b-edges induce a matching $u_{m+2}v_{m+2}, u_{m+3}v_{m+3}, \ldots, u_{3m+1}v_{3m+1}$.

The c-edges are $c_{2m+3}, c_{2m+4}, \ldots, c_{3m+1}, c_1, \ldots, c_{m+1}$, which induce two paths

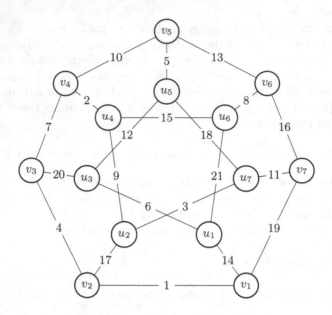

Fig. 6. CBO of $G(7,2)$

$u_{2m+3}u_{2m+5}\cdots u_{3m+1}u_2u_4\cdots u_{m+2}$ and $u_{2m+4}u_{2m+6}\cdots u_{3m}u_1u_3\cdots u_{m+3}$. It is not hard to see that all vertices have been connected by this progression. Since it has exactly $|V|-1 = 2n-1 = 6m+1$ edges, it is a spanning tree.

The second progression is

$$b_{m+2}, c_{2m+3}, a_2, b_{m+3}, c_{2m+4}, a_3, \ldots, b_{3m+1}, c_{4m+2}, a_{2m+1}, b_{3m+2}.$$

In other words, it can be obtained from the first progression by deleting the edge $a_1 = v_1v_2$ and adding the edge $b_{3m+2} = b_1 = u_1v_1$. Since the first progression induce a spanning tree, the deletion of a_1 creates an isolated vertex v_1 however the addition of b_1 will connect v_1 back. Thus the second progression also induces a spanning tree.

The third progression is

$$c_{2m+3}, a_2, b_{m+3}, c_{2m+4}, a_3, b_{m+4}, \ldots, c_{4m+2}, a_{2m+1}, b_{3m+2}, c_{4m+3}.$$

In other words, it can be obtained from the second progression by deleting the edge $b_{m+2} = u_{m+2}v_{m+2}$ and adding the edge $c_{4m+3} = c_{m+2} = u_{m+2}u_{m+4}$. Since the second progression induce a spanning tree, the deletion of b_{m+2} will disconnect this spanning tree. It is not hard to see that u_{m+2} in one component while u_{m+4} in the other component. Clearly the addition of $c_{m+2} = u_{m+2}u_{m+4}$ will connect the two components, and thus the third progression also induces a spanning tree.

3.3 The Case $n = 3m + 2$ for $m \geq 2$

Proof. Define an edge ordering $\mathcal{O}$ by

$$\mathcal{O} = (a_1, b_4, c_1, a_4, b_7, c_4, \ldots, a_{3n-5}, b_{3n-2}, c_{3n-5}, a_{3n-2}, b_{3n+1}, c_{3n-2}),$$

where the subscripts are equivalent modulo n. For example, b_{3n+1} is actually b_1. This ordering can be considered as a list of ordered triples $(a_{3i+1}, b_{3i+4}, c_{3i+1})$ for $i = 0, 1, \ldots, n - 1$.

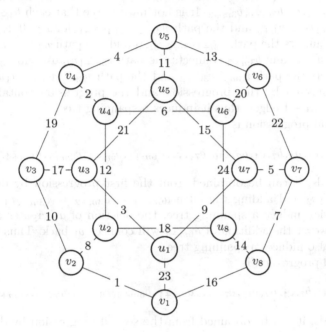

Fig. 7. CBO of $G(8,2)$

An example of $G(8,2)$ is shown in Fig. 7. The proofs for the cases of even n and odd n are the same. By symmetry, it suffices to verify each of the first three progressions induces a spanning tree. Notice that $n = 3m + 2$ and $G(n,2)$ has $2n$ vertices, and so each progression has $2n - 1 = 6m + 3$ edges.

The first progression is

$$a_1, b_4, c_1, a_4, b_7, c_4, a_7, b_{10}, c_7, \ldots, a_{6m+1}, b_{6m+4}, c_{6m+1}.$$

All the a-edges in this progression are $a_1, a_4, \ldots, a_{3m-2}, a_{3m+1}, a_{3m+4}, \ldots,$ a_{6m+1}. If the subscript k is greater than $3m + 2$, then the subscript is actually $k - (3m + 2)$. Thus the a-edges here actually are $a_1, a_4, \ldots, a_{3m-2}, a_{3m+1}, a_2, a_5, \ldots, a_{3m-1}$. In other words, the a-edges induce $m + 1$ short paths $v_1 v_2 v_3, v_4 v_5 v_6, \ldots, v_{3m-2} v_{3m-1} v_{3m}$ and $v_{3m+1} v_{3m+2}$.

All the c-edges in this progression are $c_1, c_4, \ldots, c_{3m-2}, c_{3m+1}, c_{3m+4}$, $\ldots, c_{6m+1}$. If the subscript k is greater than $3m + 2$, then the subscript is actually $k - (3m + 2)$. Thus the c-edges here actually are $c_1, c_4, \ldots, c_{3m-2}, c_{3m+1}, c_2, c_5, \ldots, c_{3m-1}$. In other words, the c-edges also induce $m + 1$ short paths, they are $u_2 u_4 u_6$, $u_5 u_7 u_9$, $u_8 u_{10} u_{12}, \ldots, u_{3m-4} u_{3m-2} u_{3m}$, $u_{3m-1} u_{3m+1} u_1 u_3$ and u_{3m+2}.

All the b-edges in this progression are $b_4, b_7, \ldots, b_{3m+1}, b_{3m+4}$, $b_{3m+7}, \ldots, b_{6m+4}$. If the subscript k is greater than $3m + 2$, then the subscript is actually $k - (3m + 2)$. Thus the b-edges in this progression actually are $b_4, b_7, \ldots, b_{3m+1}, b_2, b_5, \ldots, b_{3m+2}$. It is not hard to see that each b_{3i+2} connects the path $v_{3i+1} v_{3i+2} v_{3i+3}$ and the path $u_{3i+2} u_{3i+4} u_{3i+6}$ for $i = 0, 1, \ldots, m - 1$. Each b_{3i+1} connects the path $v_{3i+1} v_{3i+2} v_{3i+3}$ and the path $u_{3i-1} u_{3i+1} u_{3i+3}$ for $i = 1, 2, \ldots, m - 1$, and b_{3m+2} connects the path $v_{3m+1} v_{3m+2}$ and u_{3m+2}, while b_{3m+1} connects the path $v_{3m+1} v_{3m+2}$ and the path $u_{3m-1} u_{3m+1} u_1 u_3$. All vertices are connected by this progression and the progression contains exactly $|V(G)| - 1 = 2n - 1$ edges, thus it induces a spanning tree.

The second progression is

$$b_4, c_1, a_4, b_7, c_4, a_7, b_{10}, c_7, \ldots, a_{6m+1}, b_{6m+4}, c_{6m+1}, a_{6m+4}.$$

In other words, it can be obtained from the first progression by deleting the edge $a_1 = v_1 v_2$ and adding the edge $a_{6m+4} = a_{3m+2} = v_{3m+2} v_1$. Since the first progression induce a spanning tree, the deletion of a_1 creates an isolated vertex v_1 however the addition of a_{6m+4} will connect v_1 back. Thus the second progression also induces a spanning tree.

The third progression is

$$c_1, a_4, b_7, c_4, a_7, b_{10}, c_7, \ldots, a_{6m+1}, b_{6m+4}, c_{6m+1}, a_{6m+4}, b_{6m+7}.$$

In other words, it can be obtained from the second progression by deleting the edge $b_4 = u_4 v_4$ and adding the edge $b_{6m+7} = b_3 = u_3 v_3$. Since the second progression induce a spanning tree, the deletion of b_4 will disconnect this spanning tree. It is not hard to see that u_3 in one component while v_3 in the other component. Clearly the addition of $b_{6m+7} = b_3 = u_3 v_3$ will connect the two components, and thus the third progression also induces a spanning tree.

Therefore $\mathcal{O}$ is a CBO of $G(n, 2)$.

4   Conclusion

By constructing explicit cyclic base orderings, we proved the "if" part of Conjecture 1 for the generalized Petersen graphs $G(n, 1)$ and $G(n, 2)$, supporting the conjecture and providing further insight into the nature of cyclic base orderings for certain structures. Besides the theoretical advance, the systematic approach of constructing the CBOs for these graphs may prove to have real world applications.

References

1. Catlin, P.A., Grossman, J., Hobbs, A.M.: Graphs with uniform density. Congr. Numer. **65**, 281–285 (1988)
2. Catlin, P.A., Grossman, J.W., Hobbs, A.M., Lai, H.-J.: Fractional arboricity, strength and principal partitions in graphs and matroids. Discrete Appl. Math. **40**, 285–302 (1992)
3. Coxeter, H.: Self-dual configurations and regular graphs. Bull. Amer. Math. Soc. **56**, 413–455 (1950)
4. Gu, X., Horacek, K., Lai, H.-J.: Cyclic base orderings in some classes of graphs. J. Combin. Math. Combin. Comput. **88**, 39–50 (2014)
5. Gu, X., Li, J., Yang, E., Zhang, W.: Cyclic base ordering of certain degenerate graphs, submitted for publication
6. Kajitani, Y., Ueno, S., Miyano, H.: Ordering of the elements of a matroid such that its consecutive w elements are independent. Discrete Math. **72**, 187–194 (1988)
7. Li, J., Yang, E., Zhang, W.: Cyclic Base Ordering of Graphs. arXiv:2110.00892 [math.CO] (2021)
8. Watkins, M.E.: A theorem on Tait colorings with an application to the generalized Petersen graphs. J. Comb. Theor. **6**, 152–164 (1969)

The r-Dynamic Chromatic Number of Planar Graphs Without Special Short Cycles

Yuehua Bu[1,2], Ruiying Yang[1], and Hongguo Zhu[1(✉)]

[1] Department of Mathematics, Zhejiang Normal University,
Jinhua 321004, Zhejiang, China
zhuhongguo@zjnu.edu.cn

[2] Xingzhi College of Zhejiang Normal University, Jinhua 321004, Zhejiang, China

Abstract. Let k and r be two positive integers. An r-dynamic coloring of a graph G is a proper k-coloring φ such that $\mid \varphi(N_G(v)) \mid \geq \min\{d_G(v), r\}$ for each $v \in V(G)$. In this paper, we study the r-dynamic coloring of planar graphs without 3-,5-cycle, and 4-cycle is not adjacent to 7^--cycles. We prove that the upper bound of r-dynamic chromatic number of such graph is $r + 3$ if $r \geq 14$.

Keywords: r-dynamic coloring · Planar graph · Cycle · Discharging

1 Introduction

There are many generalizations and variations of ordinary graph colorings. One of the most popular is the dynamic coloring. Let G be a graph and k, r be two positive integers. A (k, r)-dynamic coloring of a graph G is mapping $\varphi : V(G) \to \{1, 2, \ldots, k\}$ k, r such that $\varphi(u) \neq \varphi(v)$ for each $uv \in E(G)$, and $\mid \varphi(N_G(v)) \mid \geq \min\{d_G(v), r\}$ for each $v \in V(G)$. The r-dynamic chromatic number, introduced by Montgomery [4] and written as $\chi_r(G)$, is the minimum k such that G has a (k, r)-dynamic coloring. In particular, we remark that the r-dynamic coloring is called 2-distance coloring when $r = \Delta(G)$.

For 2-distance coloring of planar graphs, namely Δ-dynamic coloring, Wenger [7] posed the following famous conjecture.

Conjecture 1.1. *Let G be a planar graph with maximum degree Δ. If $4 \leq \Delta \leq 7$, then $\chi_2(G) \leq \Delta + 5$. If $\Delta \geq 8$, then $\chi_2(G) \leq \lfloor \frac{3\Delta}{2} \rfloor + 8$.*

Conjecture 1.1 is still open. However, several upper bounds in terms of maximum degree Δ have been studied in [1,5,11]. For planar graphs with girth restriction condition, there are many results about $\chi_r(G)$ (see Table 1).

This work is supported by a research grant NSFC (11271334).

Table 1. The uppers of $\chi_r(G)$ for some planar graphs

Conditions	Conclusions	References
$g \geq 5, r \geq 15$	$\chi_r(G) \leq r + 5$	[10]
$g \geq 5, r \geq 24$	$\chi_r(G) \leq r + 4$	[8]
$g \geq 6, r \geq 3$	$\chi_r(G) \leq r + 5$	[6]
$g \geq 7, r \geq 16$	$\chi_r(G) \leq r + 1$	[9]
$g \geq 8, r \geq 9$	$\chi_r(G) \leq r + 1$	[3]

For planar graph G without 4,5-cycles, Zhu and Gu [12] proved that $\chi_2(G) \leq \max\{29, \Delta+3\}$. Recently, Bu and Cao [2] proved that every planar graph without 3,5-cycles and intersecting 4-cycle and $\Delta(G) \geq 15$ has $\chi_2(G) \leq \Delta + 3$. In this paper, we study the planar graphs without the special short cycles, and prove the following theorem.

Theorem 1.2. *Let G be a planar graph without 3-,5-cycles, and 4-cycle is not adjacent to 7^--cycles. If $r \geq 14$, then $\chi_r(G) \leq r + 3$.*

Let G be a simple planar graph. For a face $f \in F(G)$, f is a k(or k^+, or k^-)-face if $d(f) = k$ (or $d(f) \geq k$, or $d(f) \leq k$). For a vertex $v \in V(G)$, v is a k(or k^+, or k^-)-vertex if $d(v) = k$(or $d(v) \geq k$, or $d(v) \leq k$). For a k-vertex v of G, $N(v)$ denotes the set of vertices adjacent to v, $d(v) = |N(v)|$ denotes the degree of v, $N(v) = \{v_i | 1 \leq i \leq k\}$, $D(v) = \sum_{i=1}^{k} d(v_i)$. Let $n_2(v)$ denote the number of 2-neighbors of v. For $f \in F(G)$, we use $b(f)$ to denote the boundary walk of f and write $f = [u_1 u_2 \ldots u_n]$ if $u_1, u_2, \ldots, u_n$ are the vertices on $b(f)$ enumerated in the clockwise. Let $f_1, f_2, \ldots, f_k$ denote the incident faces of v with $f_i = [\ldots v_i v v_{i+1} \ldots](i = 1, 2, \ldots, k - 1), f_k = [\ldots v_k v v_1 \ldots]$. A 3-vertex (or 4-vertex) v is called special small if v is not incident with 4-faces. Let v be a 13^+-vertex, if v is incident with two adjacent $(6, 6^+)$-faces and two adjacent $(7, 7^+)$-faces (or at least three adjacent 8^+-faces), then v is called special large.

Let $G[V']$ be a induced subgraph of G for $V' \subseteq V$. If there exists a mapping $\varphi : V' \rightarrow [k]$ is a (k, r)-coloring of $G[V']$, then φ is called a partial (k, r)-coloring of G about $G[V']$. Suppose that φ is a partial (k, r)-coloring of G. Let $\{\varphi(v)\} = \emptyset$ for $v \in V - V'$. For each $v \in V$, we define the following color set $\varphi[v]$:

$$\varphi[v] = \begin{cases} \{\varphi(v)\}, & |\varphi(N_G(v))| \geq r; \\ \{\varphi(v)\} \bigcup \varphi(N_G(v)), & |\varphi(N_G(v))| < r. \end{cases}$$

Note that $|\varphi[v]| \leq r$ by the definition of $\varphi[v]$. Define the forbidden color set of v by $F(v) = \bigcup_{u \in N(v)} \varphi[u]$, the color set by $[k] = \{1, 2, \cdots, k\}$. We call v can be colored if $|F(v)| < k$ for each $v \in V(G)$.

2   Reducible Configurations

We prove Theorem 1.2 by contradiction. Let G be a counterexample with minimum $|V(G)|+|E(G)|$ of Theorem 1.2. That is to say, G is a planar graph without 3-,5-cycles, and 4-cycle is not adjacent to 7^--cycles, $r \geq 14$, but $\chi_r(G) \geq r + 4$. For any edge e of G, we have $\chi_r(G - e) \leq r + 3$. Then G is a connected planar graph.

Lemma 2.1. G is 2-connected.

Proof. By contradiction, suppose that v is a cut vertex of G, that is to say, there are two connected subgraphs G_1, G_2 such that $G_1 \cap G_2 = \{v\}$ and $G = G_1 \cup G_2$. By the minimality of G, G_i has a (k, r)-coloring $\varphi(i)$ for $i = 1, 2$. Suppose $\varphi_1(v) = \varphi_2(v)$. If $|\varphi_1(N_{G_1}(v)) \bigcup \varphi_2(N_{G_2}(v))| \geq \min\{d_G(v), r\}$, then we define a new coloring $\varphi \colon V(G) \rightarrow [k]$ such that $\varphi(u) = \varphi_i(u)(1 \leq i \leq 2)$ for each $u \in V(G_i)$. Obviously, φ is a (k, r)-coloring of G. If $|\varphi_1(N_{G_1}(v)) \bigcup \varphi_2(N_{G_2}(v))| < \min\{d_G(v), r\}$, then there is a vertex $v_i \in N_{G_i}(v)(i = 1, 2)$ such that $\varphi_1(v_1) = \varphi_2(v_2)$. This means that there is a color $c \in [k] - \varphi_1(N_{G_1}(v)) \bigcup \varphi_2(N_{G_2}(v)) \bigcup \{\varphi_1(v)\}$. Now we can exchange the color c with $\varphi_2(v_2)$. Repeat the steps above until $|\varphi_1(N_{G_1}(v)) \bigcup \varphi_2(N_{G_2}(v))| \geq \min\{d_G(v), r\}$. Thus, we can get a (k, r)-coloring of G, a contradiction. Suppose $\varphi_1(v) \neq \varphi_2(v)$. Now we exchange the color $\varphi_1(v)$ with $\varphi_2(v)$ in G_2 such that $\varphi_1(v) = \varphi_2(v)$ in G. As the same argument above, φ_1, φ_2 can be extended to a (k, r)-coloring of G, a contradiction. □

Lemma 2.2. There is no 2-vertex adjacent to a 2-vertex.

Proof. Otherwise, suppose there is an edge uv such that $d(u) = d(v) = 2$, $N(u) = \{v, u_1\}$ and $N(v) = \{u, v_1\}$. The graph $G' = G - uv$ has a (k, r)-coloring φ. First, we remove the colors of u, v. Since $|F(u)| \leq |\varphi[u_1] \cup \{\varphi(v_1)\}| \leq r + 1 < k$, $|F(v)| \leq |\varphi[v_1] \cup \{\varphi(u_1)\}| \leq r + 1 < k$, we can recolor u, v to get a (k, r)-coloring of G, a contradiction. □

Lemma 2.3. For each $uv \in E(G)$, if $D(u) \leq r + 2$, then $D(v) \geq r + 4$.

Proof. By contradiction, suppose $D(v) \leq r + 3$. Let $G' = G - uv$. Now G' has a (k, r)-coloring. Remove the colors of u, v. Since $|F(v)| \leq D(v) - 1 \leq r + 2$, $|F(u)| \leq D(u) - 1 \leq r + 1$, we can recolor u, v, a contradiction. □

Lemma 2.4. Let v be a 3-vertex of G.

(1) If $d(v_1) = 2$, then $d(v_2) + d(v_3) \geq r + 1$;
(2) If $n_2(v) = 2$, suppose $d(v_1) = d(v_2) = 2$, then both of v_{11}, v_{21} are r^+-vertices, and v is not incident with 4-face.

Proof. (1) By contradiction, suppose $d(v_2) + d(v_3) \leq r$. Let $G' = G - vv_1$. Then G' has a (k, r)-coloring. Remove the colors of v, v_1. Since $|F(v)| \leq D(v) - 1 \leq r + 2$, $|F(u)| \leq D(u) - 1 \leq r + 1$, we can recolor v_1, v to get a (k, r)-coloring of G, a contradiction.

(2) By contradiction, suppose $d(v_{11}) \leq r - 1$. Let $G' = G - vv_1$. Then G' has a (k,r)-coloring. Remove the colors of v_1, v, v_2. It is easy to see that $|F(v)| \leq |\varphi[v_3] \cup \{\varphi(v_{11}), \varphi(v_{21})\}| \leq r + 2$, $|F(v_2)| \leq |\varphi[v_{21}] \cup \{\varphi(v_3)\}| \leq r + 1$, $|F(v_1)| \leq |\varphi[v_{11}] \cup \{\varphi(v_3)\}| \leq r - 1 + 1 = r$. Thus, we can get a (k,r)-coloring of G by recoloring v, v_2, v_1 in turn, a contradiction. As the same argument, we can prove $d(v_{21}) \geq r$.

By contradiction, suppose v is incident with a 4-face f. There are three cases of f. First, assume $f = f_1 = [v_1 v v_2 v_{11}]$. Let $G' = G - vv_1$. Then G' has a (k,r)-coloring. Remove the colors of v_1, v, v_2. It is easy to see that $|F(v)| \leq |\varphi[v_3] \cup \{\varphi(v_{11})\}| \leq r + 1$, $|F(v_2)| \leq |\varphi[v_{11}] \cup \{\varphi(v_3)\}| \leq r + 1$. So we can recolor v, v_2. For v_1, it follows that $|F(v_1)| \leq |\varphi[v_{11}] \cup \{\varphi(v), \varphi(v_3)\}| \leq r + 2$. Therefore, we can get a (k,r)-coloring of G by recoloring v, v_2, v_1 in turn, a contradiction. If $f = f_2 = [v_2 v v_3 v_{21}]$ or $f = f_3 = [v_3 v v_1 v_{11} \cdots]$, then we can obtain a contradiction by a similar argument of the case $f = f_1$. $\qquad \square$

Lemma 2.5. *Let v be a 4-vertex of G.*

(1) If $n_2(v) = 4$, then v is not incident with 4-face, and $d(v_{i1}) \geq r (1 \leq i \leq 4)$;
(2) Suppose $n_2(v) = 3$. Let $d(v_i) = 2 (1 \leq i \leq 3)$.
  (a) If v is incident with two 4-faces, then $d(v_{11}) \geq r$, $d(v_{31}) \geq r$ and $d(v_4) \geq r - 1$;
  (b) If v is incident with exactly one 4-face, then $d(v_4) \geq r - 2$;
(3) If $n_2(v) = 2$ and v is incident with two 4-faces, then $d(v_j) + d(v_l) \geq r + 1$ for two 3^+-neighbors v_j, v_l.

Proof. (1) Let $n_2(v) = 4$. By contradiction, suppose that $f_1 = [v_1 v v_2 v_{11}]$ is a 4-face. Let $G' = G - vv_1$. Then G' admits a (k,r)-coloring. Now we remove the colors of v, v_1. It is easily can be seen that $|F(v_1)| \leq |\varphi[v_{11}] \cup \{\varphi(v_3), \varphi(v_4)\}| \leq r + 2$, $|F(v)| \leq |\varphi[v_3] \cup \varphi[v_4] \cup \{\varphi(v_2), \varphi(v_{11})\}| \leq 2 + 2 + 2 = 6$. Thus, we can recolor v_1, v to get a (k,r)-coloring of G, a contradiction.

Now we show $d(v_{i1}) \geq r (1 \leq i \leq 4)$. By contradiction, w.l.g.o, suppose $d(v_{11}) \leq r - 1$. Let $G' = G - vv_1$. Then G' admits a (k,r)-coloring. Now we remove the colors of v, v_1. Since $|F(v_1)| \leq |\varphi[v_{11}] \cup \{\varphi(v_2), \varphi(v_3), \varphi(v_4)\}| \leq r - 1 + 3 = r + 2$, $|F(v)| \leq |\varphi[v_2] \cup \varphi[v_3] \cup \varphi[v_4] \cup \{\varphi(v_{11})\}| \leq 6 + 1 = 7$, we can recolor v_1, v, $|\varphi[v_{11}]| \leq r - 3 + 1 = r - 2$.

(2) Suppose $n_2(v) = 3$. Let $d(v_i) = 2 (1 \leq i \leq 3)$.

(a) Assume that v is incident with two 4-faces $f_1 = [v_1 v v_2 v_{11}]$, $f_3 = [v_3 v v_4 v_{31}]$. By contradiction, suppose $d(v_{11}) \leq r - 1$. By the minimality of G, $G' = G - vv_1$ has a (k,r)-coloring. Now remove the colors of v, v_2, v_3, v_1. It follows from $d(v_{11}) \leq r - 1$ that $|\varphi[v_{11}]| \leq r - 3 + 1 = r - 2$. This implies that $|F(v)| \leq |\varphi[v_4] \cup \{\varphi(v_{11})\}| \leq r + 1$, $|F(v_3)| \leq |\varphi[v_{31}]| \leq r$, $|F(v_2)| \leq |\varphi[v_{11}] \cup \{\varphi(v_4)\}| \leq r - 2 + 1 = r - 1$, $|F(v_1)| \leq |\varphi[v_{11}] \cup \{\varphi(v_4)\}| \leq r - 2 + 1 = r - 1$. Therefore, we can recolor v, v_3, v_2, v_1 in turn, a contradiction. As the same argument before, it is easy to prove that $d(v_{31}) \geq r$ and $d(v_4) \leq r - 2$.

(b) If v is incident with exactly one 4-face f. By contradiction, suppose $d(v_4) \leq r - 3$. First, we consider the case when $f = f_1 = [v_1 v v_2 v_{11}]$. Let $G' = G - vv_1$. Then G' admits a (k,r)-coloring. Now we remove the colors of v, v_2, v_3, v_1.

It is easy to see that $|F(v_2)| \leq |\varphi[v_{11}] \cup \{\varphi(v_4)\}| \leq r+1$, $|F(v_3)| \leq |\varphi[v_{31}] \cup \{\varphi(v_4)\}| \leq r+1$. We can recolor v_3, v_2 in turn. Then for v_1, v, since $|F(v_1)| \leq |\varphi[v_{11}] \cup \{\varphi(v_3), \varphi(v_4)\}| \leq r+2$, $|F(v)| \leq |\varphi[v_4] \cup \varphi[v_3] \cup \varphi[v_2]| \leq r-3+2+2 = r+1$, we can recolor v_1, v in sequence, a contradiction. Now suppose $f = f_4 = [v_1vv_4v_{11}]$. We can show that $d(v_4) \geq r-2$ by a similar argument before.

(3) At first, suppose that the two 2-neighbors of v are incident with one common face. W.l.o.g, let $d(v_i) = 2(i = 1,2)$. Suppose that the two incident 4-faces of v are $f_1 = [v_1vv_2v_{11}]$ and $f_3 = [v_3vv_4v_{31}]$. By contradiction, suppose $d(v_3) + d(v_4) \leq r$. Let $G' = G - vv_1$. Then G' admits a (k,r)-coloring. Now we remove the colors of v, v_1, v_2. First, we recolor v_1 since $|F(v_1)| \leq |\varphi[v_{11}] \cup \{\varphi(v_3), \varphi(v_4)\}| \leq r+2$. Then due to $|F(v_2)| \leq |\varphi[v_{11}] \cup \{\varphi(v_3), \varphi(v_4)\}| \leq r+2$, $|F(v)| \leq |\varphi[v_1] \cup \varphi[v_3] \cup \varphi[v_4]| \leq 2 + d(v_3) + d(v_4) - 1 \leq r+1$, we can recolor v_2, v in turn, a contradiction. On the other hand, suppose that the two incident 4-faces of v are $f_2 = [v_2vv_3v_{21}]$ and $f_4 = [v_4vv_1v_{11}]$. As the same argument above, we can prove $d(v_3) + d(v_4) \geq r+1$.

Then, suppose that the two 2-neighbors of v are incident with two different faces. W.l.o.g, let $d(v_i) = 2$ for $i = 1,3$. It is easy to to prove that $d(v_3) + d(v_4) \geq r+1$ by a similar argument before. $\square$

Lemma 2.6. *If a 5(5)-vertex v is incident with two 4-faces $f_1 = [v_1vv_2v_{11}]$ and $f_3 = [v_3vv_4v_{31}]$, then $d(v_{i1}) \geq r(i = 1,3)$.*

Proof. Suppose that the lemma is not true. Assume $d(v_{11}) \leq r-1$. Let $G' = G - vv_1$. Then G' admits a (k,r)-coloring. Now we remove the colors of $v, v_1, v_2, v_3, v_4, v_5$. Note that $|F(v_3)| \leq |\varphi[v_{31}]| \leq r$, $|F(v_4)| \leq |\varphi[v_{31}]| \leq r$, $|F(v_5)| \leq |\varphi[v_{51}]| \leq r$, $|F(v_2)| \leq |\varphi[v_{11}]| \leq r-2$, $|F(v_1)| \leq |\varphi[v_{11}]| \leq r-2$, $|F(v)| \leq |\{\varphi(v_{11}), \varphi(v_{31}), \varphi(v_{51})\}| \leq 3$. This means that $v_3, v_4, v_5, v_1, v_2, v$ can be recolored in turn, a contradiction. $\square$

Lemma 2.7. *Let v be a 6-vertex of G.*

(1) *If v is a 6(6)-vertex incident with three 4-faces $f_1 = [v_1vv_2v_{11}]$, $f_3 = [v_3vv_4v_{31}]$, $f_5 = [v_5vv_6v_{51}]$, where v_{i1} are all $(r-1)^-$-vertices, then $d(v_{i1}) = r-1$ for $i = 1,3,5$;*

(2) *If v is a 6(5)-vertex incident with three 4-faces $f_1 = [v_1vv_2v_{11}]$, $f_3 = [v_3vv_4v_{31}]$, $f_5 = [v_5vv_6v_{51}]$, where $d(v_6) \leq 8$, then $d(v_{i1}) \geq r-1$ for $i = 1,3$.*

Proof. (1) Suppose that the lemma is not true. W.l.o.g., assume $d(v_{51}) \leq r-2$. Let $G' = G - vv_1$. Then G' admits a (k,r)-coloring. Now we remove the colors of $v, v_1, v_2, v_3, v_4, v_5, v_6$. Since $|F(v_1)| \leq |\varphi[v_{11}]| \leq r-1$, $|F(v_2)| \leq |\varphi[v_{11}]| \leq r-1$, $|F(v_3)| \leq |\varphi[v_{31}]| \leq r-1$, $|F(v_4)| \leq |\varphi[v_{31}]| \leq r-1$, $|F(v_5)| \leq |\varphi[v_{51}]| \leq r-3$, $|F(v_6)| \leq |\varphi[v_{51}]| \leq r-3$, $|F(v)| \leq |\{\varphi(v_{11}), \varphi(v_{31}), \varphi(v_{51})\}| = 3$, we can recolor $v_1, v_2, v_3, v_4, v_5, v_6, v$ in sequence. Thus, we can get a (k,r)-coloring of G, a contradiction.

(2) W.l.o.g., assume $d(v_{11}) \leq r-2$. Let $G' = G - vv_1$. Then G' admits a (k,r)-coloring. Now we remove the colors of $v, v_1, v_2, v_3, v_4, v_5$. Note that $|F(v_3)| \leq$

$|\varphi[v_{31}] \cup \{\varphi(v_6)\}| \leq r+1$, $|F(v_4)| \leq |\varphi[v_{31}] \cup \{\varphi(v_6)\}| \leq r+1$, $|F(v_5)| \leq |\varphi[v_{51}]| \leq r$, $|F(v_1)| \leq |\varphi[v_{11}] \cup \{\varphi(v_6)\}| \leq r-3+1 = r-2$, $|F(v_2)| \leq |\varphi[v_{11}] \cup \{\varphi(v_6)\}| \leq r-3+1 = r-2$, $|F(v)| \leq |\varphi[v_6] \cup \{\varphi(v_{11}), \varphi(v_{31})\}| \leq 8+2 = 10$. This means that we can get a (k,r)-coloring of G by recoloring $v_3, v_4, v_5, v_1, v_2, v$ in turn, a contradiction.   □

3   Proof of Theorem 1.2

In this section, we prove Theorem 1.2 by contradiction. Let G be a minimal counterexample of Theorem 1.2. By Euler's formula $|V(G)|+|F(G)|-|E(G)| = 2$, and the fact $\sum_{v \in V(G)} d(v) = \sum_{f \in F(G)} d(f) = 2|E(G)|$, we have

$$\sum_{v \in V(G)} (d(v) - 4) + \sum_{f \in F(G)} (d(f) - 4) = -8. \tag{1}$$

We define the weight function $\omega: V(G) \cup F(G) \to \mathbb{N}$ by $\omega(x) = d(x) - 4$ for each $x \in V(G)$, and $\omega(x) = d(x) - 4$ for each $x \in F(G)$, then we obtain

$$\sum_{x \in V(G) \cup F(G)} \omega(x) = -8.$$

We will design several discharging rules and reassign weight accordingly. Once the discharging is finished, a new weight function ω' is produced and we will show that for each $x \in V(G) \cup F(G)$, $\omega'(x) \geq 0$. However, the total sum of weights is unchanged. This leads to the following contradiction:

$$0 \leq \sum_{x \in V(G) \cup F(G)} \omega'(x) = \sum_{x \in V(G) \cup F(G)} \omega(x) = -8. \tag{2}$$

Let $\tau(u \mapsto v)$ denote the value transferring from u to v according to the discharging rules. We define the following discharging rules:

- **R1** Every 6^+-face sends $\frac{d(f)-4}{d(f)}$ to each incident vertex.
- **R2** Let u be a 2-vertex incident with two faces f_1, f_2, and $uv \in E(G)$. If $d(f_1) = 4$ and $d(f_2) \geq 8$, then $\tau(v \to u) = \frac{3}{4}$; If $d(f_1) = 6$ and $6 \leq d(f_2) \leq 7$, then $\tau(v \to u) = \frac{2}{3}$; If $d(f_1) = 6$ and $d(f_2) \geq 8$ then $\tau(v \to u) = \frac{7}{12}$; If $d(f_1) = 7$ and $d(f_2) \geq 7$, then $\tau(v \to u) = \frac{4}{7}$; If $d(f_1) \geq 8$ and $d(f_2) \geq 8$, then $\tau(v \to u) = \frac{1}{2}$.
- **R3** Suppose that u and v have a common 2-neighbor with $3 \leq d(u) \leq 6$. If $d(v) \geq 13$ and at least one of u, v is not special vertex, then $\tau(v \to u) = \frac{11}{52}$; If $d(v) \geq 13$ and u, v are special small vertex and special large vertex, respectively, then $\tau(v \to u) = \frac{1}{3}$; If $d(v) = 12$, then $\tau(v \to u) = \frac{1}{6}$; If $d(v) = 11$, then $\tau(v \to u) = \frac{7}{44}$.
- **R4** Let u be a 3-vertex, where $uv \in E(G)$. If $d(v) = 6$, then $\tau(v \to u) = \frac{1}{4}$; If $d(v) = 7$, then $\tau(v \to u) = \frac{3}{8}$; If $d(v) = 8$, then $\tau(v \to u) = \frac{1}{2}$; If $d(v) \geq 9$, then $\tau(v \to u) = \frac{3}{4}$.

- **R5** Let u be a 4-vertex and $uv \in E(G)$. If $6 \leq d(v) \leq 7$, then $\tau(v \to u) = \frac{1}{4}$; If $d(v) = 8$, then $\tau(v \to u) = \frac{1}{2}$; If $d(v) \geq 9$, then $\tau(v \to u) = \frac{3}{4}$.
- **R6** If u is a 5-vertex and $d(v) \geq 8$, where $uv \in E(G)$, then $\tau(v \to u) = \frac{1}{2}$.
- **R7** If u is a 6-vertex and $d(v) \geq 9$, where $uv \in E(G)$, then $\tau(v \to u) = \frac{1}{4}$.

First, we prove that $\omega'(f) = \omega(f) - \frac{d(f)-4}{d(f)} \times d(f) = 0$ by R1 for each $f \in F(G)$ since $d(f) \geq 4$. Now we check $\omega'(v) \geq 0$ for each $v \in V(G)$.

1. $d(v) = 2$, and then $\omega(v) = -2$.

Assume that v is incident with f_1 and f_2. W.l.o.g., suppose $d(f_1) \leq d(f_2)$. If f_1 is a 4-face, f_2 is a 8^+-face, then we have that $\omega'(v) \geq -2 + \frac{1}{2} + \frac{3}{4} \times 2 = 0$ by R1 and R2; If $d(f_1) = 6$ and $6 \leq d(f_2) \leq 7$, then we can obtain that $\omega'(v) \geq -2 + \frac{1}{3} \times 2 + \frac{2}{3} \times 2 = 0$ by R1 and R2; If $d(f_1) = 6$ and $d(f_2) \geq 8$, then $\omega'(v) \geq -2 + \frac{1}{2} + \frac{7}{12} \times 2 = 0$ by R1 and R2; If $d(f_1) = 7$ and $d(f_2) \geq 7$, then $\omega'(v) \geq -2 + \frac{3}{7} \times 2 + \frac{4}{7} \times 2 = 0$ by R1 and R2; If $d(f_1) \geq 8$ and $d(f_2) \geq 8$, then $\omega'(v) \geq -2 + \frac{1}{2} \times 2 + \frac{1}{2} \times 2 = 0$ by R1 and R2.

2. $d(v) = 3$ and then $\omega(v) = -1$. Note that $n_2(v) \leq 2$ by Lemma 2.4(1).

2.1. $n_2(v) = 2$.

By Lemma 2.4, v is not incident with 4-face, and $d(v_3) \geq r - 1 \geq 12$, which implies that v is special small. If $f_1 = [vv_1v_{11} \ldots v_{21}v_2v]$ is 6-face or 7-face, then the faces incident with v are all 6^+faces, and v_{11}, v_{21} are both special large. Using R3, $\tau(v_{i1} \to v) = \frac{1}{3}(i = 1, 2)$. By R1, R2, R4, we can acquire that $\omega'(v) \geq -1 - \frac{2}{3} \times 2 + \frac{1}{3} \times 3 + \frac{1}{3} \times 2 + \frac{3}{4} > 0$. If $f_1 = [vv_1v_{11} \ldots v_{21}v_2v]$ is a 8^+-face, then $\tau(v_{i1} \to v) \geq \frac{11}{52}(i = 1, 2)$ by R3. This together with R1, R2, R4 yields that $\omega'(v) \geq -1 - \frac{7}{12} \times 2 + \frac{1}{3} \times 2 + \frac{1}{2} + \frac{11}{52} \times 2 + \frac{3}{4} = \frac{9}{52} > 0$.

2.2. $n_2(v) = 1$.

It follows from Lemma 2.4(1) that $d(v_2) + d(v_3) \geq r + 1$. This implies that $(d(v_2), d(v_3)) \in \{(3, 11^+), (4, 10^+), (5, 9^+), (6, 8^+), (7^+, 7^+)\}$. So the incident faces transfer $\min\{\frac{1}{2} \times 2, \frac{1}{3} \times 3\} = 1$ to v. By R1, R2, R4, $\omega'(v) \geq -1 - \frac{3}{4} + 1 + \min\{\frac{3}{4}, \frac{1}{4} + \frac{1}{2}, \frac{3}{8} \times 2\} = 0$.

2.3. $n_2(v) = 0$.

If v is not incident with 4-face, then $\omega'(v) \geq -1 + \frac{1}{3} \times 3 = 0$ by R1. If v is incident with one 4-face, then the other two faces incident with v are 8^+-faces, which implies that $\omega'(v) \geq -1 + \frac{1}{2} \times 2 = 0$ by R1.

3. $d(v) = 4$ and then $\omega(v) = 0$.

3.1. $n_2(v) = 4$.

Note that 4(4)-vertex is not incident with 4-face by Lemma 2.5(1), which implies that v is a special small vertex. Suppose v is incident with a 6-face or 7-face $f_1 = [vv_1v_{11} \ldots v_{21}v_2v]$, the other three incident faces of v are written by f_2, f_3, f_4. Clearly, v_{11} and v_{21} are special large vertices. By R3, $\tau(v_{i1} \to v) = \frac{1}{3}(i = 1, 2)$. If $6 \leq d(f_3) \leq 7$, then v_{31} and v_{41} are special large vertices. This means that $\tau(v_{i1} \to v) = \frac{1}{3}(i = 3, 4)$ by R3. Therefore, $\omega'(v) \geq 0 - \frac{2}{3} \times 4 + \frac{1}{3} \times 4 + \frac{1}{3} \times 4 = 0$ by R1, R2, R3. If $d(f_3) \geq 8$, then we have that $\tau(v_{i1} \to v) \geq \frac{11}{52}(i = 3, 4)$ by R3. Thus, $\omega'(v) \geq 0 - \frac{2}{3} \times 2 - \frac{7}{12} \times 2 + \frac{1}{3} \times 3 + \frac{1}{2} + \frac{1}{3} \times 2 + \frac{11}{52} \times 2 > 0$ by R1, R2, R3. If the faces incident with v are all 8^+-faces, then $\omega'(v) \geq 0 - \frac{1}{2} \times 4 + \frac{1}{2} \times 4 + \frac{11}{52} \times 4 > 0$ by R1, R2.

3.2. $n_2(v) = 3$.

Suppose that the 2-neighbors of v are v_i $(1 \leq i \leq 3)$. If v is incident with two 4-faces, then $d(v_{11}) \geq r$, $d(v_{31}) \geq r$, $d(v_4) \geq r-1$ by Lemma 2.5(2). This implies that $\omega'(v) \geq 0 - \frac{3}{4} \times 3 + \frac{1}{2} \times 2 + \frac{11}{52} \times 3 + \frac{3}{4} > 0$ by $R1, R2, R3, R5$. If v is incident with one 4-face, then $d(v_4) \geq r - 2$ by Lemma 2.5(2). Using $R1, R2, R3, R5$, we have $\omega'(v) \geq 0 - \frac{3}{4} \times 2 - \frac{7}{12} + \frac{1}{2} \times 2 + \frac{1}{3} + \frac{3}{4} = 0$. If v is not incident with 4-face and $d(v_4) \geq 10$, then $\omega'(v) \geq 0 - \frac{2}{3} \times 3 + \frac{1}{3} \times 4 + \frac{3}{4} > 0$ by $R1, R2, R5$. If v is not incident with 4-face and $d(v_4) \leq 9$, then we can obtain $d(v_{i1}) \geq r(1 \leq i \leq 3)$ by Lemma 2.3. If v is incident with at least one l-face for $l = 6, 7$, then there is at least one special large vertex in $\{v_{i1}|1 \leq i \leq 3\}$, and v is a special small vertex. By $R1, R2, R3, R5$, we can acquire $\omega'(v) \geq 0 - \frac{2}{3} \times 3 + \frac{1}{3} \times 4 + \frac{1}{3} + \frac{11}{52} \times 2 > 0$. Otherwise, the incident faces of v are all 8^+-faces, which together with $R1, R2$ yields that $\omega'(v) \geq 0 - \frac{2}{3} \times 3 + \frac{1}{2} \times 4 > 0$.

3.3. $n_2(v) = 2$.

W.l.o.g., suppose that the 2-neighbors of v are v_1, v_2. If v is incident with two 4-faces, then $d(v_3) + d(v_4) \geq r + 1$ by Lemma 2.5(3). Clearly, $(d(v_3), d(v_4)) \in \{(3, 11^+), (4, 10^+), (5, 9^+), (6, 8^+), (7^+, 7^+)\}$. By $R1, R2, R5$, $\omega'(v) \geq 0 - \frac{3}{4} \times 2 + \frac{1}{2} \times 2 + \min\{\frac{3}{4}, \frac{1}{2} + \frac{1}{4}, \frac{1}{4} \times 2\} = 0$. If v is incident with one 4-face, and $d(v_3) + d(v_4) \leq r - 2$, then $d(v_{i1}) \geq r \geq 13(i = 1, 2)$ by Lemma 2.3. This means that $\tau(v_{i1} \rightarrow v) \geq \frac{11}{52}(i = 1, 2)$ by $R3$. If v is incident with one 4-face, and $d(v_3) + d(v_4) \geq r - 1 \geq 12$, then there is at least one 6^+-vertex in $\{v_3, v_4\}$. Using $R1, R2, R3, R5$, we obtain $\omega'(v) \geq 0 - \frac{3}{4} \times 2 + \frac{1}{2} \times 2 + \frac{1}{3} + \min\{\frac{1}{4}, \frac{11}{52} \times 2\} > 0$. Otherwise, v is not incident with 4-face. It is easy to see that $\omega'(v) \geq 0 - \frac{2}{3} \times 2 + \frac{1}{3} \times 4 = 0$ by $R1, R2$.

3.4. $n_2(v) \leq 1$.

If v is incident with 4-faces, then $\omega'(v) \geq 0 - \frac{3}{4} + \frac{1}{2} \times 2 > 0$ by $R1, R2$. Otherwise, v is not incident with 4-face, then $\omega'(v) \geq 0 - \frac{2}{3} + \frac{1}{3} \times 4 > 0$ by $R1, R2$.

4. $d(v) = 5$ and then $\omega(v) = 1$.

4.1. $n_2(v) = 5$.

Since $D(v) = 10 \leq r + 2$, it follows from Lemma 2.3 that $D(v_i) \geq r + 4$, $d(v_{i1}) \geq r + 4 - 5 = r - 1(1 \leq i \leq 5)$. If v is incident with two 4-faces $f_1 = [v_1vv_2v_{11}]$, $f_3 = [v_3vv_4v_{31}]$, then $d(v_{i1}) \geq r(i = 1, 3)$ by Lemma 2.6. Using $R1, R2, R3$ we have $\omega'(v) \geq 1 - \frac{3}{4} \times 4 - \frac{1}{2} + \frac{1}{2} \times 3 + \frac{11}{52} \times 4 + \frac{1}{6} > 0$. If v is incident with one 4-face, then there are at least two 8^+-faces incident with v. So $\omega'(v) \geq 1 - \frac{3}{4} \times 2 - \frac{7}{12} \times 2 - \frac{2}{3} + \frac{1}{2} \times 2 + \frac{1}{3} \times 2 + \frac{1}{6} \times 5 > 0$ by $R1, R2, R3$. Otherwise, v is not incident with 4-face. Namely, the incident faces of v are all 6^+-faces, which implies that $\omega'(v) \geq 1 - \frac{2}{3} \times 5 + \frac{1}{3} \times 5 + \frac{1}{6} \times 5 > 0$ by $R1, R2, R3$.

4.2. $n_2(v) = 4$.

Let v_5 be a 3^+-neighbor of v. If $d(v_5) \leq r - 6$, then $D(v_{i1}) \geq r + 4 - 5 = r - 1(1 \leq i \leq 4)$ by Lemma 2.3. Thus, by $R1, R2, R3$, we have that $\omega'(v) \geq 1 - \frac{3}{4} \times 4 + \frac{1}{2} \times 3 + \frac{1}{6} \times 4 > 0$. Otherwise, $d(v_5) \geq r - 5 \geq 8$, it is easy to see that $\omega'(v) \geq 1 - \frac{3}{4} \times 4 + \frac{1}{2} \times 3 + \frac{1}{2} = 0$ by $R1, R2, R6$.

4.3. $n_2(v) \leq 3$. It is easy to show that $\omega'(v) \geq 1 - \frac{3}{4} \times 3 + \frac{1}{2} \times 3 = \frac{1}{4} > 0$ by $R1, R2$.

5. $d(v) = 6$ and then $\omega(v) = 2$.

5.1. $n_2(v) = 6$.

Since $D(v) = 12 \leq r + 2$, it follows from Lemma 2.3 that $D(v_i) \geq r + 4$, $d(v_{i1}) \geq r + 4 - 6 = r - 2(1 \leq i \leq 6)$. If v is incident with three 4-faces $f_1 = [v_1vv_2v_{11}]$, $f_3 = [v_3vv_4v_{31}]$, and $f_5 = [v_5vv_6v_{51}]$, in which $d(v_{i1}) \leq r - 1$ for $i \in \{1, 3, 5\}$, then $d(v_{i1}) = r - 1$ by Lemma 2.7(1). By $R1, R2, R3$, we have that $\omega'(v) \geq 2 - \frac{3}{4} \times 6 + \frac{1}{2} \times 3 + \frac{1}{6} \times 6 = 0$. If v is incident with three 4-faces, but there is at least one r^+-vertex in $\{v_{i1} | i = 1, 3, 5\}$, then we can show that $\omega'(v) \geq 2 - \frac{3}{4} \times 6 + \frac{1}{2} \times 3 + \frac{11}{52} \times 2 + \frac{7}{44} \times 4 > 0$ by $R1, R2, R3$. If v is incident with at most two 4-faces, then it is apparent from $R1, R2, R3$ that $\omega'(v) \geq 2 - \frac{3}{4} \times 4 - \frac{7}{12} \times 2 + \frac{1}{2} \times 3 + \frac{1}{3} + \frac{7}{44} \times 6 > 0$.

5.2. $n_2(v) = 5$.

Let v_6 be a 3^+-neighbor of v. If $d(v_6) \leq 8$ and v is incident with three 4-faces $f_1 = [v_1vv_2v_{11}]$, $f_3 = [v_3vv_4v_{31}]$, and $f_5 = [v_5vv_6v_{51}]$, then $d(v_{i1}) \geq r - 1 (i = 1, 3)$ by Lemma 2.7(2). Thus, we can obtain that $\omega'(v) \geq 2 - \frac{3}{4} \times 5 - \frac{1}{4} + \frac{1}{2} \times 3 + \frac{1}{6} \times 4 = \frac{1}{6} > 0$ by $R1, R2, R3, R4$. If $d(v_6) \leq 8$ and v is incident with at most two 4-faces, then it is apparent from $R1, R2, R3, R4$ that $\omega'(v) \geq 2 - \frac{3}{4} \times 4 - \frac{7}{12} - \frac{1}{4} + \frac{1}{2} \times 3 + \frac{1}{3} = 0$. If $d(v_6) \geq 9$, then we have that $\omega'(v) \geq 2 - \frac{3}{4} \times 5 + \frac{1}{2} \times 3 + \frac{1}{4} = 0$ by $R1, R2, R7$.

5.3. $n_2(v) \leq 4$. It is easy to show that $\omega'(v) \geq 2 - \frac{3}{4} \times 4 - \frac{1}{4} \times 2 + \frac{1}{2} \times 3 = 0$ by $R1, R2, R4$.

6. $d(v) = 7$ and then $\omega(v) = 3$.

If $n_2(v) = 7$, then it follows from $R1, R2$ that $\omega'(v) \geq 3 - \frac{3}{4} \times 6 - \frac{1}{2} + \frac{1}{2} \times 4 = 0$. Otherwise, $n_2(v) \leq 6$. by $R1, R2$ and $R4$, we can obtain that $\omega'(v) \geq 3 - \frac{3}{4} \times 6 - \frac{3}{8} + \frac{1}{2} \times 4 > 0$.

7. $d(v) = 8$ and then $\omega(v) = 4$.

Suppose that the number of incident 4-faces of v is $t(t \leq 4)$. This means that v is incident with at least t 8^+-faces, at most $d(v) - 2t$ l-faces for $l = 6, 7$. From $R1 \sim R6$, we can acquire $\omega'(v) \geq 4 - \frac{3}{4} \times 2t - \frac{2}{3} \times (8 - 2t) + \frac{1}{2} \times t + \frac{1}{3} \times (8 - 2t) \geq \frac{4}{3} - \frac{1}{3}t \geq 0$.

8. $d(v) = 9$ and then $\omega(v) = 5$.

Suppose that the number of incident 4-faces of v is $t(1 \leq t \leq 4)$. This implies that v is incident with at least $t + 1$ 8^+-faces, at most $d(v) - 2t - 1$ l-faces for $l = 6, 7$. By $R1 \sim R7$, it is easy to show that $\omega'(v) \geq d(v) - 4 - \frac{3}{4} \times d(v) + \frac{1}{2} \times (t + 1) + \frac{1}{3} \times (d(v) - 2t - 1) \geq \frac{7}{12} \times d(v) - \frac{1}{6} \times t - 4 + \frac{1}{6} > 0$. On the other hand, v is not incident with 4-faces, which means that the incident faces of v are all 6^+-faces. It follows from $R1 \sim R7$ that $\omega'(v) \geq 5 - \frac{2}{3} \times 9 + \frac{1}{3} \times 9 > 0$.

9. $d(v) = 10$ and then $\omega(v) = 6$.

Suppose that the number of incident 4-faces of v is $t(t \leq 5)$. It is easy to see that v is incident with at least t 8^+-faces, at most $d(v) - 2t$ l-faces for $l = 6, 7$. Thus, we can obtain that $\omega'(v) \geq d(v) - 4 - \frac{3}{4} \times d(v) + \frac{1}{2} \times t + \frac{1}{3} \times (d(v) - 2t) \geq \frac{7}{12} d(v) - \frac{1}{6} \times t - 4 \geq 0$ by $R1 \sim R7$.

10. $d(v) = 11$ and then $\omega(v) = 7$.

Suppose there are $t(1 \leq t \leq 5)$ 4-faces incident with v. It is easy to see that v is incident with at least $t + 1$ 8^+-faces, at most $11 - 2t - 1$ l-faces for $l = 6, 7$. By $R1 \sim R7$, we can arrive at $\omega'(v) \geq 7 - (\frac{3}{4} + \frac{7}{44}) \times 2t - (\frac{2}{3} + \frac{7}{44}) \times (11 - 2t) + \frac{1}{2} \times (t + 1) + \frac{1}{3} \times (11 - 2t - 1) \geq \frac{7}{4} - \frac{1}{3}t > 0$. If v is not incident with

4-faces, then the incident faces of v are all 6^+-faces. Using $R1 \sim R7$, we have $\omega'(v) \geq 7 - (\frac{2}{3} + \frac{7}{44}) \times 11 + \frac{1}{3} \times 11 > 0$.

11. $d(v) = 12$ and then $\omega(v) = 8$.

Suppose that there are $t(1 \leq t \leq 6)$ 4-faces incident with v. Clearly, v is incident with at least t 8^+-faces, at most $12 - 2t$ l-faces for $l = 6, 7$. Thus, we can obtain that $\omega'(v) \geq 8 - (\frac{3}{4} + \frac{1}{6}) \times 2t - (\frac{2}{3} + \frac{1}{6}) \times (12 - 2t) + \frac{1}{2} \times t + \frac{1}{3} \times (12 - 2t) \geq 2 - \frac{1}{3}t \geq 0$ by $R1 \sim R7$.

12. $d(v) \geq 13$ and then $\omega(v) = d(v) - 4$.

12.1. Suppose that v is not special vertex, there are $t(1 \leq t \leq \lfloor \frac{d(v)}{2} \rfloor)$ 4-faces incident with v. If $d(v)$ is even, then v is incident with at least t 8^+-faces, at most $d(v) - 2t$ l-faces for $l = 6, 7$. Using $R1 \sim R7$, we can obtain that $\omega'(v) \geq d(v) - 4 - (\frac{3}{4} + \frac{11}{52}) \times 2t - (\frac{2}{3} + \frac{11}{52}) \times (d(v) - 2t) + \frac{1}{2} \times t + \frac{1}{3} \times (d(v) - 2t) \geq \frac{15}{52} d(v) - 4 \geq 0$. If $d(v)$ is odd, then v is incident with at least $t + 1$ 8^+-faces for $t \leq \frac{d(v)-1}{2}$, at most $d(v) - 2t - 1$ l-faces for $l = 6, 7$. This implies that $\omega'(v) \geq d(v) - 4 - (\frac{3}{4} + \frac{11}{52}) \times 2t - (\frac{2}{3} + \frac{11}{52}) \times (d(v) - 2t) + \frac{1}{2} \times (t + 1) + \frac{1}{3} \times (d(v) - 2t - 1) \geq \frac{15}{52} d(v) - \frac{11}{3} > 0$ by $R1 \sim R7$. Otherwise, v is not incident with 4-faces, then the incident faces of v are all 6^+-faces. By $R1 \sim R7$, it follows that $\omega'(v) \geq d(v) - 4 - (\frac{2}{3} + \frac{11}{52}) \times d(v) + \frac{1}{3} \times d(v) \geq \frac{71}{156} d(v) - 4 > 0$.

12.2. Suppose that v is a special vertex. It follows from the definition of special vertex that v is incident with at most $\lfloor \frac{d(v)}{2} \rfloor - 1$ 4-faces. First, assume that v is incident with $t(1 \leq t \leq \lfloor \frac{d(v)}{2} \rfloor - 1)$ 4-faces. Let v_1 be a 2-neighbor of v in which v_{11} be another neighbor of v_1. If v_1 is incident with a 4-face, then v_{11} is not a special small vertex, which implies that v transfers $\frac{3}{4} + \frac{11}{52}$ through its incident edges by $R2, R3$; If v_1 is incident with a 4-face, then v transfers at most $\frac{2}{3} + \frac{1}{3}$ through vv_1 by $R2, R3$. Let $f_1, f_2, \cdots, f_{d(v)}$ be the incident faces of v enumerated in the clockwise. Since v is a special large vertex and $t \geq 1$, there exists $i \neq j$ such that f_i and f_{i+1}, f_j and f_{j+1} are both 6^+-face and 8^+-face (or 8^+-face and 6^+-face). If $d(v)$ is even, then v is incident with at least t 8^+-faces, at most $d(v) - 2t$ l-faces for $l = 6, 7$. By $R1 \sim R7$, we have that $\omega'(v) \geq d(v) - 4 - (\frac{3}{4} + \frac{11}{52}) \times 2t - (\frac{7}{12} + \frac{1}{3}) \times 2 - (\frac{2}{3} + \frac{1}{3}) \times (d(v) - 2t - 2) + \frac{1}{2} \times t + \frac{1}{3} \times (d(v) - 2t) \geq \frac{15}{52} d(v) - \frac{146}{39} \geq 0$. If $d(v)$ is odd, then v is incident with at least $t + 1$ 8^+-faces for $t \leq \lfloor \frac{d(v)}{2} \rfloor - 1 = \frac{d(v)-3}{2}$, at most $d(v) - 2t - 1$ l-faces for $l = 6, 7$. This means that $\omega'(v) \geq d(v) - 4 - (\frac{3}{4} + \frac{11}{52}) \times 2t - (\frac{7}{12} + \frac{1}{3}) \times 2 - (\frac{2}{3} + \frac{1}{3}) \times (d(v) - 2t - 2) + \frac{1}{2} \times (t + 1) + \frac{1}{3} \times (d(v) - 2t - 1) \geq \frac{15}{52} d(v) - \frac{421}{156} > 0$ by $R1 \sim R7$. On the other hand, v is not incident with 4-faces, which means that the incident faces of v are all 6^+-faces. By $R1 \sim R7$, it follows that $\omega'(v) \geq d(v) - 4 - (\frac{2}{3} + \frac{1}{3}) \times d(v) + \frac{1}{3} \times d(v) \geq \frac{1}{3} d(v) - 4 > 0$.

Thus, we have $\omega'(x) \geq 0$ for each $x \in V(G) \cup F(G)$, which is a contradiction. This completes the proof of Theorem 1.2.

References

1. Borodin, O.V., Broersma, H.J., Glebov, A., Heuvel, J.V.D.: Stars and bunches in planar graphs. Part II: General planar graphs and colourings. CDAM researches report 2002-05 (2002)

2. Bu, Y.H., Cao, J.J.: 2-distance coloring of planar graph. Discrete Appl. Math. **13**, 2150007 (2021)
3. La, H., Montassier, M., Pinlou, A., et al.: r-hued (r+1)-coloring of planar graphs with girth at least 8 for r≥ 9. Eur. J. Comb. **91**, 103219 (2021)
4. Montgomery, B.: Dynamic Coloring of Graphs. West Virginia University, Morgantwon (2001)
5. Song, H.M., Lai, H.J.: Upper bounds of r-hued colorings of planar graphs. Discrete Appl. Math. **243**, 262–269 (2018)
6. Song, H.M., Lai, H.J., Wu, J.L.: On r-hued coloring of planar graphs with girth at least 6. Discrete Appl. Math. **198**, 251–263 (2016)
7. Wegner, G.: Graphs with given diameter and a coloring problem. Technical Report, University of Dortmund (1977)
8. Wang, X.F.: r-Dynamic Coloring of Planar Graphs. Zhejiang Noamal University, Jinhua (2020)
9. Yi, D., Zhu, J.L., Feng, L.X., et al.: Optimal r-dynamic coloring of sparse graphs. J. Comb. Optim. **38**, 545–555 (2019)
10. Zhu, J.L.: $L(2,1)$-Labeling and r-Dynamic Coloring. Zhejiang Noamal University, Jinhua (2019)
11. Zhu, J.L., Bu, Y.H.: Minimum 2-distance coloring of planar graphs and channel assignment. J. Comb. Optim. **36**, 55–64 (2018)
12. Zhu, H., Gu, Y., Sheng, J., Lv, X.: List 2-distance ($\Delta+3$)-coloring of planar graphs without 4,5-cycles. J. Comb. Optim. **36**, 1411–1424 (2018)

Distance Magic Labeling of the Halved Folded n-Cube

Yi Tian[1,2], Na Kang[3], Weili Wu[4], Ding-Zhu Du[4], and Suogang Gao[1,5(✉)]

[1] School of Mathematical Sciences, Hebei Normal University,
Shijiazhuang 050024, People's Republic of China
`sggaomail@163.com`
[2] School of Big Data Science, Hebei Finance University,
Baoding 071051, People's Republic of China
[3] School of Mathematics and Science, Hebei GEO University,
Shijiazhuang 050024, People's Republic of China
[4] Department of Computer Science, University of Texas at Dallas,
Richardson, TX 75080, USA
`{weiliwu,dzdu}@utdallas.edu`
[5] Hebei International Joint Research Center for Mathematics and Interdisciplinary
Science, Shijiazhuang 050024, People's Republic of China

Abstract. Hypercube is an important structure for computer networks. The distance plays an important role in its applications. In this paper, we study a magic labeling of the halved folded n-cube which is a variation of the n-cube. This labeling is determined by the distance. Let G be a finite undirected simple connected graph with vertex set $V(G)$, distance function ∂ and diameter d. Let $D \subseteq \{0, 1, \ldots, d\}$ be a set of distances. A bijection $l : V(G) \rightarrow \{1, 2, \ldots, |V(G)|\}$ is called a D-magic labeling of G whenever $\sum_{x \in G_D(v)} l(x)$ is a constant for any vertex $v \in V(G)$, where $G_D(v) = \{x \in V(G) : \partial(x, v) \in D\}$. A $\{1\}$-magic labeling is also called a distance magic labeling. We show that the halved folded n-cube has a distance magic labeling (resp. a $\{0, 1\}$-magic labeling) if and only if $n = 16q^2$ (resp. $n = 16q^2 + 16q + 6$), where q is a positive integer.

Keywords: D-magic labeling · Distance-regular graph · Halved folded n-cube · Network · Incomplete tournament

1 Introduction

Hypercube is an important structure for computer networks [4]. Many combinatorial structural properties are studied in order to enhance its various applications [11,13,14,25]. Especially, the distance plays an important role [7]. In this paper, we study a magic labeling of the halved folded n-cube which is a variation of the n-cube. This labeling is determined by the distance. And the magic labeling has its applications in incomplete tournament and in efficient addressing systems in communication networks, ruler models and radar pulse codes ([1,2,5,6,9,18]).

© The Author(s), under exclusive license to Springer Nature Switzerland AG 2022
Q. Ni and W. Wu (Eds.): AAIM 2022, LNCS 13513, pp. 327–338, 2022.
https://doi.org/10.1007/978-3-031-16081-3_28

Let G be a finite undirected simple connected graph with vertex set $V(G)$, distance function ∂ and diameter d. In the early 1960 s Sedláček [19] introduced the notion of magic labeling of G. The concept is motivated by the construction of magic squares [1,18]. In 1994, Vilfred [24] introduced the distance magic labeling as follows.

Definition 1. *A bijection* $l : V(G) \rightarrow \{1, 2, \ldots, |V(G)|\}$ *is called a distance magic labeling of* G *if* $\sum_{x \in N(v)} l(x)$ *is constant for every vertex* $v \in V(G)$, *where*
$N(v) = \{x \in V(G) : \partial(x, v) = 1\}$.

In 2013, O'Neal and Slaterin [17] generalized the distance magic labeling to the D-magic labeling.

Definition 2. *For a graph* G *and a set of distances* $D \subseteq \{0, 1, \ldots, d\}$, *the set* $G_D(v) = \{x \in V(G) : \partial(x, v) \in D\}$ *is called* D-*neighborhood of* v. *By a* D-*magic labeling of* G, *we mean a bijection* $l : V(G) \rightarrow \{1, 2, \ldots, |V(G)|\}$ *with the property that there exists a constant* k *such that* $w(v) = \sum_{x \in G_D(v)} l(x) = k$ *for any vertex* $v \in V(G)$, *where* $w(v)$ *is the weight of* v. *The graph* G *admitting a* D-*magic labeling is called a* D-*magic graph.*

Obviously, a $\{1\}$-magic labeling of G is precisely a distance magic labeling of G.

When studying a D-magic labeling, distance-regular graphs are a natural class of graphs to consider. For the definition of a distance-regular graph, we refer to Sect. 2. Until now, D-magic labelings of some distance-regular graphs have been studied. Simanjuntak and Anuwiksa [20] characterized strongly regular graphs which are D-magic graphs, for all possible distance sets D. Gregor and Kovář [12] proved that if $n \equiv 2 \pmod 4$, then the n-cube is a $\{j\}$-magic graph for every odd j, where $1 \leq j \leq n$. Later, Cichacz et al. [10] proved that if the n-cube is a $\{1\}$-magic graph, then $n \equiv 2 \pmod 4$. Anuwiksa et al. [3] gave the sufficient condition for the n-cube with $n \equiv 2 \pmod 4$ to be D-magic. They also proved that the n-cube has a $\{0,1\}$-magic labeling if and only if $n \equiv 1 \pmod 4$. Tian et al. [22] showed that the folded n-cube has a $\{1\}$-magic (resp. $\{0,1\}$-magic) labeling if and only if $n \equiv 0 \pmod 4$ (resp. $n \equiv 3 \pmod 4$). Recently, Miklavič and Šparl [15] provided a sufficient condition for a Hamming graph to be a $\{1\}$-magic graph and classified $\{1\}$-magic folded n-cubes by showing that the folded n-cube is a $\{1\}$-magic graph if and only if $n \equiv 0 \pmod 4$.

In this paper, we focus on a halved folded n-cube with even $n \geq 8$ which is a distance-regular graph. For its definition, we refer to Sect. 2. We show the necessary and sufficient condition for the halved folded n-cube to be a $\{1\}$-magic graph and a $\{0, 1\}$-magic graph, respectively. The difference between this paper and other references on the D-magic labeling is that one has to choose a non-square matrix such that the corresponding map is bijective, where $D = \{1\}$ or $D = \{0, 1\}$; see Notation 1. The main result is as follows.

Theorem 1. *For even* $n \geq 8$, *let* $\frac{1}{2}FQ_{n-1}$ *denote a halved folded* n-*cube. Then the following* (i), (ii) *hold:*

(i) $\frac{1}{2}FQ_{n-1}$ has a distance magic labeling if and only if $n = 16q^2$, where q is a positive integer;

(ii) $\frac{1}{2}FQ_{n-1}$ has a $\{0,1\}$-magic labeling if and only if $n = 16q^2 + 16q + 6$, where q is a positive integer.

The paper is organized as follows. Section 2 gives some definitions, basic notations and some facts used in this paper. Section 3 presents the the proof of Theorem 1.

2   Preliminaries

In this section, we review some definitions, basic notations and some facts.

Recall that G is a finite undirected simple connected graph with vertex set $V(G)$, distance function ∂ and diameter d. For $v \in V(G)$ and $i \in \{0,1,\ldots,d\}$, let $G_i(v) = \{x \in V(G) : \partial(x,v) = i\}$. We define $G_{-1}(v) = \emptyset$ and $G_{d+1}(v) = \emptyset$. Particularly, $N(v) := G_1(v)$ is called the *neighborhood* of v, that is, two vertices u and v are *adjacent* whenever $u \in N(v)$. Furthermore, $|N(v)|$ is called the *degree* of v. The graph G is *regular* if $|N(v)|$ is constant for every $v \in V(G)$. By $N[v] := N(v) \cup \{v\}$ we denote the *closed neighborhood* of v.

The graph G is called a *distance-regular graph* when for all $i \in \{0,1,\ldots,d\}$ there are constants c_i, a_i, b_i such that for any vertices x and y with $\partial(x,y) = i$, we have $|G_{i-1}(x) \cap G_1(y)| = c_i$, $|G_i(x) \cap G_1(y)| = a_i$ and $|G_{i+1}(x) \cap G_1(y)| = b_i$. Observe that G is regular with degree $b_0 = a_i + b_i + c_i$ and $|G_i(x)| = \frac{b_0 b_1 \ldots b_{i-1}}{c_1 c_2 \ldots c_i}$ which is determined by i for $0 \leq i \leq d$ (see [8, p. 127]). If $d = 2$, then G is a strongly regular graph. For more information on distance-regular graphs, we refer to [8].

Now we recall the definitions of a folded n-cube and a halved folded n-cube with even $n \geq 2$.

Let $\mathbb{F}_2^{n-1}$ be an $(n-1)$-dimensional column vector space over $\mathbb{F}_2$. Let e_i $(1 \leq i \leq n-1)$ denote the vector in $\mathbb{F}_2^{n-1}$ such that the i-th component is 1 and the others are 0, and let $\mathbf{1} = e_1 + \cdots + e_{n-1}$.

Definition 3. [21] *The folded n-cube denoted by FQ_{n-1} is a graph with vertex set consisting of all vectors in $\mathbb{F}_2^{n-1}$; two vertices u and v are adjacent whenever $u \in \{v + e_i : 1 \leq i \leq n-1\} \cup \{v + \mathbf{1}\}$.*

By [23, p. 23], FQ_{n-1} is a bipartite graph for even n and its halved graph is the *halved folded n-cube*. For the definition of the halved graph, we refer to [8, p. 25]. Inspired by [8, pp. 264–265], the halved folded n-cube can also be described in $\mathbb{F}_2^{n-1}$ as follows.

Definition 4. *Let $n \geq 2$ be an even number. The halved folded n-cube denoted by $\frac{1}{2}FQ_{n-1}$ is a graph with vertex set consisting of all vectors in $\mathbb{F}_2^{n-1}$ that contains an even number of 1's; two vertices u and v are adjacent whenever $u \in \{v + e_i + e_j : 1 \leq i < j \leq n-1\} \cup \{v + e_i + \mathbf{1} : i = 1,\ldots,n-1\}$.*

By Definition 4, we have $|V(\frac{1}{2}FQ_{n-1})| = 2^{n-2}$ and

$$N(\mathbf{0}) = \{e_1 + e_2, \ldots, e_1 + e_{n-1}, e_2 + e_3, \ldots, e_2 + e_{n-1}, \ldots,$$
$$e_{n-2} + e_{n-1}, e_1 + 1, \ldots, e_{n-1} + 1\},$$

where $\mathbf{0} := (0, \ldots, 0)^T \in V(\frac{1}{2}FQ_{n-1})$. In what follows, we use $\mathbf{v} \oplus N(\mathbf{0})$ to denote the set of all vectors $\mathbf{v} + \mathbf{u}$, where $\mathbf{u}$ runs through $N(\mathbf{0})$. Then by Definition 4, for any $\mathbf{v} \in V(\frac{1}{2}FQ_{n-1})$ we obtain

$$N(\mathbf{v}) = \mathbf{v} \oplus N(\mathbf{0}), \tag{1}$$

and thus

$$N[\mathbf{v}] = \mathbf{v} \oplus N[\mathbf{0}]. \tag{2}$$

By [8, p. 265], $\frac{1}{2}FQ_{n-1}$ with $n \geq 6$ is a distance-regular graph of diameter $d = \lfloor \frac{n}{4} \rfloor$ and its eigenvalues are $\theta_j = 2(\frac{n}{2} - 2j)^2 - \frac{n}{2}$ ($0 \leq j \leq d$). Particularly, $\frac{1}{2}FQ_{n-1}$ with $n = 6$ is the complete graph K_{16} which is obviously $\{0, 1\}$-magic, but not $\{1\}$-magic ([16]). So we assume even $n \geq 8$ for the rest of this paper.

Inspired by [3,12], we give the following definitions.

Definition 5. *A subset $A \subseteq \mathbb{F}_2^{n-2}$ is said to be balanced if for every $i \in \{1, 2, \ldots, n-2\}$*

$$|\{\mathbf{v} \in A : \mathbf{v}_i = 1\}| = \frac{|A|}{2},$$

where $\mathbf{v}_i$ denotes the i-th component of $\mathbf{v}$.

Note that balance is invariant under translation. It means that a set $A \subseteq \mathbb{F}_2^{n-2}$ is balanced if and only if $\mathbf{u} \oplus A$ is balanced, where $\mathbf{u} \in \mathbb{F}_2^{n-2}$.

Definition 6. *Suppose H is an $(n-2) \times m$ matrix with entries in $\mathbb{F}_2$. Let $r_i(H)$ denote the number of 1's in the i-th row of H for $i \in \{1, 2, \ldots, n-2\}$. The matrix H is said to be balanced whenever the set of its columns is balanced, that is, $r_i(H) = \frac{m}{2}$ for every $i \in \{1, 2, \ldots, n-2\}$.*

Definition 7. *For $\frac{1}{2}FQ_{n-1}$, let $D \subseteq \{0, 1, \ldots, \lfloor \frac{n}{4} \rfloor\}$ be a set of distances. A bijection $f : V(\frac{1}{2}FQ_{n-1}) \to \mathbb{F}_2^{n-2}$ is said to be D-neighbor balanced if the set $f(G_D(\mathbf{v})) = \{f(\mathbf{u}) : \mathbf{u} \in G_D(\mathbf{v})\}$ is balanced for every $\mathbf{v} \in V(\frac{1}{2}FQ_{n-1})$. If $D = \{1\}$ (resp. $D = \{0, 1\}$), then a D-neighbor balanced bijection is also called neighbor balanced (resp. closed neighbor balanced).*

For every regular graph, we may equivalently consider labelings with labels starting from 0 instead of 1 ([12]). In fact, we work with labels (as well as with the image of vertices of $\frac{1}{2}FQ_{n-1}$ under the above map f) in the $(n-2)$-dimensional vector space $\mathbb{F}_2^{n-2}$ over $\mathbb{F}_2$, that is, in their binary representation. Using the arguments similar to Propositions 2.1 and 3.1 for the n-cube in [12], we get the following lemma for $\frac{1}{2}FQ_{n-1}$.

Lemma 1. *For a set of distances D of $\frac{1}{2}FQ_{n-1}$, let $f : V(\frac{1}{2}FQ_{n-1}) \to \mathbb{F}_2^{n-2}$ be a bijection. If f is D-neighbor balanced, then f is a D-magic labeling of $\frac{1}{2}FQ_{n-1}$.*

Proof. For every vertex $v \in V(\frac{1}{2}FQ_{n-1})$, we obtain (with the arithmetics in $\mathbb{N}$)

$$\sum_{u \in (\frac{1}{2}FQ_{n-1})_D(v)} f(u) = \sum_{i=1}^{n-2} |\{f(u) : (f(u))_i = 1\}| 2^{i-1}$$

$$= \frac{|f((\frac{1}{2}FQ_{n-1})_D(v))|}{2}(2^{n-2} - 1) \qquad \text{(by Definition 5)}$$

$$= \frac{|(\frac{1}{2}FQ_{n-1})_D(v)|}{2}(2^{n-2} - 1).$$

Since $\frac{1}{2}FQ_{n-1}$ is a distance-regular graph, $|(\frac{1}{2}FQ_{n-1})_D(v)|$ is independent of the choice of v. Therefore, f is a D-magic labeling of $\frac{1}{2}FQ_{n-1}$.

From now on, we adopt the following notational convention.

Notation 1. *Let N be an $(n-2) \times (n-2)$ matrix with entries in $\mathbb{F}_2$ and let $M = (\mathbf{0}\ N)$ be an $(n-2) \times (n-1)$ matrix with entries in $\mathbb{F}_2$. For $\frac{1}{2}FQ_{n-1}$, we define a map*

$$f : V(\frac{1}{2}FQ_{n-1}) \to \mathbb{F}_2^{n-2}$$

by $f(v) = Mv$ for every $v \in V(\frac{1}{2}FQ_{n-1})$. For $v \in V(\frac{1}{2}FQ_{n-1})$, let v_1 be the first component of v and write $v := \begin{pmatrix} v_1 \\ \tilde{v} \end{pmatrix}$.

Lemma 2. *With reference to Notation 1, f is a bijection if and only if the rank of M is $n-2$.*

Proof. Suppose f is a bijection. Then for any $u, v \in V(\frac{1}{2}FQ_{n-1})$ if $f(u) = f(v)$, then $u = v$. It follows that the system of equations $M(u - v) = \mathbf{0}$ has only the trivial solution. Then the system of equations $N(\tilde{u} - \tilde{v}) = \mathbf{0}$ has only the trivial solution. Thus the rank of N is $n-2$, and hence the rank of M is $n-2$.

Conversely, since $|V(\frac{1}{2}FQ_{n-1})| = |\mathbb{F}_2^{n-2}| = 2^{n-2}$, it suffices to show that f is injective. Suppose $f(u) = f(v)$ for $u, v \in V(\frac{1}{2}FQ_{n-1})$. Then $M(u - v) = \mathbf{0}$. It follows that $N(\tilde{u} - \tilde{v}) = \mathbf{0}$. Since the rank of M is $n-2$, N is invertible. Thus $\tilde{u} = \tilde{v}$. Combining this with the fact that the number of 1's in both u and v is even, we obtain $u_1 = v_1$. Therefore, $u = v$, and hence f is injective.

3 Proof of Theorem 1

In this section, we give the necessary and sufficient condition for $\frac{1}{2}FQ_{n-1}$ to be a $\{1\}$-magic graph and a $\{0,1\}$-magic graph, respectively. Let B be the $(n-1) \times \frac{n(n-1)}{2}$ matrix

$$B = (e_1 + e_2, \ldots, e_1 + e_{n-1}, e_2 + e_3, \ldots, e_2 + e_{n-1}, \ldots, e_1 + 1, \ldots, e_{n-1} + 1) \qquad (3)$$

and let $B^* = (\mathbf{0}\ B)$, where $\mathbf{0} := (0, \ldots, 0)^{\mathrm{T}} \in V(\frac{1}{2}FQ_{n-1})$. Obviously, the above B (resp. B^*) corresponds to the set $N(\mathbf{0})$ (resp. $N[\mathbf{0}]$). Recall that $r_i(H)$

denotes the number of 1's in the i-th row of H for $i \in \{1, 2, \ldots, n-2\}$, where H is an $(n-2) \times m$ matrix with entries in $\mathbb{F}_2$.

Now we show the following lemma.

Lemma 3. *With reference to Notation 1, let B, B^* and r_i be as above. Suppose $r_i(M) = t$ for some $i \in \{1, \ldots, n-2\}$. Then*

$$r_i(MB^*) = r_i(MB) = \begin{cases} (t+1)(n-1-t) & \text{if } t \text{ is odd,} \\ (n-t)t & \text{otherwise.} \end{cases}$$

Proof. Since $B^* = \begin{pmatrix} \mathbf{0} & B \end{pmatrix}$, we have $MB^* = \begin{pmatrix} \mathbf{0} & MB \end{pmatrix}$. Then $r_i(MB^*) = r_i(MB)$ for every $i \in \{1, \ldots, n-2\}$.

Suppose $r_i(M) = t$ for some $i \in \{1, \ldots, n-2\}$. Let

$$M = \begin{bmatrix} 0 & a_{11} & a_{12} & \cdots & a_{1,n-2} \\ 0 & a_{21} & a_{22} & \cdots & a_{2,n-2} \\ \vdots & \vdots & \vdots & & \vdots \\ 0 & a_{n-2,1} & a_{n-2,2} & \cdots & a_{n-2,n-2} \end{bmatrix},$$

where $a_{sj} \in \mathbb{F}_2$ for any $s \in \{1, \ldots, n-2\}$ and any $j \in \{1, \ldots, n-2\}$. Clearly, the i-th row of MB is

$$(a_{i1}, a_{i2}, \ldots, a_{i,n-2}, a_{i1} + a_{i2}, a_{i1} + a_{i3}, \ldots, a_{i1} + a_{i,n-2}, a_{i2} + a_{i3}, \ldots,$$
$$a_{i2} + a_{i,n-2}, \ldots, a_{i,n-3} + a_{i,n-2}, a_{i1} + \cdots + a_{i,n-2}, a_{i2} + \cdots + a_{i,n-2},$$
$$a_{i1} + a_{i3} + \cdots + a_{i,n-2}, \ldots, a_{i1} + \cdots + a_{i,n-3}).$$

We now compute $r_i(MB)$ for the fixed i. Since $r_i(M) = t$, we have that the number of 1's in $(a_{i1}, a_{i2}, \ldots, a_{i,n-2})$ is t; the number of 1's in $(a_{i1} + a_{i2}, a_{i1} + a_{i3}, \ldots, a_{i1} + a_{i,n-2}, a_{i2} + a_{i3}, \ldots, a_{i2} + a_{i,n-2}, \ldots, a_{i,n-3} + a_{i,n-2})$ is

$$C_t^1 C_{n-2-t}^1 = t(n-2-t);$$

$a_{i1} + \cdots + a_{i,n-2} = 1$ if t is odd and $a_{i1} + \cdots + a_{i,n-2} = 0$ otherwise; the number of 1's in $(a_{i2} + \cdots + a_{i,n-2}, a_{i1} + a_{i3} + \cdots + a_{i,n-2}, \ldots, a_{i1} + \cdots + a_{i,n-3})$ is $n-2-t$ if t is odd and t otherwise. Therefore, if t is odd, then

$$r_i(MB) = t + t(n-2-t) + 1 + n - 2 - t = (t+1)(n-1-t);$$

if t is even, then $r_i(MB) = t + t(n-2-t) + 0 + t = (n-t)t$.

This completes the proof of the lemma. ∎

3.1   Proof of Theorem 1 (i)

In this subsection, we shall present the necessary and sufficient condition for $\frac{1}{2}FQ_{n-1}$ to be a $\{1\}$-magic graph. We first give the following lemma. It enables us to determine a neighbor balanced bijection by constructing the appropriate matrix.

Lemma 4. *With reference to Notation 1, the following statements are equivalent.*

(i) *f is neighbor balanced.*
(ii) *$n = 16q^2$, the rank of M is $16q^2 - 2$ and $r_i(M) \in \{8q^2 + 2q - 1, 8q^2 - 2q - 1, 8q^2 + 2q, 8q^2 - 2q\}$ for any given $i \in \{1,\ldots,n-2\}$, where q is a positive integer.*

Moreover, if (i)–(ii) hold, then f is a $\{1\}$-magic labeling of $\frac{1}{2}FQ_{n-1}$.

Proof. By Notation 1 and (1), for any $v \in V(\frac{1}{2}FQ^{n-1})$, we obtain

$$f(N(v)) = Mv \oplus MN(0). \tag{4}$$

Combining (4) with Definition 7, we know that f is neighbor balanced if and only if the set $Mv \oplus MN(0)$ is balanced for every $v \in V(\frac{1}{2}FQ^{n-1})$. Recall that the set $Mv \oplus MN(0)$ is balanced if and only if the set $MN(0)$ is balanced. Furthermore, note that the set $MN(0)$ is precisely the set of columns of the matrix MB, where B is from (3). It means that the set $MN(0)$ is balanced if and only if the matrix MB is balanced. Thus f is neighbor balanced if and only if the matrix MB is balanced.

(i) $\implies$ (ii) Since f is neighbor balanced, f is a bijection. It follows from Lemma 2 that the rank of M is $n - 2$. Fix any $i \in \{1,\ldots,n-2\}$, we assume $r_i(M) = t$. Obviously, $t > 0$. Next we show (ii) holds according to the parity of t.

Case (i) t is odd.

By Lemma 3, $r_i(MB) = (n - 1 - t)(t + 1)$. By Definition 5 and since the $(n - 2) \times \frac{n(n-1)}{2}$ matrix MB is balanced, we have $(n - 1 - t)(t + 1) = \frac{n(n-1)}{4}$. Now we let $t = 2p - 1$ $(p \in \mathbb{Z}^+)$ since t is odd. Then we have

$$n \pm \sqrt{n} = 4p. \tag{5}$$

Since both $4p$ and n are even, $\sqrt{n}$ is even. Now we let $\sqrt{n} = 2r$, where $r \geq 2$ as $n \geq 8$. Substituting it into (5), we obtain

$$2r^2 \pm r - 2p = 0. \tag{6}$$

By (6), r is even. Let $r = 2q$, where q is a positive integer. Then $n = (2r)^2 = (4q)^2 = 16q^2$, and hence the rank of M is $16q^2 - 2$. By (6), we have $p = 4q^2 \pm q$. Then $t = 2p - 1 = 2(4q^2 \pm q) - 1 = 8q^2 \pm 2q - 1$.

Case (ii) t is even.

By Lemma 3, $r_i(MB) = (n-t)t$. By Definition 5 and since the $(n-2)\times\frac{n(n-1)}{2}$ matrix MB is balanced, we have $(n - t)t = \frac{n(n-1)}{4}$. Now we let $t = 2p$ since t is even, where p is a positive integer. Then we obtain $n \pm \sqrt{n} = 4p$. By using arguments similar to the proof of case (i) above, we have $n = 16q^2$, the rank of M is $16q^2 - 2$ and $p = 4q^2 \pm q$, where q is a positive integer. Then $t = 8q^2 \pm 2q$.

By the arguments above, (ii) holds.

(ii) $\Longrightarrow$ (i) By Lemma 2 and since the rank of M is $n-2$, f is a bijection. To prove that f is neighbor balanced, it suffices to show the $(n-2) \times \frac{n(n-1)}{2}$ matrix MB is balanced. To do this, by Definition 6 and since $n = 16q^2$, it suffices to show

$$r_i(MB) = \frac{1}{2}\frac{n(n-1)}{2} = 64q^4 - 4q^2 \tag{7}$$

for every $i \in \{1, \dots, n-2\}$.

Fix any $i \in \{1, \dots, n-2\}$, we assume $r_i(M) = t$. Next we prove that (7) holds according to the parity of t.

Case (i) t is odd.

By Lemma 3, we obtain $r_i(MB) = (t+1)(n-1-t)$. It is easy to see that (7) holds if $t \in \{8q^2 + 2q - 1, 8q^2 - 2q - 1\}$.

Case (ii) t is even.

By Lemma 3, we obtain $r_i(MB) = t(n-t)$. It is easy to see that (7) holds if $t \in \{8q^2 + 2q, 8q^2 - 2q\}$.

By the arguments above, (7) holds, and hence (i) holds.

If (i)–(ii) hold, then by Lemma 1, f is a $\{1\}$-magic labeling of $\frac{1}{2}FQ_{n-1}$.

We list the following lemma which will be used in the proof of Theorem 1 (i).

Lemma 5. [20] *If G is a regular graph admitting a $\{1\}$-magic labeling, then 0 is an eigenvalue of G.*

Now we prove Theorem 1 (i).

Proof of Theorem 1 (i) Suppose that $\frac{1}{2}FQ_{n-1}$ has a $\{1\}$-magic labeling. By Lemma 5, 0 is an eigenvalue of $\frac{1}{2}FQ_{n-1}$, that is, there exists $j \in \{0, 1, \dots, \lfloor \frac{n}{4} \rfloor\}$ such that $\theta_j = 2(\frac{n}{2} - 2j)^2 - \frac{n}{2} = 0$. Thus

$$n = 4\left(\frac{n}{2} - 2j\right)^2. \tag{8}$$

We claim that $\frac{n}{2} - 2j$ is even. In fact, assume $\frac{n}{2} - 2j = p$, where p is an integer. Substituting it into (8), we have $n = 4p^2$. It follows that $\frac{n}{2} - 2j = 2p^2 - 2j$. So the claim holds. Moreover, by (8) and since $n \geq 8$, we have $\frac{n}{2} - 2j \geq 2$. Let $\frac{n}{2} - 2j = 2q$, where q is a positive integer. It follows from (8) that $n = 4(2q)^2 = 16q^2$.

Conversely, since $n = 16q^2$, where q is a positive integer, we construct a $(16q^2 - 2) \times (16q^2 - 1)$ matrix M with entries in $\mathbb{F}_2$ as follows:

$$
\begin{array}{c}
\\
8q^2 - 2q - 2 \\
8q^2 - 2q - 2 \\
2q \\
2q \\
2
\end{array}
\begin{array}{c}
1 \quad\quad 8q^2 - 2q - 2 \quad 8q^2 - 2q - 2 \quad 2q \quad 2q \quad 2 \\
\begin{bmatrix}
0 & I & J & 0 & 0 & 0 \\
0 & J & I & 0 & 0 & 0 \\
0 & J & 0 & I & 0 & 0 \\
0 & J & 0 & 0 & I & 0 \\
0 & J & 0 & 0 & 0 & I
\end{bmatrix},
\end{array}
$$

where I denotes the identity matrix and J (resp. $\mathbf{0}$) denotes the all 1's (resp. 0's) matrix. Obviously, $r_i(M) = 8q^2 - 2q - 1$ for every $i \in \{1, \ldots, n-2\}$. Moreover, by elementary row operations over $\mathbb{F}_2$, M can be transformed to the following matrix

$$\begin{bmatrix} 0 & I & 0 & 0 & 0 & 0 \\ 0 & J & I & 0 & 0 & 0 \\ 0 & J & 0 & I & 0 & 0 \\ 0 & J & 0 & 0 & I & 0 \\ 0 & J & 0 & 0 & 0 & I \end{bmatrix},$$

which has the same block representation as M and its rank is $16q^2 - 2$. So the rank of M is $16q^2 - 2$. Let $f : V(\frac{1}{2}FQ_{n-1}) \to \mathbb{F}_2^{n-2}$ be given by $f(\boldsymbol{v}) = M\boldsymbol{v}$. By Lemma 4, f is a $\{1\}$-magic labeling of $\frac{1}{2}FQ_{n-1}$. Therefore, $\frac{1}{2}FQ_{n-1}$ has a $\{1\}$-magic labeling. $\qquad\square$

3.2 Proof of Theorem 1 (ii)

In this subsection, we will present the necessary and sufficient condition for the halved folded n-cube to be a $\{0, 1\}$-magic graph. We first give the following lemma. It enables us to determine a closed neighbor balanced bijection by constructing the appropriate matrix.

Lemma 6. *With reference to Notation 1, the following statements are equivalent.*

(i) *f is closed neighbor balanced.*
(ii) *$n = 16q^2 + 16q + 6$, the rank of M is $16q^2 + 16q + 4$ and $r_i(M) \in \{8q^2 + 10q + 3, 8q^2 + 6q + 1, 8q^2 + 10q + 4, 8q^2 + 6q + 2\}$ for any given $i \in \{1, \ldots, n-2\}$, where q is a positive integer.*

Moreover, if (i)–(ii) hold, then f is a $\{0, 1\}$-magic labeling of $\frac{1}{2}FQ_{n-1}$.

Proof. By Notation 1 and (2), for any vertex $\boldsymbol{v} \in V(\frac{1}{2}FQ_{n-1})$, we obtain

$$f(N[\boldsymbol{v}]) = MN[\boldsymbol{v}] = M\boldsymbol{v} \oplus MN[\mathbf{0}]. \qquad (9)$$

Combining (9) with Definition 7, we know that f is closed neighbor balanced if and only if the set $M\boldsymbol{v} \oplus MN[\mathbf{0}]$ is balanced for every $\boldsymbol{v} \in V(\frac{1}{2}FQ^{n-1})$. Recall that the set $M\boldsymbol{v} \oplus MN[\mathbf{0}]$ is balanced if and only if the set $MN[\mathbf{0}]$ is balanced. Moreover, note that the set $MN[\mathbf{0}]$ is precisely the set of columns of the matrix MB^*. It means that the set $MN[\mathbf{0}]$ is balanced if and only if the matrix MB^* is balanced. Thus f is closed neighbor balanced if and only if the matrix MB^* is balanced.

(i) $\Longrightarrow$ (ii) Since f is closed neighbor balanced, f is a bijection. It follows from Lemma 2 that the rank of M is $n-2$. Fix any $i \in \{1, \ldots, n-2\}$, we assume $r_i(M) = t$. Obviously, $t > 0$. Next, we show (ii) holds according to the parity of t.

   Case (i) t is odd.

By Lemma 3, $r_i(MB^*) = (n - 1 - t)(t + 1)$. By Definition 5 and since the $(n - 2) \times (\frac{n(n-1)}{2} + 1)$ matrix MB^* is balanced, we have $(n - 1 - t)(t + 1) = \frac{n(n-1)+2}{4}$. Now we let $t = 2p - 1$ ($p \in \mathbb{Z}^+$) since t is odd. Then we have

$$n \pm \sqrt{n - 2} = 4p. \tag{10}$$

Since both $4p$ and n are even, $\sqrt{n - 2}$ is even. Now we let $\sqrt{n - 2} = 2r$, where $r \geq 2$ as $n \geq 8$. Substituting it into (10), we obtain

$$2r^2 \pm r + 1 - 2p = 0. \tag{11}$$

It means that r is odd. Let $r = 2q + 1$, where q is a positive integer. Then $n = (2r)^2 + 2 = (4q + 2)^2 + 2 = 16q^2 + 16q + 6$, and hence the rank of M is $16q^2 + 16q + 4$. By (11), we obtain $p = 4q^2 + 5q + 2$ or $4q^2 + 3q + 1$. Then $t = 2p - 1 = 8q^2 + 10q + 3$ or $8q^2 + 6q + 1$.

Case (ii) t is even.

By using arguments similar to the proof of case (i) above, we have $n = 16q^2 + 16q + 6$, the rank of M is $16q^2 + 16q + 4$ and $t \in \{8q^2 + 10q + 4, 8q^2 + 6q + 2\}$, where q is a positive integer.

By the arguments above, (ii) holds.

(ii) $\Longrightarrow$ (i) By Lemma 2 and since the rank of M is $n - 2$, f is a bijection. To prove that f is closed neighbor balanced, it suffices to show that the $(n - 2) \times (\frac{n(n-1)}{2} + 1)$ matrix MB^* is balanced. To do this, by Definition 6 and since $n = 16q^2 + 16q + 6$, it suffices to show

$$r_i(MB^*) = \frac{1}{2}(\frac{n(n - 1)}{2} + 1) = 64q^4 + 128q^3 + 108q^2 + 44q + 8 \tag{12}$$

for every $i \in \{1, \ldots, n - 2\}$.

Fix any $i \in \{1, \ldots, n - 2\}$, we assume $r_i(M) = t$. Next we prove that (12) holds according to the parity of t.

Case (i) t is odd. By Lemma 3, we obtain $r_i(MB^*) = (t + 1)(n - 1 - t)$. It is easy to see that (12) holds if $t \in \{8q^2 + 10q + 3, 8q^2 + 6q + 1\}$.

Case (ii) t is even. By Lemma 3, we obtain $r_i(MB^*) = t(n - t)$. It is easy to see that (12) holds if $t \in \{8q^2 + 10q + 4, 8q^2 + 6q + 2\}$.

By the arguments above, (12) holds, and hence (i) holds.

If (i)–(ii) hold, then by Lemma 1, f is a $\{0, 1\}$-magic labeling of $\frac{1}{2}FQ_{n-1}$.

To prove Theorem 1 (ii), we use the following lemma.

Lemma 7. [1] *If G is a regular graph admitting a $\{0, 1\}$-magic labeling, then -1 is an eigenvalue of G.*

Now we prove Theorem 1 (ii).

Proof of Theorem 1 (ii) Suppose that $\frac{1}{2}FQ_{n-1}$ has a $\{0, 1\}$-magic labeling. By Lemma 7, -1 is an eigenvalue of $\frac{1}{2}FQ_{n-1}$, that is, there exists $j \in \{0, 1, \ldots, \lfloor \frac{n}{4} \rfloor\}$ such that $\theta_j = 2(\frac{n}{2} - 2j)^2 - \frac{n}{2} = -1$. Thus

$$n = 4(\frac{n}{2} - 2j)^2 + 2. \tag{13}$$

We claim that $\frac{n}{2} - 2j$ is odd. In fact, assume $\frac{n}{2} - 2j = p$, where p is an integer. Substituting it into (13), we have $n = 4p^2 + 2$. It follows that $\frac{n}{2} - 2j = 2p^2 + 1 - 2j$. So the claim holds. Moreover, by (13) and since $n \geq 8$, we have $\frac{n}{2} - 2j > 1$. Let $\frac{n}{2} - 2j = 2q + 1$, where q is a positive integer. It follows from (13) that $n = 4(2q + 1)^2 + 2 = 16q^2 + 16q + 6$.

Conversely, since $n = 16q^2 + 16q + 6$, where q is a positive integer, we construct a $(16q^2 + 16q + 4) \times (16q^2 + 16q + 5)$ matrix M with entries in $\mathbb{F}_2$ as follows:

$$
\begin{array}{c}
\\
8q^2 + 4q \\
8q^2 + 4q \\
2q \\
2q \\
2q \\
2q \\
4
\end{array}
\begin{array}{c}
\begin{array}{ccccccc}
1 & 8q^2 + 4q & 8q^2 + 4q & 2q & 2q & 2q & 2q & 4
\end{array} \\
\left[
\begin{array}{cccccccc}
0 & I & J & J & 0 & 0 & 0 & 0 \\
0 & J & I & J & 0 & 0 & 0 & 0 \\
0 & J & 0 & I & J & 0 & 0 & 0 \\
0 & J & 0 & J & I & 0 & 0 & 0 \\
0 & J & 0 & J & 0 & I & 0 & 0 \\
0 & J & 0 & J & 0 & 0 & I & 0 \\
0 & J & 0 & J & 0 & 0 & 0 & I
\end{array}
\right]
\end{array},
$$

where I denotes the identity matrix and J (resp. $\mathbf{0}$) denotes the all 1's (resp. 0's) matrix. Obviously, $r_i(M) = 8q^2 + 6q + 1$ for every $i \in \{1, \ldots, n-2\}$. Moreover, by elementary row operations over $\mathbb{F}_2$, M can be transformed to the following matrix

$$
\left[
\begin{array}{cccccccc}
0 & I & 0 & 0 & 0 & 0 & 0 & 0 \\
0 & J & I & 0 & 0 & 0 & 0 & 0 \\
0 & J & 0 & I & 0 & 0 & 0 & 0 \\
0 & J & 0 & J & I & 0 & 0 & 0 \\
0 & J & 0 & J & 0 & I & 0 & 0 \\
0 & J & 0 & J & 0 & 0 & I & 0 \\
0 & J & 0 & J & 0 & 0 & 0 & I
\end{array}
\right]
$$

which has the same block representation as M and its rank is $16q^2 + 16q + 4$. So the rank of M is $16q^2 + 16q + 4$. Let $f : V(\frac{1}{2}FQ_{n-1}) \to \mathbb{F}_2^{n-2}$ be given by $f(\boldsymbol{v}) = M\boldsymbol{v}$. By Lemma 6, f is a $\{0, 1\}$-magic labeling of $\frac{1}{2}FQ_{n-1}$. Therefore, $\frac{1}{2}FQ_{n-1}$ has a $\{0, 1\}$-magic labeling. $\qquad\square$

Acknowledgements. This work was supported by the National Natural Science Foundation of China (Grant 11971146) and the National Natural Science Foundation of Hebei Province (Grant A2017403010).

References

1. Anholcer, M., Cichacz, S., Peterin, I.: Spectra of graphs and closed distance magic labelings. Discrete Math. **339**(7), 1915–1923 (2016). https://doi.org/10.1016/j.disc.2015.12.025
2. Anholcer, M., Cichacz, S., Peterin, I., Tepeh, A.: Distance magic labeling and two products of graphs. Graphs Comb. **31**(5), 1125–1136 (2014). https://doi.org/10.1007/s00373-014-1455-8
3. Anuwiksa, P., Munemasa, A., Simanjuntak, R.: D-magic and antimagic labelings of hypercubes. arXiv:1903.05005v2 [math.CO]

4. Bettayeb, S.: On the k-ary hypercube. Theor. Comput. Sci. **140**(2), 333–339 (1995). https://doi.org/10.1016/0304-3975(94)00197-Q
5. Bloom, G.S., Golomb, S.W.: Applications of numbered undirected graphs. Proc. IEEE **65**, 562–570 (1977)
6. Bloom, G.S., Golomb, S.W.: Numbered complete graphs, unusual rulers, and assorted applications. In: Alavi, Y., Lick, D.R. (eds.) Theory and Applications of Graphs. Lecture Notes in Mathematics, vol. 642, pp. 53–65. Springer, Berlin (1978). https://doi.org/10.1007/BFb0070364
7. Bose, B., Broeg, B., Kwon, Y., Ashir, Y.: Lee distance and topological properties of k-ary n-cubes. IEEE Trans. Comput. **44**(8), 1021–1030 (1995)
8. Brouwer, A.E., Cohen, A.M., Neumaier, A.: Distance-Regular Graphs. Springer-Verlag, Berlin (1989). https://doi.org/10.1007/978-3-642-74341-2
9. Chebotarev, P., Agaev, R.: Matrices of forests, analysis of networks, and ranking problems. Procedia Comput. Sci. **17**, 1134–1141 (2013)
10. Cichacz, S., Fronček, D., Krop, E., Raridan, C.: Distance magic cartesian products of graphs. Discuss. Math. Graph Theor. **36**(2), 299–308 (2016)
11. Day, K., Al-Ayyoub, A.E.: Fault diameter of k-ary n-cube networks. IEEE Trans. Parallel Distrib. Syst. **8**(9), 903–907 (1997)
12. Gregor, P., Kovář, P.: Distance magic labelings of hypercubes. Electron. Notes Discrete Math. **40**, 145–149 (2013). https://doi.org/10.1016/j.endm.2013.05.027
13. Hsieh, S.Y., Chang, Y.H.: Extraconnectivity of k-ary n-cube networks. Theor. Comput. Sci. **443**(20), 63–69 (2012). https://doi.org/10.1016/j.tcs.2012.03.030
14. Lin, C.K., Zhang, L.L., Fan, J.X., Wang, D.J.: Structure connectivity and substructure connectivity of hypercubes. Theor. Comput. Sci. **634**, 97–107 (2016)
15. Miklavič, Š., Šparl, P.: On distance magic labelings of Hamming graphs and folded hypercubes. Discuss. Math. Graph Theor. **0** (2021).https://doi.org/10.7151/dmgt.2430
16. Miller, M., Rodger, C., Simanjuntak, R.: Distance magic labelings of graphs. Australas. J. Combin. **28**, 305–315 (2003)
17. O'Neal, A., Slater, P.J.: Uniqueness of vertex magic constants. SIAM J. Discrete Math. **27**(2), 708–716 (2013). https://doi.org/10.1137/110834421
18. Prajeesh, A.V., Paramasivam, K., Kamatchi, N.: A note on handicap incomplete tournaments. Lect. Notes Comput. Sci. (including subseries Lecture Notes in Artificial Intelligence and Lecture Notes in Bioinformatics) **11638**, 1–9 (2019). https://doi.org/10.1007/978-3-030-25005-8_1
19. Sedláček, J.: Some properties of interchange graphs. Theory of Graphs and its Applications (Proc. Sympos. Smolenice, 1963), pp. 145–150. House Czech. Acad. Sci., Prague (1964)
20. Simanjuntak, R., Anuwiksa, P.: D-magic strongly regular graphs. AKCE Int. J. Graphs Comb. **17**(3), 995–999 (2020)
21. Simó, E., Yebra, J.L.A.: The vulnerability of the diameter of folded n-cubes. Discrete Math. **174**(1–3), 317–322 (1997)
22. Tian, Y., Hou, L.H., Hou, B., Gao, S.G.: D-magic labelings of the folded n-cube. Discrete Math. **344**(9), 112520 (2021)
23. van Dam, E.R., Koolen, J.H., Tanaka, H.: Distance-regular graphs. Electron. J. Combin. DS22 (2016)
24. Vilfred, V.: Sigma labelled graphs and circulant graphs. Ph.D. Thesis, University of Kerala, (1994)
25. Zhao, S.L., Yang, W.H., Zhang, S.R.: Component connectivity of hypercubes. Theor. Comput. Sci. **640**(C), 115–118 (2016). https://doi.org/10.1016/j.tcs.2016.05.035

Balanced Graph Partitioning Based on Mixed 0-1 Linear Programming and Iteration Vertex Relocation Algorithm

Zhengxi Yang[ID], Zhipeng Jiang[✉][ID], Wenguo Yang[ID], and Suixiang Gao

School of Mathematical Sciences, University of Chinese Academy of Sciences,
Beijing 100049, China
yangzhengxi20@mails.ucas.ac.cn, {jiangzhipeng,yangwg,sxgao}@ucas.ac.cn

Abstract. Graph partitioning is a classical NP problem. The goal of graphing partition is to have as few cut edges in the graph as possible. Meanwhile, the capacity limit of the shard should be satisfied. In this paper, a model for graph partitioning is proposed. Then the model is converted into a mixed 0-1 linear programming by introducing variables. In order to solve this model, we select some variables to design the vertex relocation model. This work designs a variable selection strategy according to the effect of vertex relocation on the number of local edges. For purpose of implementing graph partitioning on large scale graph, we design iterative algorithm to solve the model by selecting a small number of variables in each iteration. The algorithm relocates the shard of the vertex according to the solution of the model. In the experiment, the method in this paper is simulated and compared with BLP and its related methods in the different shard sizes on the five social network datasets. The simulation results show that the method of this paper works well.

Keywords: Graph partitioning · 0-1 mixed linear programming · Iteration algorithm

1 Introduction

The graph partitioning problem is a classical graph theory and combinatorial optimization problem. Meanwhile, the graph partitioning problem is NP-hard [3], it is difficult for direct graph partitioning methods to obtain an optimal solution in effective time. The graph partition model can be applied in many fields. In social network [2], graph partitioning algorithms are involved in friend recommendation system. Every user in the system will be served by the identified server. In order to reduce communication costs between different servers, it is expected that friends who potentially know each other are served by the same server. Friend recommendation problem can be transformed into a graph partition problem. Each user is regarded as a vertex. The edges are generated

© The Author(s), under exclusive license to Springer Nature Switzerland AG 2022
Q. Ni and W. Wu (Eds.): AAIM 2022, LNCS 13513, pp. 339–350, 2022.
https://doi.org/10.1007/978-3-031-16081-3_29

according to the user's relationship. The partition capacity limit in the graph partitioning problem is reflected in the capacity limit of the server. At the same time, it is expected that more neighboring vertices are in the same server. Graph partitioning is also used to design very large scale integrated circuit systems (VLSI) [9]. The goal of the VLSI design is to reduce the complexity by dividing the VLSI into smaller components and thus keeping the total length of all wires to be the shortest [4]. In machine learning and graph neural networks, graph partitioning algorithms are also applied. Cluster GCN [5] divides the large graph into subgraphs through the graph partitioning algorithm.

In this paper, We build a graph partitioning model with the goal of maximizing the number of local edges while maintaining the partition capacity limit. The model is converted into a mixed 0-1 linear programming by introducing variables. Then we select some variables to design the vertex relocation model. This work designs an iterative algorithm that selects some variables at each iteration and assigns new shards to the vertices corresponding to these variables. The algorithm first uses hash partitioning to obtain the initial partition of the graph. Vertices with high gain are given priority to determine the relocated shard. Then the algorithm assigns the new shard of vertex through the vertex relocation model. We compare the partitioning effect of BLP, BLP-MC, BLP-KL and our method on five social network datasets. On these datasets, we also compare the effect of different number of shards. In the experiments, our algorithms all perform better than the comparison algorithms. Our main contributions in this work are summarized as follows.

1. Construct a graph partitioning model and remove $'min'$ from the objective function by introducing variables. The model is converted into a mixed 0-1 linear programming.
2. Design an iterative algorithm that selects some variables at each iteration and relocates corresponding vertices to new shards according to the variables' value of the solution. The algorithm can solve large scale graph partitioning problems.
3. Evaluate our algorithms on real-world datasets and compare the results with some other methods such as BLP, BLP-MC and BLP-KL.

The paper is organized as follows: Sect. 2 summarizes related work on graph partitioning problem; Sect. 3 proposes a graph partitioning model based on local edge maximization and converts it into a mixed 0-1 linear programming by introducing variables; Sect. 4 gives the iteration algorithm; Sect. 5 shows the numerical results obtained on datasets of different size and topology. Finally, we draw some conclusions in Sect. 6.

2   Related Work

There is a large number of literature on methods that solve graph partitioning problem, including spectral partitioning [14], geometric partitioning [8], streaming graph partition [1], linear programing [13] and semi-define programing [12].

For each vertex, the hash partitioning [1] determines the shard in which the vertex is located based on the vertex number and shard number. Hash partitioning can be defined as the mapping function $f(v) = hash(v) \bmod (k)$.

The KL algorithm [10] proposed by Kernighan and Lin is a local search method. The selection strategy finds the swap of vertex assignments that yields the largest decrease in the total number of cut edge. The algorithm considers node swaps between $\frac{1}{2}K(K-1)$ shard pairs. M. Fiduccia and M. Mattheyses propose the Fiduccia-Mattheyses (FM) algorithm [7] to improve the KL algorithm. The difference with the KL algorithm is that the FM algorithm uses a single vertex movement and introduces the bucket list data structure to reduce the time complexity.

Johan Ugander and Lars Backstrom propose the balanced label propagation algorithm (BLP) [15] based on linear programming. The idea of this method is inspired by the label propagation principle, which is vertices tend to move to the shards with more neighboring vertices. The algorithm first determines the vertices that tend to be migrated to another shard. Then the vertices with high gains were prioritized for migration. It transforms a maximally concave optimization problem into a linear programming problem. The constraints of linear programming limit the number of adjustable vertices in the shard. Finally, the node migration strategy is optimized through the solution of the model.

Zishi Deng, Torsten Suel [6] studied the combination of graph partitioning initialization algorithm and BLP improvement algorithm. They propose three methods of interruption, probability-based disruption, clustered constrained relocation, and round-robin Kernighan–Lin swaps, to improve the BLP algorithm. They name the combination of BLP and clustered constrained relocation as BLP-MC and name the combination of BLP and round-robin Kernighan–Lin swaps as BLP-KL. In addition, random, SBM and Metis are considered as initialization methods for graph partitioning. Then they analyze the improvement degree of BLP and BLP improvement method to the initial partition.

3 Mixed 0-1 Linear Programming Model for Graph Partitioning

Problem Definition
Balanced graph partitioning of a graph $G = (V, E)$ is a partition $\{V_1, \ldots, V_K\}$ of $V(G)$, where K is the shard number and $|V| = n$. Each shard (subset) in the partition is subject to the following conditions:

1. $V_i \cap V_j = \emptyset, \forall i \neq j, V_1 \cup V_2 \cup \ldots \cup V_k = V$
2. $|V_i| = |V_j|, i = 1, 2, \ldots, K, j = 1, 2, \ldots, K$

The balanced graph partitioning with slack condition is:
2′. $lm \leq |V_i| \leq um, i = 1, 2, \ldots, k$.
where $lm = (1 - \epsilon)\frac{|V|}{K}, um = (1 + \epsilon)\frac{|V|}{K}$.
The goal of balanced graph partitioning is that the number of local edges (with endpoints in the same shards) is maximized.

Based on the definition of balanced graph partitioning, this work builds a graph partitioning model based on local edge maximization and converts it into a mixed 0-1 linear programming by introducing variables.

Notation

x_{it}: Binary variable. Indicating whether the vertex i in shard V_t.

$$x_{it} = \begin{cases} 1 & i \in V_t \\ 0 & \text{otherwise} \end{cases}$$

n: Constant. The number of vertices, $n = |V|$.
K: Constant. The number of shards.
$N(i)$: Constant. The ordered set of the neighbor vertices of vertex i. $N(i) = \{j | j > i, e(i,j) \in E(G)\}$.
lm: Constant. Lower limit for the number of vertices in a shard, $lm = (1 - \epsilon)\frac{|V|}{K}$.
um: Constant. Upper limit on the number of vertices in a shard, $um = (1 + \epsilon)\frac{|V|}{K}$.

The vertex can only stay in one shard. For $i = 1, 2, \ldots, n$,

$$\sum_{t=1}^{K} x_{it} = 1 \tag{1}$$

Balanced graph partitioning requires an equal number of vertices in each shard. For $t = 1, 2, \ldots, K$,

$$lm \leq \sum_{i=1}^{n} x_{it} \leq um \tag{2}$$

According to the definition of the graph partitioning problem, our objective function is to maximize the local edges in the graph. We can get the following model,

$$\max \sum_{i=1}^{n} \sum_{t=1}^{K} \sum_{j \in N(i)} \min\{x_{it}, x_{jt}\}$$

$$\text{s.t.} \sum_{t=1}^{K} x_{it} = 1, \qquad\qquad i = 1, 2, \ldots, n \tag{3}$$

$$lm \leq \sum_{i=1}^{n} x_{it} \leq um, \qquad\qquad t = 1, 2, \ldots, K$$

$$x_{it} \in \{0, 1\}, \qquad\qquad i = 1, 2, \ldots, n, t = 1, 2, \ldots, K$$

In order to remove $'min'$ from the objective function, some variables are introduced. $y_{it} = \sum_{j \in N(i)} \min\{x_{it}, x_{jt}\}$, By a simple derivation we can obtain

$y_{it} = \min\left\{|N(i)|x_{it}, \sum_{j\in N(i)} x_{jt}\right\}$, y_{it} is equivalent to the following four constraints,

$$y_{it} \le |N(i)|x_{it}, i = 1, 2, \ldots, n, t = 1, 2, \ldots, K \tag{4}$$

$$y_{it} \le \sum_{j\in N(i)} x_{jt}, i = 1, 2, \ldots, n, t = 1, 2, \ldots, K \tag{5}$$

$$y_{it} \ge 0, i = 1, 2, \ldots, n, t = 1, 2, \ldots, K \tag{6}$$

$$y_{it} \ge |N(i)|x_{it} + \sum_{j\in N(i)} x_{jt} - |N(i)|, i = 1, 2, \ldots, n, t = 1, 2, \ldots, K \tag{7}$$

Naturally, the following model 2 is obtained.

$$\max \quad \sum_{i=1}^{n}\sum_{t=1}^{K} y_{it}$$

$$\begin{aligned}
s.t. \sum_{t=1}^{K} x_{it} &= 1, & i &= 1, 2, \ldots, n\\
lm \le \sum_{i=1}^{n} x_{it} &\le um, & t &= 1, 2, \ldots, K\\
y_{it} &\le |N(i)|x_{it}, & i &= 1, 2, \ldots, n, t = 1, 2, \ldots, K\\
y_{it} &\le \sum_{j\in N(i)} x_{jt}, & i &= 1, 2, \ldots, n, t = 1, 2, \ldots, K\\
y_{it} &\ge 0, & i &= 1, 2, \ldots, n, t = 1, 2, \ldots, K\\
y_{it} &\ge |N(i)|x_{it} + \sum_{j\in N(i)} x_{jt} - |N(i)|, & i &= 1, 2, \ldots, n, t = 1, 2, \ldots, K\\
x_{it} &\in \{0, 1\}, & i &= 1, 2, \ldots, n, t = 1, 2, \ldots, K
\end{aligned}$$
$$\tag{8}$$

Due to the difficulty in solving large scale integer programming, we cannot solve Model 2 using the existing integer programming solver. So we consider to design the iteration algorithm.

4   Iteration Algorithm

Model 2 is a 0-1 mixed linear programming model which is NP hard, cannot be solved in polynomial time. The number of variables is $2K|V|$, where the number of 0-1 variables is $K|V|$, the number of constraints is $K + |V| + 4K|V|$. When the total number of vertices of the graph is large, the number of variables and constraints are numerous, it is not feasible to solve in time. Designing the iterative algorithm, in each round, For all x_{it}, choose a small number of them as decision variables. Determine y_{it} according to the chosen x_{it}.

Notation

S_x: The set of variables selected. $S_x = \{x_{it}|x_{it} \quad is \quad selected\}$.

n_s: The number of selected variables. $n_s = |S_x|$

S_{xc}: The set of variables corresponding to the selected vertices and its current shard. $S_{xc} = \{x_{is}|i \in V_s, x_{it} \in S_x\}$.

S_v: The set of vertices associated with variables in S_x. $S_v = \{i|\exists t, x_{it} \in S_x\}$.

S_s: The set of shards associated with variables in S_x. $S_s = \{V_s|x_{it} \in S_x, i \in V_s\}$.

$S_{y_{it}}$: The set of selected variables corresponding to the neighbor vertices of vertex i in shard V_t. $S_{y_{it}} = \{x_{jt}|j \in N(i)\} \cap (S_x \cup S_{xc})$.

The goal of our model is to relocate vertices according to the chosen variables such that the number of local edges is maximized. The vertices associated with the selected variables are referred to as the selected vertices. Y_1 is the number of local edges with the selected vertices as endpoints.

$$Y_1 = \sum_{x_{it} \in S_x \cup S_{xc}} y_{it} \tag{9}$$

Y_2 is the number of local edges when the selected vertex is an endpoint and the other endpoint is not selected.

$$Y_2 = \sum_{x_{it} \in S_x \cup S_{xc}} \sum_{j \in N(i) \cap V_t, x_{jt} \notin S_{xc}} x_{it} + \sum_{i \notin S_v} \sum_{x_{jt} \in S_x \cup S_{xc}, j \in N(i)} x_{jt} \tag{10}$$

Y_3 is the number of local edges with unselected vertices as endpoints.

$$Y_3 = \sum_{i \notin S_v} \sum_{j \in N(i) \cap V_t, j \notin S_v} 1 \tag{11}$$

$Y_1 + Y_2 + Y_3$ is equal to the sum of all the local edges in the graph. The goal of balanced graph partitioning is that the number of local edges is maximized. Based on the selection of partial variables, we can derive the following model 3,

$$\max \quad Y_1 + Y_2 + Y_3$$

$$s.t. \quad \sum_{x_{it} \in S_x \cup S_{xc}} x_{it} = 1, \qquad\qquad\qquad i \in S_v$$

$$lm - |V_t| \le \sum_{x_{it} \in S_x} x_{it} - \sum_{x_{ju} \in S_x, j \in V_t, u \neq t} x_{ju} \le um - |V_t|, \quad t \in S_s$$

$$y_{it} \le |S_{y_{it}}|x_{it}, \qquad\qquad\qquad x_{it} \in S_x, S_{y_{it}} \neq \emptyset$$

$$y_{it} \le \sum_{x_{jt} \in S_{y_{it}}} x_{jt}, \qquad\qquad\qquad x_{it} \in S_x, S_{y_{it}} \neq \emptyset$$

$$y_{it} \ge 0, \qquad\qquad\qquad x_{it} \in S_x, S_{y_{it}} \neq \emptyset$$

$$y_{it} \ge |S_{y_{it}}|x_{it} + \sum_{x_{jt} \in S_{y_{it}}} x_{jt} - |S_{y_{it}}|, \qquad x_{it} \in S_x, S_{y_{it}} \neq \emptyset$$

$$x_{it} \in \{0, 1\}, \qquad\qquad\qquad x_{it} \in S_x, S_{z_{it}} \neq \emptyset \tag{12}$$

4.1   Variables Selection Strategy

We use the $gain_{it}$ to represent the number of local edges increased after vertex i relocated to shard V_t. In order to ensure that vertex relocation is considered among all shard pairs. The selection strategy for variable x_{it} is as follows: The variable selection strategy is to preferentially select the variable with a large gain, and then consider all shard pairs. If $i \in V_s$, then (V_s, V_t) is said to be the shard pair associated with the variable x_{it}. We give the shard pair set associated with the selected variables set S_x,

$$SP = \{(V_s, V_t) \,|\, \forall x_{it} \in S_x, i \in V_s\} \tag{13}$$

If $i \in V_s$, then x_{it} is called the variable corresponding to the shard pair (V_s, V_t). We denote the set of variables associated with a shard pair is

$$SPV_{st} = \{x_{it} | i \in V_s, x_{it} \in S_x\} \tag{14}$$

Algorithm 1. Variables Selection

Input $gain$, Number of partitions K, Number of selected variables n
Output Selected variables set S_x
1: $m = \lceil \frac{n}{K(K-1)} \rceil$
2: Sort $gain$ from largest to smallest
3: $S_x = \emptyset$, $j = 0$, $SP = \emptyset$, $SPV_{st} = \emptyset$, $\forall s, t$
4: **while** $|SP| \leq K(K-1)$ **do**
5:     Find the x_{it} corresponding to $gain[j]$, where $i \in V_s$
6:     $M = m$
7:     **if** $|S_x| <= n$ **then**
8:         $M = 2m$
9:     **end if**
10:    **if** $|SPV_{st}| < M$ **then**
11:        Add x_{it} to SPV_{st}
12:        Add x_{it} to S_x
13:    **else if** $(V_s, V_t) \notin SP$ **then**
14:        Add (V_s, V_t) to SP
15:    **end if**
16:    j++
17: **end while**

4.2   Update Parameter

With each iteration, we need to update the vertex gain, the upper and lower bounds on the number of vertices in the shard. It only need to recalculate the gain of the relocated vertices and their neighbor vertices, and update the lm, um values of the shard involved in the vertex relocation.

Algorithm 2. Update Parameters

Input $lm, um, gain$
Output Updated $lm, um, gain$
1: **repeat**
2:    **if** $x_{it} = 1$ and $i \in V_s$ **then**
3:        $Shard \quad V_s : um = um + 1, lm = lm + 1$
4:        $Shard \quad V_t : um = um - 1, lm = lm - 1$
5:    **end if**
6: **until** All x_{it} are iterated
7: **Update** $gain$

4.3   Iteration Vertex Relocation Algorithm

We use hash partitioning to assign each vertex an initial location. Then we calculate the number of vertices in each shard of the graph in initial state, the gain from vertex relocation, and the number of local edges. This paper select some of the variables according to the variable selection strategy and build the 0-1 mixed linear programming. On the based of the results of the model, the vertices corresponding to the variables are relocated. We design the following vertex relocation algorithm (IVRA).

Algorithm 3. Iteration Vertex Relocation Algorithm (IVRA)

Input Graph $G(V, E)$, Relaxation factor ϵ, Partition number K, Number of variables n
Output Partition$\{V_1, V_2, \ldots, V_K\}$
1: Hash Partition
2: Initial $um, lm, gain, global\_opt\_local\_edge\_num$
3: $improvement_flag = $ True
4: **while** $improvement\_flag$ **do**
5:    $improvement_flag = $ False
6:    **Variables Selection**
7:    Construct and solve model 3
8:    Vertex relocation according to the solution variables x_{it} of model 3
9:    **if** The solution objective value of model 3 more than
    $global\_opt\_local\_edge\_num$ **then**
10:        $improvement_flag = $ True
11:        $global_opt_local_edge_num = $ The solution objective value of model 3
12:    **end if**
13:    **Update Parameters**
14: **end while**

5 Experiments

We performed an evaluation over a number of social networks, with varying numbers of shards. Social network real dataset Table 1 is from the Stanford SNAP collection [11]. Due to the ring edges do not become cut edges, this work removes the ring edges (edges where two endpoints overlap) from the datasets FB Athletes and FB Companies. All the graphs are undirected graphs. Our experiments were implemented in C++ and Microsoft Visual Studio Professional 2019. Our algorithm solves the model 3 with CPLEX Optimization Studio 20.1.

Table 1. Graph dataset from the Stanford SNAP collection

Graph name	Vertices	Edges
FB combined	4,039	88,234
FB Athletes	13,866	86,858
FB Companies	14,113	52,310
Youtube	1,134,890	2,987,624
LiveJournal	3,997,962	34,681,189

In the FB combined, we select 100 variables in each iteration i.e. $n_s = 100$. In the FB Athletes and FB Companies, we select 200 variables in each iteration i.e. $n_s = 200$. In the LiveJournal and Youtube, we select 2000 variables in each iteration i.e. $n_s = 2000$. We use CPLEX to solve the 0-1 mixed linear programming in each iteration.

5.1 Experimental Effect and Comparison

We test the performance of our algorithm on Table 1. The partition results of livejournal are reported in Fig. 1. In this figure, five different graph partitioning models, including hash, BLP, BLP-MC, BLP-KL and our method are utilized for the experiments. BLP-MC and BLP-KL are improved methods of the BLP algorithm proposed by Zishi Deng, Torsten Suel [6]. BLP, BLP-MC, BLP-KL use hash partitioning as initialization. Compared with BLP, BLP-MC and BLP-KL algorithms, our algorithm (IVRA) has the largest improvement on the partition result of hash initialization. Especially for 90 shards, IVRA is 28.45% better than BLP algorithm. Considering the shard number of 10, 30, 50, 70 and 90, the average improvement of IVRA compared to BLP is 22.89%.

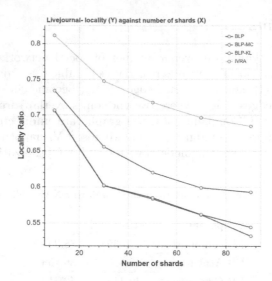

Fig. 1. Performance comparison between the BLP and our algorithm in different shard size of LiveJournal, $n_s = 2000$.

Table 2 shows the local edge ratio of different methods in Facebook_combined. Especially for 11 shards, IVRA is 11.34% better than BLP algorithm. Considering the shard number of 3, 5, 7, 9 and 11, the average improvement of IVRA compared to BLP is 7.32%. Table 3 shows the local edge ratio in Facebook Athletes. Especially for 7 shards, IVRA is 14.04% better than BLP algorithm. Considering the shard number of 3, 5, 7, 9 and 11, the average improvement of IVRA compared to BLP is 11%. Table 4 shows the local edge ratio in Facebook Companies. The largest improvement is obtained with a shard number of 9, IVRA is 7.5% better than BLP algorithm. Considering the shard number of 3, 5, 7, 9 and 11, the average improvement of IVRA compared to BLP is 6.43%. Table 5 shows the local edge ratio in Youtube. The largest improvement is obtained with a shard number of 10, IVRA is 24.97% better than BLP algorithm. Considering the shard number of 10, 30, 50, 70 and 90, the average improvement of IVRA compared to BLP is 19.98%. Our algorithm outperforms BLP and its improved algorithms on all five datasets.

Table 2. Performance comparison in Facebook_combined, $n_s = 100$.

Method/shard number	3	5	7	9	11
Hash	0.333964	0.199016	0.139969	0.111034	0.090067
BLP	0.935898	0.883435	0.874549	0.877485	0.824002
BLP-MC	0.935898	0.88348	0.87464	0.877485	0.824195
BLP-KL	0.93592	0.909581	0.912857	0.899404	0.875955
IVRA	**0.957409**	**0.950858**	**0.951515**	**0.93481**	**0.917651**

Table 3. Performance comparison in Facebook Athletes, $n_s = 200$.

Method/shard number	3	5	7	9	11
Hash	0.331567	0.200134	0.141847	0.111183	0.091128
BLP	0.773174	0.742487	0.687724	0.656691	0.670261
BLP-MC	0.773842	0.74289	0.688023	0.657106	0.670433
BLP-KL	0.784901	0.745217	0.69846	0.678255	0.702929
IVRA	**0.851113**	**0.782528**	**0.784278**	**0.743581**	**0.752485**

Table 4. Performance comparison in Facebook Companies, $n_s = 200$.

Method/shard number	3	5	7	9	11
Hash	0.332464	0.200092	0.141542	0.107873	0.09126
BLP	0.780186	0.728734	0.71377	0.678395	0.689944
BLP-MC	0.783218	0.730288	0.714097	0.678395	0.690538
BLP-KL	0.788321	0.746211	0.719161	0.698519	0.695104
IVRA	**0.830277**	**0.774278**	**0.757012**	**0.729271**	**0.730806**

Table 5. Performance comparison in Youtube, $n_s = 2000$.

Method/shard number	10	30	50	70	90
Hash	0.098640	0.032643	0.019643	0.014011	0.010824
BLP	0.605539	0.569444	0.513764	0.507491	0.474939
BLP-MC	0.612002	0.576813	0.520701	0.516207	0.48472
BLP-KL	0.688525	0.630836	0.597185	0.563218	0.51894
IVRA	**0.756734**	**0.658552**	**0.606803**	**0.594505**	**0.589026**

6   Conclusion

The graph partitioning problem is a classical NP problem with many application contexts. We propose a model of vertex relocation based on 0-1 mixed linear programming. We design an iterative algorithm that selects some variables at each iteration and assigns new shards to the vertices corresponding to these variables. We compare the partitioning effect of BLP, BLP-MC, BLP-KL and our method on five social network datasets. On these datasets, we also compare the effect of different number of shards. In the experiments, compared with BLP and its improved algorithm, the effect of our method has been significantly improved.

The iterative algorithm solves the model by selecting a small number of variables for optimization in each round. The choice of variables has a crucial impact on the partition results. For this reason, our future work will concern more reasonable method for variables selection strategy. For instance this could be obtained considering the structure information of different networks.

References

1. Abbas, Z., Kalavri, V., Carbone, P., Vlassov, V.: Streaming graph partitioning: an experimental study. Proc. VLDB Endow. **11**(11), 1590–1603 (2018)
2. Boyd, D.M., Ellison, N.B.: Social network sites: definition, history, and scholarship. J. Comput. Mediat. Commun. **13**(1), 210–230 (2007)
3. Bui, T.N., Jones, C.: Finding good approximate vertex and edge partitions is NP-hard. Inf. Process. Lett. **42**(3), 153–159 (1992)
4. Buluç, A., Meyerhenke, H., Safro, I., Sanders, P., Schulz, C.: Recent advances in graph partitioning. In: Kliemann, L., Sanders, P. (eds.) Algorithm Engineering. LNCS, vol. 9220, pp. 117–158. Springer, Cham (2016). https://doi.org/10.1007/978-3-319-49487-6_4
5. Chiang, W.L., Liu, X., Si, S., Li, Y., Bengio, S., Hsieh, C.J.: Cluster-GCN: an efficient algorithm for training deep and large graph convolutional networks. In: Proceedings of the 25th ACM SIGKDD International Conference on Knowledge Discovery & Data Mining, pp. 257–266 (2019)
6. Deng, Z., Suel, T.: Optimizing iterative algorithms for social network sharding. In: 2021 IEEE International Conference on Big Data (Big Data), pp. 400–408. IEEE (2021)
7. Fiduccia, C.M., Mattheyses, R.M.: A linear-time heuristic for improving network partitions. In: 19th Design Automation Conference, pp. 175–181. IEEE (1982)
8. Hungershöfer, J., Wierum, J.-M.: On the quality of partitions based on space-filling curves. In: Sloot, P.M.A., Hoekstra, A.G., Tan, C.J.K., Dongarra, J.J. (eds.) ICCS 2002. LNCS, vol. 2331, pp. 36–45. Springer, Heidelberg (2002). https://doi.org/10.1007/3-540-47789-6_4
9. Kahng, A.B., Lienig, J., Markov, I.L., Hu, J.: VLSI Physical Design: From Graph Partitioning to Timing Closure. Springer, Cham (2011). https://doi.org/10.1007/978-90-481-9591-6
10. Kernighan, B.W., Lin, S.: An efficient heuristic procedure for partitioning graphs. Bell Syst. Tech. J. **49**(2), 291–307 (1970)
11. Leskovec, J., Krevl, A.: Snap datasets (2022). https://snap.stanford.edu/data/
12. Lisser, A., Rendl, F.: Graph partitioning using linear and semidefinite programming. Math. Program. **95**(1), 91–101 (2003)
13. Nip, K., Shi, T., Wang, Z.: Some graph optimization problems with weights satisfying linear constraints. J. Comb. Optim. **43**(1), 200–225 (2022)
14. Pothen, A., Simon, H.D., Liou, K.P.: Partitioning sparse matrices with eigenvectors of graphs. SIAM J. Matrix Anal. Appl. **11**(3), 430–452 (1990)
15. Ugander, J., Backstrom, L.: Balanced label propagation for partitioning massive graphs. In: Proceedings of the Sixth ACM International Conference on Web Search and Data Mining, pp. 507–516 (2013)

Partial Inverse Min-Max Spanning Tree Problem Under the Weighted Bottleneck Hamming Distance

Qingzhen Dong, Xianyue Li$^{(\boxtimes)}$, and Yu Yang

School of Mathematics and Statistics, Lanzhou University,
Lanzhou 730000, Gansu, People's Republic of China
lixianyue@lzu.edu.cn

Abstract. Given a undirected connected weighted graph G and a forest F of G, the partial inverse min-max spanning tree problem is to adjust weight function with minimum cost such that there is a min-max spanning tree with respect to the new weight function containing F. In this paper, we study this problem under the weighted bottleneck Hamming distance. Firstly, we consider this problem with value of optimal tree restriction, and present a polynomial time algorithm to solve it. Then, by characterizing the properties of the value of optimal tree, we present a strongly polynomial algorithm for this problem with time complexity $O(m^2 \log m)$, where m is the number of edges of G. Moreover, we show that these algorithms can be generalized to solve these problems with capacitated constraint.

Keywords: Min-max spanning tree · Partial inverse problem · Weighted bottleneck Hamming distance · Strongly polynomial time algorithm

1 Introduction

Given a combinatorial optimization problem A and a feasible solution B, the inverse problem of A is to adjust the parameters so that B becomes an optimal solution with respect to the new parameters, and the change of parameters is required to be least. The change of parameters is usually measured by l_p-norm or Hamming distance. In 1992, Burton and Toint [1] introduced the inverse problem into the field of combinatorial optimization. They pointed out that the problems of traffic planning and seismic tomography can be transformed into the inverse shortest path problem. Heuberger [5] and Demange and Monnot [4] surveyed the inverse combinatorial optimization problems.

A natural generalization of inverse problem is partial inverse problem. The partial inverse combinatorial optimization problem is to replace the feasible solution B in the inverse problem with the partial solution C (a partial solution is a

Supported by National Natural Science Foundation of China (Nos. 11871256, 12071194), and the Basic Research Project of Qinghai (No. 2021-ZJ-703).

Q. Ni and W. Wu (Eds.): AAIM 2022, LNCS 13513, pp. 351–362, 2022.
https://doi.org/10.1007/978-3-031-16081-3_30

part of the feasible solution), and adjust the parameters so that these is an optimal solution with respect to the new parameters contains the partial solution, and the change of the parameters is required to be least.

Spanning tree is a classical combinatorial optimization problem. A lot of results have been studied on the partial inverse minimum/maximum spanning tree problems. In 2008, Cai et al. [2] studied the partial inverse minimum spanning tree problem in which the weight can only be decreased, and gave a polynomial time algorithm to solve it under any norm. In 2016, Li et al. [6] considered the partial inverse maximum spanning tree problem in which the weight can only be decreased and presented a polynomial time algorithm for this problem under l_∞-norm. In 2018, Li et al. [7] considered the partial inverse maximum spanning tree in which the weight function can only be decreased under l_p-norm, and proved that when $|E'| \geq 2$, the problem is APX-Hard; when $|E'| = 1$, the problem is polynomial solvable, where E' is the edge set of the partial solution. In 2019, Li et al. [8] studied the partial inverse maximum spanning tree under the weighted sum Hamming distance, they showed that the problem is APX-Hard when the partial solution has at least two edges, and polynomial solvable when the partial solution has only one edge. In the same paper, they also proved that this problem is polynomial solvable under the weighted bottleneck Hamming distance. In 2020, Li et al. [9] gave the approximation algorithms for partial inverse maximum spanning tree problem. In 2021, Li et al. [10] showed that partial inverse maximum spanning tree problem under the l_∞-norm is polynomial solvable.

Similar to the minimum/maximum spanning tree, *min-max spanning tree* is also a major problem of spanning tree. Let $G = (V, E)$ be a connected graph, in which $V = \{v_1, v_2, \ldots, v_n\}$ is the vertex set, $E = \{e_1, e_2, \ldots, e_m\}$ is the edge set, and a weight function w defined on E. The min-max spanning tree problem is to find a spanning tree T^* of $\mathcal{T}$ such that $w(T^*) = min\{w(T)|T \in \mathcal{T}\}$, where $\mathcal{T}$ is the set of all spanning trees and $w(T) = max_{e \in T} w(e)$. It is known that this problem can be solved in $O(m)$ [3]. For the *inverse min-max spanning tree problem* (abbreviated as IMMST), Yang et al. [14] firstly showed that IMMST under the l_1-norm and l_∞-norm can be solved in strongly polynomial time. In 2008, Liu et al. [11] studied IMSST under the weighted sum-type Hamming distance, they presented a strongly polynomial algorithm to solve it with time complexity $O(n^4)$. In 2009, Liu et al. [12] showed that IMSST under the weighted bottleneck-type Hamming distance is also polynomial solvable. For the *partial inverse min-max spanning tree problem* (abbreviated as PIMMST), Tayyebi and Sepasian [13] presented two polynomial algorithms to solve this problem under the l_1-norm, this is the first study of PIMMST.

For PIMMST under other measure functions, we haven't found any results yet. Thus, it is very meaningful to study whether they have polynomial time algorithms or they are NP-Hard. In this parer, we consider PIMMST under the weighted bottleneck Hamming distance and present polynomial time algorithm to solve it.

This paper is organized as follows. In Sect. 2, we give some basic definitions and properties, and present an algorithm for determining whether a given weight function w is feasible. In Sect. 3, we show the main results. In detail, we firstly present a polynomial algorithm for PIMMST with value of optimal tree restriction under the weighted bottleneck Hamming distance. Then, by characterizing the properties of the value of optimal tree, we present a polynomial algorithm of PIMMST under the weighted bottleneck Hamming distance. In Sect. 4, we study the *capacitated partial inverse min-max spanning tree problem* (abbreviated as CPIMMST) under the weighted bottleneck Hamming distance and present a polynomial algorithm for this problem. In Sect. 5, we make a conclusion.

2 Preliminary

At the beginning of this section, the definition of PIMMST under the weighted bottleneck Hamming distance is introduced as follows.

Definition 1. *Given an undirected connected graph $G = (V, E)$, in which V is the vertex set, E is the edge set, a weight function w defined on E, a norm function c defined on E, and a forest F of G, the goal of PIMMST under the weighted bottleneck Hamming distance is to find a new weight function w^* satisfying:*

(1) there exists a min-max spanning tree T of G with respect to w^* such that $E(F) \subseteq E(T)$;
(2) $\|w^* - w\|_{WBH} = max_{e \in E}\{c(e) \cdot H(w(e), w^*(e))\}$ is minimum,

where $H(w(e), w^*(e))$ is the Hamming distance between $w(e)$ and $w^*(e)$, that is, if $w^*(e) = w(e)$, $H(w(e), w^*(e)) = 0$; otherwise, $H(w(e), w^*(e)) = 1$. Furthermore, we call such a tree T in the first condition as optimal tree, and a weight function w' is feasible for PIMMST if it satisfies the first condition.

The following lemma is necessary and sufficient conditions for a spanning tree T of G to be min-max. For the convenience of the following description, for any weight function w and any real number p, we define $E_{\bar{w},p}^{\geq} = \{e \in E \mid w(e) \geq p\}$, that is, $E_{\bar{w},p}^{\geq}$ is the set of edges whose weights are greater than or equal to p.

Lemma 1. *A spanning tree T of G is a min-max spanning tree with respect to weight function w if and only if $G - E_{w,w(T)}^{\geq}$ is disconnected.*

Proof. Let $G' = G - E_{w,w(T)}^{\geq}$. We firstly prove the necessity by contradiction. If G' is connected, then $w(T') < w(T)$ for any spanning tree T' of G'. Clearly, T' is also a spanning tree of G, which contradicts that T is min-max.

Next, we will prove the sufficiency. Since G' is disconnected, it implies that the weight of each spanning tree of G is greater than or equal to $w(T)$. Thus, T is a min-max spanning tree of G with respect to w.   □

Next, we will explore the properties of an optimal solution of PIMMST under the weighted bottleneck Hamming distance. Firstly, the definition of fundamental cut is introduced.

Definition 2. *Let $G = (V, E)$ be a connected graph, and T be a spanning tree of G. For any edge $e \in E(T)$, $T - e$ has exactly two components. The set of edges connecting these two components is called the* fundamental cut *with respect to T and e, and denoted by $K(T, e)$.*

Clearly, for any edge $e' \in K(T, e)$, $T - e + e'$ is also a spanning tree of G.

Theorem 1. *For any instance $I = (G, w, c, F)$ of PIMMST under the weighted bottleneck Hamming distance, there exist an optimal solution w^* and an optimal tree T^* with respect to w^*, such that*

(1) for any $e \notin E(F)$, $w^*(e) \geq w(e)$;
(2) for any $e \in E(F)$, if $w^*(e) \neq w(e)$, then $w^*(e) = p^*$, where $p^* = w(T^*)$.

Proof. For part (1), we firstly prove that there exist an optimal solution w^* and an optimal tree T^* such that for any $e \in E \backslash E(T^*)$, $w^*(e) \geq w(e)$. Let w^* be any optimal solution of I, and T^* be an optimal tree with respect to w^*. Suppose that there exists an edge $e' \in E \backslash E(T^*)$ with $w^*(e') < w(e')$. Let

$$w'(e) = \begin{cases} w(e'), & e = e'; \\ w^*(e), & otherwise. \end{cases}$$

It is obvious that the only difference between w^* and w' is on the edge e', thus $w^*(T^*) = w'(T^*) = p^*$. If $w^*(e') \geq p^*$, then $w'(e') = w(e') > w^*(e') \geq p^*$. It implies that $E^{\geq}_{w^*, p^*} \subseteq E^{\geq}_{w', p^*}$. By Lemma 1, $G - E^{\geq}_{w^*, p^*}$ is disconnected, so $G - E^{\geq}_{w', p^*}$ is also disconnected. By Lemma 1, w' is a feasible solution of I. However, we have

$$\|w' - w\|_{WBH} \leq \|w^* - w\|_{WBH}.$$

If $\|w' - w\|_{WBH} < \|w^* - w\|_{WBH}$, which contradicts with the optimality of w^*. Hence, w' is another optimal solution of I with $w'(e') = w(e')$. By repeating the above process, we can obtain an optimal solution w^* of I, such that T^* is also an optimal tree and $w^*(e) \geq w(e)$ for any edge $e \in E \backslash E(T^*)$.

Next, we will prove that there exist an optimal solution w^* and an optimal tree T^* such that for any $e \in E(T^*) \backslash E(F)$, $w^*(e) \geq w(e)$. Let w^* be the weight function obtain by the above process. Suppose that there exists an edge $e' \in E(T^*) \backslash E(F)$ with $w^*(e') < w(e')$. Let

$$w'(e) = \begin{cases} w(e'), & e = e'; \\ w^*(e), & otherwise. \end{cases}$$

Let $e'' = \arg min\{w'(e) | e \in K(T^*, e')\}$, and $T' = T^* - e' + e''$. We will prove that T' is a min-max spanning tree with respect to w'. Since $K(T^*, e') \cap F = \emptyset$, T' is a spanning tree containing F. If $w'(e'') = w'(T') = max_{e \in T'} w'(e)$, by the choice of e'', we can see that $K(T^*, e') \subseteq E^{\geq}_{w', w'(T')}$. By Lemma 1, T' is a min-max spanning tree with respect to w'. Otherwise, $w'(e'') < w'(T') \leq p^*$. For any edge $e \in E^{\geq}_{w^*, p^*}$, if $e = e'$, then $w'(e') = w(e') > w^*(e') = p^* \geq w'(T')$; else $e \neq e'$, then $w'(e) = w^*(e) \geq p^* \geq w'(T')$. It implies that $E^{\geq}_{w^*,p^*} \subseteq E^{\geq}_{w',w'(T')}$. Hence, by Lemma 1, w' is a feasible solution of I. However, we have

$$\|w' - w\|_{WBH} \leq \|w^* - w\|_{WBH}.$$

If $\|w' - w\|_{WBH} < \|w^* - w\|_{WBH}$, which contradicts with the optimality of w^*. Hence, w' is another optimal solution of I with $w'(e') = w(e')$. Besides, for any $e \in E\backslash E(T')$, $w'(e) \geq w(e)$ since $T' = T^* - e' + e''$. Hence, by repeating the above process, we can obtain an optimal solution w^* of I, such that T^* is an optimal tree and $w^*(e) \geq w(e)$ for any edge $e \in E(T^*)\backslash E(F)$.

The results of the second part are obvious. □

Remark 1. The one of main differences between partial inverse minimum spanning tree problem (abbreviated as PIMST) and PIMMST is that PIMMST does not satisfies the separation property. Li et al. [8] showed that for any instance I of PIMST under the weighted bottleneck Hamming distance, there exists an optimal solution w^*, such that for any edge $e \in E(F)$, $w^*(e) \leq w(e)$ and for any edge $e \notin E(F)$, $w^*(e) \geq w(e)$. However, Theorem 1 can only ensure that PIMMST under the weighted bottleneck Hamming distance satisfies the second part of separation property. Figure 1 illustrates a counter-example of the first part. In this instance, the partial solution $F = \{v_1v_2, v_3v_4\}$ (dash edges), the two parameters on each edge are $w(e)$ and $c(e)$ (for example, $w(v_1v_2) = 4$, $c(v_1v_2) = 1$). Obviously, the optimal solution is to increase $w(v_1v_2)$ from 4 to 6 and maintain the weights of other edges. Therefore, we can not use the research methods of PIMST to study PIMMST.

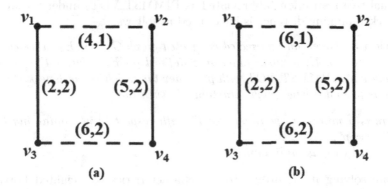

Fig. 1. (a) An instance I of PIMMST under the weighted bottleneck Hamming distance; (b) the optimal solution of I.

At the end of this section, we present the following algorithm to determine whether a given weight function is a feasible solution of this problem.

Remark 2. The main calculation steps of Algorithm 1 are lines 1, 2, 3, which all take $O(m)$ time [3].

Algorithm 1: Determine whether a given weight function w is feasible

---

  Input: A connected graph G, a forest F and a weight function w;
  Output: "True" or "False";
 1 Find a min-max spanning tree T with respect to w of G;
 2 Construct a new graph G' by contracting all edges in F;
 3 Find a min-max spanning tree T' of G';
 4 Set $T'' := T' \cup F$;
 5 **if** $w(T'') = w(T)$ **then**
 6 | return "Ture";
 7 **else**
 8 | return "False";
 9 **end**

---

3 PIMMST Under the Weighted Bottleneck Hamming Distance

In this section, we study the PIMMST under the weighted bottleneck Hamming distance. Before solving this problem, we firstly consider a restricted version of this problem in Subsect. 3.1.

3.1 PIMMST with Value of Optimal Tree Restriction Under the Weighted Bottleneck Hamming Distance

At the beginning of this subsection, the formal definition of PIMMST with value of optimal tree restriction (abbreviated as PIMMST_VOT) under the weighted bottleneck Hamming distance is introduced as follows.

Definition 3. *Given an undirected connected graph $G = (V, E)$, a weight function w defined on E, a norm function c defined on E, a forest F of G and a real number p^*, PIMMST_VOT with p^* under the weighted bottleneck Hamming distance is to find a new weight function w', such that*

(1) there is a min-max spanning tree T' with respect to w' containing F with $w'(T') = p^*$;
(2) $\|w' - w\|_{WBH}$ is minimum.

Before solving it, according to the characteristics of weighted bottleneck Hamming distance, we firstly give the decision version of this problem and propose the following important results.

Given an instance $I = (G, w, c, F; p^*)$ of PIMMST_VOT, and a constant C, the **decision version** of PIMMST_VOT under the weighted bottleneck Hamming distance is to judge whether there exists a new weight function w', such that

(1) there is a min-max spanning tree T' with respect to w' containing F with $w'(T') = p^*$;
(2) $\|w' - w\|_{WBH} \leq C$.

Theorem 2. *Let*

$$w''(e) = \begin{cases} p^*, & c(e) \leq C; \\ w(e), & otherwise. \end{cases}$$

Then, the answer of decision version of PIMMST_VOT under the weighted bottleneck Hamming distance is "Ture" if and only if there is a min-max spanning tree T' with respect to w'' containing F with $w''(T') = p^*$.

Proof. Let's prove the sufficiency firstly. By the definition of w'', we have $\|w'' - w\|_{WBH} \leq C$, which satisfies condition (2). So, the answer of decision version of PIMMST_VOT under the weighted bottleneck Hamming distance is "Ture".

Next, we prove the necessity. Let w' be a feasible solution of PIMMST_VOT, and T' be a min-max spanning tree with respect to w' containing F with $w'(T') = p^*$. Since $w'(T') = p^*$, there is an edge $e' \in E(T')$, $w'(e') = p^*$, and for any other edge $e \in E(T')$, $w'(e) \leq p^*$. If $c(e') \leq C$, we have $w''(e') = p^*$; otherwise, $w''(e') = w(e') = w'(e') = p^*$. Thus, $w''(e') = p^*$. For any other edge $e \in E(T')$, if $c(e) \leq C$, we have $w''(e) = p^*$; otherwise, we have $w''(e) = w(e) = w'(e) \leq p^*$. Thus, $w''(e) \leq p^*$. Therefore, $w''(T') = p^*$.

In the following, we will prove that T' is also a min-max spanning tree with respect to w'' containing F. Clearly, $E(F) \subseteq E(T')$. For any edge $e \in E_{w',p^*}$, if $c(e) \leq C$, then $w''(e) = p^*$; otherwise we have $w''(e) = w(e) = w'(e) \geq p^*$. It implies that $E_{w',p^*} \subseteq E_{w'',p^*}$. By Lemma 1, $G - E^{\geq}_{w',p^*}$ is disconnected, so $G - E^{\geq}_{w'',p^*}$ is also disconnected. By Lemma 1, T' is a min-max spanning tree with respect to w''. Therefore, T' is a min-max spanning tree with respect to w'' containing F. $\square$

By Theorem 2, we present the following algorithm for the decision version of PIMMST_VOT under the weighted bottleneck Hamming distance.

Algorithm 2: Algorithm for the decision version of PIMMST_VOT under the weighted bottleneck Hamming distance

---

Input: An intance $I = (G, w, c, F; p^*)$ of PIMMST_VOT, a constant $C \geq 0$.
Output: "Ture" or "False".
1 Set

$$w'(e) = \begin{cases} p^*, & c(e) \leq C; \\ w(e), & otherwise; \end{cases} \qquad (1)$$

2 Calculate E_{w',p^*};
3 Set $G' := G - E^{\geq}_{w',p^*}$;
4 **if** G' *is connected* **then**
5 | return "False" ;
6 **else**
7 | Execute Algorithm 1 on (G, w', F);
8 **end**

Remark 3. In Algorithm 2, we can use Depth First Search or Breadth First Search to judge whether G' is connected, which takes $O(m)$ time. Combined this with the running time of Algorithm 1, 2 takes $O(m) + O(m) = O(m)$ time.

Clearly, the optimal value of PIMMST_VOT under the weighted bottleneck Hamming distance must be $c(e_i)$ for some edge e_i. Hence, combining Binary search method with Algorithm 2, we give an algorithm for PIMMST_VOT under the weighted bottleneck Hamming distance.

Algorithm 3: Algorithm for PIMMST_VOT under the weighted bottleneck Hamming distance

---

Input: An intance $I = (G, w, c, F; p^*)$ of PIMMST_VOT.

Output: An optimal solution $w_{p^*}^*$ and optimal value $C_{min}^{p^*}$.

1 Order the edges in E as $c(e_1) \leq c(e_2) \leq \ldots \leq c(e_m)$;
2 Use Binary search to find the minimum $c(e_i)$ such that Algorithm 2 on $(G, w, c, F; p^*, c(e_i))$ returns "Ture";
3 Set $C_{min}^{p^*} := c(e_i)$ and

$$w_{p^*}^*(e) = \begin{cases} p^*, & c(e) \leq C_{min}^{p^*}; \\ w(e), & otherwise; \end{cases} \qquad (2)$$

4 **return** $w_{p^*}^*$, $C_{min}^{p^*}$.

Remark 4. In Algorithm 3, sort the weights of edges takes $O(m \log m)$ time. Binary search takes $O(\log m)$ iterations and step 2 takes $O(m \log m)$ times. So, the time complexity of Algorithm 3 is $O(m \log m)$.

3.2 PIMMST Under the Weighted Bottleneck Hamming Distance

According to Algorithm 3, if we can give a candidate set of values of optimal trees in polynomial scale for PIMMST under the weighted bottleneck Hamming distance, we can solve this problem in strongly polynomial time. The following theorem gives the range of optimal tree value.

Theorem 3. *For any instance $I = (G, w, c, F)$ of PIMMST under the weighted bottleneck Hamming distance, there exists an optimal solution w^* with its optimal tree T^* such that $w^*(T^*) = w(e_h)$, for some $e_h \in E$.*

Proof. Let w^* be an optimal solution of I and T^* be an optimal tree with respect to w^* such that $w^*(T^*) = p^*$. If $p^* = w(e_h)$ for some $e_h \in E$, then Theorem 3 already holds. Otherwise, we sort the edges with $w(e_1) \leq w(e_2) \leq \ldots \leq w(e_m)$ and there are three cases of p^*.

Case 1. $w(e_i) < p^* < w(e_{i+1})$, for some $1 \leq i \leq m - 1$.

Now, let

$$\bar{w}(e) = \begin{cases} w(e_i), & w(e_i) < w^*(e) \leq p^*; \\ w^*(e), & otherwise. \end{cases}$$

We will firstly prove that $\bar{w}(T^*) = w(e_i)$. For any $e \in T^*$, if $w^*(e) \leq w(e_i)$, then $\bar{w}(e) = w^*(e) \leq w(e_i)$; if $w(e_i) < w^*(e) \leq p^*$, then $\bar{w}(e) = w(e_i)$. Thus, $\bar{w}(T^*) = w(e_i)$.

Next, we will prove that $G - E^{\geq}_{\bar{w}, \bar{w}(T^*)}$ is disconnected. For any edge $e \in E^{\geq}_{w^*, p^*}$, if $w^*(e) = p^*$, then $\bar{w}(e) = w(e_i) = \bar{w}(T^*)$; if $w^*(e) > p^*$, then $\bar{w}(e) = w^*(e) > p^* > \bar{w}(T^*)$. Thus, $E^{\geq}_{w^*, p^*} \subseteq E^{\geq}_{\bar{w}, \bar{w}(T^*)}$. By Lemma 1, $G - E^{\geq}_{w^*, p^*}$ is disconnected. So, $G - E^{\geq}_{\bar{w}, \bar{w}(T^*)}$ is also disconnected.

By Lemma 1, we can see that $\bar{w}$ is a feasible solution of I. On the other hand, for any edge e with $w(e_i) < w^*(e) \leq p^* < w(e_{i+1})$, $H(w(e), w^*(e)) = 1 \geq H(w(e), \bar{w}(e))$. Hence,

$$\|\bar{w} - w\|_{WBH} \leq \|w^* - w\|_{WBH}.$$

By the optimality of w^*, $\bar{w}$ is another optimal solution of I, and T^* is an optimal tree with respect to $\bar{w}$ such that $\bar{w}(T^*) = w(e_i)$.

Case 2. $p^* > w(e_m)$.

Let

$$\bar{w}(e) = \begin{cases} w(e_m), & w^*(e) > w(e_m); \\ w^*(e), & otherwise. \end{cases}$$

Similar to case 1, we can also prove that $\bar{w}$ is another optimal solution of I, and T^* is optimal with respect to $\bar{w}$ such that $\bar{w}(T^*) = w(e_m)$.

Case 3. $p^* < w(e_1)$.

Let

$$\bar{w}(e) = \begin{cases} w(e_1), & w^*(e) < w(e_1); \\ w^*(e), & otherwise. \end{cases}$$

Clearly, for any edge $e \in T^*$, since $w^*(e) \leq p^* < w(e_1)$, $\bar{w}(e) = w(e_1)$. Thus, $\bar{w}(T^*) = w(e_1)$.

For any edge $e \in E$, we have $\bar{w}(e) \geq w(e_1)$. Thus, $E^{\geq}_{\bar{w}, \bar{w}(T^*)} = E$. It implies that $G - E^{\geq}_{\bar{w}, \bar{w}(T^*)}$ is disconnected. By Lemma 1, $\bar{w}$ is a feasible solution of I. On the other hand, for any edge $w^*(e) < w(e_1)$, $H(w(e), w^*(e)) = 1 \geq H(w(e), \bar{w}(e))$. Hence,

$$\|\bar{w} - w\|_{WBH} \leq \|w^* - w\|_{WBH}.$$

By the optimality of w^*, $\bar{w}$ is another optimal solution of I, and T^* is optimal with respect to $\bar{w}$ such that $\bar{w}(T^*) = w(e_1)$. $\square$

By Theorem 3, we present the following algorithm to solve PIMMST under the weighted bottleneck Hamming distance.

Algorithm 4: Algorithm for PIMMST under the weighted bottleneck Hamming distance

---

Input: An instance $I = (G, w, c, F)$ of PIMMST.
Output: An optimal solution w^*.
1 Sorting the set of weights of all edges in E with $p_1 < p_2 < \cdots < p_t$;
2 **for** $i = 1$ *to* t **do**
3 $\quad$ | Execute Algorithm 3 with $(G, w, c, F; p_i)$;
4 **end**
5 Set $C^* := \min C_{min}^{p_i}$, $p^* = \arg\min C_{min}^{p_i}$;
6 Set

$$w^*(e) = \begin{cases} p^*, & c(e) \leq C^*; \\ w(e), & otherwise; \end{cases} \tag{3}$$

$\quad$ **return** w^*.

Remark 5. Algorithm 4 needs to execute Algorithm 3 at most m times, so the time complexity of Algorithm 4 is $O(m^2 \log m)$.

4 CPIMMST Under the Weighted Bottleneck Hamming Distance

In this section, we consider the capacitated PIMMST (abbreviated as CPIMMST) under the weighted bottleneck Hamming distance. CPIMMST is to add a capacity constraint on PIMMST. That is, there are upper and lower non-negative bound functions u and l, so that

$$-l(e) \leq w^*(e) - w(e) \leq u(e), \text{ for any edge } e \in E.$$

Similar to Sect. 3, we firstly consider CPIMMST with value of optimal tree restriction under the weighted bottleneck Hamming distance.

4.1 CPIMMST with Value of Optimal Tree Restriction Under the Weighted Bottleneck Hamming Distance

At the beginning of this subsection, the formal definition of CPIMMST with value of optimal tree restriction (abbreviated as CPIMMST_VOT) under the weighted bottleneck Hamming distance is introduced as follows.

Definition 4. *Given an undirected connected graph $G = (V, E)$, a weight function w defined on E, a norm function c defined on E, a forest F of G, upper and lower bound functions u and l and a real number p^*, CPIMMST_VOT with p^* under the weighted bottleneck Hamming distance is to find a new weight function w', such that*

(1) there is a min-max spanning tree T' with respect to w' containing F with $w'(T') = p^*$;
(2) $-l(e) \le w'(e) - w(e) \le u(e)$, for any edge $e \in E$;
(3) $\|w' - w\|_{WBH}$ is minimum.

By the same discussion on Subsect. 3.1, we can obtain the following theorem for the decision version of CPIMMST_VOT.

Theorem 4. *Let*

$$w''(e) = \begin{cases} max\{p^*, w(e) - l(e)\}, & c(e) \le C \ and \ w(e) > p^*; \\ min\{p^*, w(e) + u(e)\}, & c(e) \le C \ and \ w(e) < p^*; \\ w(e), & otherwise. \end{cases}$$

Then the answer of decision version of CPIMMST_VOT with parameter C under the weighted bottleneck Hamming distance is "Ture" if and only if there is a min-max spanning tree T'' with respect to w'' containing F with $w''(T'') = p^*$.

4.2   CPIMMST Under the Weighted Bottleneck Hamming Distance

Similar to Subsect. 3.2, we can also give the candidate set of p^* in polynomial scale for CPIMMST under the weighted bottleneck Hamming distance. The following theorem give the range of optimal tree values.

Theorem 5. *For any instance $I = (G, w, c, l, u, F)$ of CPIMMST under the weighted bottleneck Hamming distance, there exists an optimal solution w^* with its optimal tree T^* such that $w^*(T^*)$ must be one of the value in $W = \{w(e_h)|e_h \in E\} \cup \{w(e_h) - l(e_h)|e_h \in E\} \cup \{w(e_h) + u(e_h)|e_h \in E\} \cap [\underline{w}, \infty)$, where $\underline{w} = max\{w(e) - l(e)|e \in E(F)\}$.*

Remark 6. Based on the above results, we can generalize the algorithms for PIMMST in Sect. 3 to solving CPIMMST directly. In detail, for the decision version of CPIMMST_VOT, we just need to replace Eq. (1) in Algorithm 2 with

$$w'(e) = \begin{cases} max\{p^*, w(e) - l(e)\}, & c(e) \le C \ and \ w(e) > p^*; \\ min\{p^*, w(e) + u(e)\}, & c(e) \le C \ and \ w(e) < p^*; \\ w(e), & otherwise. \end{cases}$$

For CPIMMST_VOT, it just need to replace Eq. (2) in Algorithm 3 with

$$w'(e) = \begin{cases} max\{p^*, w(e) - l(e)\}, & c(e) \le C_{min}^{p^*} \ and \ w(e) > p^*; \\ min\{p^*, w(e) + u(e)\}, & c(e) \le C_{min}^{p^*} \ and \ w(e) < p^*; \\ w(e), & otherwise. \end{cases}$$

For CPIMMST, it just to replace Line 1 in Algorithm 4 with "Sorting the set of values in W with $p_1 < p_2 < \cdots < p_t$", and replace Eq. (3) in Algorithm 4 with

$$w'(e) = \begin{cases} max\{p^*, w(e) - l(e)\}, & c(e) \le C^* \ and \ w(e) > p_i; \\ min\{p^*, w(e) + u(e)\}, & c(e) \le C^* \ and \ w(e) < p_i; \\ w(e), & otherwise. \end{cases}$$

5 Concluding

In this paper, we study the partial inverse min-max spanning tree problem under the weighted bottleneck Hamming distance, and present an algorithm to solve this problem with time complexity $O(m^2 \log m)$. Moreover, we show that the algorithm can also solve capacitated partial inverse min-max spanning tree problem under the weighted bottleneck Hamming distance. In the future research, it is very meaningful to continue to study partial inverse min-max spanning tree problem under other measurement standards.

References

1. Burton, D., Toint, P.L.: On an instance of the inverse shortest paths problem. Math. Program. **53**(1–3), 45–61 (1992)
2. Cai, M.-C., Duin, C.W., Yang, X., Zhang, J.: The partial inverse minimum spanning tree problem when weight increasing is forbidden. Eur. J. Oper. Res. **188**, 348–353 (2008)
3. Camerini, P.M.: The min-max spanning tree problem and some extensions. Inf. Process. Lett. **7**(1), 10–14 (1978)
4. Demange, M., Monnot, J.: An Introduction to Inverse Combinatorial Problems, 2nd edn. In: Paschos, V.T. (ed.) Paradigms of Combinatorial Optimization. Wliey, Hoboken (2014)
5. Heuberger, C.: Inverse combinatorial optimization: a survey on problems, methods, and results. J. Comb. Optim. **8**(3), 329–361 (2004)
6. Li, S., Zhang, Z., Lai, H.-J.: Algorithms for constraint partial inverse matroid problem with weight increase forbidden. Theor. Comput. Sci. **640**, 119–124 (2016)
7. Li, X., Zhang, Z., Du, D.-Z.: Partial inverse maximum spanning tree in which weight can only be decreased under l_p-norm. J. Global Optim. **70**, 677–685 (2018)
8. Li, X., Shu, X., Huang, H., Bai, J.: Capacitated partial inverse maximum spanning tree under the weighted Hamming distance. J. Comb. Optim. **38**, 1005–1018 (2019)
9. Li, X., Zhang, Z., Yang, R., Zhang, H., Du, D.-Z.: Approximation algorithms for capacitated partial inverse maximum spanning tree problem. J. Global Optim. **77**(2), 319–340 (2020)
10. Li, X., Yang, R., Zhang, H., Zhang, Z.: Capacitated partial inverse maximum spanning tree under the weighted l_∞-norm. In: Du, D.-Z., Du, D., Wu, C., Xu, D. (eds.) COCOA 2021. LNCS, vol. 13135, pp. 389–399. Springer, Cham (2021). https://doi.org/10.1007/978-3-030-92681-6_31
11. Liu, L., Yao, E.: Inverse min-max spanning tree problem under the weighted sum-type Hamming distance. Theoret. Comput. Sci. **196**, 28–34 (2008)
12. Liu, L., Wang, Q.: Constrained inverse min-max spanning tree problems under the weighted Hamming distance. J. Global Optim. **43**, 83–95 (2009)
13. Tayyebi, J., Sepasian, A.R.: Partial inverse min-max spanning tree problem. J. Comb. Optim. **40**, 1–17 (2020)
14. Yang, X., Zhang, J.: Some inverse min-max network problems under weighted l_1 and l_∞ norms with bound constraints on changes. J. Comb. Optim. **13**(2), 123–135 (2007)

Mixed Metric Dimension of Some Plane Graphs

Na Kang, Zhiquan Li, Lihang Hou, and Jing Qu[✉]

School of Mathematics and Science, Hebei GEO University,
Shijiazhuang 050031, People's Republic of China
qjhj8079@126.com

Abstract. Let G be a finite undirected simple connected graph with
vertex set $V(G)$ and edge set $E(G)$. A vertex $u \in V(G)$ resolves two
elements (vertices or edges) $v, w \in V(G) \cup E(G)$ if $d(u,v) \neq d(u,w)$.
A subset S_m of vertices in G is called a mixed metric generator for G
if every two distinct elements (vertices and edges) of G are resolved by
some vertex of S_m. The minimum cardinality of a mixed metric generator
for G is called the mixed metric dimension and is denoted by $dim_m(G)$.
In this paper, we study the mixed metric dimension for the plane graph
of web graph $\mathbb{W}_n$ and convex polytope $\mathbb{D}_n$.

Keywords: Mixed metric dimension · Mixed metric generator · Plane
graph

1 Introduction

The concept of the metric dimension of graph G was introduced independently
by Slater [18] and Harary and Melter [6]. After these two seminal papers, several
works concerning applications, as well as some theoretical properties, of this
invariant were published. For instance, applications to the navigation of robots
in networks were discussed in [14] and applications to chemistry were discussed
in [2,3,10,11].

Let G be a finite undirected simple connected graph with vertex set $V(G)$
and edge set $E(G)$. The distance $d(u,v)$ between two vertices $u,v \in V(G)$ is
the number of edges in a shortest path between them in G. A vertex $x \in V(G)$
resolves or distinguishes two vertices $u,v \in V(G)$ if $d(u,x) \neq d(v,x)$. A set
$S \subset V(G)$ is a *metric generator* for G if every two distinct vertices of G can be
distinguished by some vertex in S. A *metric basis* of G is a metric generator of
minimum cardinality. The cardinality of a metric basis, denoted by $dim(G)$ is
called the *metric dimension* of G.

Similar to metric dimension, edge metric dimension was introduced by [12]
which uniquely identifies the edges related to a graph. The distance between the
vertex u and edge $e = vw$ is defined as $d(e, u) = min\{d(v, u), d(w, u)\}$. The
vertex $u \in V(G)$ resolves or distinguishes two edges of a graph $e_1, e_2 \in E(G)$
if $d(e_1, u) \neq d(e_2, u)$. A set $S_e \subset V(G)$ is an edge metric generator for G if

© The Author(s), under exclusive license to Springer Nature Switzerland AG 2022
Q. Ni and W. Wu (Eds.): AAIM 2022, LNCS 13513, pp. 363–375, 2022.
https://doi.org/10.1007/978-3-031-16081-3_31

every two distinct edges of G can be distinguished by some vertex of S_e. An *edge metric basis* of G is an edge metric generator of minimum cardinality. The cardinality of an edge metric basis, denoted by $dim_e(G)$ is called the *edge metric dimension* of G. Recently, this variant has been investigated by [15,19,20].

The mixed metric dimension is the combination of well studied metric and edge metric dimension. It was introduced by Kelenc et al. [11]. A vertex $v \in V(G)$ resolves or distinguishes two elements (vertices or edges) $a, b \in V(G) \cup E(G)$ if $d(v, a) \neq d(v, b)$. A set $S_m \subset V(G) \cup E(G)$ is a *mixed metric generator* for G if every two distinct elements (vertices or edges) of G can be distinguished by some vertex in S_m. A *mixed metric basis* of G is a mixed metric generator of minimum cardinality. The cardinality of a mixed metric basis, denoted by $dim_m(G)$ is called the *mixed metric dimension* of G. Let $S_m = \{v_1, v_2, \ldots, v_k\}$ be an ordered subset of $V(G)$. Let a be an element (vertex or edge) of G. The k-tuple $r(a|S_m) = (d(a, v_1), d(a, v_2), \ldots, d(a, v_k))$ is called mixed metric representation of a with respect to S_m. Clearly, S_m is a mixed metric generator if and only if for every two distinct elements (vertices or edges) a, b of G we have $r(a|S_m) \neq r(b|S_m)$. Calculation of the mixed metric dimension of a graph G can be found in [4,5, 16,17].

In this paper, we study the mixed metric dimension of two classes of plane graphs: web graph $\mathbb{W}_n$, plane graph (convex polytope) $\mathbb{D}_n$. We show that the mixed metric dimension of $\mathbb{W}_n$ is not constant and the mixed metric dimension of $\mathbb{D}_n$ is constant. For $\mathbb{W}_n$, $dim_m(\mathbb{W}_n) = n + 1$ when $n \geqslant 3$. For $\mathbb{D}_n$, $dim_m(\mathbb{D}_n) = 4$ when $n \geqslant 3$.

The organization of the paper is as follows. In the following section, we recall some results concerning plane graphs: $\mathbb{W}_n$, $\mathbb{D}_n$. In Sect. 3, we study the mixed metric dimension of $\mathbb{W}_n$. In Sect. 4, we study the mixed metric dimension of $\mathbb{D}_n$. In the last section, we conclude this paper.

2   Preliminaries

The web graph $\mathbb{W}_n$ [13] (Fig. 1) has $3n$ vertices and $4n$ edges. We have the vertex set
$$V(\mathbb{W}_n) = \{a_i, b_i, c_i | 1 \leqslant i \leqslant n\},$$
and all edges $E(\mathbb{W}_n) = \{a_i a_{i+1}, a_i b_i, b_i b_{i+1}, b_i c_i | 1 \leqslant i \leqslant n\}$ $(a_{n+1} = a_1, b_{n+1} = b_1)$.

The plane graph (convex polytope) $\mathbb{D}_n$ [1] (Fig. 2) has $4n$ vertices and $6n$ edges. We have the vertex set
$$V(\mathbb{D}_n) = \{a_i, b_i, c_i, d_i | 1 \leqslant i \leqslant n\},$$
and all edges $E(\mathbb{D}_n) = \{a_i a_{i+1}, a_i c_i, c_i d_i, c_{i+1} d_i, b_i d_i, b_i b_{i+1} | 1 \leqslant i \leqslant n\}$ $(a_{n+1} = a_1, b_{n+1} = b_1, c_{n+1} = c_1)$. Let $A = \{a_i : 1 \leqslant i \leqslant n\}$, $B = \{b_i : 1 \leqslant i \leqslant n\}$, $C = \{c_i : 1 \leqslant i \leqslant n\}$, $D = \{d_i : 1 \leqslant i \leqslant n\}$.

Lemma 1. [8] *For $n \geqslant 5$, let $\mathbb{W}_n$ be a web graph. Then $dim(\mathbb{W}_n)$ is equal to 2 if n is odd and 3 if n is even.*

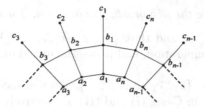

Fig. 1. The web graph $\mathbb{W}_n$

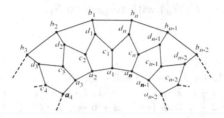

Fig. 2. The plane graph $\mathbb{D}_n$

Lemma 2. [7] *Let $\mathbb{D}_n$ be the graph of convex polytope with $n \geqslant 4$, then we have* $dim(\mathbb{D}_n) = 3$.

Lemma 3. [21] *For the web graph $\mathbb{W}_n$ with $n \geqslant 3$, we have $dim_e(\mathbb{W}_n) = 3$.*

Lemma 4. [21] *For the graph of convex polytope $\mathbb{D}_n$ with $n \geqslant 3$, then we have* $dim_e(\mathbb{D}_n) = 3$.

3   The Mixed Metric Dimension of Web Graph $\mathbb{W}_n$

In this section we intend to present the mixed metric dimension of web graph $\mathbb{W}_n$ (Fig. 1).

Lemma 5. *Let $\mathbb{W}_n$ be the web graph, where $n \geqslant 3$. Let $W = \{c_1, c_2, \ldots, c_n\}$ be a subset of $V(\mathbb{W}_n)$. For arbitrary mixed metric generators S_m of $\mathbb{W}_n$, we have $W \subseteq S_m$.*

Proof. Suppose that $c_i \notin S_m$. Then we have $r(b_i c_i | S_m) = r(b_i | S_m)$, which is a contradiction to the fact that S_m is a mixed metric generator. Therefore we have $W \subseteq S_m$.

Lemma 6. *Let $\mathbb{W}_n$ be the web graph, where $n \geqslant 3$. Then $dim_m(\mathbb{W}_n) \geqslant n + 1$.*

Proof. Let S_m be any mixed metric generators for $\mathbb{W}_n$. By Lemma 5, we get $|S_m| \geqslant n$. If $|S_m| = n$, then we have $S_m = \{c_1, c_2, \ldots, c_n\}$. Note that $r(a_i b_i | S_m) = r(b_i | S_m)$, which is a contradiction to the fact that S_m is a mixed metric generator. Thus, we have $dim_m(\mathbb{W}_n) \geqslant n + 1$.

Theorem 1. *Let* $\mathbb{W}_n$ *be the web graph, where* $n \geqslant 3$. *Then* $dim_m(\mathbb{W}_n) = n+1$.

Proof. For $3 \leqslant n \leqslant 5$, we find that $\{a_1, c_1, c_2, \ldots, c_n\}$ is the mixed metric basis of $\mathbb{W}_n$ by total enumeration, and hence the mixed metric dimension of $\mathbb{W}_n$ is $n+1$.

For $n \geqslant 6$, let $S_m = \{a_1, c_1, c_2, \ldots, c_n\}$. We will show that S_m is a mixed metric generator of $\mathbb{W}_n$ in Cases (I) and (II), respectively.

Case (I) n is odd. In this case, we can write $n = 2l + 1$, where $l \geqslant 3$ is an integer. Let $S_1 = \{a_1, c_1, c_3, c_{l+3}\}$. We give mixed metric representations of any element of $V(\mathbb{W}_n) \cup E(\mathbb{W}_n)$ with respect to S_1.

$$r(a_i|S_1) = \begin{cases} (i-1,\ i+1,\ 5-i,\ l+i), & 1 \leqslant i \leqslant 2; \\ (i-1,\ i+1,\ i-1,\ l+5-i), & 3 \leqslant i \leqslant l+1; \\ (2l+2-i,\ 2l+4-i,\ i-1,\ l+5-i), & l+2 \leqslant i \leqslant l+3; \\ (2l+2-i,\ 2l+4-i,\ 2l+6-i,\ i-l-1),\ l+4 \leqslant i \leqslant 2l+1. \end{cases}$$

$$r(b_i|S_1) = \begin{cases} (i,\ i,\ 4-i,\ l+i-1), & 1 \leqslant i \leqslant 2; \\ (i,\ i,\ i-2,\ l-i+4), & 3 \leqslant i \leqslant l+1; \\ (2l-i+3,\ 2l-i+3,\ i-2,\ l-i+4), & l+2 \leqslant i \leqslant l+3; \\ (2l+3-i,\ 2l+3-i,\ 2l+5-i,\ i-l-2),\ l+4 \leqslant i \leqslant 2l+1. \end{cases}$$

$$r(c_i|S_1) = \begin{cases} (2,\ 0,\ 4,\ l+1), & i = 1; \\ (3,\ 3,\ 3,\ l+2), & i = 2; \\ (4,\ 4,\ 0,\ l+2), & i = 3; \\ (i+1,\ i+1,\ i-1,\ l+5-i), & 4 \leqslant i \leqslant l+1; \\ (l+2,\ l+2,\ l+1,\ 3), & i = l+2; \\ (l+1,\ l+1,\ l+2,\ 0), & i = l+3; \\ (2l+4-i,\ 2l+4-i,\ 2l+6-i,\ i-l-1),\ l+4 \leqslant i \leqslant 2l+1. \end{cases}$$

$$r(a_i a_{i+1}|S_1) = \begin{cases} (i-1,\ i+1,\ 4-i,\ l+i), & 1 \leqslant i \leqslant 2; \\ (i-1,\ i+1,\ i-1,\ l-i+4), & 3 \leqslant i \leqslant l+1; \\ (l-1,\ l+1,\ l+1,\ 2), & i = l+2; \\ (2l+1-i,\ 2l+3-i,\ 2l+5-i,\ i-l-1),\ l+3 \leqslant i \leqslant 2l+1. \end{cases}$$

$$r(a_i b_i|S_1) = \begin{cases} (i-1,\ i,\ 4-i,\ l-1+i), & 1 \leqslant i \leqslant 2; \\ (i-1,\ i,\ i-2,\ l+4-i), & 3 \leqslant i \leqslant l+1; \\ (2l+2-i,\ 2l+3-i,\ i-2,\ l+4-i), & l+2 \leqslant i \leqslant l+3; \\ (2l+2-i,\ 2l+3-i,\ 2l+5-i,\ i-l-2),\ l+4 \leqslant i \leqslant 2l+1. \end{cases}$$

$$r(b_i b_{i+1}|S_1) = \begin{cases} (i,\ i,\ 3-i,\ l-1+i), & l \leqslant i \leqslant 2; \\ (i,\ i,\ i-2,\ l+3-i), & 3 \leqslant i \leqslant l+1; \\ (l,\ l,\ l,\ 1), & i = l+2; \\ (2l-i+2,\ 2l-i+2,\ 2l+4-i,\ i-l-2),\ l+3 \leqslant i \leqslant 2l+1. \end{cases}$$

$$r(b_i c_i|S_1) = \begin{cases} (1,\ 0,\ 3,\ l), & i = 1; \\ (2,\ 2,\ 2,\ l+1), & i = 2; \\ (3,\ 3,\ 0,\ l+1), & i = 3; \\ (i,\ i,\ i-2,\ l+4-i), & 4 \leqslant i \leqslant l+1; \\ (l+1,\ l+1,\ l,\ 2), & i = l+2; \\ (l,\ l,\ l+1,\ 0), & i = l+3; \\ (2l+3-i,\ 2l+3-i,\ 2l+5-i,\ i-l-2),\ l+4 \leqslant i \leqslant 2l+1. \end{cases}$$

Note that when $1 \leqslant i \leqslant n$ and $i \neq 1, 3, l+3$, we have $r(b_i c_i | S_1) = r(b_i | S_1)$. In other cases, all mixed metric representations with respect to S_1 are pairwise different. Therefore, in other cases, all mixed metric representations with respect to S_m are pairwise different. However, when $1 \leqslant i \leqslant n$ and $i \neq 1, 3, l+3$, we have $r(b_i c_i | S_1 \cup c_i) \neq r(b_i | S_1 \cup c_i)$. It follows that $r(b_i c_i | S_m) \neq r(b_i | S_m)$ for $1 \leqslant i \leqslant n$. Hence S_m is a mixed metric generator and therefore $dim_m(\mathbb{W}_n) \leqslant n+1$. By Lemma 6 we have $dim_m(\mathbb{W}_n) \geqslant n+1$. Thus, we obtain that $dim_m(\mathbb{W}_n) = n+1$.

Case (II) n is even. In this case, we can write $n = 2l$, where $l \geqslant 3$ is an integer. Let $S_1 = \{a_1, c_1, c_3, c_{l+2}\}$. We give mixed metric representations of any element of $V(\mathbb{W}_n) \cup E(\mathbb{W}_n)$ with respect to S_1.

$$r(a_i|S_1) = \begin{cases} (i-1,\ i+1,\ 5-i,\ l+i), & 1 \leqslant i \leqslant 2; \\ (i-1,\ i+1,\ i-1,\ l+4-i), & 3 \leqslant i \leqslant l+1; \\ (l-1,\ l+1,\ l+1,\ 2), & i = l+2; \\ (2l-i+1,\ 2l-i+3,\ 2l-i+5,\ i-l), & l+3 \leqslant i \leqslant 2l. \end{cases}$$

$$r(b_i|S_1) = \begin{cases} (i,\ i,\ 4-i,\ l-1+i), & 1 \leqslant i \leqslant 2; \\ (i,\ i,\ i-2,\ l+3-i), & 3 \leqslant i \leqslant l+1; \\ (l,\ l,\ l,\ 1), & i = l+2; \\ (2l-i+2,\ 2l-i+2,\ 2l-i+4,\ i-l-1), & l+3 \leqslant i \leqslant 2l. \end{cases}$$

$$r(c_i|S_1) = \begin{cases} (2,\ 0,\ 4,\ l+1), & i = 1; \\ (3,\ 3,\ 3,\ l+2), & i = 2; \\ (4,\ 4,\ 0,\ l+1), & i = 3; \\ (i+1,\ i+1,\ i-1,\ l+4-i), & 4 \leqslant i \leqslant l+1; \\ (l+1,\ l+1,\ l+1,\ 0), & i = l+2; \\ (2l-i+3,\ 2l-i+3,\ 2l-i+5,\ i-l), & l+3 \leqslant i \leqslant 2l. \end{cases}$$

$$r(a_i a_{i+1}|S_1) = \begin{cases} (0,\ 2,\ 3,\ l+1), & i = 1; \\ (1,\ 3,\ 2,\ l+1), & i = 2; \\ (i-1,\ i+1,\ i-1,\ l+3-i), & 3 \leqslant i \leqslant l; \\ (l-1,\ l+1,\ l,\ 2), & i = l+1; \\ (l-2,\ l,\ l+1,\ 2), & i = l+2; \\ (2l-i,\ 2l-i+2,\ 2l-i+4,\ i-l), & l+3 \leqslant i \leqslant 2l. \end{cases}$$

$$r(a_i b_i|S_1) = \begin{cases} (i-1,\ i,\ 4-i,\ l-1+i), & 1 \leqslant i \leqslant 2; \\ (i-1,\ i,\ i-2,\ l+3-i), & 3 \leqslant i \leqslant l+1; \\ (l-1,\ l,\ l,\ 1), & i = l+2; \\ (2l-i+1,\ 2l-i+2,\ 2l-i+4,\ i-l-1), & l+3 \leqslant i \leqslant 2l. \end{cases}$$

$$r(b_i b_{i+1}|S_1) = \begin{cases} (1,\ 1,\ 2,\ l), & i = 1; \\ (2,\ 2,\ 1,\ l), & i = 2; \\ (i,\ i,\ i-2,\ l+2-i), & 3 \leqslant i \leqslant l; \\ (2l-i+1,\ 2l-i+1,\ i-2,\ 1), & l+1 \leqslant i \leqslant l+2; \\ (2l-i+1,\ 2l-i+1,\ 2l-i+3,\ i-l-1), & l+3 \leqslant i \leqslant 2l. \end{cases}$$

$$r(b_i c_i|S_1) = \begin{cases} (1,\ 0,\ 3,\ l), & i = 1; \\ (2,\ 2,\ 2,\ l+1), & i = 2; \\ (3,\ 3,\ 0,\ l), & i = 3; \\ (i,\ i,\ i-2,\ l+3-i), & 4 \leqslant i \leqslant l+1; \\ (l,\ l,\ l,\ 0), & i = l+2; \\ (2l-i+2,\ 2l-i+2,\ 2l-i+4,\ i-l-1), & l+3 \leqslant i \leqslant 2l. \end{cases}$$

Note that when $1 \leqslant i \leqslant n$ and $i \neq 1, 3, l+2$, we have $r(b_i c_i | S_1) = r(b_i | S_1)$. In other cases, all mixed metric representations with respect to S_1 are pairwise different. Thus, in other cases, all mixed metric representations with respect to S_m are pairwise different. However, when $1 \leqslant i \leqslant n$ and $i \neq 1, 3, l+2$, we have $r(b_i c_i | S_1 \cup c_i) \neq r(b_i | S_1 \cup c_i)$. It follows that $r(b_i c_i | S_m) \neq r(b_i | S_m)$ for $1 \leqslant i \leqslant n$. Hence S_m is a mixed metric generator and therefore $dim_m(\mathbb{W}_n) \leqslant n+1$. By Lemma 6 we have $dim_m(\mathbb{W}_n) \geqslant n+1$. Thus, we obtain that $dim_m(\mathbb{W}_n) = n+1$.

Therefore, for $n \geqslant 3$ we have $dim_m(\mathbb{W}_n) = n+1$.

4   The Mixed Metric Dimension of Plane Graph (Convex Polytope) $\mathbb{D}_n$

In this section, we intend to present the mixed metric dimension of plane graph (convex polytope) $\mathbb{D}_n$ (Fig. 2).

Lemma 7. *Let $\mathbb{D}_n$ be the plane graph (convex polytope), where $n \geqslant 10$. Then $dim_m(\mathbb{D}_n) \leqslant 4$.*

Proof. We consider two cases.

Case (I) n is odd. In this case, we can write $n = 2l+1$, where $l \geqslant 5$ is an integer. Let $S_m = \{a_1, a_{l+1}, b_2, b_{l+2}\}$. We will show that S_m is a mixed metric generator of $\mathbb{D}_n$. We give mixed metric representations of any element of $V(\mathbb{D}_n) \cup E(\mathbb{D}_n)$ with respect to S_m.

$$r(a_i | S_m) = \begin{cases} (i-1, \ l-i+1, \ 5-i, \ l+i+1), & 1 \leqslant i \leqslant 2; \\ (i-1, \ l-i+1, \ i, \ l-i+5), & 3 \leqslant i \leqslant l+1; \\ (l, \ 1, \ l+2, \ 3), & i = l+2; \\ (2l-i+2, \ i-l-1, \ 2l-i+6, \ i-l), & l+3 \leqslant i \leqslant 2l+1. \end{cases}$$

$$r(b_i | S_m) = \begin{cases} (3, \ l+2, \ 1, \ l), & i = 1; \\ (i+2, \ l-i+3, \ i-2, \ l-i+2), & 2 \leqslant i \leqslant l; \\ (2l-i+4, \ i-l+2, \ i-2, \ l-i+2), & l+1 \leqslant i \leqslant l+2; \\ (2l-i+4, \ i-l+2, \ 2l-i+3, \ i-l-2), & l+3 \leqslant i \leqslant 2l+1. \end{cases}$$

$$r(c_i | S_m) = \begin{cases} (i, \ l-i+2, \ 4-i, \ l+i), & 1 \leqslant i \leqslant 2; \\ (i, \ l-i+2, \ i-1, \ l-i+4), & 3 \leqslant i \leqslant l+1; \\ (l+1, \ 2, \ l+1, \ 2), & i = l+2; \\ (2l-i+3, \ i-l, \ 2l-i+5, \ i-l-1), & l+3 \leqslant i \leqslant 2l+1. \end{cases}$$

$$r(d_i | S_m) = \begin{cases} (2, \ l+1, \ 2, \ l+1), & i = 1; \\ (i+1, \ l-i+2, \ i-1, \ l-i+3), & 2 \leqslant i \leqslant l; \\ (2l-i+3, \ i-l+1, \ i-1, \ l-i+3), & l+1 \leqslant i \leqslant l+2; \\ (2l-i+3, \ i-l+1, \ 2l-i+4, \ i-l-1), & l+3 \leqslant i \leqslant 2l+1. \end{cases}$$

$$r(a_i a_{i+1} | S_m) = \begin{cases} (i-1, \ l-1, \ 3, \ l+2), & 1 \leqslant i \leqslant 2; \\ (i-1, \ l-i, \ i, \ l-i+4), & 3 \leqslant i \leqslant l; \\ (2l-i+1, \ i-l-1, \ i, \ 3), & l+1 \leqslant i \leqslant l+2; \\ (2l-i+1, \ i-l-1, \ 2l-i+5, \ i-l), & l+3 \leqslant i \leqslant 2l+1. \end{cases}$$

$$r(a_i c_i | S_m) = \begin{cases} (i-1,\ l-i+1,\ 4-i,\ l+i), & 1 \leqslant i \leqslant 2; \\ (i-1,\ l-i+1,\ i-1,\ l-i+4), & 3 \leqslant i \leqslant l+1; \\ (l,\ 1,\ l+1,\ 2), & i = l+2; \\ (2l-i+2,\ i-l-1,\ 2l-i+5,\ i-l-1), & l+3 \leqslant i \leqslant 2l+1. \end{cases}$$

$$r(c_i d_i | S_m) = \begin{cases} (1,\ l+1,\ 2,\ l+1), & i = 1; \\ (i,\ l-i+2,\ i-1,\ l-i+3), & 2 \leqslant i \leqslant l+1; \\ (l+1,\ 2,\ l+1,\ 1), & i = l+2; \\ (2l-i+3,\ i-l,\ 2l-i+4,\ i-l-1), & l+3 \leqslant i \leqslant 2l+1. \end{cases}$$

$$r(c_{i+1} d_i | S_m) = \begin{cases} (2,\ l,\ 2,\ l+1), & i = 1; \\ (i+1,\ l-i+1,\ i-1,\ l-i+3), & 2 \leqslant i \leqslant l; \\ (2l-i+2,\ i-l+1,\ i-1,\ l-i+3), & l+1 \leqslant i \leqslant l+2; \\ (2l-i+2,\ i-l+1,\ 2l-i+4,\ i-l-1), & l+3 \leqslant i \leqslant 2l; \\ (1,\ l+1,\ 3,\ l), & i = 2l+1. \end{cases}$$

$$r(b_i d_i | S_m) = \begin{cases} (2,\ l+1,\ 1,\ l), & i = 1; \\ (i+1,\ l-i+2,\ i-2,\ l-i+2), & 2 \leqslant i \leqslant l; \\ (2l-i+3,\ i-l+1,\ i-2,\ l-i+2), & l+1 \leqslant i \leqslant l+2; \\ (2l-i+3,\ i-l+1,\ 2l-i+3,\ i-l-2), & l+3 \leqslant i \leqslant 2l+1. \end{cases}$$

$$r(b_i b_{i+1} | S_m) = \begin{cases} (3,\ l+1,\ 0,\ l), & i = 1; \\ (i+2,\ l-i+2,\ i-2,\ l-i+1), & 2 \leqslant i \leqslant l-1; \\ (l+2,\ 3,\ i-2,\ l+1-i), & l \leqslant i \leqslant l+1; \\ (2l-i+3,\ i-l+2,\ 2l-i+2,\ i-l-2), & l+2 \leqslant i \leqslant 2l; \\ (3,\ l+2,\ 1,\ l-1), & i = 2l+1. \end{cases}$$

Note that all mixed metric representations with respect to S_m are pairwise different. We deduce that S_m is a mixed metric generator for $\mathbb{D}_n$.

Case (II) n is even. In this case, we can write $n = 2l$, where $l \geqslant 5$ is an integer. Let $S_m = \{a_1,\ a_{l+1},\ b_2,\ b_{l+2}\}$. We will show that S_m is a mixed metric generator of $\mathbb{D}_n$. We give mixed metric representations of any element of $V(\mathbb{D}_n) \cup E(\mathbb{D}_n)$ with respect to S_m.

$$r(a_i | S_m) = \begin{cases} (i-1,\ l-i+1,\ 5-i,\ l+i), & 1 \leqslant i \leqslant 2; \\ (i-1,\ l-i+1,\ i,\ l-i+5), & 3 \leqslant i \leqslant l+1; \\ (l-1,\ 1,\ l+2,\ 3), & i = l+2; \\ (2l-i+1,\ i-l-1,\ 2l-i+5,\ i-l), & l+3 \leqslant i \leqslant 2l. \end{cases}$$

$$r(b_i | S_m) = \begin{cases} (i+2,\ l-i+3,\ 2-i,\ l+i-2), & 1 \leqslant i \leqslant 2; \\ (i+2,\ l-i+3,\ i-2,\ l-i+2), & 3 \leqslant i \leqslant l; \\ (2l-i+3,\ i-l+2,\ i-2,\ l-i+2), & l+1 \leqslant i \leqslant l+2; \\ (2l-i+3,\ i-l+2,\ 2l-i+2,\ i-l-2), & l+3 \leqslant i \leqslant 2l. \end{cases}$$

$$r(c_i | S_m) = \begin{cases} (i,\ l-i+2,\ 4-i,\ l+i-1), & 1 \leqslant i \leqslant 2; \\ (i,\ l-i+2,\ i-1,\ l-i+4), & 3 \leqslant i \leqslant l+1; \\ (l,\ 2,\ l+1,\ 2), & i = l+2; \\ (2l-i+2,\ i-l,\ 2l-i+4,\ i-l-1), & l+3 \leqslant i \leqslant 2l. \end{cases}$$

$$r(d_i | S_m) = \begin{cases} (2,\ l+1,\ 2,\ l), & i = 1; \\ (i+1,\ l-i+2,\ i-1,\ l-i+3), & 2 \leqslant i \leqslant l; \\ (2l-i+2,\ i-l+1,\ i-1,\ l-i+3), & l+1 \leqslant i \leqslant l+2; \\ (2l-i+2,\ i-l+1,\ 2l-i+3,\ i-l-1), & l+3 \leqslant i \leqslant 2l. \end{cases}$$

$$r(a_i a_{i+1} | S_m) = \begin{cases} (i-1,\ l-1,\ 3,\ l+i), & 1 \leqslant i \leqslant 2; \\ (i-1,\ l-i,\ i,\ l-i+4), & 3 \leqslant i \leqslant l; \\ (2l-i,\ i-l-1,\ i,\ l-i+4), & l+1 \leqslant i \leqslant l+2; \\ (2l-i,\ i-l-1,\ 2l-i+4,\ i-l), & l+3 \leqslant i \leqslant 2l. \end{cases}$$

$$r(a_i c_i | S_m) = \begin{cases} (i-1,\ l-i+1,\ 4-i,\ l+i-1), & 1 \leqslant i \leqslant 2; \\ (i-1,\ l-i+1,\ i-1,\ l-i+4), & 3 \leqslant i \leqslant l+1; \\ (l-1,\ 1,\ l+1,\ 2), & i = l+2; \\ (2l-i+1,\ i-l-1,\ 2l-i+4,\ i-l-1), & l+3 \leqslant i \leqslant 2l. \end{cases}$$

$$r(c_i d_i | S_m) = \begin{cases} (1,\ l+1,\ 2,\ l), & i = 1; \\ (i,\ l-i+2,\ i-1,\ l-i+3), & 2 \leqslant i \leqslant l+1; \\ (l,\ 2,\ l+1,\ 1), & i = l+2; \\ (2l-i+2,\ i-l,\ 2l-i+3,\ i-l-1), & l+3 \leqslant i \leqslant 2l. \end{cases}$$

$$r(c_{i+1} d_i | S_m) = \begin{cases} (2,\ l,\ 2,\ l), & i = 1; \\ (i+1,\ l-i+1,\ i-1,\ l-i+3), & 2 \leqslant i \leqslant l; \\ (2l-i+1,\ i-l+1,\ i-1,\ l-i+3), & l+1 \leqslant i \leqslant l+2; \\ (2l-i+1,\ i-l+1,\ 2l-i+3,\ i-l-1), & l+3 \leqslant i \leqslant 2l. \end{cases}$$

$$r(b_i d_i | S_m) = \begin{cases} (2,\ l+1,\ 1,\ l-1), & i = 1; \\ (i+1,\ l-i+2,\ i-2,\ l-i+2), & 2 \leqslant i \leqslant l; \\ (2l-i+2,\ i-l+1,\ i-2,\ l-i+2), & l+1 \leqslant i \leqslant l+2; \\ (2l-i+2,\ i-l+1,\ 2l-i+2,\ i-l-2), & l+3 \leqslant i \leqslant 2l. \end{cases}$$

$$r(b_i b_{i+1} | S_m) = \begin{cases} (3,\ l+1,\ 0,\ l-1), & i = 1; \\ (i+2,\ l-i+2,\ i-2,\ l-i+1), & 2 \leqslant i \leqslant l-1; \\ (l+2,\ 3,\ l-2,\ 1), & i = l; \\ (l+1,\ 3,\ l-1,\ 0), & i = l+1; \\ (2l-i+2,\ i-l+2,\ 2l-i+1,\ i-l-2), & l+2 \leqslant i \leqslant 2l-1; \\ (3,\ l+2,\ 1,\ l-2), & i = 2l. \end{cases}$$

Note that all mixed metric representations with respect to S_m are pairwise different. We deduce that S_m is a mixed metric generator for $\mathbb{D}_n$.

Therefore, for $n \geqslant 10$ we have $dim_m(\mathbb{D}_n) \leqslant 4$.

Lemma 8. *Let $\mathbb{D}_n$ be the plane graph (convex polytope), where $n \geqslant 10$. Let $C_i = \{c_i, c_{i+1}, d_i\} \subset C \cup D$, $D_i = \{d_{i-1}, d_i, c_i\} \subset C \cup D$. Then the following* (i) *and* (ii) *hold.*

(i) $r(b_i | B \cup C \cup D \setminus C_i) = r(b_i d_i | B \cup C \cup D \setminus C_i)$ *for* $1 \leqslant i \leqslant n$;
(ii) $r(a_i | A \cup C \cup D \setminus D_i) = r(a_i c_i | A \cup C \cup D \setminus D_i)$ *for* $1 \leqslant i \leqslant n$.

Proof. (i) We consider the subsequent two cases depending upon n.

Case (I) n is odd. In this case, we can write $n = 2l + 1$, where $l \geqslant 5$ is an integer. Now, we calculate the distance between the vertexs b_i and x_j, and the distance between the edges $b_i d_i$ and the vertex x_j, where $x_j \in B \cup C \cup D$,

$1 \leqslant i, j \leqslant n.$

$$\begin{cases} d(b_i, b_j) = d(b_i d_i, b_j) = |i - j|, & |i - j| \leqslant l; \\ d(b_i, b_j) = d(b_i d_i, b_j) = n - |i - j|, & l + 1 \leqslant |i - j| \leqslant 2l. \end{cases}$$

$$\begin{cases} d(b_i, c_j) = 2, d(b_i d_i, c_j) = 1, & j = i, j = i+1; \\ d(b_i, c_j) = d(b_i d_i, c_j) = |i - j| + 1, & |i - j| \leqslant l \text{ and } j > i, j \leqslant i, j \leqslant i+1; \\ d(b_i, c_j) = d(b_i d_i, c_j) = |i - j| + 2, & |i - j| \leqslant l \text{ and } j < i, j \leqslant i, j \leqslant i+1; \\ d(b_i, c_j) = d(b_i d_i, c_j) = n - |i - j| + 2, & l + 1 \leqslant |i - j| \leqslant 2l \text{ and } i \leqslant l; \\ d(b_i, c_j) = d(b_i d_i, c_j) = n - |i - j| + 1, & l + 1 \leqslant |i - j| \leqslant 2l \text{ and } l + 1 \leqslant i \leqslant 2l + 1. \end{cases}$$

$$\begin{cases} d(b_i, d_j) = 1, d(b_i d_i, d_j) = 0, & j = i; \\ d(b_i, d_j) = d(b_i d_i, d_j) = |i - j| + 1, & |i - j| \leqslant l \text{ and } j \leqslant i; \\ d(b_i, d_j) = d(b_i d_i, d_j) = n - |i - j| + 1, & l + 1 \leqslant |i - j| \leqslant 2l. \end{cases}$$

In this case, it is not hard to see that $r(b_i | B \cup C \cup D \setminus C_i) = r(b_i d_i | B \cup C \cup D \setminus C_i)$.

Case (II) n is even. Similar to the proof of Case (I) we may obtain $r(b_i | B \cup C \cup D \setminus C_i) = r(b_i d_i | B \cup C \cup D \setminus C_i)$.

So (i) holds.

(ii) We consider the subsequent two cases depending upon n.

Case (I) n is odd. In this case, we can write $n = 2l + 1$, where $l \geqslant 5$ is an integer. Now, we calculate the distance between the vertexs a_i and the vertex x_j, and the distance between the edges $a_i c_i$ and the vertex x_j, where $x_j \in A \cup C \cup D$, $1 \leqslant i, j \leqslant n$.

$$\begin{cases} d(a_i, a_j) = d(a_i c_i, a_j) = |i - j|, & |i - j| \leqslant l; \\ d(a_i, a_j) = d(a_i c_i, a_j) = n - |i - j|, & l + 1 \leqslant |i - j| \leqslant 2l. \end{cases}$$

$$\begin{cases} d(a_i, c_j) = 1, d(a_i c_i, c_j) = 0, & j = i; \\ d(a_i, c_j) = d(a_i c_i, c_j) = |i - j| + 1, & |i - j| \leqslant l \text{ and } j \leqslant i; \\ d(a_i, c_j) = d(a_i c_i, c_j) = n - |i - j| + 1, & l + 1 \leqslant |i - j| \leqslant 2l. \end{cases}$$

$$\begin{cases} d(a_i, d_j) = 2, d(a_i c_i, d_j) = 1, & j = i, j = i-1; \\ d(a_i, d_j) = d(a_i c_i, d_j) = |i - j| + 2, & |i - j| \leqslant l \text{ and } j > i, j \leqslant i, j \leqslant i-1; \\ d(a_i, d_j) = d(a_i c_i, d_j) = |i - j| + 1, & |i - j| \leqslant l \text{ and } j < i, j \leqslant i, j \leqslant i-1; \\ d(a_i, d_j) = d(a_i c_i, d_j) = n - |i - j| + 1, & l + 1 \leqslant |i - j| \leqslant 2l \text{ and } i \leqslant l; \\ d(a_i, d_j) = d(a_i c_i, d_j) = n - |i - j| + 2, & l + 1 \leqslant |i - j| \leqslant 2l \text{ and } l + 1 \leqslant i \leqslant 2l + 1. \end{cases}$$

In this case, it is not hard to see that $r(a_i | A \cup C \cup D \setminus D_i) = r(a_i c_i | A \cup C \cup D \setminus D_i)$.

Case (II) n is even. Similar to the proof of Case (I) we may obtain $r(a_i | A \cup C \cup D \setminus D_i) = r(a_i c_i | A \cup C \cup D \setminus D_i)$.

So (ii) holds.

Lemma 9. Let $\mathbb{D}_n$ be the plane graph (convex polytope), where $n \geqslant 10$. Then each mixed metric basis S_m of $\mathbb{D}_n$ contains at least one vertex of A and one vertex of B.

Proof. We first show that S_m contains at least one vertex of A. Suppose on the contrary that S_m does not contain any vertex of A. Then $S_m \subset B \cup C \cup D$. By Lemma 8(i), we have $r(b_i|B \cup C \cup D \setminus C_i) = r(b_i d_i|B \cup C \cup D \setminus C_i)$, where $C_i = \{c_i, c_{i+1}, d_i\} \subset C \cup D$. This means that S_m contains at least one vertex of C_i. Also, we observe that

$$|C_i \cap C_j| = \begin{cases} 1, & |i-j| = 1; \\ 0, & |i-j| \neq 1. \end{cases}$$

From which it follows that S_m contains at least $\lceil \frac{n}{2} \rceil$ vertices of $C \cup D$. Since $n \geqslant 10$, then $dim_m(\mathbb{D}_n) \geqslant 5$. But, $dim_m(\mathbb{D}_n) \leqslant 4$ by Lemma 7. This is a contradiction.

Secondly, we show that S_m contains at least one vertex of B. Suppose on the contrary that S_m does not contain any vertex of B. Then $S_m \subset A \cup C \cup D$. By Lemma 8(ii), we have $r(a_i|A \cup C \cup D \setminus D_i) = r(a_i c_i|A \cup C \cup D \setminus D_i)$, where $D_i = \{d_{i-1}, d_i, c_i\} \subset C \cup D$. This means that S_m contains at least one vertex of D_i. Also, we observe that

$$|D_i \cap D_j| = \begin{cases} 1, & |i-j| = 1; \\ 0, & |i-j| \neq 1. \end{cases}$$

From which it follows that S_m contains at least $\lceil \frac{n}{2} \rceil$ vertices of $C \cup D$. Since $n \geqslant 10$, then $dim_m(\mathbb{D}_n) \geqslant 5$. But, $dim_m(\mathbb{D}_n) \leqslant 4$ by Lemma 7. This is a contradiction.

Thus, each mixed metric basis S_m of $\mathbb{D}_n$ contains at least one vertex of A and one vertex of B. □

Theorem 2. *Let $\mathbb{D}_n$ be the plane graph (convex polytope), where $n \geqslant 3$. Then $dim_m(\mathbb{D}_n) = 4$.*

Proof. For $n = 3$, we find that $\{a_1, a_2, d_3, b_2\}$ is the mixed metric basis of $\mathbb{D}_n$ by total enumeration, and hence the mixed metric dimension of $\mathbb{D}_n$ is 4. For $n = 4$, we find that $\{a_1, a_2, d_3, b_2\}$ is the mixed metric basis of $\mathbb{D}_n$ by total enumeration, and hence the mixed metric dimension of $\mathbb{D}_n$ is 4. For $5 \leqslant n \leqslant 9$, we find that $\{a_1, a_{l+1}, b_2, b_{l+2}\}$ is the mixed metric basis of $\mathbb{D}_n$ by total enumeration, and hence the mixed metric dimension of $\mathbb{D}_n$ is 4. For $n \geqslant 10$, we consider the following two cases.

Case (I) n is odd. We show that $dim_m(\mathbb{D}_n) \neq 3$. By Lemma 9 we know that a mixed metric basis S_m for $\mathbb{D}_n$ contains at least one vertex of A and one vertex of B. Since the vertices of graph $\mathbb{D}_n$ are symmetric, without loss of generality, we assume that a_1 and b_i are these two vertices, where $1 \leqslant i \leqslant l+1$. By calculating, there are following four possibilities to be discussed.
(1) If $S_m = \{a_1, b_i, a_j\}$, where $1 \leqslant i \leqslant l+1$ and $2 \leqslant j \leqslant 2l+1$, then we obtain

$$\begin{cases} r(c_{l+1}|S_m) = r(c_{l+1} d_{l+1}|S_m), & 1 \leqslant i \leqslant l \text{ and } 2 \leqslant j \leqslant 2l+1; \\ r(c_{2l+1}|S_m) = r(c_{2l+1} d_{2l+1}|S_m), & i = l+1 \text{ and } 2 \leqslant j \leqslant 2l+1. \end{cases}$$

(2) If $S_m = \{a_1, b_i, b_j\}$, where $1 \leqslant i \leqslant l+1$ and $2 \leqslant j \leqslant 2l+1$, then we obtain $r(d_{2l+1}|S_m) = r(c_{2l+1} d_{2l+1}|S_m)$.

(3) If $S_m = \{a_1, b_i, c_j\}$, where $1 \leqslant i \leqslant l+1$ and $1 \leqslant j \leqslant 2l+1$, then we obtain

$$
\begin{cases}
r(a_1|S_m) = r(a_1 a_{2l+1}|S_m), & 1 \leqslant i \leqslant l \text{ and } 1 \leqslant j \leqslant l+1; \\
r(a_1|S_m) = r(a_1 a_2|S_m), & 1 \leqslant i \leqslant l \text{ and } l+2 \leqslant j \leqslant 2l+1; \\
r(d_{2l+1}|S_m) = r(c_{2l+1} d_{2l+1}|S_m), & i = l+1 \text{ and } 1 \leqslant j \leqslant l; \\
r(c_{2l+1}|S_m) = r(c_{2l+1} d_{2l+1}|S_m), & i = l+1 \text{ and } l+1 \leqslant j \leqslant 2l+1.
\end{cases}
$$

(4) If $S_m = \{a_1, b_i, d_j\}$, where $1 \leqslant i \leqslant l+1$ and $1 \leqslant j \leqslant 2l+1$, then we obtain

$$
\begin{cases}
r(b_1|S_m) = r(b_1 b_{2l+1}|S_m), & i = 1 \text{ and } 1 \leqslant j \leqslant l+1; \\
r(b_1|S_m) = r(b_1 b_2|S_m), & i = 1 \text{ and } l+2 \leqslant j \leqslant 2l+1; \\
r(d_{l+2}|S_m) = r(c_{l+2} d_{l+2}|S_m), & 2 \leqslant i \leqslant l \text{ and } 1 \leqslant j \leqslant 2l+1 \text{ and } j \neq l, l+1; \\
r(c_{l+2}|S_m) = r(c_{l+2} d_{l+2}|S_m), & 2 \leqslant i \leqslant l \text{ and } j = l, l+1.
\end{cases}
$$

By the above we see that there is no resolving set with three vertices for $V(\mathbb{D}_n)$, then $dim_m(\mathbb{D}_n) \geqslant 4$. By Lemma 7 we know that $dim_m(\mathbb{D}_n) \leqslant 4$, so $dim_m(\mathbb{D}_n) = 4$ holds.

Case (II) n is even. We show that $dim_m(\mathbb{D}_n) \neq 3$. By Lemma 9 we know that a mixed metric basis S_m for $\mathbb{D}_n$ contains at least one vertex of A and one vertex of B. Since the vertices of graph $\mathbb{D}_n$ are symmetric, without loss of generality, we assume that a_1 and b_i are these two vertices, where $1 \leqslant i \leqslant l$. By calculating, there are following four possibilities to be discussed.

(1) If $S_m = \{a_1, b_i, a_j\}$, where $1 \leqslant i \leqslant l$ and $2 \leqslant j \leqslant 2l$, then we obtain

$$
\begin{cases}
r(c_l|S_m) = r(c_l d_l|S_m), & 1 \leqslant i \leqslant l-1 \text{ and } 2 \leqslant j \leqslant 2l; \\
r(c_{2l}|S_m) = r(c_{2l} d_{2l}|S_m), & i = l \text{ and } 2 \leqslant j \leqslant 2l.
\end{cases}
$$

(2) If $S_m = \{a_1, b_i, b_j\}$, where $1 \leqslant i \leqslant l$ and $2 \leqslant j \leqslant 2l$, then we obtain $r(d_{2l}|S_m) = r(c_{2l} d_{2l}|S_m)$.

(3) If $S_m = \{a_1, b_i, c_j\}$, where $1 \leqslant i \leqslant l$ and $1 \leqslant j \leqslant 2l$, then we obtain

$$
\begin{cases}
r(d_{2l}|S_m) = r(c_{2l} d_{2l}|S_m), & 1 \leqslant i \leqslant l \text{ and } 1 \leqslant j \leqslant l; \\
r(d_l|S_m) = r(c_l d_l|S_m), & 1 \leqslant i \leqslant l \text{ and } l+1 \leqslant j \leqslant 2l.
\end{cases}
$$

(4) If $S_m = \{a_1, b_i, d_j\}$, where $1 \leqslant i \leqslant l$ and $1 \leqslant j \leqslant 2l$, then we obtain

$$
\begin{cases}
r(a_1|S_m) = r(a_1 a_{2l}|S_m), & i = 1 \text{ and } 1 \leqslant j \leqslant l; \\
r(a_1|S_m) = r(a_1 a_2|S_m), & i = 1 \text{ and } l+1 \leqslant j \leqslant 2l; \\
r(d_{l+2}|S_m) = r(c_{l+2} d_{l+2}|S_m), & 2 \leqslant i \leqslant l \text{ and } 1 \leqslant j \leqslant 2l \text{ and } j \neq l, l+1; \\
r(c_{l+2}|S_m) = r(c_{l+2} d_{l+2}|S_m), & 2 \leqslant i \leqslant l \text{ and } j = l, l+1.
\end{cases}
$$

From the above we know that there is no resolving set with three vertices for $V(\mathbb{D}_n)$, then $dim_m(\mathbb{D}_n) \geqslant 4$. By Lemma 7 we know that $dim_m(\mathbb{D}_n) \leqslant 4$, so $dim_m(\mathbb{D}_n) = 4$ holds.

5   Conclusion

In this paper, we studied the mixed metric dimension for two families of plane graphs (web graphs and convex polytopes) in metric graph theory. For web graphs, a lower bound for the mixed metric dimension was proved and a mixed metric basis was then obtained to determine the mixed metric dimension. For convex polytopes, an upper bound for the mixed metric dimension was discovered and the above bound was then proved to be tight. The future research can be thought of as finding the mixed metric dimension for other families of plane graphs, especially rotationally symmetric ones.

Acknowledgements. This work was supported by the NSF of China (No. 11971146) and the Doctoral Scientific Research of Shijiazhuang University of Economics of China (No. BQ201517).

References

1. Bača, M.: Labellings of two classes of convex polytopes. Util. Math. **34**, 24–31 (1988)
2. Chartrand, G., Poisson, C., Zhang, P.: Resolvability and the upper dimension of graphs. Comput. Math. Appl. **39**, 19–28 (2000)
3. Chartrand, G., Eroh, L., Johnson, M.A., Oellermann, O.R.: Resolvability in graphs and the metric dimension of a graph. Discrete. Appl. Math. **105**, 99–113 (2000)
4. Danas, M.M.: The mixed metric dimension of flower snarks and wheels. Open Math. **19**(1), 629–640 (2021)
5. Danas, M.M., Kratica, J., Savić, A., Maksimović, Z.L.: Some new general lower bounds for mixed metric dimension of graphs (2020). arXiv:2007.05808
6. Harary, F., Melter, R.A.: On the metric dimension of a graph. Ars Combin. **2**, 191–195 (1976)
7. Imran, M., Baig, A.Q., Ahmad, A.: Families of plane graphs with constant metric dimension. Util. Math. **88**, 43–57 (2012)
8. Imran, M., Baig, A.Q., Bokhary, S.A.: On the metric dimension of rotationally-symmetric graphs. Ars Combin. **124**, 111–128 (2016)
9. Imran, M., Siddiqui, M.K., Naeem, R.: On the metric dimension of generalized Petersen multigraphs. IEEE. Access. **6**, 74328–74338 (2018)
10. Johnson, M.A.: Structure-activity maps for visualizing the graph variables arising in drug design. J. Biopharm. Stat. **3**, 203–236 (1993)
11. Kelenc, A., Kuziak, D., Taranenko, A., Yero, I.G.: Mixed metric dimension of graphs. Appl. Math. Comput. **314**(1), 429–438 (2017)
12. Kelenc, A., Tratnik, N., Yero, I.G.: Uniquely identifying the edges of a graph: the edge metric dimension. Discrete. Appl. Math. **251**, 204–220 (2018)
13. Koh, K.M., Rogers, D.G., Teo, H.K., Yap, K.Y.: Graceful graphs: some further results and problems. Congr. Numer. **29**, 559–571 (1980)
14. Khuller, S., Raghavachari, B., Rosenfeld, A.: Landmarks in graphs. Discret. Appl. Math. **70**, 217–229 (1996)
15. Peterin, I., Yero, I.G.: Edge metric dimension of some graph operations. Bull. Malays. Math. Sci. Soc. **43**(3), 2465–2477 (2019). https://doi.org/10.1007/s40840-019-00816-7
16. Raza, H., Liu, J.B., Qu, S.: On mixed metric dimension of rotationally symmetric graphs. IEEE Access. **8**, 188146–188153 (2020)
17. Raza, H., Ji, Y.: Computing the mixed metric dimension of a generalized Petersen graph $P(n, 2)$. Front. Phys. 8: 211. https://doi.org/10.3389/fphy.2020.00211
18. Slater, P.J.: Leaves of trees. Congr. Numer. **14**, 549–559 (1975)

19. Zhu, E., Taranenko, A., Shao, Z., Xu, J.: On graphs with the maximum edge metric dimension. Discret. Appl. Math. **257**, 317–324 (2019)
20. Zubrilina, N.: On the edge dimension of a graph. Discret. Math. **341**(7), 2083–2088 (2018)
21. Zhang, Y.Z., Gao, S.G.: On the edge metric dimension of convex polytopes and its related graphs. J. Comb. Optim. **39**, 334–350 (2020)

On the Transversal Number of k-Uniform Connected Hypergraphs

Zian Chen[1], Bin Chen[1], Zhongzheng Tang[2], and Zhuo Diao[3(✉)]

[1] Center for Discrete Mathematics, Fuzhou University, Fuzhou 350003, Fujian, China
[2] School of Science, Beijing University of Posts and Telecommunications,
Beijing 100876, China
tangzhongzheng@amss.ac.cn
[3] School of Statistics and Mathematics, Central University of Finance
and Economics, Beijing 100081, China
diaozhuo@amss.ac.cn

Abstract. For $k \geq 3$, let H be a k-uniform connected hypergraph on n vertices and m edges. The transversal number $\tau(H)$ is the minimum number of vertices that intersect every edge. We prove the following inequality: $\tau(H) \leq \frac{(k-1)m+1}{k}$. Furthermore, the extremal hypergraphs with equality holds are exactly hypertrees with perfect matching. Based on the proofs, some combinatorial algorithms on the transversal number are designed.

Keywords: Transversal · k-uniform · Extremal hypergraph · Perfect matching

1 Introduction

A simple hypergraph is a hypergraph without loops and multiple edges. Let $H(V, E)$ be a simple hypergraph with vertex set V and edge set E. As for a graph, the order of H, denoted by n, is the number of vertices. The number of edges will be denoted by m.

For each vertex $v \in V$, the degree $d(v)$ is the number of edges in E that contains v. We say v is an isolated vertex of H if $d(v) = 0$. Hypergraph H is k-regular if each vertex's degree is k ($d(v) = k, \forall v \in V$). Hypergraph H is k-uniform if each edge contains exactly k vertices ($| e | = k, \forall e \in E$). Hypergraph H is called *linear* if any two distinct edges have at most one common vertex. ($| e_1 \cap e_2 | \leq 1, \forall e_1, e_2 \in E$). The rank is $r(H) = max_{e \in E} | e |$.

For any vertex set $S \subseteq V$, we write $H \backslash S$ for the sub-hypergraph of H obtained from H by deleting all vertices in S and all edges incident with some

---

Supported by National Natural Science Foundation of China under Grant No. 11901605, No. 12101069, the disciplinary funding of Central University of Finance and Economics, the Emerging Interdisciplinary Project of CUFE, the Fundamental Research Funds for the Central Universities and Innovation Foundation of BUPT for Youth (500422309).

Q. Ni and W. Wu (Eds.): AAIM 2022, LNCS 13513, pp. 376–387, 2022.
https://doi.org/10.1007/978-3-031-16081-3_32

vertices in S. For any edge set $A \subseteq V$, we write $H \backslash A$ for the sub-hypergraph of H obtained from H by deleting all edges in A and keeping vertices. If S is a singleton set s, we write $H \backslash s$ instead of $H \backslash \{s\}$.

Let $k \geq 2$ be an integer. A cycle of length k, denoted as k-cycle, is a vertex-edge sequence $C = v_1 e_1 v_2 e_2 \cdots v_k e_k v_1$ with: (1) $\{e_1, e_2, \ldots, e_k\}$ are distinct edges of H. (2) $\{v_1, v_2, \ldots, v_k\}$ are distinct vertices of H. (3) $\{v_i, v_{i+1}\} \subseteq e_i$ for each $i \in [k]$, here $v_{k+1} = v_1$. We consider the cycle C as a sub-hypergraph of H with vertex set $\{v_i, i \in [k]\}$ and edge set $\{e_j, j \in [k]\}$.

Similarily, a path of length k, denoted as k-path, is a vertex-edge sequence $P = v_1 e_1 v_2 e_2 \cdots v_k e_k v_{k+1}$ with: (1) $\{e_1, e_2, \ldots, e_k\}$ are distinct edges of H. (2) $\{v_1, v_2, \ldots, v_{k+1}\}$ are distinct vertices of H. (3) $\{v_i, v_{i+1}\} \subseteq e_i$ for each $i \in [k]$. We consider the path P as a sub-hypergraph of H with vertex set $\{v_i, i \in [k+1]\}$ and edge set $\{e_j, j \in [k]\}$.

A hypergraph $H = (V, E)$ is called *connected* if any two of its vertices are linked by a path in H. For two distinct vertices u and v, the distance between u and v is the length of a shortest path connecting u and v, denoted by $d(x, y)$. A hypergraph $H = (V, E)$ is called a *hypertree* if H is connected and acyclic, not containing any cycles, denoted by $T(V, E)$.

Given a hypergraph $H(V, E)$, a set of vertices $S \subseteq V$ is a transversal if every edge is incident with at least a vertex in S. The transversal number is the minimum cardinality of a transversal, denoted by $\tau(H)$. A set of edges $A \subseteq E$ is a matching if every two distinct edges have no common vertex. The matching number is the maximum cardinality of a matching, denoted by $\nu(H)$. In this paper, we consider the transversal number in k-uniform connected hypergraphs.

1.1 Known Results

Hypergraphs are systems of sets which are conceived as natural extensions of graphs. A subset S of vertices in a hypergraph H is a transversal (also called vertex cover or hitting set in many papers) if S has a nonempty intersection with every edge of H. The transversal number $\tau(H)$ of H is the minimum size of a transversal in H. Transversals in hypergraphs are well studied in the literature [3,6,7,9–13].

Chvátal and McDiarmid [3] established the following upper bound on the transversal number of a uniform hypergraph in terms of its order and size.

Theorem 1. *For $k \geq 2$, if H is a k-uniform hypergraph on n vertices with m edges, then $\tau(H) \leq \frac{n + \lfloor \frac{k}{2} \rfloor m}{\lfloor \frac{3k}{2} \rfloor}$.*

Henning and Yeo [5] proposed the following question:

Conjecture 1. For $k \geq 2$, let H be a k-uniform hypergraph on n vertices with m edges. If H is linear, then $(k + 1)\tau(H) \leq n + m$ holds for all $k \geq 2$?

The Chvátal and McDiarmid theorem implies that $(k + 1)\tau(H) \leq n + m$ holds for $k \in \{2, 3\}$ even without the linearity constraint imposed on H. Henning and Yeo [5] remarked that if H is not linear, then Conjecture 1 is not always true, showing an example by taking $k = 4$ and letting $\overline{F}_7$ be the complement of the Fano plane F_7. Henning and Yeo [5] proved the following theorem which verified Conjecture 1 for linear hypergraphs with maximum degree two:

Theorem 2. *For $k \geq 2$, let H be a k-uniform linear hypergraph satisfying $\Delta(H) \leq 2$. Then, $(k+1)\tau(H) \leq n(H) + m(H)$ with equality if and only if each component of H consists of a single edge or is the dual of a complete graph of order $k + 1$ and k is even.*

Henning and Yeo [8] proposed the following conjecture in another paper:

Conjecture 2. $\tau(H) \leq \frac{n}{k} + \frac{m}{6}$ holds for all uniform hypergraphs with maximum degree at most 3.

Henning and Yeo [8] showed that $\tau(H) \leq \frac{n}{k} + \frac{m}{6}$ holds when $k = 2$ and characterized the hypergraphs for which equality holds. Chvátal and McDiarmid [3] showed that $\tau(H) \leq \frac{n}{k} + \frac{m}{6}$ holds when $k = 3$. Henning and Yeo characterized the extremal hypergraphs. Henning and Yeo [8] showed that $\tau(H) \leq \frac{n}{k} + \frac{m}{6}$ holds when $\Delta(H) \leq 2$ and characterized the hypergraphs for which it holds with equality in that case.

1.2   Our Results

Chen, Diao, Hu and Tang [2] proved the following theorem on 3-uniform connected linear hypergraphs:

Theorem 3. *Let $H(V, E)$ be a 3-uniform connected linear hypergraph with m edges. Then $\tau(H) \leq \frac{2m+1}{3}$.*

Diao [4] proved the same inequality on 3-uniform connected hypergraphs and characterized the extremal hypergraphs:

Theorem 4. *Let $H(V, E)$ be a 3-uniform connected hypergraph with m edges. Then $\tau(H) \leq \frac{2m+1}{3}$. Furthermore, $\tau(H) = \frac{2m+1}{3}$ holds if and only if H is a hypertree with perfect matching.*

In this paper, we generalize the above results for k-uniform connected hypergraphs, as stated in the following theorem:

Theorem 5. *For $k \geq 3$, let $H(V, E)$ be a k-uniform connected hypergraph with m edges. Then $\tau(H) \leq \frac{(k-1)m+1}{k}$. Furthermore, $\tau(H) = \frac{(k-1)m+1}{k}$ holds if and only if H is a hypertree with perfect matching.*

For k-uniform connected hypergraphs, Conjecture 1 and Theorem 5 are related. If Conjecture 1 holds, combined with $n \leq (k-1)m + 1$, we have

$$(k+1)\tau(H) \leq n + m, n \leq (k-1)m + 1 \Rightarrow \tau(H) \leq \frac{n+m}{k+1} \leq \frac{km+1}{k+1},$$

which is a weaker result of Theorem 5.

The main content of the article is organized as follows:

- In Sect. 2, we introduce a *breaking-cycle* operation and prove $\tau(H) \leq [(k-1)m+1]/k$ holds for k-uniform connected hypergraphs. Based on the proof, a polynomial time algorithm is designed to compute a transversal set with cardinality at most $[(k-1)m+1]/k$.
- In Sect. 3, we characterize the extremal k-uniform connected hypergraphs with $\tau(H) = [(k-1)m+1]/k$. By structure analysis, we prove the extremal hypergraphs are exactly hypertrees with perfect matching.

2   The Upper Bound for Transversal Number

In this section, we prove $\tau(H) \leq \frac{(k-1)m+1}{k}$ holds for k-uniform connected hypergraphs, as stated in Theorem 5.

The content is organized as follows:

- The conception and property of *breaking-cycle* operations are introduced by Definition 1 and Lemma 1.
- The upper bound for transversal number is proved in Lemma 2, Lemma 3 and Theorem 7.
- Computing the transversal with cardinality no more than the upper bound is shown in Algorithm 2.

Definition 1. *For a k-uniform connected hypergraph $H(V, E)$, let $C = v_1 e_1 v_2 e_2 \cdots v_k e_k v_1$ be a cycle in H. The breaking-cycle operation of C is breaking an adjacent vertex v_i in the cycle C, as shown in Fig. 1.*

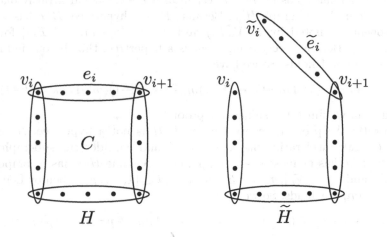

Fig. 1. The schematic diagram of *breaking-cycle* operation

Lemma 1. *The breaking-cycle operation does not decrease the transversal number.*

Proof. Denote $\widetilde{H}$ as the hypergraph obtained by the *breaking-cycle* operation of C. Obviously, $\widetilde{H}$ is k-uniform and connected. v_i is the breaking adjacent vertex in H and $\widetilde{v}_i$ is the corresponding leaf in $\widetilde{H}$; $e_i(v_i, v_{i+1})$ is the breaking edge in H and $\widetilde{e}_i(\widetilde{v}_i, v_{i+1})$ is the corresponding edge in $\widetilde{H}$, as shown in Fig. 1.

Take a minimum transversal $\widetilde{S}$ of $\widetilde{H}$. If $\widetilde{v}_i \in \widetilde{S}$, then replace $\widetilde{v}_i$ by v_{i+1}. Thus there is a minimum transversal $\widetilde{S}$ of $\widetilde{H}$ and $\widetilde{v}_i \notin \widetilde{S}$. Take $S = \widetilde{S}$ and S is a transversal of H. Thus $\tau(H) \leq \tau(\widetilde{H})$.

Lemma 2. *For k-uniform connected hypergraph $H(V, E)$, $n \leq (k-1)m + 1$.*

Proof. We prove this lemma by induction on m. When $m = 0$, $H(V, E)$ is an isolate vertex, $n \leq (k-1)m + 1$ holds. Assume this lemma holds for $m \leq t$. When $m = t + 1$, take arbitrarily one edge e and consider the subgraph $H \backslash e$. Obviously, $H \backslash e$ has at most k components. Assume $H \backslash e$ has p components $H_i(V_i, E_i)$ and $n_i = |\, V_i\, |$, $m_i = |\, E_i\, |$ for each $i \in \{1, \ldots, p\}$. Then by our induction, $n_i \leq (k-1)m_i + 1$ holds. So we have

$$n = n_1 + \cdots + n_p \leq (k-1)m_1 + \cdots + (k-1)m_p + p = (k-1)(m-1) + p$$

$$= (k-1)m + p - k + 1 \leq (k-1)m + 1.$$

By induction, we finish our proof.

Lemma 3. *For k-uniform connected hypergraph $H(V, E)$, $n = (k-1)m + 1$ if and only if H is a hypertree.*

Proof. Sufficiency: if H is a hypertree, we prove $n = (k-1)m + 1$ by induction on m. When $m = 0$, $H(V, E)$ is an isolate vertex, $n = (k-1)m + 1$ holds. Assume this lemma holds for $m \leq t$. When $m = t + 1$, take arbitrarily one edge e and consider the subgraph $H \backslash e$. Because H is a hypertree, $H \backslash e$ has exactly k components, denoted by $H_i(V_i, E_i)$ and $n_i = |\, V_i\, |, m_i = |\, E_i\, |$ for each $i \in \{1, \ldots, k\}$. Because every component is a hypertree, thus by our induction, $n_i = (k-1)m_i + 1$ holds. So we have

$$n = n_1 + \cdots + n_k = (k-1)m_1 + \cdots + (k-1)m_k + k = (k-1)(m-1) + k = (k-1)m + 1.$$

By induction, we finish the sufficiency proof.

Necessity: We prove by contradiction. If H is not a hypertree, H contain a cycle C. Take arbitrarily one edge e in C and consider the subgraph $H \backslash e$. obviously, $H \backslash e$ has at most $k - 1$ components. Assume $H \backslash e$ has p components $H_i(V_i, E_i)$ and $n_i = |\, V_i\, |, m_i = |\, E_i\, |$ for each $i \in \{1, \ldots, p\}$. Then by Lemma 2, $n_i \leq (k-1)m_i + 1$ holds. So we have

$$n = n_1 + \cdots + n_p \leq (k-1)m_1 + \cdots + (k-1)m_p + p = (k-1)(m-1) + p$$

$$= (k-1)m + p - k + 1 \leq (k-1)m < (k-1)m + 1,$$

which is a contradiction with $n = (k-1)m + 1$. Thus H is a hypertree and we finish our necessity proof.

Transversals and matchings are related in a prime-dual way. The property $\tau(H) = \nu(H)$ is called *König Property*. The next theorem is useful in our proof:

Theorem 6 [1]. $T(V, E)$ *is a hypertree,* $\tau(T) = \nu(T)$.

Theorem 7. *For $k \geq 3$, let $H(V, E)$ be a k-uniform connected hypergraph with m edges. Then $\tau(H) \leq \frac{(k-1)m+1}{k}$.*

Proof. Take arbitrarily a sequences of *breaking-cycle* operations. There is a sequence of k-uniform connected hypergraphs $\widetilde{H}_1, \ldots, \widetilde{H}_t$. All these hypergraphs have m edges. $\widetilde{H}_t$ is a hypertree and the number of vertices is n_t. According to Lemma 1, we have the following inequalities:

$$\tau(H) \leq \tau(\widetilde{H}_1) \leq \cdots \leq \tau(\widetilde{H}_t).$$

$\widetilde{H}_t$ is a hypertree. According to Theorem 6, we have the following inequalities:

$$\tau(\widetilde{H}_t) = \nu(\widetilde{H}_t) \leq \frac{n_t}{k} = \frac{(k-1)m+1}{k},$$

which means $\tau(H) \leq \frac{(k-1)m+1}{k}$.

The proof of Theorem 7 also implies a combinatorial algorithm for computing a transversal of k-uniform connected m-edge hypergraph H with cardinality at most $\frac{(k-1)m+1}{k}$.

First, if H is a hypertree, then the optimal transversal can be found in polynomial time as follows.

Algorithm 1. Transversal of Hypertrees

Input: A hypertree $T(V, E)$.

Output: ALG1(T), an optimal transversal of H.

1: **if** $| E |= 0$ **then**

2:     **return** $\varnothing$

3: **if** $| E |= 1$ **then**

4:     Let u be an arbitrary vertex in V.

5:     **return** $\{u\}$

6: Find $e \in E$ that contains exactly one vertex v with $d(v) \geq 2$.

7: Let $T_1, T_2, \ldots, T_l$ be all components of $T \setminus v$.

8: **return** $\{v\} \cup \bigcup_{i=1}^{l} \text{ALG1}(T_i)$

Remark 1. For any hypertree T with $m \geq 2$ edges, the longest path in T is $P = v_1 e_1 v_2 e_2 \cdots e_t v_{t+1}$ with $t \geq 2$. T is acyclic and P is the longest path in T, thus e_1 has exactly one vertex v_2 with degree greater than 1 and e_t has exactly one vertex v_t with degree greater than 1. The vertices in Step 6 of Algorithm 1 exist.

Theorem 8. *For a hypertree $T(V, E)$, Algorithm 1 outputs a minimum transversal set of T.*

Proof. $T(V, E)$ is a hypertree. In the process of Algorithm 1, there is a series of acyclic hypergraphs, denoted as $\{T_i, 1 \leq i \leq k\}$. T_{i+1} is the subhypergraph by deleting v_i in T_i. T_1 is T and T_k is a trivial hyperforest, whose components are isolated vertices or single edges. For $1 \leq i \leq k$, Algorithm 1 constructs an vertex set S_i in T_i. We will prove S_i is the minimum transversal set of T_i by backward induction on i.

- For $i = k$, T_k is a trivial hyperforest. S_k is the set of vertices by taking arbitrarily a vertex for each single edge. Obviously, S_k is the minimum transversal set of T_k.
- Let S_{i+1} be the minimum transversal set of T_{i+1}. We need to show that S_i is the minimum transversal set of T_i. $T_{i+1} = T_i \backslash v_i$, $S_i = S_{i+1} \cup \{v_i\}$. On one side, for any minimum transversal set S of T_i, there is a vertex $v \in e_i$ in S and $S \backslash \{v\}$ is a transversal set of $T_i \backslash v$. $T_{i+1} = T_i \backslash v_i$ is a subhypergraph of $T_i \backslash v$. Thus $S \backslash \{v\}$ is also a transversal set of T_{i+1} and the following inequalities hold:

$$\tau(T_i) = | S |, \quad | S \backslash \{v\} | = | S | - 1 \geq \tau(T_{i+1}) \Rightarrow \tau(T_i) \geq \tau(T_{i+1}) + 1.$$

On the other side, S_{i+1} is the minimum transversal set of T_{i+1} and $S_i = S_{i+1} \cup \{v_i\}$ is a transversal set of T_i. Thus the following inequalities hold:

$$| S_i | = | S_{i+1} \cup \{v_i\} | = | S_{i+1} | + 1 = \tau(T_{i+1}) + 1 \geq \tau(T_i).$$

Above all, $\tau(T_i) = \tau(T_{i+1}) + 1$ and S_i is the minimum transversal set of T_i.

Take $S = S_1$, $T = T_1$ and Algorithm 1 outputs a minimum transversal set S of the hypertree T.

Next, we show Algorithm 2, which will call Algorithm 1 as a subprocedure.

Theorem 9. *For a k-uniform connected hypergraph $H(V, E)$ with m edges, Algorithm 2 outputs a transversal set S of H with $| S | \leq \frac{(k-1)m+1}{k}$.*

Proof. In the process of breaking-cycle operations, there is a series of k-uniform connected hypergraphs, denoted as $\{\tilde{H}_i, 1 \leq i \leq t\}$. $\tilde{H}_{i+1}$ is generated by breaking-cycle C_i in $\tilde{H}_i$, as shown in Fig. 1. $\tilde{H}_1$ is H and $\tilde{H}_k$ is a hypertree T with the number of vertices n_t. All these hypergraphs have m edges. According to Lemma 3 and Theorem 6, we have the following inequalities:

$$\tau(\tilde{H}_t) = \nu(\tilde{H}_t) \leq \frac{n_t}{k} = \frac{(k-1)m+1}{k}.$$

Algorithm 2. Transversal of k-Uniform Connected Hypergraphs

Input: A k-uniform connected hypergraph H with m edges.

Output: A transversal of H with cardinality at most $\frac{(k-1)m+1}{k}$.

1: Set $\widetilde{H}_1 = H$ and $i = 1$.

2: **while** $\widetilde{H}_i$ contains some cycle C_i **do**

3:     Do breaking-cycle operation on vertex v_i in cycle C_i.

4:     Denote the resulting hypergraph as $\widetilde{H}_{i+1}$.

5:     Set $i = i + 1$.

6: Compute an optimal transversal S_i of the hypertree $\widetilde{H}_i$ by Algorithm 1.

7: **for** $j = i$ to 2 **do**

8:     **if** $\widetilde{v}_{j-1} \in S_j$ **then**

9:         Take $S_{j-1} = S_j \setminus \{\widetilde{v}_{j-1}\} \cup \{u\}$ where u is an arbitrarily vertex in $\widetilde{e}_{j-1} \setminus \widetilde{v}_{j-1}$.

10:     **else**

11:         Take $S_{j-1} = S_j$.

12: **return** S_1

Algorithm 1 is an exact algorithm to compute transversal number for hypertrees, which outputs a transversal set $\mid S_t \mid = \tau(\widetilde{H}_t) \leq \frac{(k-1)m+1}{k}$.

$\widetilde{H}_{i+1}$ is generated by breaking-cycle C_i in $\widetilde{H}_i$, as shown in Fig. 1. S_{i+1} is a transversal set of $\widetilde{H}_{i+1}$, we can construct a transversal set S_i of $\widetilde{H}_i$ as follows:

- $\widetilde{v}_i \in S_{i+1}$, take $S_i = S_{i+1} \setminus \{\widetilde{v}_i\} \cup \{u\}$, here u is an arbitrarily vertex in $\widetilde{e}_i \setminus \widetilde{v}_i$;
- $\widetilde{v}_i \notin S_{i+1}$, take $S_i = S_{i+1}$.

According to the rules, we have $\mid S_i \mid \leq \mid S_{i+1} \mid$. Thus we have a series of transversal set S_i of $\widetilde{H}_i$ with

$$\mid S_1 \mid \leq \mid S_2 \mid \leq \cdots \leq \mid S_t \mid \leq \frac{(k-1)m+1}{k}.$$

Take $S = S_1, H = \widetilde{H}_1$ and Algorithm 1 outputs a transversal set S of H with $\mid S \mid \leq \frac{(k-1)m+1}{k}$.

Remark 2. Whenever an operation is executed, the value $\sum_{v:d(v)\geq 2} d(v)$ is decreased. The number of operations executed is at most km. Thus, Algorithm 2 runs in polynomial time.

Remark 3. Let H be a connected hypergraph with rank k. If H is not k-uniform, we can construct a connected k-uniform hypergraph H' by adding new vertices to each edge. The simple operation keeps the transversal number. Thus we have

$$\tau(H) = \tau(H') \leq \frac{(k-1)m' + 1}{k} = \frac{(k-1)m + 1}{k},$$

which states the bound $\tau(H) \leq \frac{(k-1)m+1}{k}$ also holds in rank k hypergraphs.

3　Extremal k-Uniform Connected Hypergraphs

In this subsection, we characterize the extremal hypergraphs achieving the bound in Theorem 5.

Theorem 10. *Let $H(V, E)$ be a connected k-uniform hypergraph with m edges. Then $\tau(H) = \frac{(k-1)m+1}{k}$ if and only if $H(V, E)$ is a hypertree with perfect matching.*

Proof. Sufficiency: If $H(V, E)$ is a hypertree with perfect matching, then according to Lemma 3 and Theorem 6, we have next equalities:

$$\tau(H) = \nu(H) = \frac{n}{k} = \frac{(k-1)m+1}{k}.$$

Necessity: When $\tau(H) = \frac{(k-1)m+1}{k}$, we need to prove $H(V, E)$ is a hypertree with perfect matching. It is enough to prove $H(V, E)$ is acyclic. Actually, if $H(V, E)$ is acyclic, according to Lemma 3 and Theorem 6, we have next inequalities:

$$\tau(H) = \nu(H) \leq \frac{n}{k} = \frac{(k-1)m+1}{k}.$$

Combined with $\tau(H) = \frac{(k-1)m+1}{k}$, we have next equalities, which says $H(V, E)$ is a hypertree with perfect matching.

$$\tau(H) = \nu(H) = \frac{n}{k} = \frac{(k-1)m+1}{k}.$$

By contradiction, let us take out a counterexample $H(V, E)$ with minimum edges. Then $\tau(H) = \frac{(k-1)m+1}{k}$ and $H(V, E)$ contains cycles. We have a series of claims:

Claim 1 . *Every two distinct cycles in H share common edges.*

Actually, for every two distinct cycles C_1 and C_2, if $E(C_1) \cap E(C_2) = \varnothing$, then we can partition the set of edges $E(H)$ into two parts $E(H_1)$ and $E(H_2)$ such that $E(C_1) \subseteq E(H_1), E(C_2) \subseteq E(H_2)$ and the edge-induced subhypergraphs H_1 and H_2 are both connected. Because $H(V, E)$ is a counterexample with minimum edges, we have next inequalities, a contradiction with the assumption $\tau(H) = \frac{(k-1)m+1}{k}$.

$$\tau(H_1) \leq \frac{(k-1)m_1}{k}, \quad \tau(H_2) \leq \frac{(k-1)m_2}{k}$$

$$\Rightarrow \tau(H) \leq \tau(H_1) + \tau(H_2) \leq \frac{(k-1)m_1}{k} + \frac{(k-1)m_2}{k} = \frac{(k-1)m}{k}.$$

Let us take out a shortest cycle C. Because $\tau(C) \leq \frac{m_c+1}{2} < \frac{(k-1)m+1}{k}$, we know $E(H) \backslash E(C) \neq \varnothing$. Furthermore, according to Claim 1, we know $E(H) \backslash E(C)$ induces some hypertrees. The next claim is essential.

Claim 2 . *Every hypertree induced by $E(H) \backslash E(C)$ must be an edge.*

We assume that there exists a hypertree T with $\mid E(T) \mid \geq 2$. Then let us take arbitrarily a vertex $v \in T \cap C$ and denote the farthest vertex from v in T as v'. We have next two cases.

Case 1: distance $d(v, v') = 1$ in T, we have a partial structure in Fig. 2. Now we can take $\{e_1, e_2\}$ as E_1 and other edges as E_2. It is easy to see that the edge-induced subhypergraphs H_1 and H_2 are both connected. Thus we have next inequalities, which is contradiction with $\tau(H) = \frac{(k-1)m+1}{k}$.

$$\tau(H) \leq \tau(H_1) + \tau(H_2) \leq 1 + \frac{(k-1)(m-2)+1}{k} = \frac{(k-1)m-k+3}{k} \leq \frac{(k-1)m}{k}.$$

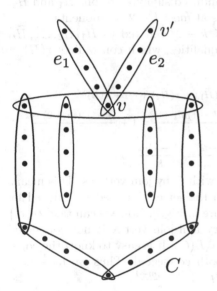

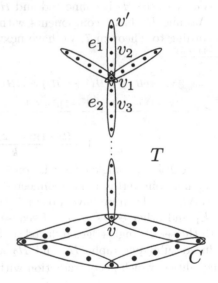

Fig. 2. distance $d(v, v') = 1$ in T **Fig. 3.** distance $d(v, v') \geq 2$ in T

Case 2: distance $d(v, v') \geq 2$ in T, we have a partial structure in Fig. 3. Because v' is the farthest vertex from v in T, for any vertex $v_i \in e_1 \backslash v_1$, there must be $d(v_i) = 1$ in T, which says e_1 is the unique edge containing v_i in T. We can take the edges incident with v_1 in T as E_1 and other edges as E_2. It is easy to know H_1 is connected and H_2 has at most $k - 1$ components. Furthermore, H_1 contains e_1, e_2, thus $m_1 \geq 2$.

Assume H_2 has t components with $t \leq k - 1$, denoted as $\widetilde{H}_1, \widetilde{H}_2, \ldots, \widetilde{H}_t$, and $\widetilde{H}_1$ contains the cycle C. Because $H(V, E)$ is a counterexample with minimum edges, we have next inequalities, also a contradiction with the assumption $\tau(H) = \frac{(k-1)m+1}{k}$.

$$\tau(\widetilde{H}_1) \leq \frac{(k-1)\widetilde{m}_1}{k}, \tau(\widetilde{H}_2) \leq \frac{(k-1)\widetilde{m}_2 + 1}{k}, \ldots, \tau(\widetilde{H}_t) \leq \frac{(k-1)\widetilde{m}_t + 1}{k}$$

$$\Rightarrow \tau(H) \leq \tau(H_1) + \tau(H_2) = \tau(H_1) + \tau(\widetilde{H}_1) + \tau(\widetilde{H}_2) + \cdots + \tau(\widetilde{H}_t)$$

$$\leq 1 + \frac{(k-1)\widetilde{m}_1}{k} + \frac{(k-1)\widetilde{m}_2 + 1}{k} + \cdots + \frac{(k-1)\widetilde{m}_t + 1}{k} = 1 + \frac{(k-1)(m - m_1) + t - 1}{k}$$

$$\leq 1 + \frac{(k-1)(m-2) + k - 2}{k} = \frac{(k-1)m}{k}.$$

Above all, in whatever case, there always exists a contradiction. Thus our assumption that there exists a hypertree T with $| E(T) | \geq 2$ doesn't hold on and every hypertree induced by $E(H) \backslash E(C)$ must be an edge.

Finally, let us consider the set of single edges induced by $E(H) \backslash E(C)$.

Case 1: there exists a single edge e connected with C by a non-join vertex. Then we have a partial structure in Fig. 4. Now we can take $\{e, e'\}$ as E_1 and other edges as E_2. Let us consider the edge-induced subhypergraphs H_1 and H_2. It is easy to see H_1 is connected and H_2 has at most $k - 2$ components.

Assume H_2 has t components with $t \leq k - 2$, denoted as $\widetilde{H}_1, \widetilde{H}_2, \ldots, \widetilde{H}_t$. According to Theorem 7, we have next inequalities, which contradicts $\tau(H) = \frac{(k-1)m+1}{k}$.

$$\tau(H) \leq \tau(H_1) + \tau(H_2) = \tau(H_1) + \tau(\widetilde{H}_1) + \tau(\widetilde{H}_2) + \cdots + \tau(\widetilde{H}_t)$$

$$\leq 1 + \frac{(k-1)\widetilde{m}_1 + 1}{k} + \frac{(k-1)\widetilde{m}_2 + 1}{k} + \cdots + \frac{(k-1)\widetilde{m}_t + 1}{k} = 1 + \frac{(k-1)(m-2) + t}{k}$$

$$\leq 1 + \frac{(k-1)(m-2) + k - 2}{k} = \frac{(k-1)m}{k}.$$

Case 2: Every single edge e is connected with C by join vertices. This means every non-join vertex is not connected with the set of single edges induced by $E(H) \backslash E(C)$. Then we have a partial structure in Fig. 5. Now we can take $\{e, e'\}$ as E_1 and other edges as E_2. Because every non-join vertex is not connected with the set of single edges induced by $E(H) \backslash E(C)$. It is easy to know the edge-induced subhypergraphs H_1 and H_2 are both connected. Thus we have next inequalities, which is contradiction with $\tau(H) = \frac{(k-1)m+1}{k}$.

$$\tau(H) \leq \tau(H_1) + \tau(H_2) \leq 1 + \frac{(k-1)(m-2) + 1}{k} = \frac{(k-1)m - k + 3}{k} \leq \frac{(k-1)m}{k}.$$

Above all, in whatever case, there always exists a contradiction. Thus our initial assumption that $H(V, E)$ contains cycles doesn't hold on. Thus $H(V, E)$ is a hypertree with perfect matching.

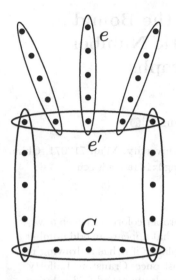

Fig. 4. partial structure in Case 1   **Fig. 5.** partial structure in Case 2

References

1. Berge, C.: Hypergraphs. North-Holland, Paris (1989)
2. Chen, X., Diao, Z., Hu, X., Tang, Z.: Covering triangles in edge-weighted graphs. Theory Comput. Syst. **62**(6), 1525–1552 (2018). https://doi.org/10.1007/s00224-018-9860-7
3. Chvátal, V., Mcdiarmid, C.: Small transversals in hypergraphs. Combinatorica **12**(1), 19–26 (1992). https://doi.org/10.1007/BF01191201
4. Diao, Z.: On the vertex cover number of 3-uniform hypergraphs. J. Oper. Res. Soc. China **9**, 427–440 (2021). https://doi.org/10.1007/s40305-019-00284-7
5. Dorfling, M., Henning, M.A.: Linear hypergraphs with large transversal number and maximum degree two. Eur. J. Comb. **36**, 231–236 (2014)
6. Henning, M.A., Löwenstein, C.: Hypergraphs with large transversal number and with edge sizes at least four. Discrete Appl. Math. **10**(3), 1133–1140 (2012)
7. Henning, M.A., Yeo, A.: Total domination in 2-connected graphs and in graphs with no induced 6-cycles. J. Graph Theory **60**(1), 55–79 (2010)
8. Henning, M.A., Yeo, A.: Hypergraphs with large transversal number. Discrete Math. **313**(8), 959–966 (2013)
9. Henning, M.A., Yeo, A.: Lower bounds on the size of maximum independent sets and matchings in hypergraphs of rank three. J. Graph Theory **72**(2), 220–245 (2013)
10. Henning, M.A., Yeo, A.: Transversals and matchings in 3-uniform hypergraphs. Eur. J. Comb. **34**(2), 217–228 (2013)
11. Lai, F.C., Chang, G.J.: An upper bound for the transversal numbers of 4-uniform hypergraphs. J. Comb. Theory Ser. B **50**(1), 129–133 (1990)
12. Thomassé, S., Yeo, A.: Total domination of graphs and small transversals of hypergraphs. Combinatorica **27**(4), 473–487 (2007). https://doi.org/10.1007/s00493-007-2020-3
13. Tuza, Z.: Covering all cliques of a graph. Discrete Math. **86**(1–3), 117–126 (1990)

An Improvement of the Bound on the Odd Chromatic Number of 1-Planar Graphs

Bei Niu[ID] and Xin Zhang[(✉)][ID]

School of Mathematics and Statistics, Xidian University, Xi'an 710071, China
beiniu@stu.xidian.edu.cn, xzhang@xidian.edu.cn

Abstract. An odd coloring of a graph is a proper coloring in such a way that every non-isolated vertex has some color that appears an odd number of times on its neighborhood. A graph is 1-planar if it has a drawing in the plane so that each edge is crossed at most once. Cranston, Lafferty, and Song showed that every 1-planar graph admits an odd 23-coloring [arXiv:2202.02586v4]. In this paper, we improve their bound to 16.

Keywords: 1-planar graph · Odd coloring · Discharging

1 Introduction

Throughout the paper, all graphs are finite, simple and undirected. By $V(G)$, $E(G)$, and $\delta(G)$, we denote the set of vertices, the set of edges, and the minimum degree of a graph G, respectively. If G is a plane graph, then $F(G)$ denotes the set of faces of G, The *neighborhood* $N_G(v)$ of a vertex v is the set of vertices adjacent to v in G. The *degree* of a vertex v in G, denoted by $d_G(v)$, is the size of $N_G(v)$, and the *degree* of a face f in a plane graph G, denoted by $d_G(f)$, is the the number of edges that are incident with f in G, where cut-edges are counted twice. A k-, k^+-, and k^--*vertex* (resp. face) is a vertex (resp. face) of degree k, at least k and at most k, respectively. For other undefined notation, we refer the readers to the book [1].

A coloring of vertices of a hypergraph is *conflict-free* if at least one vertex in each (hyper-)edge has a unique color, see [6]. Its research was initially motivated by a frequency assignment problem in cellular networks. Such networks consist of fixed-position base stations and roaming clients, each base station is assigned a certain frequency and transmits data in this frequency within some given region. Roaming clients have a range of communication and come under the influence of different subsets of base stations. This situation can be modeled by means of a hypergraph whose vertices correspond to the base stations. The range of communication of a mobile agent, that is, the set of base stations it can

Supported by the Fundamental Research Funds for the Central Universities (QTZX22053) and the National Natural Science Foundation of China (11871055).

communicate with, is represented by a hyperedge $e \in E$. A conflict-free coloring of such a hypergraph implies an assignment of frequencies, to the base stations, which enables clients to connect to a base station holding a unique frequency in the client's range, thus avoiding interferences.

Recently, Petruševski and Škrekovski [9] introduced the notion of odd coloring, which is a relaxation of conflict-free coloring. Formally, an *odd c-coloring* of a graph is a proper c-coloring with the additional constraint that each vertex of positive degree has a color appearing an odd number of times among its neighborhood. A graph G is *odd c-colorable* if it has an odd c-coloring. The *odd chromatic number* of a graph G, denoted $\chi_o(G)$, is the minimum c such that G has an odd c-coloring. Petruševski and Škrekovski [9] put forward the following conjecture:

Conjecture 1. ([9]). *Every planar graph admits an odd 5-coloring.*

The best progress towards this conjecture is due to Petr and Portier [8], who proved that the odd chromatic number of every planar graph is at most 8, improving the preceding bound 9 of Petruševski and Škrekovski [9]. Supporting this conjecture, Cranston [4] showed that every planar graph of girth at least 7 is odd 5-colorable, Caro, Petruševski and Škrekovski [2] also proved that every outerplanar graph admits an odd 5-coloring. Qi and Zhang [10] later verified Conjecture 1 for another two subclasses of planar graphs, saying outer-1-planar graphs and 2-boundary planar graphs, generalizing the result of Petruševski and Škrekovski [9]. Note that the bound 5 in Conjecture 1 would be sharp as $\chi_o(C_5) = 5$.

There are normally two ways to generalize the planarity. One way is to allow a drawing without crossings in a surface, such as a torus, rather than a plane. In view of this, Metrebian [7] showed that every torodial graph admits an odd 9-coloring. Another generalization can be established in the way of allowing bounded number of crossings per edge. Ringel [11] introduced the notion of 1-planarity in 1965. Precisely, a graph is *1-planar* if it can be drawn in the plane so that each edge is crossed by at most one other edge. Recently, Cranston, Lafferty, and Song [5] showed that every 1-planar graph admits an odd 23-coloring (the first bound for the odd chromatic number of 1-planar graphs was 47, due to the first version of [5]).

The aim of this paper is to find a better upper bound for the odd chromatic number of 1-planar graphs by showing the following.

Theorem 2. *Every 1-planar graph admits an odd 16-coloring.*

The proof of Theorem 2 is relied on the proof of the odd 23-colorability of Cranston, Lafferty, and Song [5]. Readers will see that we borrow all structural lemmas there. However, with our new discharging rules, the counting of final charges becomes easier and surprisingly the bound descends.

2   The Proof of Theorem 2

Suppose for a contradiction that G is a minimal counterexample (in terms of $|V(G)| + |E(G)|$) to this theorem. The *associated plane graph* $G^\times$ of G is the plane graph obtained from G by turning all crossings of G into new vertices of degree four. Those new 4-vertices are *false* vertices of $G^\times$, and the original vertices of G are *true* vertices of $G^\times$. A face of $G^\times$ is *false* if it is incident with at least one false vertex, and *true* otherwise. For each vertex $v \in V(G)$, let $d_2(v)$ denote the number of 2-vertices adjacent to v in G. An *odd vertex* of G is a vertex having odd degree. If $v \in V(G)$ has even degree at most 6 then we call it *small vertex*, and if $v \in V(G)$ has degree at least 8 then we call it *big vertex*.

Claim 1. *[5, Claim 1]* $\delta(G) \geq 2$.

Claim 2. *[5, Claim 2] Every odd vertex in G has degree at least 9.*

Claim 3. *[5, Claim 3] No two small vertices are adjacent in G.*

Claim 4. *[5, Claim 4] Every edge incident to a small vertex in G has a crossing.*

Claim 5. *[3, Lemma 2.1] If v is a vertex with $d_2(v) \geq 1$, then $2d(v) \geq d_2(v)+16$.*

Claim 6. *[5, Claim 6] The graph $G^\times$ has no loop or 2-face, and every 3-face in $G^\times$ is incident to either three big vertices or two big vertices and one false 4-vertex.*

Claim 7. *[5, Claim 7] Every 2-vertex in $G^\times$ is incident to a 5^+-face and to another 4^+-face.*

Claim 8. *[5, in the proof of Claim 9] For a 4-face zz_1vz_2z and a 6-face $vz_1uz_3wz_2v$ of $G^\times$, if z_1, z_2, z_3 are false vertices, then at most two vertices among u, v, w are 2-vertices.*

Note that Claims 2 and 5 are different from their original forms. However, we can prove them using the same arguments only with certain numbers changed. Moreover, although we change the definitions of small and big vertices, comparing to the ones in [5], the proofs of Claims 3, 4, and 7 work in the same logic as in [5].

We apply the discharging method to $G^\times$. Formally, for each vertex $v \in V(G^\times)$, let $\mathrm{ch}(v) := d_{G^\times}(v)-4$ be its initial charge, and for each face $f \in F(G^\times)$, let $\mathrm{ch}(f) := d_{G^\times}(f) - 4$ be its initial charge. Clearly,

$$\sum_{x \in V(G^\times) \cup F(G^\times)} \mathrm{ch}(x) = -8 < 0$$

by the well-known Euler's formula.

For convenience we use $d(v)$ and $d(f)$ instead of $d_{G^\times}(v)$ and $d_{G^\times}(f)$ if v is a true vertex and f is a face in $G^\times$, respectively. The discharging rules are defined as follows.

R1 Every big vertex sends 1/4 to each of its incidence 2-vertices;
R2 Every big vertex sends 1/3 to each of its incidence true 3-faces;
R3 Every big vertex sends 1/2 to each of its incidence false faces;

R4 If f is a 4^+-face with positive charge after applying **R1–R3** and f is incident with at least one 2-vertex, then f redistribute its positive charge to each of its incidence 2-vertices equally.

Let $\text{ch}^*(x)$ be the charge of $x \in V(G^\times) \cup F(G^\times)$ after applying the above rules. Since our rules only move charge around, and do not affect the sum, we have

$$\sum_{x \in V(G^\times) \cup F(G^\times)} \text{ch}^*(x) = \sum_{x \in V(G^\times) \cup F(G^\times)} \text{ch}(x) < 0.$$

Next, we prove that $\text{ch}^*(x) \geq 0$ for each $x \in V(G^\times) \cup F(G^\times)$ by Propositions 1, 2 and 3. This gives

$$\sum_{x \in V(G^\times) \cup F(G^\times)} \text{ch}^*(x) \geq 0,$$

a contradiction. Note that every true vertex of $G^\times$ is either small or big by Claim 2 and the final charge of any false vertex of $G^\times$ is trivially 0.

Proposition 1. *The final charge of every face of $G^\times$ is non-negative.*

Proof. By Claim 6, every face of $G^\times$ has degree at least 3. If f is a true 3-face, then f is incident only with big vertices by Claim 6 and thus $\text{ch}^*(f) = 3-4+3\times\frac{1}{3} = 0$ by **R2**. If f is a false 3-face, then f is incident with two big vertices by Claim 6 and thus $\text{ch}^*(f) = 3 - 4 + 2 \times \frac{1}{2} = 0$ by **R3**. If f is 4^+-face incident with 2-vertices, then $\text{ch}^*(f) = 0$ by **R4**. If f is 4^+-face incident with no 2-vertex, then no rules will be applied to f and thus $\text{ch}^*(f) = \text{ch}(f) = d(f) - 4 \geq 0$. $\quad\square$

Proposition 2. *The final charge of every small vertex of $G^\times$ is non-negative.*

Proof. It is sufficient to prove this result for an arbitrary arbitrarily 2-vertex v, as $\text{ch}^*(v) = \text{ch}(v)$ for $d(v) = 4, 6$ by the discharging rules. Assume $N_G(v) = \{x, y\}$. By Claim 4, vx and vy are crossed. Assume that vx is crossed by $u_1 u_2$ at a false vertex z_1, and vy is crossed by $w_1 w_2$ at a false vertex z_2, such that u_1, z_1, v, z_2, w_1 are on one face, say f_1, and u_2, z_1, v, z_2, w_2 are on another face, say f_2. It may be possible that $u_1 = w_1$ or $u_2 = w_2$. Assume, without loss of generality, that $d(f_1) \leq d(f_2)$. Since v is incident to a 5^+-face and to another 4^+-face by Claim 7, we consider two cases.

Case 1. $d(f_1) = 4$.

This situation implies $u_1 = w_1$ and $u_2 \neq w_2$. For convenience we let $z = u_1 = w_1$. Note that $d(f_2) \geq 5$.

Subcase 1.1. z is a big vertex, see Fig. 1(a).

Now f_1 sends $1/2$ to v by **R3** and **R4**. Next we look at f_2. Besides z_1, v, z_2, there are at most $\lceil \frac{d(f_2)-3}{2} \rceil$ 2-vertices on f_2 by Claim 3. Hence at most $\lceil \frac{d(f_2)-1}{2} \rceil$ 2-vertices exist on f_2. By **R4**, f_2 sends to v at least

$$\alpha := \frac{d(f_2) - 4}{\left\lceil \frac{d(f_2)-1}{2} \right\rceil}.$$

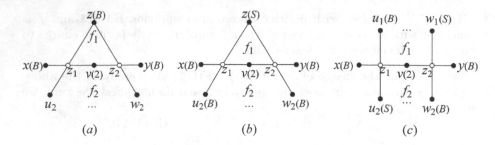

Fig. 1. Illustration for the proof of Proposition 2.

If $d(f_2) \geq 7$, then $\alpha \geq 1$.

If $d(f_2) = 6$, then f_2 is incidence with at most two 2-vertices by Claim 8. Hence f_2 sends to v at least $\frac{6-4}{2} = 1$, too.

If $d(f_2) = 5$, then $u_2 w_2 \in E(G)$. This implies that u_2 and w_2 are both big by Claim 4. Hence f_2 has charge $5 - 4 + 2 \times \frac{1}{2} = 2$ after applying **R3**, which will be sent to v by **R4**.

In each case, v receives at least 1 from f_2, 1/2 from f_1, and 1/4 from each of x and y by **R1**. This gives ch$^*(v) \geq 2 - 4 + 1 + \frac{1}{2} + 2 \times \frac{1}{4} = 0$.

Subcase 1.2. z is a small vertex, see Fig. 1(b).

This situation implies u_2 and w_2 are big by Claim 3.

Now we mainly look at f_2. Besides u_2, z_1, v, z_2, w_2, there are at most $\lfloor \frac{d(f_2)-5}{2} \rfloor$ 2-vertices on f_2 by Claims 3 and 4. Hence at most $\lfloor \frac{d(f_2)-3}{2} \rfloor$ 2-vertices exist on f_2. By **R3** and **R4**, f_2 sends to v at least

$$\frac{d(f_2) - 4 + 2 \times \frac{1}{2}}{\left\lfloor \frac{d(f_2)-3}{2} \right\rfloor} \geq 2$$

and thus ch$^*(v) \geq 2 - 4 + 2 = 0$.

Case 2. $d(f_1) \geq 5$.

If u_i and w_i are big for some $i \in \{1,2\}$, then by a similar argument as in Subcase 1.2, we conclude that f_i sends at least 2 to v and thus ch$^*(v) \geq 2 - 4 + 2 = 0$. Hence we assume u_1 is big and w_1 is small. By Claim 3, w_2 is big. So we further assume u_2 is small, see Fig. 1(c).

Now besides u_1, z_1, v, z_2, there are at most $\lceil \frac{d(f_1)-4}{2} \rceil$ 2-vertices on f_1 by Claim 3. Hence at most $\lceil \frac{d(f_1)-2}{2} \rceil$ 2-vertices exist on f_1. By **R3** and **R4**, f_1 sends to v at least

$$\frac{d(f_1) - 4 + \frac{1}{2}}{\left\lceil \frac{d(f_1)-2}{2} \right\rceil} \geq \frac{3}{4}$$

as $d(f_1) \geq 5$. By symmetry, f_2 sends to v at least 3/4.

Since both x and y sends 1/4 to v by **R1**, ch$^*(v) \geq 2 - -4 + 2 \times \frac{3}{4} + 2 \times \frac{1}{4} = 0$.

□

Proposition 3. *The final charge of every big vertex of $G^\times$ is non-negative.*

Proof. Let v be a big vertex. If $d_2(v) = 0$, then $\mathrm{ch}^*(v) \geq d(v) - 4 - \frac{1}{2}d(v) \geq 0$ by **R2** and **R3** as $d(v) \geq 8$. If $d_2(v) \geq 1$, then $d_2(v) \leq 2d(v) - 16$ by Claim 5. Hence

$$\mathrm{ch}^*(v) \geq d(v) - 4 - \frac{1}{4}d_2(v) - \frac{1}{2}d(v)$$

$$\geq d(v) - 4 - \frac{1}{4}(2d(v) - 16) - \frac{1}{2}d(v)$$

$$= 0$$

by **R1–R3**. □

References

1. Bondy, J.A., Murty, U.S.R.: Graph Theory. GTM 244, Springer, New York (2008)
2. Caro, Y., Petruševski, M., Škrekovski, R.: Remarks on odd colorings of graphs. (2022). arXiv: 2201.03608v1
3. Cho, E.-K., Choi, I., Kwon, H., Park, B.: Odd coloring of sparse graphs and planar graphs. (2022). arXiv: 2202.11267v1
4. Cranston, D.W.: Odd Colorings of Sparse Graphs (2022).arXiv:2201.01455v1
5. Cranston, D.W., Lafferty, M., Song, Z.-X.: A note on odd coloring of 1-planar graphs (2022). arXiv: 2202.02586v4
6. Even, G., Lotker, Z., Ron, D., Smorodinsky, S.: Conflict-free colorings of simple geometric regions with applications to frequency assignment in cellular networks. SIAM J. Comput. **33**, 94–136 (2003)
7. Metrebian, H.: Odd coloring on the torus (2022). arXiv: 2205.05398v1
8. Petr, J., Portier, J.: The odd chromatic number of a planar graph is at most 8 (2022). arXiv: 2201.12381v2
9. Petruševski, M., Škrekovski, R.: Coloring with neighborhood parity condition (2022). arXiv: 2112.13710v2
10. Qi, M.K., Zhang, X.: Odd coloring of two subclasses of planar graphs (2022). arXiv: 2205.09317v1
11. Ringel, G.: Ein Sechsfarbenproblem auf der Kugel, Abhandlungen aus dem Mathematischen Seminar der Universität Hamburg, **29** 107–117 (1965)

Fast Searching on k-Combinable Graphs

Yuan Xue, Boting Yang$^{(\boxtimes)}$, and Sandra Zilles

Department of Computer Science, University of Regina, Regina, Canada
{xue228,boting,zilles}@cs.uregina.ca

Abstract. Finding an optimal fast search strategy for graphs is challenging, sometimes even when graphs have very small treewidth, like cacti, cartesian product of a tree and an edge, etc. However, it may be easier to find an optimal fast search strategy for some critical subgraphs of the given graph. Although fast searching is not subgraph-closed, this observation still motivates us to establish relationships between optimal fast search strategies for a graph and its subgraphs. In this paper, we introduce the notion of k-combinable graphs and propose a new method for computing their fast search number. Assisted by the new method, we investigate the fast search number of cacti graphs and the cartesian product of a tree and an edge. Algorithms for producing fast search strategies for the above graphs, along with rigorous proofs, are given in this paper.

1 Introduction

Inspired by an article of Breisch [3] who considered the problem of finding a lost explorer in dark complex caves, Parsons [9] first introduced the graph search problem in which both searchers and fugitive move continuously along edges of a graph. Motivated by applied problems in the real world and theoretical issues in computer science and mathematics, graph searching has become a hot topic. It has many models, such as edge searching, node searching, mixed searching, fast searching, etc. These models are basically defined by the class of graphs, the actions of searchers and fugitives, visibility of fugitives, and conditions on what constitutes capture [1,2,6,8].

Given a graph that contains an invisible fugitive, the fast search problem is to find the fast search number, i.e., the minimum number of searchers to capture the fugitive in the fast search model. This model was first introduced by Dyer, Yang and Yaşar [5] in 2008. Let G denote an undirected graph. In the fast search model, a fugitive hides either on vertices or on edges of G. The fugitive can move at a great speed at any time from one vertex to another along a path that contains no searchers. We call an edge *contaminated* if it may contain the fugitive, and we call an edge *cleared* if we are certain that it does not contain the fugitive. In order to capture the fugitive, one launches a set of searchers on some vertices of the graph; these searchers then clear the graph edge by edge while at the same time guarding the already cleared parts of the graph. There are two actions for searchers: placing and sliding. An edge is cleared by a sliding action and every edge must be traversed exactly once. A *fast search strategy* for a graph is a sequence of actions of searchers that clear all contaminated edges of the graph. The *fast search number* of G, denoted by $\mathrm{fs}(G)$, is the smallest number of searchers needed to capture the fugitive in G. For more details about the model setting, please refer to [5].

Q. Ni and W. Wu (Eds.): AAIM 2022, LNCS 13513, pp. 394–405, 2022.
https://doi.org/10.1007/978-3-031-16081-3_34

Dyer et al. [5] proposed a linear time algorithm for computing the fast search number of trees. Stanley and Yang [10] gave a linear time algorithm for computing the fast search number of Halin graphs and their extensions. They also presented a quadratic time algorithm for computing the fast search number of cubic graphs, while the problem of finding the node search number of cubic graphs is NP-complete [7]. Yang [13] proved that the problem of finding the fast search number of a graph is NP-complete; and it remains NP-complete for Eulerian graphs. He also proved that the problem of determining whether the fast search number of G is a half of the number of odd vertices in G is NP-complete; and it remains NP-complete for planar graphs with maximum degree 4. Dereniowski et al. [4] characterized graphs for which 2 or 3 searchers are sufficient in the fast search model. They proved that the fast searching problem is NP-hard for multigraphs. Dyer et al. [5] considered complete bipartite graphs $K_{m,n}$, and computed the fast search number of $K_{m,n}$ when m is even. They also presented lower and upper bounds on the fast search number of $K_{m,n}$ when m is odd. Xue et al. [12] provided lower bounds and upper bounds on the fast search number of complete k-partite graphs. They also solved the open problem of determining the fast search number of complete bipartite graphs. In [11], Xue and Yang provided lower bounds on the fast search number, and gave formulas for the fast search number of the cartesian product of an Eulerian graph and a path, as well as variants of the cartesian product.

In this paper, we introduce the notion of k-combinable graphs, and develop a new method for computing their fast search number. The method can be seen as a general method for finding lower bounds on the fast search number. Using this new method, we examine the fast search number of several classes of graphs including cacti graphs and cartesian product of a tree and an edge.

2 K-Combinable Graphs

We first introduce a class of graphs named k-*combinable graphs*. Then we describe our method for finding an optimal fast search strategy for k-combinable graphs. Let G be a connected graph and let E'_G be the set of all pendant edges of G. The *profile* of G is an ordered tuple $\pi_G = (\pi_1, \ldots, \pi_z)$ of positive integers, which is defined as follows:

1. If $E'_G = \emptyset$, then $z = 1$ and $\pi_1 = \mathrm{fs}(G)$.
2. If $E'_G \neq \emptyset$ and $|E'_G| = k$, then $z = k!2^k$ and each component π_i of π_G is associated with a specific permutation σ and a specific orientation of each edge in E'_G. In particular, π_i is the smallest number of searchers with which a fast search strategy can clear G if it traverses the edges in E'_G in the order of σ and in the directions as given by the chosen orientations.

Let G_1 be a connected graph that has $k_1 \geq 1$ pendant edges, and let G_2 be a connected graph having $k_2 \geq 1$ pendant edges. We choose k to be a constant satisfying that $1 \leq k \leq \min\{k_1, k_2\}$. Let $\overrightarrow{e_1} = (u_1 u'_1, \ldots, u_k u'_k)$, where $u_i u'_i \in E(G_1)$ and u'_i is a leaf node. Let $\overrightarrow{e_2} = (v_1 v'_1, \ldots, v_k v'_k)$, where $v_i v'_i \in E(G_2)$ and v'_i is a leaf node. Let H be the graph obtained from G_1 and G_2 by performing the following operations on G_1 and G_2 with respect to $\overrightarrow{e_1}$ and $\overrightarrow{e_2}$:

1. remove edges $u_i u_i'$ and $v_i v_i'$, for $1 \le i \le k$;
2. remove vertices u_i' and v_i', for $1 \le i \le k$;
3. connect u_i and v_i by adding a new edge, for $1 \le i \le k$.

Note that the above operations depend on the choice of the sequences $\overrightarrow{e_1}$ and $\overrightarrow{e_2}$ which we will henceforth call edge pairing sequences. If we permute either of the edge pairing sequences, this would create a different result. Hence, we define the *align operation* on G_1 and G_2 with respect to $\overrightarrow{e_1}$ and $\overrightarrow{e_2}$, denoted as $(G_1, \overrightarrow{e_1}) \triangle (G_2, \overrightarrow{e_2})$, to be the graph obtained by performing the above operations.

Definition 1. Let $m \ge 2$. Let $G_1, \ldots, G_m$ be connected graphs. The sequence $(G_1, \ldots, G_m)$ is k-combinable if there are edge sequences $\overrightarrow{e_1}, \ldots, \overrightarrow{e_m}, \overrightarrow{e_{1,2}}, \ldots, \overrightarrow{e_{1,m-1}}$ such that:

1. For $1 \le i \le m$, $\overrightarrow{e_i}$ is a sequence of pendant edges of G_i.
2. For $2 \le i \le m - 1$, $\overrightarrow{e_{1,i}}$ is a sequence of pendant edges of H_i, where $H_2 = (G_1, \overrightarrow{e_1}) \triangle (G_2, \overrightarrow{e_2})$, and $H_{i+1} = (H_i, \overrightarrow{e_{1,i}}) \triangle (G_{i+1}, \overrightarrow{e_{i+1}})$.
3. For $1 \le i \le m$, the set of all edges of G_i, which occur in $\overrightarrow{e_1}, \ldots, \overrightarrow{e_m}$ and $\overrightarrow{e_{1,2}}, \ldots, \overrightarrow{e_{1,m-1}}$, has size at most k.
4. For $2 \le j \le m - 1$, the set of all edges of H_j, which occur in $\overrightarrow{e_1}, \ldots, \overrightarrow{e_m}$ and $\overrightarrow{e_{1,2}}, \ldots, \overrightarrow{e_{1,m-1}}$, has size at most k.

Further, we call H_m a k-combination of $(G_1, \ldots, G_m)$, in particular, this is the k-combination of $(G_1, \ldots, G_m)$ with respect to $\overrightarrow{e_1}, \ldots, \overrightarrow{e_m}, \overrightarrow{e_{1,2}}, \ldots, \overrightarrow{e_{1,m-1}}$.

Obviously, there may exist more than one graph that is a k-combination of $(G_1, G_2, \ldots, G_m)$. Further, for each k-combination G of $(G_1, G_2, \ldots, G_m)$, there exist specific $\overrightarrow{e_{1,2}}, \ldots, \overrightarrow{e_{1,m-1}}$ and $\overrightarrow{e_1}, \ldots, \overrightarrow{e_m}$ for obtaining G. In the remainder of this section, we always assume that every time an algorithm handles profiles of graphs, it implicitly associates the profiles with corresponding $\overrightarrow{e_{1,i}}$ and $\overrightarrow{e_j}$, where $2 \le i \le m - 1$ and $1 \le j \le m$.

Theorem 1. *There exists an algorithm that, given the profiles and edge pairing sequences of G_1 and G_2 such that G is the k-combination of (G_1, G_2) with respect to the edge pairing sequences, runs in $O((k_1 + k_2 - k)! 2^{k_1 + k_2 - k})$ time to compute the profile of G. Here k_i refers to the number of pendant edges of G_i, where $1 \le i \le 2$.*

Proof. We briefly introduce the idea of how to compute the profile of G. Since G_1 and G_2 have k_1 and k_2 pendant edges respectively, the sizes of profiles of G_1 and G_2 are $k_1! 2^{k_1}$ and $k_2! 2^{k_2}$. Let $\overrightarrow{e_1}$ and $\overrightarrow{e_2}$ denote the edge pairing sequences of G_1 and G_2 respectively. Consider all the edges in $\overrightarrow{e_1}$ and $\overrightarrow{e_2}$. If we are given a set of rules instructing how these edges are cleared in a strategy, then in accordance with the rules, we can figure out the number of searchers that need to be placed on the non-leaf vertices in $V(G_1)$ and $V(G_2)$. For each parameter in the profile of G, it takes $O(k! 2^k)$ time to compute its value. Further, we know the size of the profile of G is $(k_1 + k_2 - 2k)! 2^{k_1 + k_2 - 2k}$. Hence, the time complexity for computing the profile of G is $O((k_1 + k_2 - k)! 2^{k_1 + k_2 - k})$. ☐

From Theorem 1, it is easy to see that our method can be applied to find an optimal fast search strategy for quite complicated graphs, if the graph can be split into two smaller graphs for which fast search strategies are easy to find. Moreover, if we are given G that is a k-combination of $(G_1, \ldots, G_m)$ where $m \geq 3$, by repeatedly applying the procedure presented in the proof of Theorem 1, we can find an optimal fast search strategy for G as stated in Theorem 2. This novel method reveals an interesting property of fast searching that has not been exploited systematically in the literature to date.

Theorem 2. *Let G be a k-combination of $(G_1, \ldots, G_m)$ with respect to $\overrightarrow{e_1}, \ldots, \overrightarrow{e_m}$, $\overrightarrow{e_{1,2}}, \ldots, \overrightarrow{e_{1,m-1}}$, where $G_1, \ldots, G_m$ are connected graphs and k is a constant. There exists an algorithm which, given (1) the profiles of $G_1, G_2, \ldots G_m$ in sequence, and (2) $\overrightarrow{e_1}, \ldots, \overrightarrow{e_m}$ and $\overrightarrow{e_{1,2}}, \ldots, \overrightarrow{e_{1,m-1}}$, runs in polynomial time to compute the profile of G. Furthermore, the fast search number of G can be found in polynomial time.*

In the next section, we will apply Theorem 2 to the finding of optimal fast search strategy for cacti graphs; further, we also apply the theorem to the finding of optimal fast search strategies for cartesian product of a tree and an edge. We will show that (1) how to split a graph into smaller subgraphs, and (2) how to apply Theorem 2 to obtain an optimal fast search strategy, upon knowing the profiles of all the subgraphs in (1).

3   Cacti Graphs

A connected graph is a *cactus* if and only if each of its edges is contained in at most one cycle. In this section, we use G to denote a cactus graph. Let $v \in V(G)$ and let $\mathcal{G}_1, \ldots, \mathcal{G}_k$ be all the connected components from G by deleting v and all its incident edges. We use $\mathcal{G}_v^i$ to denote the subgraph of G induced by $V(\mathcal{G}_i) \cup \{v\}$, where $1 \leq i \leq k$. $\mathcal{G}_v^1, \ldots, \mathcal{G}_v^k$ are called *sub-cacti* of G with respect to vertex v. Note that $\mathcal{G}_v^1, \ldots, \mathcal{G}_v^k$ must satisfy:

(i) $V(\mathcal{G}_v^1) \cup \cdots \cup V(\mathcal{G}_v^k) = V(G)$,
(ii) $V(\mathcal{G}_v^i) \cap V(\mathcal{G}_v^j) = v$, where $1 \leq i \neq j \leq k$, and
(iii) $u_1, u_2 \in V(G)$ are adjacent, only if there exists i such that $u_1, u_2 \in V(\mathcal{G}_v^i)$.

Consider $\mathcal{G}_v^i$, where $1 \leq i \leq k$. Note that v has degree at most two in $\mathcal{G}_v^i$. If v is a leaf node in $\mathcal{G}_v^i$, then let u be a vertex in $V(\mathcal{G}_v^i)$ satisfying $u \sim v$. We use $\pi_I(\mathcal{G}_v^i)$ to denote the minimum number of searchers placed on $V(\mathcal{G}_v^i) \setminus \{v\}$ in a strategy for $\mathcal{G}_v^i$, in which vu is cleared by sliding a searcher from v to u. An *I-strategy* for $\mathcal{G}_v^i$ is a strategy in which (1) vu is cleared by sliding a searcher from v to u, and (2) $\pi_I(\mathcal{G}_v^i)$ searchers are placed on $V(\mathcal{G}_v^i) \setminus \{v\}$. Note that if vu is cleared by sliding a searcher from v to u in a strategy, then a searcher must be placed on v at the beginning of the strategy. We use $\pi_O(\mathcal{G}_v^i)$ to denote the minimum number of searchers placed on $V(\mathcal{G}_v^i) \setminus \{v\}$ in a strategy for $\mathcal{G}_v^i$, in which vu is cleared by sliding a searcher from u to v. An *O-strategy* for $\mathcal{G}_v^i$ is a strategy for $\mathcal{G}_v^i$ in which (1) vu is cleared by sliding a searcher from u to v, and (2) $\pi_O(\mathcal{G}_v^i)$ searchers are placed on $V(\mathcal{G}_v^i) \setminus \{v\}$.

If v has degree two in $\mathcal{G}_v^i$, then let u_1 and u_2 be the two vertices in $V(\mathcal{G}_v^i)$ satisfying that $u_1 \sim v$ and $u_2 \sim v$. For $i \in \{1, 2\}$, we say vu_i is cleared by a *slide-in* action if a searcher slides from v to u_i along vu_i, and we say vu_i is cleared by a *slide-out* action if a searcher slides from u_i to v along vu_i. We use $\pi_{I,I}(\mathcal{G}_v^i)$ to denote the minimum number of searchers placed on $V(\mathcal{G}_v^i) \setminus \{v\}$ in a strategy for $\mathcal{G}_v^i$, in which vu_1 and vu_2 are both cleared by slide-in actions. We use $\pi_{O,O}(\mathcal{G}_v^i)$ to denote the minimum number of searchers placed on $V(\mathcal{G}_v^i) \setminus \{v\}$ in a strategy for $\mathcal{G}_v^i$, in which vu_1 and vu_2 are both cleared by slide-out actions. We use $\pi_{I,O}(\mathcal{G}_v^i)$ to denote the minimum number of searchers placed on $V(\mathcal{G}_v^i)\setminus\{v\}$ in a strategy for $\mathcal{G}_v^i$, in which vu_1 or vu_2 is cleared by a slide-in action, and later the other edge is cleared by a slide-out action. We use $\pi_{O,I}(\mathcal{G}_v^i)$ to denote the minimum number of searchers placed on $V(\mathcal{G}_v^i) \setminus \{v\}$ in a strategy for $\mathcal{G}_v^i$, in which vu_1 or vu_2 is cleared by a slide-out action, and later the other edge is cleared by a slide-in action. A strategy for $\mathcal{G}_v^i$ is an *II-strategy*, in which (1) $\pi_{I,I}(\mathcal{G}_v^i)$ searchers are placed on $V(\mathcal{G}_v^i) \setminus \{v\}$, and (2) vu_1 and vu_2 are both cleared by slide-in actions. In a similar way, we define *IO-strategy*, *OI-strategy* and *OO-strategy* for $\mathcal{G}_v^i$ respectively.

Definition 2. Consider a sub-cactus of G with respect to vertex v, i.e., $\mathcal{G}_v^i$.

1. If v has exactly one incident edge in $\mathcal{G}_v^i$, then the profile of $\mathcal{G}_v^i$ is defined as the pair $(\pi_I(\mathcal{G}_v^i), \pi_O(\mathcal{G}_v^i))$.
2. If v has exactly two incident edges in $\mathcal{G}_v^i$, then the profile of $\mathcal{G}_v^i$ is defined as the 4-tuple $(\pi_{I,I}(\mathcal{G}_v^i), \pi_{I,O}(\mathcal{G}_v^i), \pi_{O,I}(\mathcal{G}_v^i), \pi_{O,O}(\mathcal{G}_v^i))$.

For cactus graph G and $v \in V(G)$, we use G_v' to denote the graph obtained by adding either one or two pendant edges to v. There are two possibilities for G_v':

(1) v has one added pendant edge in G_v', say vu. Let $\pi_I(G_v')$ be the minimum number of searchers placed on $V(G_v') \setminus \{u\}$ in a strategy for G_v', in which vu is cleared by sliding a searcher from u to v. An *I-strategy* for G_v' is a strategy, in which (a) $\pi_I(G_v')$ searchers are placed on $V(G_v') \setminus \{u\}$, and (b) vu is cleared by sliding a searcher from u to v. In a similar way, we define $\pi_O(G_v')$ and *O-strategy* for G_v'. The *profile of G_v'* is defined as the pair $(\pi_I(G_v'), \pi_O(G_v'))$.

(2) v has two added pendant edges in G_v'. Notice that there are four distinct ways to clear the two added pendant edges of v. In a similar way, we define (1) $\pi_{I,I}(G_v')$, $\pi_{I,O}(G_v')$, $\pi_{O,I}(G_v')$ and $\pi_{O,O}(G_v')$ for G_v', and (2) *II-strategy*, *IO-strategy*, *OI-strategy* and *OO-strategy* for G_v'. The *profile of G_v'* is defined as 4-tuple $(\pi_{I,I}(G_v'), \pi_{I,O}(G_v'), \pi_{O,I}(G_v'), \pi_{O,O}(G_v'))$.

Definition 3. For a strategy $\mathcal{S}$ for G, let the *reversed strategy* for $\mathcal{S}$ be obtained from $\mathcal{S}$ by making the following modifications:

1. Remove all placing actions from $\mathcal{S}$.
2. For each vertex $v \in V(G)$ that contains searchers at the end of $\mathcal{S}$, insert a placing action at the beginning that places the same number of searchers on v.

3. For each edge $e \in E(G)$, reverse the sliding action on e by letting searcher move in the opposite way to clear it.
4. Reverse the order of all sliding actions.

Clearly, the reversed strategy for S uses the same number of searchers to clear G. Hence, we have $\pi_{I,I}(\mathcal{G}_v^i) = \pi_{O,O}(\mathcal{G}_v^i) - 2$, and $\pi_I(\mathcal{G}_v^i) = \pi_O(\mathcal{G}_v^i) - 1$.

Lemma 1. $\mathcal{G}_v^i$ *must have one of the following properties:*

1. $\pi_{I,I}(\mathcal{G}_v^i) = \pi_{I,O}(\mathcal{G}_v^i) = \pi_{O,I}(\mathcal{G}_v^i) = \pi_{O,O}(\mathcal{G}_v^i) - 2;$
2. $\pi_{I,I}(\mathcal{G}_v^i) = \pi_{I,O}(\mathcal{G}_v^i) = \pi_{O,I}(\mathcal{G}_v^i) - 1 = \pi_{O,O}(\mathcal{G}_v^i) - 2;$
3. $\pi_{I,I}(\mathcal{G}_v^i) = \pi_{I,O}(\mathcal{G}_v^i) = \pi_{O,I}(\mathcal{G}_v^i) - 2 = \pi_{O,O}(\mathcal{G}_v^i) - 2;$
4. $\pi_{I,I}(\mathcal{G}_v^i) = \pi_{I,O}(\mathcal{G}_v^i) - 1 = \pi_{O,I}(\mathcal{G}_v^i) - 1 = \pi_{O,O}(\mathcal{G}_v^i) - 2;$
5. $\pi_{I,I}(\mathcal{G}_v^i) = \pi_{I,O}(\mathcal{G}_v^i) - 1 = \pi_{O,I}(\mathcal{G}_v^i) - 2 = \pi_{O,O}(\mathcal{G}_v^i) - 2;$
6. $\pi_{I,I}(\mathcal{G}_v^i) = \pi_{I,O}(\mathcal{G}_v^i) - 2 = \pi_{O,I}(\mathcal{G}_v^i) - 2 = \pi_{O,O}(\mathcal{G}_v^i) - 2;$
7. $\pi_I(\mathcal{G}_v^i) = \pi_O(\mathcal{G}_v^i) - 1.$

For convenience, we say $\mathcal{G}_v^i$ satisfies $(^i)$ if it has the i-th property in Lemma 1, where $1 \leq i \leq 7$. Consider $\mathcal{G}_v^1, \ldots, \mathcal{G}_v^k$. Let χ_v^i be the number of sub-cacti that satisfy $(^i)$, where $1 \leq i \leq 7$. Obviously, we have $0 \leq \chi_v^i \leq k$. Two strategies for G are said to be *equivalent* if they use the same number of searchers to clear G.

For any cactus graph G, algorithm FASTSEARCHCACTUS (See Algorithm 1) computes the minimum number of searchers required for clearing G.

Algorithm 1: FASTSEARCHCACTUS(G)

---

1 **Input:** A cactus graph G.
2 **Output:** The fast search number of G.
  1: Arbitrarily select a cut vertex v in $V(G)$, whose removal results in $k \geq 2$ connected components $H_1, \ldots, H_k$. Let G_i denote the subgraph of G induced by $V(H_i) \cup \{v\}$, where $1 \leq i \leq k$. Let E_{cut} denote the edge set consisting of all edges connecting v and vertices in $V(H_1)$. Let G' be the subgraph of G induced by $V(H_2) \cup \cdots \cup V(H_k) \cup V(E_{\text{cut}})$.
  2: Let $\mathcal{P}_{G_i}$ be the output of CLEARCACTI1(G_i, v), where $1 \leq i \leq k$.
  3: Let $\mathcal{P}_{G'}$ be the output of CLEARCACTI3($G', E_{\text{cut}}, v, \{\mathcal{P}_{G_2}, \ldots, \mathcal{P}_{G_k}\}$).
  4: List all the possible combinations of the profiles from $\mathcal{P}_{G_1}$ and $\mathcal{P}_{G'}$ respectively with respect to sliding actions on all the edges in E_{cut}.
  5: **return** the minimum number of searchers in all the combinations.

In algorithm FASTSEARCHCACTUS, we define G_i, where $1 \leq i \leq k$. Algorithm CLEARCACTI1 (See Algorithm 2) computes the profiles of G_i. The input of the algorithm includes G_i, along with the cut vertex $v \in V(G)$. The output is the profile of G_i.

Algorithm 2: CLEARCACTI1(G_i, v)

1: If G_i is a tree, then let $\pi_I(G_i)$ be the number of searchers that are placed on $V(G_i) \setminus \{v\}$ in the I-strategy produced by FS(G_i) in [5]. Let $\pi_O(G_i) \leftarrow \pi_I(G_i) + 1$. Let $(\pi_I(G_i), \pi_O(G_i))$ be the profile of G_i.

2: If G_i is a simple cycle, then let $\pi_{I,I}(G_i) \leftarrow 0, \pi_{I,O}(G_i) \leftarrow 0, \pi_{O,I}(G_i) \leftarrow 2,$ and $\pi_{O,O}(G_i) \leftarrow 2$. Let $(\pi_{I,I}(G_i), \pi_{I,O}(G_i), \pi_{O,I}(G_i), \pi_{O,O}(G_i))$ be the profile of G_v^i.

3: If G_i is neither a tree nor a simple cycle, then there are two subcases:
   (i) if v is contained in a cycle of G_i, then let the output of CLEARCACTI2(G_i, v) be the profile of G_i;
   (ii) if v is a leaf node of G_i, then let $u \in V(G_i)$ be the vertex such that $v \sim u$; let the output of CLEARCACTI1$(G_i - \{uv\}, u)$ be the profile of G_i.

4: **return** the profile of G_i.

Algorithm CLEARCACTI2 (See Algorithm 3) is used to compute the profile of a sub-cactus in which v is contained in a cycle. The input of the algorithm includes a sub-cactus G_i and the cut vertex v. The output of the algorithm is the profile of G_i.

Algorithm 3: CLEARCACTI2(G_i, v)

1: Let $\mathcal{C} = vu_1 \ldots u_{k'}v$ be the shortest cycle in G_i that contains v. Let $\mathcal{H}_{u_1}, \ldots, \mathcal{H}_{u_{k'}}$ denote the k' connected components that contain $u_1, \ldots, u_{k'}$ respectively, which are obtained by deleting all edges in $E(\mathcal{C})$ from G_i. Let $E_j \subset E(\mathcal{C})$ be the set containing the two incident edges of u_j, where $1 \leq j \leq k'$. Let $\mathcal{G}_{u_j}$ denote the connected subgraph obtained from $\mathcal{H}_{u_j}$ by adding two edges in E_j to u_j.

2: For $j \leftarrow 1, \ldots, k'$:
   (2.1) Let $H_1, \ldots, H_m$ denote all the connected components of $\mathcal{H}_{u_j}$ after removing the vertex u_j. Let H'_ℓ be the subgraph of $\mathcal{H}_{u_j}$ induced by $V(\mathcal{H}_\ell) \cup \{u_j\}$, where $1 \leq \ell \leq m$.
   (2.2) Let $\mathcal{P}_{H'_\ell}$ be the output of CLEARCACTI1(H'_ℓ, u_j), where $1 \leq \ell \leq m$.
   (2.3) Let the output of CLEARCACTI3$(\mathcal{G}_{u_j}, E_j, u_j, \{\mathcal{P}_{H'_1}, \ldots, \mathcal{P}_{H'_m}\})$ be the profile of $\mathcal{G}_{u_j}$.

3: Let $W \leftarrow \mathcal{G}_{u_1}$. Let $j \leftarrow 2$.

4: Note that W and $\mathcal{G}_{u_i}$ have one edge in common. A strategy for $W \cup \mathcal{G}_{u_j}$ can be obtained from strategies for W and $\mathcal{G}_{u_j}$ by reaching an accord on the sliding action on the common edge of W and $\mathcal{G}_{u_j}$. Note that in the graph $W \cup \mathcal{G}_{u_j}$, u_1 and u_j have one pendent edge in $E(\mathcal{C})$ respectively. Compute the profile of $W \cup \mathcal{G}_{u_j}$ with respect to the sliding actions on the pendent edges of u_1 and u_j, which consists of
   $$\pi_{I,I}(W \cup \mathcal{G}_{u_j}), \pi_{I,O}(W \cup \mathcal{G}_{u_j}), \pi_{O,I}(W \cup \mathcal{G}_{u_j}), \pi_{O,O}(W \cup \mathcal{G}_{u_j}).$$

5: Let $W \leftarrow W \cup \mathcal{G}_{u_j}$. If $j = k'$, then go to Step 6; otherwise, let $j \leftarrow j + 1$ and go to Step 4.

6: **return** the profile of W.

Algorithm CLEARCACTI3 (see Algorithm 4) is used for computing the profile of G'_v, which is obtained by adding either one or two pendant edges to the cut vertex v. The input of the algorithm includes G', E_{cut}, v and $\mathcal{P}$. The output of the algorithm is the profile of G'.

Algorithm 4: CLEARCACTI3(G', E_{cut}, v, $\mathcal{P}$)

1: If $|E_{\text{cut}}| = 1$, then let $\pi_I(G')$ be obtained from the output of
   CLEARCACTI4(G', $1, 1, \mathcal{P}$). Let $\pi_O(G') \leftarrow \pi_I(G') + 1$.
2: If $|E_{\text{cut}}| = 2$, then:
   (i) let $\pi_{I,I}(G')$ be the minimum number of searchers required for clearing
      G', where edges in E_{cut} are cleared by slide-in actions.
   (ii) let $\pi_{O,O}(G') \leftarrow \pi_{I,I}(G') + 2$.
   (iii) let $\pi_{I,O}(G')$ be the minimum number of searchers required for clearing
      G', where one edge in E_{cut} is cleared by a slide-in action, followed by the
      other edge in E_{cut} being cleared by a slide-out action.
   (iv) let $\pi_{O,I}(G')$ be the minimum number of searchers required for clearing
      G', where one edge in E_{cut} is cleared by a slide-out action, followed by
      the other edge in E_{cut} being cleared by a slide-in action.
3: **return** the profile of G'.

Algorithm CLEARCACTI4(G', $\sigma_1, \sigma_2, \mathcal{P}$) (which is omitted due to space limit) is called by CLEARCACTI3 as a subroutine, which computes the total number of searchers for clearing G' under some specific setting. Let $\mathcal{P}$ be the set containing the profiles of all the sub-cacti of $G' - E_{\text{cut}}$ with respect to vertex v. We use σ_1 to record the number of available searchers on v which could be used in an II-strategy or an I-strategy for a sub-cactus. We use σ_2 to denote the maximum number of searchers residing on v at some moment in a strategy for G'. For simplicity, σ_2 is set to 2 if there exists some moment in a strategy for G' at which v contains two or more searchers.

Lemma 2. *Consider all the sub-cacti of G with respect to v. For any strategy for G, there exists an equivalent strategy such that all the sub-cacti are cleared in the following order:*

1. *all the sub-cacti that are cleared by an O-strategy or an OO-strategy;*
2. *all the sub-cacti that are cleared by an OI-strategy (for each sub-cactus, perform all actions of searchers in its strategy until one of v's incident edges is cleared);*
3. *all the sub-cacti that are cleared by an IO-strategy;*
4. *all the sub-cacti that are cleared by an OI-strategy (for each sub-cactus, perform all actions of searchers in its strategy after one of v's incident edges is cleared);*
5. *all the sub-cacti that are cleared by an I-strategy or an II-strategy.*

Lemma 3. *Consider all the sub-cacti of G with respect to v, denoted as $G_1, \ldots, G_k$. For any strategy for G, there exists an equivalent strategy in which:*

1. *if G_i satisfies (^{1}), then it is cleared by an OI-strategy or an OO-strategy;*
2. *if G_i satisfies (^{2}), then it is cleared by an IO-strategy, an OI-strategy or an OO-strategy;*

3. *if G_i satisfies $(^3)$, then it is cleared by an IO-strategy or an OO-strategy;*
4. *if G_i satisfies $(^4)$, then it is cleared by an II-strategy, an OI-strategy or an OO-strategy;*
5. *if G_i satisfies $(^5)$, then it is cleared by an II-strategy, an IO-strategy or an OO-strategy;*
6. *if G_i satisfies $(^6)$, then it is cleared by an II-strategy, or an OO-strategy.*

Definition 4. A strategy is called a standard strategy for G with respect to v, where $v \in V(G)$, if (1) all the sub-cacti with respect to v are cleared in the order given in Lemma 2, and (2) each sub-cactus G_v^i, where $1 \leq i \leq k$, is cleared by a strategy in accordance with Lemma 3.

In the remainder of this section, we assume that every strategy for G_v' is a standard strategy with respect to v without subscripts.

Theorem 3. *For any cactus graph G, the fast search number of G can be computed in linear time by algorithm FASTSEARCHCACTUS.*

Proof. The algorithm FASTSEARCHCACUTS runs in linear time, as we can verify the time complexity as follows:

1. the profile of the sub-cactus with respect to each vertex in $V(G)$ has constant size;
2. the profile of the sub-cactus with respect to each vertex in $V(G)$ is computed at most once;
3. the profile of the sub-cactus with respect to each vertex in $V(G)$ is passed as parameter at most once when computing the profile of other sub-cactus;
4. the computation of the profile of the sub-cactus with respect to a vertex in $V(G)$ takes constant time.

Obviously, the algorithm FASTSEARCHCACUTS computes the fast search number of G in linear time.   □

Theorem 4. *For any cactus graph G, we can obtain an optimal fast search strategy in linear time using FASTSEARCHCACTUS.*

Proof. This can be achieved by first using a back-track method to record how every edge of G is cleared after calling FASTSEARCHCACTUS. In addition, we can record the vertices of G on which searchers are placed throughout FASTSEARCHCACTUS. Based on these records, we can easily obtain an optimal fast search strategy for G by letting those searchers move along edges following the prescribed directions.   □

4   Cartesian Product of a Tree and an Edge

Given two graphs G and H, the *cartesian product* of G and H, denoted $G \square H$, is the graph whose vertex set is the cartesian product $V(G) \times V(H)$, and in which two vertices (u, v) and (u', v') are adjacent if and only if $u = u'$ and v is adjacent to v' in H, or $v = v'$ and u is adjacent to u' in G.

In what follows, we apply Theorem 2 to find an optimal fast search strategy for $T \square P_2$, where T has at least three vertices. Let S_n, where $n \geq 3$, denote a star graph of n vertices. Let H_n denote the graph obtained by connecting the center vertices of two copies of S_n. Without loss of generality, let S_n^1 and S_n^2 denote the two copies of S_n in H_n. For any pair of edges that are from $E(S_n^1)$ and $E(S_n^2)$ respectively, there are four distinct ways to clear the two edges in a fast search strategy for H_n:

1. both edges are cleared by sliding a searcher from leaf to center node;
2. one of the two edges is cleared by sliding a searcher from leaf to center node, followed by the other edge being cleared by sliding a searcher from center node to leaf;
3. one of the two edges is cleared by sliding a searcher from center node to leaf, followed by the other edge being cleared by sliding a searcher from leaf to center node;
4. both edges are cleared by sliding a searcher from center node to leaf.

For convenience, we use II, IO, OI and OO to represent the above four ways respectively in the remainder of this section. Note that there are two layers in $T \square P_2$. Let T_1 and T_2 be the two layers in $T \square P_2$. Let $v_c^1 \in V(T_1)$ be a vertex of degree $k \geq 3$. Let $v_c^2 \in V(T_2)$ be the vertex where $v_c^2 \sim v_c^1$. Let V_c' be the subset of $V(T \square P_2)$, which consists of v_c^1, v_c^2 and all their adjacent vertices in $V(T \square P_2)$. Let E_c' be the subset of $E(T \square P_2)$, in which v_c^1 or v_c^2 is an end point of each edge. We use $\mathcal{G}_c'$ to denote the connected subgraph of $T \square P_2$, whose vertex set is V_c' and edge set is E_c'. It is easy to see that $\mathcal{G}_c'$ is the same as H_k. Let $\mathcal{G}_1, \ldots, \mathcal{G}_{k-1}$ be the connected components after deleting all edges in E_c' and all isolated vertices from $T \square P_2$. We use $\mathcal{G}_i'$, where $1 \leq i \leq k-1$, to denote the subgraph of $T \square P_2$, which is obtained from $\mathcal{G}_i$ by adding two pendant edges in $E(T \square P_2)$ that connect vertices from $\{v_c^1, v_c^2\}$ and $V(\mathcal{G}_i)$. Note that there are four ways to clear the two pendant edges of $\mathcal{G}_i'$. We use $s_1(\mathcal{G}_i')$ to denote the minimum numbers of searchers needed to be placed on $V(\mathcal{G}_i)$ in a strategy for $\mathcal{G}_i'$, in which the two pendant edges are cleared by II. In a similar way, we define $s_2(\mathcal{G}_i')$, $s_3(\mathcal{G}_i')$ and $s_4(\mathcal{G}_i')$.

Lemma 4. $\mathcal{G}_i'$ *must have one of the following properties:*

1. $s_1(\mathcal{G}_i') = s_2(\mathcal{G}_i') = s_3(\mathcal{G}_i') = s_4(\mathcal{G}_i') - 2;$
2. $s_1(\mathcal{G}_i') = s_2(\mathcal{G}_i') = s_3(\mathcal{G}_i') - 1 = s_4(\mathcal{G}_i') - 2;$
3. $s_1(\mathcal{G}_i') = s_2(\mathcal{G}_i') = s_3(\mathcal{G}_i') - 2 = s_4(\mathcal{G}_i') - 2;$
4. $s_1(\mathcal{G}_i') = s_2(\mathcal{G}_i') - 1 = s_3(\mathcal{G}_i') - 1 = s_4(\mathcal{G}_i') - 2;$
5. $s_1(\mathcal{G}_i') = s_2(\mathcal{G}_i') - 1 = s_3(\mathcal{G}_i') - 2 = s_4(\mathcal{G}_i') - 2.$
6. $s_1(\mathcal{G}_i') = s_2(\mathcal{G}_i') - 2 = s_3(\mathcal{G}_i') - 2 = s_4(\mathcal{G}_i') - 2.$

Let $\mathcal{G}'$ be the graph with vertex set $V(\mathcal{G}_1) \cup \cdots \cup V(\mathcal{G}_{k-2}) \cup V_c'$ and edge set $E(\mathcal{G}_1) \cup \cdots \cup E(\mathcal{G}_{k-2}) \cup E_c'$. Given the profiles of $\mathcal{G}_i'$, where $1 \leq i \leq k-2$, we can compute the minimum number of searchers required for clearing $\mathcal{G}'$.

Lemma 5. *For each connected component $\mathcal{G}_i'$, where $1 \leq i \leq k-2$, if we know $s_1(\mathcal{G}_i')$, $s_2(\mathcal{G}_i')$, $s_3(\mathcal{G}_i')$ and $s_4(\mathcal{G}_i')$, then we can compute $s_1(\mathcal{G}')$, $s_2(\mathcal{G}')$, $s_3(\mathcal{G}')$ and $s_4(\mathcal{G}')$.*

Note that $E(\mathcal{G}')$ and $E(\mathcal{G}'_j)$ have exactly two common edges. Given the profiles of $\mathcal{G}'$ and $\mathcal{G}'_j$, we can list all the possible combinations of the profiles with respect to the sliding actions on the two edges. The fast search number of $T \Box P_2$ is the minimum number of searchers in all the combinations. From Lemma 5, we have the following result:

Lemma 6. *For each connected component $\mathcal{G}'_i$, where $1 \le i \le k-1$, if we know $s_1(\mathcal{G}'_i)$, $s_2(\mathcal{G}'_i)$, $s_3(\mathcal{G}'_i)$ and $s_4(\mathcal{G}'_i)$, then we can compute the minimum number of searchers for clearing $T \Box P_2$.*

Theorem 5. *An optimal fast search strategy for $T \Box P_2$ can be found in polynomial time.*

Proof. We briefly describe a strategy below for finding an optimal fast search strategy for $T \Box P_2$.

1. Arbitrarily select a pair of vertices v and v', where v' is the corresponding vertex of v in $T \Box P_2$.
2. For each of the connected components of $T \Box P_2$ with respect to v and v', compute its profile.
3. Compute the optimal fast search number of $T \Box P_2$ based on the profiles of all the connected components.
4. Use a back-track method to record how every edge of $T \Box P_2$ is cleared, as well as all vertices that are placed searchers. Based on these records, produce an optimal fast search strategy for $T \Box P_2$ by letting those searchers move along edges following the prescribed directions.

Clearly, the above strategy can find an optimal fast search strategy for $T \Box P_2$ in polynomial time.   □

References

1. Bienstock, D.: Graph searching, path-width, tree-width and related problems (a survey). DIMACS Ser. Discrete Math. Theoret. Comput. Sci. **5**, 33–49 (1991)
2. Bonato, A., Yang, B.: Graph searching and related problems. In: Pardalos, P.M., Du, D.-Z., Graham, R.L. (eds.) Handbook of Combinatorial Optimization, pp. 1511–1558. Springer, New York (2013). https://doi.org/10.1007/978-1-4419-7997-1_76
3. Breisch, R.: An intuitive approach to speleotopology. Southwestern Cavers **6**(5), 72–78 (1967)
4. Dereniowski, D., Diner, Ö., Dyer, D.: Three-fast-searchable graphs. Discret. Appl. Math. **161**(13), 1950–1958 (2013)
5. Dyer, D., Yang, B., Yaşar, Ö.: On the fast searching problem. In: Fleischer, R., Xu, J. (eds.) AAIM 2008. LNCS, vol. 5034, pp. 143–154. Springer, Heidelberg (2008). https://doi.org/10.1007/978-3-540-68880-8_15
6. Fomin, F.V., Thilikos, D.M.: An annotated bibliography on guaranteed graph searching. Theoret. Comput. Sci. **399**(3), 236–245 (2008)
7. Makedon, F.S., Papadimitriou, C.H., Sudborough, I.H.: Topological bandwidth. SIAM J. Algebraic Discrete Methods **6**(3), 418–444 (1985)

8. Megiddo, N., Hakimi, S.L., Garey, M.R., Johnson, D.S., Papadimitrioum, C.H.: The complexity of searching a graph. J. ACM **35**(1), 18–44 (1988)
9. Parsons, T.: Pursuit-evasion in a graph. In: Proceedings of the International Conference on the Theory and Applications of Graphs, pp. 426–441. Springer-Verlag (1976). https://doi.org/10.1007/BFb0070400
10. Stanley, D., Yang, B.: Fast searching games on graphs. J. Comb. Optim. **22**(4), 763–777 (2011)
11. Xue, Y., Yang, B.: The fast search number of a cartesian product of graphs. Discret. Appl. Math. **224**, 106–119 (2017)
12. Xue, Y., Yang, B., Zhong, F., Zilles, S.: The fast search number of a complete k-partite graph. Algorithmica **80**(12), 3959–3981 (2018)
13. Yang, B.: Fast edge searching and fast searching on graphs. Theoret. Comput. Sci. **412**(12), 1208–1219 (2011)

Class Ramsey Numbers Involving Induced Graphs

Yan Li[1] and Ye Wang[2]($\boxtimes$)

[1] University of Shanghai for Science and Technology, Shanghai 200093, China
[2] Harbin Engineering University, Harbin 150001, China
ywang@hrbeu.edu.cn

Abstract. For graphs F, G and H, the class Ramsey number involving induced graph $r(G; H, F - ind)$ is defined to be the minimal n such that any red/blue edge coloring of K_n contains a red G, a blue H or a blue induced F. In this note, we shall show a general sharp lower bound for $r(G; H, F - ind)$, and determine some class Ramsey numbers involving induced graphs.

Keywords: Class Ramsey number · Induced graph · Induced Turán number

1 Introduction

For graph G with vertex set $V(G)$ and edge set $E(G)$, we write $|G| = |V(G)|$ and $e(G) = |E(G)|$. For vertex disjoint graphs G and H, let $G \cup H$ be the disjoint union of G and H, and $G + H$ the graph obtained by preserving the edges of G and H and adding new edges to connect G and H completely. Particularly, denote by nG the union of n disjoint copies of G. For $S \subseteq V(G)$, we denote by $G \setminus S$ the subgraph of G induced by $V(G) \setminus S$. For an edge coloring of F by red and blue and a vertex v of F, we denote the red and blue neighbors of v in F by $N_F^R(v)$ and $N_F^B(v)$, respectively. Let $d_F^R(v) = |N_F^R(v)|$ and $d_F^B(v) = |N_F^B(v)|$.

For a graph H and a class $\mathcal{G} = \{G_1, G_2, \ldots, G_n\}$ of graphs, let $H \xrightarrow{k} \mathcal{G}$ denote that any edge coloring of H with k colors contains at least a monochromatic $G_i \in \mathcal{G}$. We write $H \xrightarrow{k} G$ for $H \xrightarrow{k} \{G\}$ if $\mathcal{G} = \{G\}$ is a singleton. Thus the Ramsey number is defined as $R_k(G) = \min\{N \mid K_N \xrightarrow{k} G\}$, and the class Ramsey number is defined as $R_k(\mathcal{G}) = \min\{N \mid K_N \xrightarrow{k} \mathcal{G}\}$.

Forbidden subgraph is a basic topic in extremal graph theory. In addition to Ramsey number, the Turán number $ex(n; G)$ is well known as the largest number of edges of graphs on n vertices that contains no G. A related topic is asking what happened if the forbidden subgraph is induced. Recently, Loh, Tait, Timmons and Zhou introduced the following definition.

Supported in part by Natural Science Foundation of Heilongjiang Province of China (LH2021A004).

Definition 1 [7]. *For graphs H and F, the induced Turán number*

$$ex(n, \{H, F - ind\}) = \max\{e(G) \,|\, G \text{ contains neither an } H \text{ nor an induced } F \text{ with } |G| = n\}.$$

For a reasonable version of the following definition in Ramsey number, we shall note the fact that if the edges of K_n are blue completely, there is neither a red G nor a blue induced F when F is not complete. In order to define the class Ramsey number involving induced graphs, let

$$K_n \to (G; H, F - ind)$$

signify that any red/blue edge coloring of K_n contains a red G, a blue H or a blue induced F.

Definition 2. *For graphs G, H and F, the class Ramsey number involving induced graphs is defined to be*

$$r(G; H, F - ind) = \min\left\{n \,|\, K_n \to (G; H, F - ind)\right\}.$$

It is easy to have the following relation.

Lemma 1. *For graphs G, H and F, it holds*

$$r(G, \{H, F\}) \le r(G; H, F - ind) \le r(G, H).$$

Let $\alpha(H)$ be the independent number of H, $\chi(H)$ the chromatic number of H, and $\sigma(H)$ the chromatic surplus of H, which is the minimum number of vertices in a color class among vertex coloring of H by $\chi(H)$ colors. If G is connected of order $n \ge \sigma(H)$, then

$$r(G, H) \ge (\chi(H) - 1)(n - 1) + \sigma(H). \tag{1}$$

Burr [1] defined G to be H-good if the inequality (1) holds as an equality. In this note, we shall show a general sharp lower bound for $r(G; H, F - ind)$, and determine some $r(G; H, F - ind)$ as follows.

Theorem 1. *Let G, H and F be connected graphs with $n = |G| \ge \sigma(H)$. If F satisfies one of the following conditions:*

(1) F is a non-complete k-partite graph with $k \le \chi(H)$;
(2) $\chi(F) \ge \chi(H) + 1$;
(3) $\alpha(F) \ge n$,

then

$$r(G; H, F - ind) \ge (\chi(H) - 1)(n - 1) + \sigma(H). \tag{2}$$

Furthermore, if G is H-good, then the inequality (2) becomes an equality.

By Theorem 1, we have some particular equalities. For example, as T_n is K_m-good (see [3]),

$$r(T_n; K_m, K_{s,t} - ind) = (n-1)(m-1) + 1$$

for any $m, n \geq 1$ and $t \geq n$. As C_n is H-good for large n (see [2]), and thus

$$r(C_n; H, F - ind) = (n-1)(\chi(H) - 1) + \sigma(H)$$

for F with $\alpha(F) \geq n$. When $H = K_3$ and $F = K_{s,t}$, we shall determine all $r(C_n; K_3, K_{s,t} - ind)$ as follows.

Theorem 2. *Let n, s and t be positive integers with $n \geq 5$ and $t \geq s+1$. Then*

$$r(C_n; K_3, K_{s,t} - ind) = \begin{cases} 2n - 1 & \text{if } t \geq n, \\ \max\{n + s - 1, 2t\} & \text{otherwise.} \end{cases}$$

In this note, we also have the following results, where P_n is a path on n vertices.

Theorem 3. *Let n, s and t be positive integers with $t \geq s$. Then*

$$r(P_n; K_3, K_{s,t} - ind) = \begin{cases} 2n - 1 & \text{if } t \geq n, \\ \max\{n + s - 1, 2t - 1\} & \text{otherwise.} \end{cases}$$

Theorem 4. *Let n, m, s and t be positive integers with $n \geq m$ and $t \geq s \geq 2$. Then*

$$r(nK_2; mK_2, K_{s,t} - ind) = 2n + m - 1.$$

Theorem 5. *Let n, m, s and t be positive integers with $st \geq 2$. Then*

$$r(K_n; mK_2, K_{s,t} - ind) = n + 2m - 2.$$

2   Proofs of Main Results

Proof of Theorem 1. Let $\chi = \chi(H)$, $\sigma = \sigma(H)$ and $N = (\chi-1)(n-1)+\sigma-1$. Color the edges of K_N red and blue such that the red graph is isomorphic to $(\chi - 1)K_{n-1} \cup K_{\sigma-1}$, so it contains no G. The chromatic number of the blue graph is χ, and the smallest vertex color class has size $\sigma-1$, so it contains neither H nor induced F.

Before proceeding to the proof of Theorem 2, we need some preliminary lemmas. □

Lemma 2 [6]. *For all $n \geq 4$, C_n is K_3-good, namely, $r(C_n, K_3) = 2n - 1$.*

Lemma 3. *For $n \geq 3$, if the edges of K_n are colored by red and blue such that there is neither a red hamiltonian path nor a blue K_3, then this K_n can be divided into two red cliques which are adjacent to each other by blue completely.*

Proof. Suppose that the longest red path in K_n is $P = v_1 v_2 \ldots v_s$ with $s \leq n-1$. As its maximality, any vertex in $V(K_n) \setminus V(P)$ is adjacent to v_1 and v_s by blue. Since there is no blue K_3, $v_1 v_s$ is red. Then $v_1 v_2 \ldots v_s v_1$ is a red cycle such that v_i is adjacent to $V(K_n) \setminus V(P)$ by blue completely, where $i \in [s]$. As K_n contains no blue K_3, we get two red cliques and the edges between them are blue completely, completing the proof. $\square$

For the figures in this note, red edges are solid and blue edges are dashed. If all the edges between two cliques are red (or blue), a thick solid (or dashed) line is drawn between the cliques.

Proof of Theorem 2. For $t \geq n$, as C_n is K_3-good and $K_{s,t}$ satisfies the condition (3) in Theorem 1, we obtain that $r(C_n; K_3, K_{s,t} - ind) = 2n - 1$. Now we consider the case that $t \leq n - 1$.

For the lower bound, color the edges of K_{n+s-2} red and blue such that the red graph is isomorphic to $K_{s-1} \cup K_{n-1}$ and the blue graph is isomorphic to $K_{s-1,n-1}$. We also color the edges of K_{2t-1} red and blue such that the red graph is isomorphic to $K_1 + 2K_{t-1}$ and the blue graph is isomorphic to $K_{t-1,t-1}$. It is easy to see that neither K_{n+s-2} nor K_{2t-1} contains a red C_n, a blue K_3 or a blue induced $K_{s,t}$.

For the upper bound, let $N = \max\{n + s - 1, 2t\}$. For any red/blue edge coloring of K_N, suppose K_N contains no blue K_3, and we shall show that K_N must contain either a red C_n or a blue induced $K_{s,t}$. Consider the following two cases.

Case 1. Suppose K_N contains no red hamiltonian path. As K_N contains neither a red hamiltonian path nor a blue K_3, by Lemma 3, K_N can be divided into two red cliques Q_1 and Q_2 with $|Q_1| \leq |Q_2|$, which are adjacent to each other by blue completely. As K_N contains no red C_n, we have $|Q_2| \leq n - 1$. Then

$$|Q_2| \geq \lceil N/2 \rceil \geq t, \ |Q_1| = N - |Q_2| \geq N - n + 1 \geq s,$$

which implies a blue induced $K_{s,t}$.

Case 2. Suppose K_N contains a red hamiltonian path $P = v_1 v_2 \ldots v_N$. As K_N contains neither a red C_n nor a blue K_3 and $N \geq n$, $v_1 v_n$ is blue, and then either $v_1 v_i$ or $v_i v_n$ must be red, for $i \in \{2, \ldots, n-1\}$. We may assume $v_1 v_j$ is red with

$$j = \max\{i \mid v_1 v_i \text{ is red}, 2 \leq i \leq n - 1\},$$

then $v_{j-1} v_n$ is blue, otherwise there is a red C_n. Since $v_1 v_n$ and $v_{j-1} v_n$ are both blue, $v_1 v_{j-1}$ is red. Similarly, $v_{j-1} v_n, v_{j-2} v_n, \ldots, v_2 v_n$ and $v_1 v_n$ are all blue. Then $\{v_1, v_2, \ldots, v_{j-1}\}$ induces a red clique. Because of the maximality of j, $v_1 v_{j+1}, v_1 v_{j+2}, \ldots, v_1 v_n$ are all blue, and then $\{v_{j+1}, v_{j+2}, \ldots, v_n\}$ induces a red clique.

Subcase 2.1. $j = 2$. Then $v_1 v_{n+1}$ is blue, otherwise we can omit an internal vertex in P to yield a red C_n. Since $v_1 v_3, v_1 v_4, \ldots, v_1 v_{n+1}$ are all blue, $\{v_3, v_4, \ldots, v_{n+1}\}$ induces a red clique. Continue this procedure, then $\{v_3, v_4, \ldots, v_N\}$ induces a red clique of order $N - 2 \geq t$, which is adjacent to v_1 by blue completely. Suppose that there is no red C_n, then $N \leq n + 1$, which implies $s \leq 2$. If $s = 1$, $\{v_1, v_3, v_4, \ldots, v_N\}$ induces a blue induced $K_{s,t}$. If $s = 2$, v_2 must be adjacent to $\{v_4, v_5, \ldots, v_N\}$ by blue completely, otherwise we get a red cycle C_n. As $N - 3 \geq 2t - 3 \geq t$, $\{v_1, v_2, v_4, \ldots, v_N\}$ induces a blue induced $K_{s,t}$.

Subcase 2.2. $3 \leq j \leq n - 2$. Then any edge between $\{v_1, v_2, \ldots, v_{j-1}\}$ and $\{v_{j+2}, v_{j+3}, \ldots, v_n\}$ is blue. Suppose to the contrary that there is a red edge ab with $a \in \{v_1, v_2, \ldots, v_{j-1}\}$ and $b \in \{v_{j+2}, v_{j+3}, \ldots, v_n\}$. Then we obtain a red C_n with $C_n = abv_n \ldots v_{j+1}v_jv_1 \ldots a$ or $C_n = abv_n \ldots v_{j+1}v_jv_{j-1} \ldots a$. For $N = n$, $\{v_1, v_2, \ldots, v_{j-1}\}$ and $\{v_{j+2}, v_{j+3}, \ldots, v_N\}$ induce two red cliques with the edges between them being blue completely. For $N = n + 1$, v_1v_{n+1} is blue, otherwise we can omit an internal vertex in P to yield a red C_n. Since $v_1v_{j+1}, v_1v_{j+2}, \ldots, v_1v_{n+1}$ are all blue, $\{v_{j+1}, v_{j+2}, \ldots, v_{n+1}\}$ induces a red clique. If v_{n+1} is adjacent to any vertex $a \in \{v_1, v_2, \ldots, v_{j-1}\}$ by red, we obtain a red C_n with $C_n = av_{n+1} \ldots v_{j+1}v_jv_1 \ldots a$ or $C_n = av_{n+1} \ldots v_{j+1}v_jv_{j-1} \ldots a$. Thus $K_N \setminus \{v_j, v_{j+1}\}$ can be divided into two red cliques with the edges between them being blue completely. Continue this procedure. For $N \geq n$, we will get two red cliques Q_1 and Q_2 with the edges between them being blue completely with $V(Q_1) = \{v_1, v_2, \ldots, v_{j-1}\}$ and $V(Q_2) = \{v_{j+2}, v_{j+3}, \ldots, v_N\}$, shown in Fig. 1.

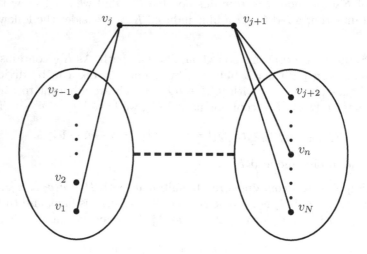

Fig. 1. The edge coloring of K_N for $3 \leq j \leq n - 2$.

If $|Q_2| = 1$, then $N = n = j + 2$, which implies $s = 1$. As

$$N - 3 \geq \max\{2t - 3, n - 3\} = \max\{2t - 3, j - 1\} \geq \max\{2t - 3, 2\} \geq t,$$

we obtain that $\{v_1, v_2, \ldots, v_{j-1}, v_N\}$ induces a blue induced $K_{1,t}$.

If $|Q_2| \geq 2$, then $d_{Q_1}^R(v_j) \geq 2$ and $d_{Q_2}^R(v_{j+1}) \geq 2$. Without loss of generality, suppose that $|Q_2| \geq |Q_1|$ (if $|Q_1| > |Q_2|$, the proof is similar). As K_N contains no red C_n, we have

$$n - 1 \geq |Q_2| \geq |Q_1| \geq N - 2 - n + 1 \geq s - 2.$$

Suppose that $s \leq t - 2$. For $|Q_1| = s - 2$, we have $|Q_2| = N - 2 - |Q_1| \geq n - 1$. Along with vertex v_{j+1}, we obtain a red C_n. For $|Q_1| = s - 1$, we have $|Q_2| = N - 2 - |Q_1| \geq \max\{n - 2, t\}$. As there is neither a red C_n nor a blue K_3, v_j is adjacent to each vertex in Q_2 and Q_1 by blue and red, respectively. Then $\{v_1, v_2, \ldots, v_j\}$ induces a red K_s. Note that $|Q_2| \geq t$, and we obtain a blue induced $K_{s,t}$. For $s \leq |Q_1| \leq t - 2$, we have $|Q_2| = N - 2 - |Q_1| \geq t$, yielding a blue induced $K_{s,t}$. For $|Q_1| \geq t - 1$, we have $|Q_2| \geq |Q_1| \geq t - 1$. Note that $d_{Q_2}^R(v_j) = 0$ or $d_{Q_1}^R(v_{j+1}) = 0$, otherwise there will be a red C_n. Without loss of generality, let $d_{Q_2}^R(v_j) = 0$. Since the edges between Q_1 and Q_2 are all blue, v_j is adjacent to any vertex in Q_1 by red, yielding a blue induced $K_{s,t}$.

If $s = t - 1$, similarly, we will obtain a blue induced $K_{s,t}$.

Subcase 2.3. $j = n-1$. Then $\{v_1, v_2, \ldots, v_{n-2}\}$ induces a red clique. For $N = n$, we have $s = 1$. Since $n \geq 5$, we have $n - 2 \geq \max\{2, 2t - 2\} \geq t$, which yielding a blue induced $K_{s,t}$. Similarly, for $N > n$, we have $K_N \setminus \{v_{n-1}, v_n\}$ can be divided into two red cliques Q_1 and Q_2 with the edges between them being blue completely with $V(Q_1) = \{v_1, v_2, \ldots, v_{n-2}\}$ and $V(Q_2) = \{v_{n+1}, \ldots, v_N\}$. If $|Q_2| \geq 2$, similarly as in Subcase 2.2, there is a blue induced $K_{s,t}$. If $|Q_2| = 1$, then $N = n + 1$, which implies $s \leq 2$. Note that v_n must be adjacent to $V(Q_1)$ by blue completely, otherwise we get a red cycle C_n, and

$$n - 2 \geq \max\{3, N - 3\} \geq \max\{3, 2t - 3\} \geq t.$$

So $\{v_{n+1}\} \cup V(Q_1)$ and $\{v_n, v_{n+1}\} \cup V(Q_1)$ induce blue induced $K_{1,t}$ and $K_{2,t}$, respectively. $\square$

Lemma 4 [9]. *For positive integers n and m, $r(P_n, K_m) = (n - 1)(m - 1) + 1$.*

Proof of Theorem 3. For $t \geq n$, as P_n is K_3-good and $K_{s,t}$ satisfies the condition (3) in Theorem 1, we obtain that $r(P_n; K_3, K_{s,t} - ind) = 2n - 1$. Now we consider the case that $t \leq n - 1$.

For the lower bound, color the edges of K_{n+s-2} red and blue such that the red graph is isomorphic to $K_{s-1} \cup K_{n-1}$ and the blue graph is isomorphic to $K_{s-1,n-1}$. We also color the edges of K_{2t-2} red and blue such that the red graph is isomorphic to $2K_{t-1}$ and the blue graph is isomorphic to $K_{t-1,t-1}$. It is easy to see that neither K_{n+s-2} nor K_{2t-2} contains a red P_n, a blue K_3 or a blue induced $K_{s,t}$.

For the upper bound, let $N = \max\{n + s - 1, 2t - 1\}$. For any red/blue edge coloring of K_N, suppose K_N contains neither a red P_n nor a blue K_3, by Lemma 3, K_N can be divided into two red cliques K_a and K_b with $a \leq b$, which are adjacent to each other by blue completely. As K_N contains no red P_n, we have $b \leq n - 1$. Then

$$b \geq \lceil N/2 \rceil \geq t, \ a = N - b \geq N - n + 1 \geq s,$$

which implies a blue induced $K_{s,t}$.   □

Lemma 5 [4,5,8]. *For* $n \geq m \geq 1$, $r(nK_2, mK_2) = 2n + m - 1$.

Proof of Theorem 4. By Lemma 1 and Lemma 5, we obtain that $r(nK_2; mK_2, K_{s,t} - ind) \leq 2n + m - 1$. For the lower bound, color the edges of K_{2n+m-2} red and blue such that the red graph is isomorphic to K_{2n-1} and the blue graph is isomorphic to $(2n - 1)K_1 + K_{m-1}$. It is easy to see that K_{2n+m-2} contains neither a red nK_2, a blue mK_2 nor a blue induced $K_{s,t}$, which implies the lower bound $r(nK_2; mK_2, K_{s,t} - ind) \geq 2n + m - 1$.   □

Lemma 6 [8]. *For* $n \geq 2$ *and* $m \geq 1$, $r(K_n, mK_2) = n + 2m - 2$.

Proof of Theorem 5. By Lemma 1 and Lemma 6, we obtain that $r(K_n; mK_2, K_{s,t} - ind) \leq n + 2m - 2$. For the lower bound, color the edges of K_{n+2m-3} red and blue such that the red graph is isomorphic to $K_{n-2} + (2m - 1)K_1$ and the blue graph is isomorphic to K_{2m-1}. If $K_{s,t} \neq K_{1,1}$, it is easy to see that K_{n+2m-3} contains neither a red K_n, a blue mK_2 nor a blue induced $K_{s,t}$, which implies the lower bound $r(K_n; mK_2, K_{s,t} - ind) \geq n + 2m - 2$.   □

References

1. Burr, S.: Ramsey numbers involving graphs with long suspended paths. J. London Math. Soc. **24**(3), 405–413 (1981)
2. Burr S., Erdős P.: On the magnitude of generalized Ramsey numbers of graphs. In: Infinite and Finite Sets, vol. 2, Colloquia Mathematica Societatis Janos Bolyai, 10, pp. 214–240. North-Holland, Amsterdam/London (1973)
3. Chvátal, V.: Tree-complete graph Ramsey numbers. J. Graph Theory **1**(1), 93 (1977)
4. Cockayne, E.J., Lorimer, P.J.: On Ramsey graph numbers for stars and stripes. Canad. Math. Bull. **18**(1), 31–34 (1975)
5. Cockayne, E.J., Lorimer, P.J.: The Ramsey number for stripes. J. Aust. Math. Soc. Ser. A. **19**, 252–256 (1975)
6. Faudree, R.J., Schelp, R.H.: All Ramsey numbers for cycles in graphs. Discrete Math. **8**, 313–329 (1974)
7. Loh, P., Tait, M., Timmons, C., Zhou, R.: Induced Turán numbers. Comb. Probabil. Comput. **27**(2), 274–288 (2018)
8. Lorimer, P.J.: The Ramsey numbers for stripes and one complete graph. J. Graph Theory **8**(1), 177–184 (1984)
9. Parsons, T.D.: The Ramsey numbers $r(P_m, K_n)$. Discret. Math. **6**, 159–162 (1973)

Injective Edge Coloring of Power Graphs and Necklaces

Yuehua Bu[1,2], Wenwen Chen[1,2], and Junlei Zhu[3(✉)] ⓘ

[1] Department of Mathematics, Zhejiang Normal University, Jinhua 321004, China
[2] Zhejiang Normal University Xingzhi College, Jinhua 321004, China
[3] College of Data Science, Jiaxing University, Jiaxing 314001, China
zhujl-001@163.com

Abstract. A k-*injective-edge coloring of a graph* G is an edge coloring $c :$ $E(G) \to \{1, 2, \cdots, k\}$ such that $c(e_1) \neq c(e_3)$ for any three consecutive edges e_1, e_2, e_3 of a path or a 3-cycle. The minimum integer k such that G has a k-injective-edge coloring is called the *injective chromatic index of* G, denoted by $\chi_i'(G)$. In this paper, we determined the exact injective chromatic index of power graphs of path and necklaces.

Keywords: Injective edge coloring · Power graph · Necklace

1 Introduction

All graphs in this paper are simple and finite. For a planar graph G, we denote its vertex set, edge set, maximum degree and minimum degree by $V(G), E(G), \Delta(G)$ and $\delta(G)$ (simply V, E, Δ and δ) respectively. For $v \in V(G)$, let $d_G(v)$ (simply $d(v)$) denote the degree of v in G. A vertex of degree k (resp. at least k, at most k) is called a k-vertex (resp. k^+-vertex, k^--vertex). For graph G, the k-power graph G^k of G is defined as $V(G^k) = V(G), E(G^k) = E(G) \cup \{uv|d(u,v) = k\}$. 2-power graph is also called power graph.

A k-injective-edge coloring of a graph G is a mapping $c : E(G) \to \{1, 2, \cdots, k\}$ such that $c(e_1) \neq c(e_3)$ for any three consecutive edges e_1, e_2, e_3 of a path or a 3-cycle. The minimum integer k such that G has a k-injective-edge coloring is called the injective chromatic index of G, denoted by $\chi_i'(G)$. A k-strong-edge-coloring of a graph G is a mapping $f : E(G) \to \{1, 2, \cdots, k\}$ such that two distinct edges are colored differently if their distance is at most two. The strong chromatic index of G, written $\chi_s'(G)$, is the minimum k such that G has a k-strong-edge-coloring. Note that an injective edge coloring is not necessarily a proper edge coloring and a strong edge coloring is a proper edge coloring. Moreover, $\chi_i'(G) \leq \chi_s'(G)$.

The notion of injective edge coloring was introduced in 2015 by Cardoso et al. [3] motivated by the Packet Radio Network problem. Independently, this notion was studied as induced star arboricity in 2019 by Axenovich et al. [1].

Supported by National Science Foundation of China under Grant No.11901243.

Cardoso et al. [3] proved that it is NP-hard to compute the injective chromatic index of a graph. In [3], authors gave exact values of injective chromatic index for some classes of graphs such as paths, cycles, wheels, Petersen graph, complete graphs and complete bipartite graphs.

Theorem 1. *(Cardoso et al. [3])*

1. *If $n \geq 4$, then $\chi_i'(P_n) = 2$.*
2. *If $n \equiv 0(mod4)$, then $\chi_i'(C_n) = 2$. Otherwise, $\chi_i'(C_n) = 3$.*
3. *$\chi_i'(K_{p,q}) = \min\{p, q\}$.*

For the case of trees, they proved that the injective chromatic index of a tree is at most 3. The algorithmic complexity of the injective edge-coloring problem has been further studied in [5].

Bu and Qi [2] considered upper bounds on injective chromatic index for subcubic graphs with maximum average degree restriction. Some of their results have been improved by Ferdjallah et al. [4] and Kostochka et al. [7].

A Halin graph $G = T \cup C$ is a planar graph G constructed as follows. Let T be a tree of order at least 4. All vertices of T are either of degree 1 or of degree at least 3. Let C be a cycle connecting the leaves of T in such a way that C forms the boundary of the unbounded face. The tree T and the cycle C are called the characteristic tree and the adjoint cycle of G, respectively.

The necklace is a particular Halin graph whose characteristic tree is a caterpillar. A caterpillar is a tree whose interior vertices are all of degree 3 and the removal of the leaves together with their incident edges becomes a path. For any positive integer h, a necklace, denoted Ne_h, is a cubic Halin graph whose characteristic tree T_h consists of the path $v_0, v_1, \cdots, v_h, v_{h+1}$ and leaves $v_{1'}, v_{2'}, \cdots, v_{h'}$ such that the unique neighbor of $v_{i'}$ in T_h is v_i for $1 \leq i \leq h$ and vertices $v_0, v_{1'}, \cdots, v_{h'}, v_{h+1}$ are connected in order to form the adjoint cycle C_{h+2}.

In 2006, Shiu and Tam [8] determined the strong chromatic index for a necklace (see Fig. 1) as follows.

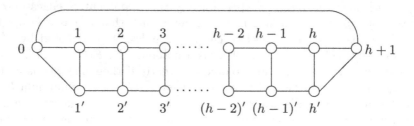

Fig. 1. Necklace Ne_h

Theorem 2. *(Shiu and Tam [8])*
Let $h \geq 1$, then

(1) If h is odd, then $\chi'_s(Ne_h) = 6$;
(2) If h is even, then $\chi'_s(Ne_h) = 7$;
(3) If $h = 4$, then $\chi'_s(Ne_h) = 8$;
(4) If $h = 2$, then $\chi'_s(Ne_h) = 9$;

2   Main Results

Note that $\chi'_i(G) \leq \chi'_s(G)$, the strong chromatic index of necklaces in Theorem 2 must be the upper bound of injective chromatic index of necklaces. In this paper, we shall determine the exact injective chromatic index of necklaces. Moreover, we shall determine the exact injective chromatic index of power graphs. The following two theorems are the main results we obtained.

Theorem 3. *Let P_n be a path with n vertex.*

(1) If $n = 3$ or $n = 4$, then $\chi'_i(P_n^2) = 3$.
(2) If $n = 5$, then $\chi'_i(P_n^2) = 4$.
(3) If $n = 6$, then $\chi'_i(P_n^2) = 5$.
(4) If $n \geq 7$, then $\chi'_i(P_n^2) = 6$.

Theorem 4. *Let $h \geq 1$, then we have*

(1) If $h = 1, 2$, then $\chi'_i(Ne_h) = 6$;
(2) If $h = 3$, then $\chi'_i(Ne_h) = 5$;
(3) If $h = 4$, then $\chi'_i(Ne_h) = 4$;
(4) If $h \geq 5$, then $\chi'_i(Ne_h) \leq 5$;

3   Proof of Theorem 3

(1) Note that $P_3^2 = K_3$, we have that $\chi'_i(P_3^2) = 3$.
    Note that $v_1 v_2 v_3 v_1$ is a 3-cycle in P_4^2, we have that $\chi'_i(P_4^2) \geq 3$. On the other hand, we can obtain a 3-injective-edge-coloring of P_4^2 as in Fig. 2. Thus, $\chi'_i(P_4^2) = 3$.

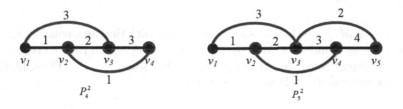

Fig. 2. Power graphs of P_4 and P_5

(2) Since $v_1v_2v_3v_1$ is a 3-cycle, the edges of this 3-cycle should be colored with distinct colors. Note that the distance between edge v_4v_5 and any edge in 3-cycle $v_1v_2v_3v_1$ is two, we have that $\chi'_i(P_5^2) \geq 4$. On the other hand, we can check that $\chi'_i(P_5^2) \leq 4$, see Fig. 2. Thus, $\chi'_i(P_5^2) = 4$.

(3) For P_6^2, we can see Fig. 3. Note the analysis of P_5^2, the edges v_1v_2, v_1v_3, v_2v_3, v_4v_5 should be colored with distinct colors. For edges v_4v_6 and v_4v_5, they are in the same 3-cycle. For edges v_4v_6 and any edge in $v_1v_2v_3$, they are at distance two. Thus the colors used on v_4v_5 and $v_1v_2v_3$ are forbidden for edge v_4v_6. Hence, $\chi'_i(P_6^2) \geq 5$. On the hand, we can obtain a 5-injective-edge-coloring as in Fig. 3. Hence, $\chi'_i(P_6^2) = 5$.

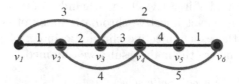

Fig. 3. 5-injective-edge-coloring of P_6^2

(4) First, we shall prove $\chi'_i(P_n^2) \geq 6$.

For P_7^2, we see Fig. 4. By the analysis of P_6^2, it is easy to see that $\chi'_i(P_7^2) \geq 5$. Assume that $\chi'_i(P_7^2) = 5$ and let c be a 5-injective-edge-coloring of P_7^2. Since $|E(P_7^2)| = 11$, there must be at least three edges, denoted by e_1, e_2, e_3, colored with the same color. Let $M = \{e_1, e_2, e_3\}$. Since in P_6^2, $v_1v_2, v_1v_3, v_2v_3, v_4v_5, v_4v_6$ must be colored with distinct colors, it cannot be hold that $M \subseteq E(P_6^2)$. Thus, $v_5v_7 \in M$ or $v_6v_7 \in M$.

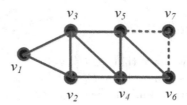

Fig. 4. P_7^2

Let $v_5v_7 \in M$. Note that edges can be colored with the same color on v_5v_7 can only be v_1v_2, v_3v_5 and v_4v_5. Moreover, $c(v_1v_2) \neq c(v_3v_5)$, $c(v_1v_2) \neq c(v_4v_5)$. And $c(v_3v_5) \neq c(v_4v_5)$ since v_3v_5 and v_4v_5 are in the same 3-cycle. Thus, $|M| \leq 2$, which is a contradiction. Hence, $v_5v_7 \notin M$.

Let $v_6v_7 \in M$. Note that edges can be colored with the same color on v_6v_7 can only be v_1v_2, v_1v_3, v_2v_3 and v_4v_6. Since $v_1v_2v_3v_1$ is a 3-cycle, at most one edge of v_1v_2, v_1v_3, v_2v_3 belongs to M and thus $|M| \leq 3$. Note that the distance between v_4v_6 and any edge in 3-cycle $v_1v_2v_3v_1$ is two, we have that $|M| \leq 2$, which is a contradiction. Hence, $v_6v_7 \notin M$.

By analysis above, the assumption that $\chi'_i(P_7^2) = 5$ cannot be hold. Thus, $\chi'_i(P_7^2) \geq 6$. Since $P_7^2 \subseteq P_n^2(n \geq 7)$, we have that $\chi'_i(P_n^2) \geq 6$.

Next, we shall prove $\chi'_i(P_n^2) \leq 6$.

Let $c : E(P_n^2) \to C = \{1, 2, 3, 4, 5, 6\}$. We color the edges of P_n^2 according to the following rules.

We color edges $v_iv_{i+1}(i = 1, 2, \cdots, n-1)$ with colors 1, 2, 3, 4, 5, 6, 1, 2, 3, 4, 5, 6, $\cdots$, color edges $v_iv_{i+2}(i = 1, 2, \cdots, n-2)$ with colors 3, 1, 2, 6, 4, 5, 3, 1, 2, 6, 4, 5, $\cdots$.

In the following, we shall prove that c is an injective-edge-coloring of power graph P_n^2. See Fig. 5.

Let $M_i = \{e | e \in E(P_n^2), c(e) = i\}$. By the definition of c, we have that $M_i = \{v_iv_{i+1}, v_{i+6}v_{i+7}, \cdots, v_{i+1}v_{i+3}, v_{i+1+6}v_{i+3+6}, \cdots\}$, $i = 1, 2, 4, 5$, $M_j = \{v_jv_{j+1}, v_{j+6}v_{j+7}, \cdots, v_{j-2}v_j, v_{j-2+6}v_{j+6}, \cdots\}$, $j = 3, 6$. It is easy to see that for any two edges $a, b \in M_i$ ($1 \leq i \leq 6$), there is no edge $c \in E(P_n^2)$ such that a, c, b are consecutive in P_n^2. Thus, c is an injective-edge-coloring of P_n^2. Hence, $\chi'_i(P_n^2) \leq 6$.

Together with $\chi'_i(P_n^2) \geq 6$, we have that $\chi'_i(P_n^2) = 6$.

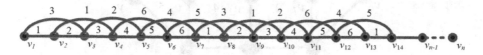

Fig. 5. 6-injective-edge-coloring of P_n^2

4   Proof of Theorem 4

(1) Note that any two edges in Ne_1 (see Fig. 6) should be colored with distinct colors, we have that $\chi'_i(Ne_1) = 6$.

For Ne_2, we can see Fig. 6. Since v_2v_3, $v_2v'_2$, v'_2v_3 are in the same 3-cycle, we should color them with distinct colors. And the distance between any edge in 3-cycle $v_2v'_2v_3v_2$ and any edge in 3-cycle $v_0v_1v'_1v_0$ is two, we need six colors to color these edges in 3-cycles $v_2v'_2v_3v_2$ and $v_0v_1v'_1v_0$. Thus, $\chi'_i(Ne_2) \geq 6$. On the other hand, we can obtain a 6-injective-edge-coloring of Ne_2 as in Fig. 6. Thus, $\chi'_i(Ne_2) = 6$.

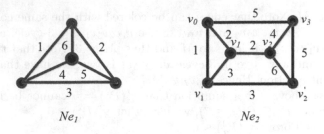

Fig. 6. 6-injective-edge-coloring of Ne_1 and Ne_2

(2) For Ne_3, we can see Fig. 7. Note that $v_0v_1v_1'v_0$ and $v_3v_3'v_4v_3$ are two 3-cycles and the distance between v_3v_4 and v_0v_1 or v_0v_1' is two, we need at least four distinct colors to color them to get an injective-edge-coloring of Ne_3. And since the distance between v_2v_2' and any edge in $\{v_0v_1, v_0v_1', v_1v_1', v_3v_4, v_3v_3', v_3'v_4\}$ is two, we need at least five distinct colors to color v_0v_1, v_0v_1', v_1v_1', v_3v_4, $v_3'v_4$, v_3v_3', v_2v_2' and thus $\chi_i'(Ne_3) \geq 5$. On the other hand, we can obtain a 5-injective-edge-coloring of Ne_3 as in Fig. 7. Thus, $\chi_i'(Ne_3) = 5$.

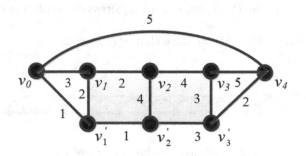

Fig. 7. 5-injective-edge-coloring of Ne_3

(3) For Ne_4, we can see Fig. 8. Since $v_0v_1v_1'v_0$ is a 3-cycle, we need three distinct colors to color it. And since the distance between v_2v_2' and any edge in 3-cycle $v_0v_1v_1'v_0$ is two, we need a new color to color v_2v_2'. Thus, $\chi_i'(Ne_4) \geq 4$. On the other hand, we can obtain a 4-injective-edge-coloring of Ne_4 as in Fig. 8. Thus, $\chi_i'(Ne_4) = 4$.

(4) For $h \geq 5$, let $G' = G - v_{h+1}$. Let $c : E(G) \to C = \{1, 2, 3, 4, 5\}$.

Step 1: Color G' according to the following rules:

Let $C_3 = v_0v_1v_1'$. Since the distance between v_2v_2' and any edge in C_3 is two, we need a fourth color to color edge v_2v_2' and thus $\chi_i'(G') \geq 4$. See Fig. 9, let $c(v_0v_1') = 1$, $c(v_1v_1') = 2$, $c(v_0v_1) = 3$, $c(v_1v_2) = c(v_2v_2') = c(v_2v_3) = 4$, $c(v_1'v_2') = 2$. We then color edges $v_iv_{i+1} (i = 3, 4, \cdots, h - 1)$ with colors 1, 1, 3, 3, 2, 2, 1, 1, 3, 3, 2, 2, $\cdots$, color edges $v_i'v_{i+1}' (i = 2, 3, \cdots, h - 1)$ with colors 3, 3, 2, 2, 1, 1, 3, 3, 2, 2, 1, 1, $\cdots$ and color edges $v_iv_i' (i = 3, 4, \cdots, h - 1)$ with colors 3, 1, 2, 3, 1, 2, $\cdots$. Then a 4-injective-edge-coloring of G' is obtained.

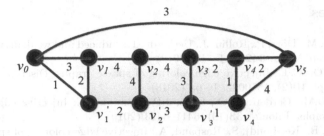

Fig. 8. 4-injective-edge-coloring of Ne_4

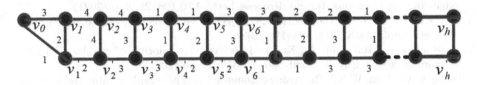

Fig. 9. 4-injective-edge-coloring of G'

Step 2: Add a vertex v_{h+1} to G' and add three edges $v_h v_{h+1}, v'_h v_{h+1}, v_0 v_{h+1}$ to get Ne_h, i.e. $Ne_h = G' + v_{h+1} + v_h v_{h+1} + v'_h v_{h+1} + v_0 v_{h+1}$. Next, we color these three edges. Let $c(v_h v_{h+1}) = c(v_0 v_{h+1}) = 5, c(v_{h+1} v'_h) = 4$, see Fig. 10.

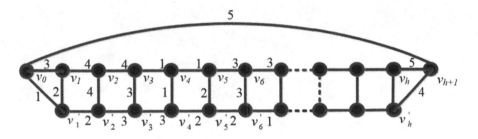

Fig. 10. 5-injective-edge-coloring of Ne_h, where $h \geq 5$

Note that there are only four colors, 1,2,3,4, have been used for a 4-injective-edge-coloring of G', we can get a 5-injective-edge-coloring of $G - v'_h v_{h+1}$ by coloring edges $v_h v_{h+1}, v_0 v_{h+1}$ with color 5. It's easy to see that $N_2(v'_h v_{h+1}) = \{v_0 v_1, v_0 v'_1, v'_{h-2} v'_{h-1}, v_{h-1} v'_{h-1}, v_{h-1} v_h\}$. By the coloring rules in step 1, only colors 1, 2 and 3 used for coloring the edges in $N_2(v'_h v_{h+1})$. Thus, we can color $v'_h v_{h+1}$ with color 4. Hence, Ne_h has a 5-injective-edge-coloring.

By analysis above, we complete the proof of Theorem 4.

Acknowledgements. Thank you to anonymous reviewers for comments that improved this paper.

References

1. Axenovich, M., Dörr, P., Rollin, J., Ueckerdt, T.: Induced and weak induced arboricities. Discrete Math. **342**, 511–519 (2019)
2. Bu, Y.H., Qi, C.T.: Injective edge coloring of sparse graphs. Discrete Math. Algorithms Appl. **10**(2), 1850022, 16 pp (2018)
3. Cardoso, D.M., Cerdeira, J.O., Cruz, J.P., Dominic, C.: Injective edge chromatic index of graphs. Filomat **33**(19), 6411–6423 (2019)
4. Ferdjallah, B., Kerdjoudj, S., Raspaud, A.: Injective edge-coloring of sparse graphs. arXiv:1907.09838
5. Foucaud, F., Hocquard, H., Lajou, D.: Complexity and algorithms for injective edge-coloring in graphs. Inform. Process. Lett., **170**,106121, 9pp (2021)
6. Lv, J.B., Zhou, J.X., Nian, N.H.: List injective edge-coloring of subcubic graphs. Discrete Appl. Math. **302**, 163–170 (2021)
7. Kostochka, A., Raspaud, A., Xu, J.: Injective edge-coloring of graphs with given maximum degree. Eur. J. Comb. **96**, 103355, 12 pp (2021)
8. Shiu, W.C., Tam, W.K.: The strong chromatic index of complete cubic Halin graphs. Appl. Math. Lett. **22**, 754–758 (2009)

Total Coloring of Planar Graphs Without Some Adjacent Cycles

Liting Wang and Huijuan Wang$^{(\boxtimes)}$

School of Mathematics and Statistics, Qingdao University, Qingdao 266071, China
sduwhj@163.com

Abstract. Let $G = (V, E)$ be a graph. If x, $y \in V \cup E$ are two adjacent or incident elements, then a k-total-coloring of graph G is a mapping φ from $V \cup E$ to $\{1, 2, \ldots, k\}$ on condition that $\varphi(x) \neq \varphi(y)$. In this paper, we define G to be a planar graph with maximum degree $\Delta \geq 8$. We prove that if for each vertex $v \in V(G)$, there exist two integers i_v, $j_v \in \{3, 4, 5, 6, 7, 8\}$ on condition that v is not incident with adjacent i_v-cycles and j_v-cycles, then G has a $(\Delta + 1)$-total-coloring.

Keywords: Total coloring · Planar graph · Short cycle

1 Introduction

In this paper, all graphs mentioned are finite, simple and undirected. Undefined notions and terminologies can be referred to [1]. Suppose G is a graph, then V and $d(v)$ are used to denote the vertex set and the degree of v. We use F, $d(f)$ and E to denote the face set, the degree of f and the edge set respectively. Then $\Delta = \max\{d(v) | v \in V\}$ is the maximum degree of a graph and $\delta = \min\{d(v) | v \in V\}$ is the minimum degree. We use n-vertex, n^+-vertex, or n^--vertex to denote the vertex v when $d(v) = n$, $d(v) \geq n$, or $d(v) \leq n$ respectively. A n-face, n^+-face, or n^--face are analogously defined. We use $(n_1, n_2, \ldots, n_k)$ to denote a k-face and its boundary vertices are n_i-vertex ($i = 1, 2 \ldots, k$). Similarly, we can define a $(n_1{}^+, n_2{}^-, \ldots, n_k)$-face. For instance, a (l, m^+, n^-)-face is a 3-face whose boundary vertices are l-vertex, m^+-vertex and n^--vertex respectively. If two cycles or faces have at least one common edge, then we call they are adjacent. We use $n_k(f)$ to denote the number of k-vertices that is incident with f. The number of k^+-face incident with f is denoted as $n_{k+}(f)$ and the number of k^--face incident with f is denoted as $n_{k-}(f)$. We use $n_k(v)$ to denote the number of k-vertices adjacent to v and use $f_k(v)$ to denote the number of k-faces incident with v. If G has a k-total-coloring, then we say that G can be totally colored by k colors. For the convenience of description, we say that G is total-k-colorable when G can be totally colored by k colors. If G can be totally colored by at least k colors, then k is the total chromatic number of G that is defined as χ''. It is easy to know that $\chi''(G) \geq \Delta + 1$. For the upper bound of χ'', Behzad [2] and Vizing [3] put forth the Total Coloring Conjecture (for short, TCC):

© The Author(s), under exclusive license to Springer Nature Switzerland AG 2022
Q. Ni and W. Wu (Eds.): AAIM 2022, LNCS 13513, pp. 421–432, 2022.
https://doi.org/10.1007/978-3-031-16081-3_37

Conjecture 1. For any graph, $\Delta + 1 \leq \chi''(G) \leq \Delta + 2$.

TCC has attracted lots of researchers' attention. However, this conjecture remains open even for planar graphs. In 1971, Rosenfeld [4] and Vijayaditya [5] confirmed TCC for all graphs with $\Delta \leq 3$ independently. Kostochka [6] proved that $\chi''(G) \leq \Delta + 2$ when $\Delta \leq 5$. For a planar graph, TCC is unsolved only when $\Delta = 6$ (see [6,18]). With the advances in research, some researchers found that $\chi''(G)$ of some specific graphs have an exact upper bound $\Delta + 1$. In 1989, Sánchez-Arroyo [7] demonstrated that it is a NP-complete problem to determine whether $\chi''(G) = \Delta + 1$ for a specified graph G. Moreover, for every fixed $k \geq 3$, McDiarmid and Sánchez-Arroyo [8] demonstrated that to determine whether a specific k-regular bipartite graph is total-$(\Delta + 1)$-colorable or not is also a NP-complete problem. However, it is possible to prove that $\chi''(G) = \Delta + 1$ when G is a planar graph having large maximum degree. It has been proved that $\chi''(G) = \Delta + 1$ on condition that G is a planar graph when $\Delta(G) \geq 11$ [9], $\Delta(G) = 10$ [10] and $\Delta(G) = 9$ [11]. It is still open to determine whether a planar graph is total-$(\Delta + 1)$-colorable when $\Delta = 6$, 7 and 8. If G is a planar graph and $\Delta(G) = 8$, then there are some relevant results obtained by adding some restrictions. For instance, for a planar graph with $\Delta(G) \geq 8$, it is proved that G is total-$(\Delta + 1)$-colorable if G does not contain k-cycles ($k = 5, 6$) [13], or adjacent 3-cycles [12], or adjacent 4-cycles [14]. Wang et al. [15] proved $\chi''(G) = \Delta + 1$ if there exist two integers i, $j \in \{3, 4, 5\}$ such that G does not contain adjacent i-cycles and j-cycles. Recently, a result has been proved in [20] for a planar graph with $\Delta(G) = 8$, that is, if for each vertex $v \in V$, there exist two integers i_v, $j_v \in \{3, 4, 5, 6, 7\}$ on condition that v is not incident with adjacent i_v-cycles and j_v-cycles, then G is total-$(\Delta + 1)$-colorable. Now we improve some former results and get the following theorem.

Theorem 1. *Suppose G is a planar graph with maximum degree $\Delta \geq 8$. If for each vertex $v \in V$, there exist two integers i_v, $j_v \in \{3, 4, 5, 6, 7, 8\}$ on condition that v is not incident with adjacent i_v-cycles and j_v-cycles. Then G is total-$(\Delta + 1)$-colorable.*

2   Reducible Configurations

Theorem 1 has been proved for $\Delta \geq 9$ in [11]. So we presume that $\Delta = 8$ in the rest of this paper. Suppose $G = (V, E)$ is a minimal counterexample to Theorem 1, that is, $|V| + |E|$ is as small as possible. In other words, G cannot be totally colored by $\Delta + 1$ colors, but every proper subgraph of G can be totally colored with $\Delta + 1$ colors. In this section, we give some information of configurations for our minimal counterexample G. A configuration is called to be reducible if it cannot occur in the minimal counterexample G. Firstly, we show some known properties of G.

Lemma 1. ([9]). *(a) G is 2-connected.*

(b) Suppose $v_1 v_2$ is an edge of G. If $d(v_1) \leq 4$, then $d(v_1) + d(v_2) \geq \Delta + 2 = 10$.

(c) Suppose G_{28} is a subgraph of G that is induced by the edges joining 2-vertices to 8-vertices. Then G_{28} is a forest.

Lemma 2. ([16]). G has no subgraph isomorphic to the configurations depicted in Fig. 1, where $7-v$ is used to denote the vertex of degree of seven. If a vertex is marked by •, then it has no more neighbors that are not depicted in G.

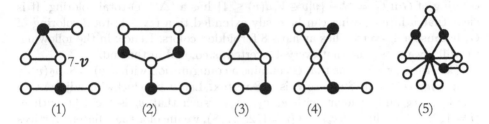

(1)   (2)   (3)   (4)   (5)

Fig. 1. Reducible configurations of Lemma 2

Lemma 3. ([19]). Suppose $v \in V$, $d(v) = d$ and $d \geq 6$. Let v be clockwise adjacent to $v_1, \ldots, v_d$ and incident with $f_1, f_2, \ldots, f_d$ such that v_i is the common vertex of f_{i-1} and f_i ($i \in \{1, 2, \ldots, d\}$). Notice that f_0 and f_d denote a same face. Let $d(v_1) = 2$ and $N(v_1) = \{v, u_1\}$. Then G contains none of the following configurations.(see Fig. 2):

(1) there exists an integer k ($2 \leq k \leq d-1$) such that $d(v_{k+1}) = 2$, $d(v_i) = 3$ ($2 \leq i \leq k$) and $d(f_j) = 4$ ($1 \leq j \leq k$).
(2) there exist two integers k and t ($2 \leq k < t \leq d-1$) such that $d(v_k) = 2$, $d(v_i) = 3$ ($k+1 \leq i \leq t$), $d(f_t) = 3$ and $d(f_j) = 4$ ($k \leq j \leq t-1$).
(3) there exist two integers k and t ($3 \leq k \leq t \leq d-1$) such that $d(v_i) = 3$ ($k \leq i \leq t$), $d(f_{k-1}) = d(f_t) = 3$ and $d(f_j) = 4$ ($k \leq j \leq t-1$).
(4) there exists an integer k ($2 \leq k \leq d-2$) such that $d(v_d) = d(v_i) = 3$ ($2 \leq i \leq k$), $d(f_k) = 3$ and $d(f_j) = 4$ ($0 \leq j \leq k-1$).

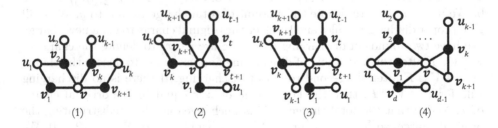

(1)   (2)   (3)   (4)

Fig. 2. Reducible configurations of Lemma 3

Lemma 4. ([20]). *Suppose u is a 6-vertex. If u is incident with one 3-cycle which is incident with a 4-vertex, then $n_{5+}(u) = 5$.*

Lemma 5. ([17]). *G contains no $(6, 6, 4^+)$-cycles.*

Lemma 6. *Suppose $v \in G$. If $d(v) = 8$ and $n_2(v) \geq 1$, then $n_{5+}(v) \geq 1$.*

Proof. Suppose G' is a subgraph of G. The mapping φ is said to be a nice coloring of G if $G' = G - \{v | v \in V, d(v) \leq 4\}$ has a $(\Delta + 1)$-total-coloring. It is clear that a nice coloring can be easily extended to a $(\Delta + 1)$-total-coloring of G, because a 4^--vertex has at most 8 forbidden colors. Hence, in the following, we will always assume that every 4^--vertex is colored in the end.

Contrarily, we assume that G contains a configuration with $d(v) = 8$, $n_2(v) \geq 1$, and $n_{5+}(v) = 0$. Suppose v is a 8-vertex. Let v be clockwise adjacent to $v_1, v_2, \ldots, v_8$ and incident with $e_1, e_2, \ldots, e_8$ such that v_i is incident with e_i $(i = 1, 2, \ldots, 8)$. Since $d(v_i) \leq 4$ $(i = 1, 2, \ldots, 8)$, we uncolor the adjacent vertices of v and color them in the end. We may assume that $d(v_1) = 2$. Then the one edge incident with v_1 is e_1, and the other edge incident with v_1 is denoted as e_9. Because of the minimality of G, $H = G - e_1$ has a nice coloring. Firstly, suppose $\varphi(e_9) = 9$. Otherwise, we color e_1 with 9 to get a nice coloring of G, which is a contradiction, so $\varphi(e_9) = 9$. We recolor v with 9, and color e_1 with 1 to get a nice coloring of G, which is a contradiction.   □

3   Discharging

In this section, we will accomplish the proof of Theorem 1 by using discharging method. The discharging method is a familiar and important way to solve coloring problems for a planar graph. By Euler's formula $|V| - |E| + |F| = 2$, we have

$$\sum_{v \in V}(2d(v) - 6) + \sum_{f \in F}(d(f) - 6) = -6(|V| - |E| + |F|) = -12 < 0$$

We define $w(x)$ of $x \in V \cup F$ to be the original charge function. Let $w(v) = 2d(v) - 6$ for every $v \in V$ and $w(f) = d(f) - 6$ for every $f \in F$. So $\sum_{v \in V \cup F} w(x) < 0$. We use $\omega(x \rightarrow y)$ to denote the amount of total charge from x to y. We shall give proper discharging rules and transfer the original charge to get a new charge. We have two rounds of discharging rules. We use $w^*(x)$ to denote the charge of $x \in V \cup F$ after the first round of discharging and use $w'(x)$ to denote the charge of $x \in V \cup F$ after the second round of discharging. If there is no discharging rule for $x \in V \cup F$, then the last charge of x is equal to the original charge of x. Notice that the total charge of G is unchangeable after redistributing the original charge, so $\sum_{x \in V \cup F} w'(x) = \sum_{x \in V \cup F} w(x) = -6\chi(\Sigma) = -12 < 0$. We will get an obvious contradiction by proving that $\sum_{x \in V \cup F} w'(x) \geq 0$.

These are the discharging rules:

R1. Suppose v is a 2-vertex. If u is adjacent to v, then $\omega(u \rightarrow v) = 1$.

R2. Let f be a face which is incident with v. Suppose $d(v) = 4$ or 5. If $d(f) = 4$, then $\omega(v \to f) = \frac{1}{2}$. If $d(f) = 5$, then $\omega(v \to f) = \frac{1}{3}$. Finally v sends the surplus charge to 3-faces incident with it evenly.

R3. If a 3-face is incident with 6-vertices and 7^+-vertices, then it receives $\frac{5}{4}$ from 7^+-vertices.

R4. Every 7^+-face sends $\frac{d(f)-6}{d(f)}$ to its adjacent 3-faces.

If $w^*(f) < 0$ of a 5^--face after the first round discharging, then we have the second round discharging:

R5. If $w^*(f) < 0$, then f receives $\left|\frac{w^*(f)}{n_{6+}(v)}\right|$ from every 6^+-vertices incident it which do not give any charge to f.

Lemma 7. *Suppose f is a face which is incident with v.*

1. If $d(v) = 6$, then

$$
\omega(v \to f) \leq
\begin{cases}
\frac{5}{4}, & \text{if } d(f) = 3 \text{ and } n_4(f) = 1, \\
\frac{11}{10}, & \text{if } d(f) = 3 \text{ and } n_5(f) \geq 1, \\
1, & \text{if } d(f) = 3 \text{ and } n_{6+}(f) = 3, \\
\frac{7}{8}, & \text{if } d(f) = 3, n_{5-}(f) = 0 \text{ and } n_{7+}(f) = 1, \\
\frac{1}{2}, & \text{if } d(f) = 3 \text{ and } n_{7+}(f) = 2, \\
\frac{2}{3}, & \text{if } d(f) = 4 \text{ and } n_{3-}(f) = 1, \\
\frac{1}{2}, & \text{if } d(f) = 4 \text{ and } n_{3-}(f) = 0, \\
\frac{1}{3}, & \text{if } d(f) = 5.
\end{cases}
$$

2. If $d(v) \geq 7$, then

$$
\omega(v \to f) \leq
\begin{cases}
\frac{3}{2}, & \text{if } d(f) = 3 \text{ and } n_{3-}(f) = 1, \\
\frac{5}{4}, & \text{if } d(f) = 3 \text{ and } n_{3-}(f) = 0, \\
1, & \text{if } d(f) = 4 \text{ and } n_{3-}(f) = 2, \\
\frac{3}{4}, & \text{if } d(f) = 4, n_{3-}(f) = 1 \text{ and } n_4(f) = 1, \\
\frac{2}{3}, & \text{if } d(f) = 4, n_{3-}(f) = 1 \text{ and } n_{5+}(f) = 3, \\
\frac{1}{2}, & \text{if } d(f) = 4 \text{ and } n_{3-}(f) = 0, \\
\frac{1}{3}, & \text{if } d(f) = 5.
\end{cases}
$$

Proof. Suppose v is incident with a 4^+-face f. Then it is clear that Lemma 7 is correct by R2 and R5. Now we think about that f is a 3-face that is incident with v. If $d(v) = 6$, then there exist no 3^--vertices adjacent to v by Lemma 1(b). If there exists a 4-vertex incident with f, then f is incident with a 7^+-vertex by Lemma 5. So $\omega(v \to f) \leq 3 - \frac{5}{4} - \frac{1}{4} = \frac{5}{4}$. If $n_5(f) = 1$ and $n_{6+}(v) = 2$, then $\omega(v \to f) \leq \frac{3-\frac{4}{5}}{2} = \frac{11}{10}$. Suppose $n_5(f) = 2$. If there exists one 5-vertex incident with five 3-faces, then the other 5-vertex is incident with at least two 6^+-faces. So $\omega(v \to f) \leq 3 - \frac{4}{5} - \frac{4}{3} \leq \frac{11}{10}$. If there exists one 5-vertex incident with four

3-faces, then all of the two 5-vertices are incident with at least one 6^+-face. So $\omega(v \to f) \le 3 - 1 \times 2 \le \frac{11}{10}$. Suppose $n_{6+}(f) = 3$. Then $\omega(v \to f) \le \frac{3}{3} = 1$. If $n_{5-}(f) = 0$ and $n_{7+}(f) = 1$, then the 7^+-vertex sends $\frac{5}{4}$ to f by R4, so $\omega(v \to f) \le \frac{3 - \frac{5}{4}}{2} = \frac{7}{8}$. If $n_{7+}(f) = 2$, then $\omega(v \to f) \le 3 - \frac{5}{4} \times 2 = \frac{1}{2}$. If $d(v) \ge 7$, then there exists at most one 3^--vertex adjacent to v, so $\omega(v \to f) \le \frac{3}{2}$. If $n_{3-}(f) = 0$, then $\omega(v \to f) \le \frac{3 - \frac{1}{2}}{2} = \frac{5}{4}$. $\qquad\square$

Lemma 8. *Suppose $d(v) = 8$. Let v be clockwise adjacent to $v_1, v_2, \ldots, v_n$ ($n \ge 3$) and incident with $f_1, f_2, \ldots, f_{n-1}$ such that f_j is incident with v_j and v_{j+1}. Clearly, f_0 and f_d denote a same face. If $d(v_1) = d(v_n) = 2$ and $d(v_i) \ge 3$ ($i = 2, 3, \ldots, n-1$), then $\sum_{i=1}^{n-1} \omega(v \to f_i) \le \frac{5}{4}n - \frac{9}{4}$.*

Proof. By Lemma 2, we know that $d(f_1) \ge 4$ and $d(f_{n-1}) \ge 4$. Firstly, suppose $d(f_1) = 4$ and $d(f_{n-1}) = 4$. If $\min\{d(f_2), d(f_3), \ldots, d(f_{n-2})\} \ge 5$, then $n \ge 4$, so $\sum_{i=1}^{n-1} \omega(v \to f_i) \le 1 \times 2 + \frac{1}{3}(n-3) \le \frac{5}{4}n - \frac{9}{4}$. If $\min\{d(f_2), d(f_3), \ldots, d(f_{n-2})\} = 4$ and $\max\{d(f_2), d(f_3), \ldots, d(f_{n-2})\} = 5$, then $\sum_{i=1}^{n-1} \omega(v \to f_i) \le n - 2 + \frac{1}{3} \le \frac{5}{4}n - \frac{9}{4}$. If $d(f_2) = d(f_3) = \ldots = d(f_{n-2}) = 4$, then $\sum_{i=1}^{n-1} \omega(v \to f_i) \le n - 3 + \frac{3}{4} \times 2 \le \frac{5}{4}n - \frac{9}{4}$ by Lemma 3. Suppose $\min\{d(f_2), d(f_3), \ldots, d(f_{n-2})\} = 3$ and $\max\{d(f_2), d(f_3), \ldots, d(f_{n-2})\} = 4$. If $d(f_2) = 4$ or $d(f_{n-2}) = 4$, then $\omega(v \to f_1) + \omega(v \to f_2) \le \max\{1 \times 2, \frac{3}{4} + \frac{5}{4}\} = 2$ and $\omega(v \to f_{n-2}) + \omega(v \to f_{n-1}) \le \max\{1 \times 2, \frac{3}{4} + \frac{5}{4}\} = 2$. Moreover, v sends more charge to 3-faces than 4-faces, so we assume that v is incident with 3-faces as more as possible. Hence, $\sum_{i=1}^{n-1} \omega(v \to f_i) \le 2 \times 2 + \frac{5}{4} \times (n-5) \le \frac{5}{4}n - \frac{9}{4}$. Suppose $d(f_2) = d(f_3) = \ldots = d(f_{n-2}) = 3$, then f_j ($2 \le j \le n-2$) receives at most $\frac{5}{4}$ from v by Lemma 3. Hence, $\sum_{i=1}^{n-1} \omega(v \to f_i) \le \frac{3}{4} \times 2 + \frac{5}{4} \times (n-3) \le \frac{5}{4}n - \frac{9}{4}$. Secondly, suppose $\min\{d(f_1), d(f_{n-1})\} = 4$ and $\max\{d(f_1), d(f_{n-1})\} \ge 5$. If $d(f_2) = d(f_3) = \ldots = d(f_{n-2}) = 3$, then $\sum_i^{n-1} \omega(v \to f_i) \le \frac{3}{4} + \frac{1}{3} + \frac{3}{2} + \frac{5}{4} \times (n-4) \le \frac{5}{4}n - \frac{9}{4}$. If $\max\{d(f_2), d(f_3), \ldots, d(f_{n-2})\} = 4$, then $\sum_{i=1}^{n-1} \omega(v \to f_i) \le 1 \times 2 + \frac{1}{3} + \frac{3}{2} + \frac{5}{4} \times (n-5) \le \frac{5}{4}n - \frac{9}{4}$. Finally, suppose $\min\{d(f_1), d(f_{n-1})\} \ge 5$. Then $\sum_{i=1}^{n-1} \omega(v \to f_i) \le \frac{1}{3} \times 2 + \frac{3}{2} \times 2 + \frac{5}{4} \times (n-5) \le \frac{5}{4}n - \frac{9}{4}$. $\qquad\square$

In the rest of this paper, we can check that $w'(x) \ge 0$ for every $x \in V \cup F$ which is a contradiction to our assumption. Let $f \in F$. If $d(f) \ge 7$, then $w'(f) \ge w(f) - \frac{d(f) - 6}{d(f)} \times d(f) = 0$ by R4. If f is a 6-face, then $w'(f) = w(f) = 0$. Suppose $d(f) \le 5$. If $n_{6+}(f) \ge 1$, then $w'(f) \ge 0$ by R5. If $n_{6+}(f) = 0$, then $n_5(f) = d(f)$. Suppose $d(f) = 3$ and the boundary vertices of f are consecutively v_1, v_2 and v_3. Then $d(v_i) = 5$ ($i = 1, 2, 3$). By R2, 4^+-face receives at most $\frac{1}{2}$ from incident 4-vertices or 5-vertices. Suppose $f_3(v_i) \le 3$ ($i = 1, 2, 3$). Then $\omega(v_i \to f) \ge 1$, so $w'(f) \ge 3 - 6 + 1 \times 3 = 0$. Suppose there exists $f_3(v_i) \ge 4$. Without loss of generality, assume that $f_3(v_3) \ge 4$. Then we have $f_3(v_1) \le 4$ and $f_3(v_2) \le 4$. Otherwise, $f_3(v_1) = 5$ or $f_3(v_2) = 5$, then for any integers $j, k \in \{3, 4, 5, 6, 7, 8\}$, there exists a vertex incident with adjacent j-cycles and k-cycles. So we get a contradiction to the condition of Theorem 1. If $f_3(v_1) = 4$, then v_1 is incident with a 9^+-face and v_2 is incident with at least two 6^+-faces, so $\omega(v_1 \to f) \ge 1$

and $\omega(v_2 \to f) \geq 1$. Consequently, $w'(f) \geq 3 - 6 + \frac{4}{5} + 1 + \frac{4}{3} > 0$. Similarly, we know that if $f_3(v_2) = 4$, then $w'(f) > 0$. Suppose $f_3(v_1) = f_3(v_2) = 3$. Then v_1 and v_2 is incident with at least one 6^+-face, so $\omega(v_i \to f) \geq \frac{4 - \frac{1}{2}}{3} = \frac{7}{6}$, $(i = 1, 2)$. Consequently, $w'(f) \geq 3 - 6 + \frac{4}{5} + \frac{7}{6} \times 2 > 0$. If $d(f) = 4$, then $w'(f) \geq 4 - 6 + \frac{1}{2} \times 4 = 0$ by R2. If $d(f) = 5$, then $w'(f) \geq 5 - 6 + \frac{1}{3} \times 5 > 0$ by R2. So for every $f \in F$, we prove that $w'(f) \geq 0$. Next, we consider that $v \in V$. Suppose $d(v) = 2$. Then it is clear that $w(v) = -2$, so $w'(v) = -2 + 1 \times 2 = 0$ by R1. If $d(v) = 3$, then $w'(v) = w(v) = 0$. Suppose $d(v) = 4$ or $d(v) = 5$. Then $w'(v) = 0$ by R2.

If v is a 6^+-vertex of G. Let v be clockwise adjacent to $v_1, \ldots, v_d$ and incident with $f_1, f_2, \ldots, f_d$ such that v_i is the common vertex of f_{i-1} and f_i $(i \in \{1, 2, \ldots, d\})$. Notice that f_0 and f_d denote the same face. Suppose $d(v) = 6$. Then there exist no 3^--vertices incident with v by Lemma 1 (b). Clearly, $w(v) = 2d(v) - 6 = 6$. By Lemma 4, there exist at most two 3-faces incident with a 4-vertex. Hence, if $f_3(v) \leq 3$, then $w'(v) \geq 6 - \frac{5}{4} \times 2 - \frac{11}{10} \times 1 - \frac{2}{3} \times 3 > 0$ by R4. Suppose $f_3(v) = 4$. If $f_{5+}(v) \geq 1$, then $w'(v) \geq 6 - \frac{5}{4} \times 2 - \frac{11}{10} \times 2 - \frac{2}{3} - \frac{1}{3} > 0$. If $f_4(v) = 2$, then there exist three boundary vertices of the two 4-faces adjacent v, that is, all of the two 4-faces are incident with four 4^+-vertices. Hence, $w'(v) \geq 6 - \frac{5}{4} \times 2 - \frac{11}{10} \times 2 - \frac{1}{2} \times 2 > 0$. Suppose $f_3(v) \geq 5$. If v is adjacent to a 5-vertex v_0 and f is a 3-face incident with v and v_0, then $f_3(v_0) \leq 3$, so $\omega(v_0 \to f) \geq 1$ and $\omega(v \to f) \leq 1$. Suppose $f_3(v) = 5$. If $f_{5+}(v) = 1$, then $w'(v) \geq 6 - \frac{5}{4} \times 2 - 1 \times 3 - \frac{1}{3} > 0$. If $f_4(v) = 1$, then there exist three boundary vertices of the 4-faces adjacent to v, that is, the 4-face is incident with four 4^+-vertices. Hence, $w'(v) \geq 6 - \frac{5}{4} \times 2 - 1 \times 3 - \frac{1}{2} = 0$.

Suppose $f_3(v) = 6$, that is, $d(f_i) = 3$ $(i = 1, 2, \ldots, 6)$. By Lemma 4, v is incident with at most one 4-vertex. So we may assume that $d(v_6) = 4$, then $d(v_1) \geq 7$ and $d(v_5) \geq 7$ by Lemma 5. Suppose $f_{6+}(v_6) = 2$. Then $\omega(v_6 \to f_5) \geq 1$ and $\omega(v_6 \to f_6) \geq 1$, so $\omega(v \to f_5) \leq 1$ and $\omega(v \to f_6) \leq 1$. Therefore, $w'(v) \geq 6 - 1 \times 6 = 0$. Otherwise, $f_{5-}(v) \geq 3$. Let f_x be the 5^--face incident with v_6 except f_5 and f_6. Suppose $d(f_x) = 5$. Then we get a contradiction to the condition of Theorem 1. Suppose $d(f_x) = 4$. Then v_6 is adjacent to v_4 and v_1 is adjacent to v_3. So we know that $f_{6+}(v_6) = 1$ and $\omega(v_6 \to f_i) \geq \frac{2 - \frac{1}{2}}{2} = \frac{3}{4}$ $(i = 5, 6)$. Therefore, $\omega(v \to f_i) \leq 3 - \frac{5}{4} - \frac{3}{4} \leq 1$ $(i = 5, 6)$, and $w'(v) \geq 6 - 1 \times 6 = 0$. Suppose $d(f_x) = 3$. Then each of the boundary vertices of f is adjacent to v. If v_6 is adjacent to v_4 and v_1 is adjacent to v_4, then $d(v_4) \geq 7$ by Lemma 5. So $\omega(v_4 \to f_4) = \frac{5}{4}$ and $\omega(v_5 \to f_4) = \frac{5}{4}$, then $\omega(v \to f_4) \leq \frac{1}{2}$ and $w'(v) \geq 6 - \frac{5}{4} \times 2 - 1 \times 3 - \frac{1}{2} = 0$. If v_6 is adjacent to v_3 and v_1 is adjacent to v_3, then $d(v_3) \geq 7$ by Lemma 5. Suppose $d(v_2) \geq 6$ and $d(v_4) \geq 6$. Then $\omega(v \to f_i) \leq \frac{3 - \frac{5}{4}}{2} = \frac{7}{8}$ $(i = 1, 2, 3, 4)$. Hence, $w'(v) \geq 6 - \frac{5}{4} \times 2 - \frac{7}{8} \times 4 = 0$. Suppose $d(v_2) = 5$ or $d(v_4) = 5$. Without of generality, assume that $d(v_4) = 5$. Then $\omega(v_4 \to f_3) \geq 1$ and $\omega(v_4 \to f_4) \geq 1$. So $\omega(v \to f_3) \leq 3 - 1 - \frac{5}{4} = \frac{3}{4}$ and $\omega(v \to f_4) \leq 3 - 1 - \frac{5}{4} = \frac{3}{4}$. Therefore, $w'(v) \geq 6 - \frac{5}{4} \times 2 - 1 \times 2 - \frac{3}{4} \times 2 = 0$.

Suppose $d(v) = 7$. Then it is easy to know that $f_3(v) \leq 6$ and v is not adjacent to a 2^--vertices by Lemma 1 (b). Clearly, $w(v) = 2d(v) - 6 = 8$. Suppose there exist no 3-faces incident with a 3-vertex. If $f_3(v) = 6$, then $f_{9+}(v) = 1$, so $w'(v) \geq 8 - \frac{5}{4} \times 6 > 0$ by Lemma 7. If $f_3(v) = 5$, then there exist no 4-faces incident with two 3^--vertex. So $w'(v) \geq 8 - \frac{5}{4} \times 5 - \frac{3}{4} \times 2 > 0$ by Lemma 7. If $f_3(v) \leq 4$, then $w'(v) \geq 8 - \frac{5}{4} \times 4 - 1 \times 3 = 0$. Now we presume that there exists at least one 3-face that is incident with a 3-vertex. Then all of the 4-faces are incident with at most one 3^--vertex. By Lemma 2, there exist at most two 3-faces incident with a 3-vertex. If $f_3(v) = 6$, then $f_{9+}(v) = 1$, so $w'(v) \geq 8 - \frac{3}{2} \times 2 - \frac{5}{4} \times 4 = 0$ by Lemma 7. Suppose $f_3(v) = 5$. If v is incident with at least one 5^+-face, then $w'(v) \geq 8 - \frac{3}{2} \times 2 - \frac{5}{4} \times 3 - \frac{2}{3} - \frac{1}{3} > 0$ by Lemma 7. Otherwise, $f_4(v) = 2$, then there exist three boundary vertices of the 4-face adjacent to v, so $w'(v) \geq 8 - \frac{3}{2} \times 2 - \frac{5}{4} \times 3 - \frac{3}{4} - \frac{1}{2} = 0$. Suppose $f_3(v) \leq 4$. Then $w'(v) \geq 8 - \frac{3}{2} \times 2 - \frac{5}{4} \times 2 - \frac{3}{4} \times 3 > 0$ by Lemma 7. If $d(v) = 8$, then we know that $w(v) = 2 \times 8 - 6 = 10$, $f_3(v) \leq 6$ and $n_2(v) \leq 7$ by Lemma 6. By Lemma 7 and Lemma 8, we shall consider the following cases by discussing the number of $n_2(v)$.

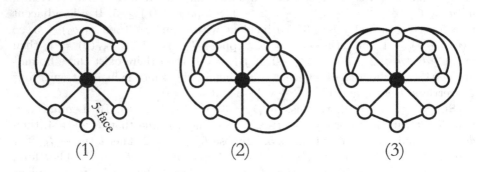

(1)   (2)   (3)

Fig. 3. $n_2(v) = 0$ and $f_3(v) = 6$

Case 1. $n_2(v) = 0$. Suppose $f_3(v) = 6$. If $f_{6+}(v) \geq 1$ or $f_{5+}(v) \geq 2$, then $w'(v) \geq 10 - \frac{3}{2} \times 6 - 1 = 0$ by Lemma 7. Otherwise, $f_{6+}(v) = 0$ and $f_{5+}(v) \leq 1$. Suppose $f_4(v) = 1$ and $f_5(v) = 1$. According to the condition of Theorem 1, there is only one case in which the location of the faces satisfying the condition of 1. We depict this case in Fig. 3 (1). It is clear that there exist three boundary vertices of the 4-faces adjacent to v, and there is at least one 3-face which is not incident with a 3-vertex by Lemma 2. If the 4-face is incident with at most one 3-vertex, then $w'(v) \geq 10 - \frac{3}{2} \times 5 - \frac{5}{4} - \frac{3}{4} - \frac{1}{3} > 0$. Otherwise, there exist two 3-vertex incident with the 4-face, then there exist at least two 3-faces that are not incident with a 3-vertex by Lemma 2. Hence, $w'(v) \geq 10 - \frac{3}{2} \times 4 - \frac{5}{4} \times 2 - \frac{3}{4} - \frac{1}{3} > 0$. Suppose $f_4(v) = 2$. There are only two cases satisfying the condition of Theorem 1. We depict these cases in Fig. 3(2) and (3). In Fig. 3(2), there exist at least four 3-faces all of which are adjacent to a 8^+-face. By R4, if there exists a 8^+-face adjacent to a 3-face, then 8^+-face sends $\frac{1}{4}$ to the 3-face, so each of the 3-face adjacent to a 8^+-face receives at most $\frac{3-\frac{1}{4}}{2} = \frac{11}{8}$ from the boundary vertices. There exist at most one 4-face incident with two 3-vertices in Fig. 3(2). By Lemma 2, there exist at least one 3-face that is not incident with a 3-vertex, so $w'(v) \geq 10 - \frac{3}{2} - \frac{11}{8} \times 4 - \frac{5}{4} - 1 - \frac{3}{4} = 0$. In Fig. 3(3), there exist at least four 3-faces all of which are adjacent to a 8^+-face. By Lemma 2, there is at most one 4-face incident with two 3-vertices. If all of the two 4-faces are incident with at most one 3-vertex, then $w'(v) \geq 10 - \frac{3}{2} \times 2 - \frac{11}{8} \times 4 - \frac{3}{4} \times 2 = 0$. Otherwise, there exists one 4-face that is incident with two 3-vertices, then there exist at least three 3-faces that are not incident with a 3-vertex by Lemma 2. Hence, $w'(v) \geq 10 - \frac{3}{2} \times 3 - \frac{5}{4} \times 3 - 1 - \frac{3}{4} = 0$. Suppose $f_3(v) = 5$. Then by the condition of Theorem 1, we know that $f_{5+}(v) \geq 1$, so $w'(v) \geq 10 - \frac{3}{2} \times 5 - 1 \times 2 - \frac{1}{3} \times 2 > 0$.

Case 2. $n_2(v) = 1$. Then $2 \times 8 - 6 - 1 = 9$.

Case 2.1. Let the 2-vertex be incident with a 3-cycle. It is clear that $f_3(v) \leq 6$ and there exist no 3-faces incident with a 3-vertex by Lemma 2. So v is incident with at most one 3-face that receives $\frac{3}{2}$ from v. If $f_3(v) = 6$, then by the condition of Theorem 1, we know that $f_{6+}(v) \geq 1$ or $f_{5+}(v) \geq 2$, so $w'(v) \geq 9 - \frac{3}{2} - \frac{5}{4} \times 5 > 0$ by Lemma 7. Suppose $f_3(v) = 5$. If $f_4(v) = 3$, then there are at least two $(8, 4^+, 4^+, 2^+)$-faces between the three 4-faces by Lemma 2. Hence, $w' \geq 9 - \frac{3}{2} - \frac{5}{4} \times 4 - 1 - \frac{3}{4} \times 2 = 0$. If $f_4(v) \leq 2$, then we have $w'(v) \geq 9 - \frac{3}{2} - \frac{5}{4} \times 4 - 1 \times 2 - \frac{1}{3} > 0$. Suppose $f_3(v) = 4$. If $f_4(v) = 4$, then there exist at least two $(8, 4^+, 4^+, 2^+)$-faces between the four 4-faces by Lemma 2. Hence, $w' \geq 9 - \frac{3}{2} - \frac{5}{4} \times 3 - 1 \times 2 - \frac{3}{4} \times 2 > 0$. If $f_4(v) \leq 3$, then $w'(v) \geq 9 - \frac{3}{2} - \frac{5}{4} \times 3 - 1 \times 3 - \frac{1}{3} > 0$. If $f_3(v) \leq 3$, then $w'(v) \geq 9 - \frac{3}{2} - \frac{5}{4} \times 2 - 1 \times 5 = 0$.

Case 2.2. Let the 2-vertex not be incident with a 3-cycle. Then $f_3(v) \leq 6$. Suppose $f_3(v) = 6$. Then the six 3-faces are adjacent and $f_{9+}(v) = 1$, so there exist at least four $(8, 4^+, 4^+)$-faces between the six 3-faces by Lemma 3. Therefore, $w'(v) \geq 9 - \frac{3}{2} \times 2 - \frac{5}{4} \times 4 - 1 \times 1 > 0$ by Lemma 7. Suppose $f_3(v) = 5$. It is easy to know that $f_{6+}(v) \geq 1$ by the condition of Theorem 1. If $f_4(v) = 2$, then there exist three the boundary vertices of the two 4-faces adjacent to v. So v is incident with at least two $(8, 4^+, 4^+)$-faces and one $(8, 4^+, 4^+, 2^+)$-face by Lemma 3. Hence, $w'(v) \geq 9 - \frac{3}{2} \times 3 - \frac{5}{4} \times 2 - 1 \times 1 - \frac{3}{4} \times 1 > 0$. If $f_4(v) = 1$, then there exists at least one $(8, 4^+, 4^+)$-face between the five 3-faces. by Lemma 3. Hence, $w'(v) \geq 9 - \frac{3}{2} \times 4 - \frac{5}{4} \times 1 - 1 \times 1 - \frac{1}{3} \times 2 > 0$. If $f_4(v) = 0$, then $w'(v) \geq 9 - \frac{3}{2} \times 5 - \frac{1}{3} \times 3 > 0$. Suppose $f_3(v) = 4$. Then we have $f_4(v) \leq 3$ by the condition of Theorem 1. If $f_4(v) = 3$, then there exist at least two $(8, 4^+, 4^+)$-faces between four the 3-faces. Hence, $w'(v) \geq 9 - \frac{3}{2} \times 2 - \frac{5}{4} \times 2 - 1 \times 3 - \frac{1}{3} \times 1 > 0$. If $f_4(v) \leq 2$, then $w'(v) \geq 9 - \frac{3}{2} \times 4 - 1 \times 2 - \frac{1}{3} \times 2 > 0$. Suppose $f_3(v) = 3$. If there exists a 5^+-face incident with v, then $w'(v) \geq 9 - \frac{3}{2} \times 3 - 1 \times 4 - \frac{1}{3} > 0$. Otherwise, $f_4(v) = 5$, then there exist at least three $(8, 4^+, 4^+, 2^+)$-faces between the five 4-faces. Hence, $w'(v) \geq 9 - \frac{3}{2} \times 3 - 1 \times 2 - \frac{3}{4} \times 3 > 0$. If $f_3(v) \leq 2$, then $w'(v) \geq 9 - \frac{3}{2} \times 2 - 1 \times 6 = 0$.

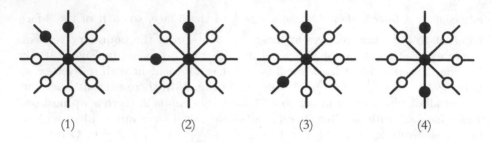

Fig. 4. $n_2(v) = 2$

Case 3. $n_2(v) = 2$. Then $2 \times 8 - 6 - 2 = 8$ and there are four cases in which 2-vertices are located. We depict these cases in Fig. 4. In Fig. 4 (1), $w'(v) \geq 8 - (\frac{5}{4} \times 8 - \frac{9}{4}) > 0$ by Lemma 8. In Fig. 4 (2), $w'(v) \geq 8 - (\frac{5}{4} \times 7 - \frac{9}{4}) - (\frac{5}{4} \times 3 - \frac{9}{4}) = 0$. In Fig. 4 (3), $w'(v) \geq 8 - (\frac{5}{4} \times 6 - \frac{9}{4}) - (\frac{5}{4} \times 4 - \frac{9}{4}) = 0$. In Fig. 4 (4), $w'(v) \geq 8 - (\frac{5}{4} \times 5 - \frac{9}{4}) \times 2 = 0$ by Lemma 8.

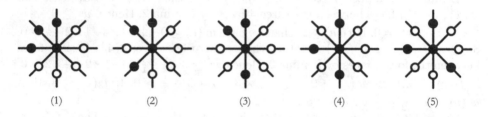

Fig. 5. $n_2(v) = 3$

Case 4. $n_2(v) = 3$. Then $2 \times 8 - 6 - 3 = 7$ and there are five cases in which 2-vertices are located. We depict these cases in Fig. 5. In Fig. 5(1), $w'(v) \geq 7 - (\frac{5}{4} \times 7 - \frac{9}{4}) > 0$ by Lemma 8. In Fig. 5(2), $w'(v) \geq 7 - (\frac{5}{4} \times 6 - \frac{9}{4}) - (\frac{5}{4} \times 3 - \frac{9}{4}) > 0$. In Fig. 5(3), $w'(v) \geq 7 - (\frac{5}{4} \times 5 - \frac{9}{4}) - (\frac{5}{4} \times 4 - \frac{9}{4}) > 0$. In Fig. 5(4), $w'(v) \geq 7 - (\frac{5}{4} \times 5 - \frac{9}{4}) - (\frac{5}{4} \times 3 - \frac{9}{4}) \times 2 = 0$. In Fig. 5(5), $w'(v) \geq 7 - (\frac{5}{4} \times 3 - \frac{9}{4}) - (\frac{5}{4} \times 4 - \frac{9}{4}) \times 2 = 0$ by Lemma 8.

Case 5. $n_2(v) = 4$. Then $2 \times 8 - 6 - 4 = 6$ and there are eight cases in which 2-vertices are located. We depict these cases in Fig. 6. In Fig. 6(1), $w'(v) \geq 6 - (\frac{5}{4} \times 6 - \frac{9}{4}) > 0$ by Lemma 8. In Fig. 5(2) and (4), $w'(v) \geq 6 - (\frac{5}{4} \times 5 - \frac{9}{4}) - (\frac{5}{4} \times 3 - \frac{9}{4}) > 0$. In Fig. 6(3) and (7), $w'(v) \geq 6 - (\frac{5}{4} \times 4 - \frac{9}{4}) \times 2 > 0$. In Fig. 6(5) and (6), $w'(v) \geq 6 - (\frac{5}{4} \times 3 - \frac{9}{4}) \times 2 - (\frac{5}{4} \times 4 - \frac{9}{4}) > 0$. In Fig. 6(8), $w'(v) \geq 6 - (\frac{5}{4} \times 3 - \frac{9}{4}) \times 4 = 0$ by Lemma 8.

Case 6. $n_2(v) \geq 5$. Suppose $n_2(v) = 5$. Then $2 \times 8 - 6 - 5 = 5$ and $f_3(v) \leq 2$. Suppose $f_3(v) = 2$. Then $f_{6+}(v) \geq 4$ by Lemma 2. Consequently, $w'(v) \geq 5 - \frac{3}{2} \times 2 - 1 \times 2 = 0$ by Lemma 7. If $f_3(v) = 1$, then $f_{6+}(v) \geq 3$ and $f_4(v) \leq 4$. If $f_4(v) = 4$. then all of the four 4-faces are $(8, 4^+, 4^+, 2^+)$-faces. Hence, $w'(v) \geq$

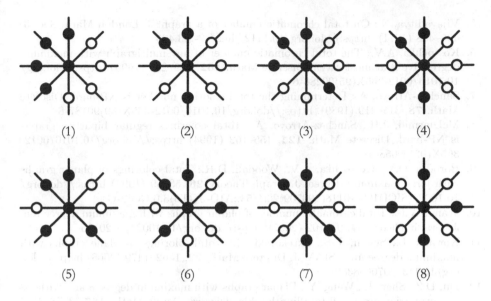

(1)   (2)   (3)   (4)

(5)   (6)   (7)   (8)

Fig. 6. $n_2(v) = 4$

$5 - \frac{3}{2} \times 1 - \frac{3}{4} \times 4 > 0$. If $f_4(v) \leq 3$, then $w'(v) \geq 5 - \frac{3}{2} \times 1 - 1 \times 3 - \frac{1}{3} > 0$. Suppose $f_3(v) = 0$. Then $f_{6+}(v) \geq 2$. If $f_4(v) = 6$, then all of the six 4-faces are $(8, 4^+, 4^+, 2^+)$-faces. Hence, $w'(v) \geq 5 - \frac{3}{4} \times 6 > 0$. If $f_4(v) = 5$, then there exist at least four $(8, 4^+, 4^+, 2^+)$-faces between the five 4-faces. Hence, $w'(v) \geq 5 - 1 \times 1 - \frac{3}{4} \times 4 - \frac{1}{3} \times 1 > 0$. If $f_4(v) \leq 4$, then $w'(v) \geq 5 - 1 \times 4 - \frac{1}{3} \times 2 > 0$. Suppose $n_2(v) = 6$. Then $2 \times 8 - 6 - 6 = 4$ and $f_3(v) \leq 1$. If $f_3(v) = 1$, then $f_{6+}(v) \geq 5$ and $f_4(v) \leq 2$. So $w'(v) \geq 4 - \frac{3}{2} - 1 \times 2 > 0$. If $f_3(v) = 0$, then $f_{6+}(v) \geq 4$. Hence, $w'(v) \geq 4 - 1 \times 4 = 0$. Suppose $n_2(v) = 7$. Then by Lemma 2 we know that $f_{6+}(v) \geq 6$ by and $f_3(v) = 0$, so $w'(v) \geq 10 - 7 - 1 \times 2 > 0$.

In summary, we know that $w'(x) \geq 0$ for every $x \in V \cup F$, so $\sum_{x \in V \cup F} w'(x) \geq 0$. Hence, we get the desired contradiction and finish the proof of Theorem 1.

Acknowledgements. We thanks for the support by Shandong Provincial Natural Science Foundation of China under Grant ZR2020MA045.

References

1. Bondy, J.A., Murty, U.S.R.: Graph Theory with Applications. North-Holland, NewYork (1982)
2. Behzad, M.: Graphs and their chromatic numbers. Ph.D. Thesis, Michigan State University (1965)
3. Vizing, V.G.: Some unsolved problems in graph theory. UspekhiMat. Nauk **23**, 117–134 (1968)
4. Roesnfeld, M.: On the total colorings of certain graphs. Israel J. Math **9**, 396–402 (1971). https://doi.org/10.1007/BF02771690

5. Vijayaditya, N.: On total chromatic number of a graph. J. London Math. Soc. **3**, 405–408 (1971). https://doi.org/10.1112/jlms/s2-3.3.405

6. Kostochka, A.V.: The total chromatic number of any multigraph with maximum degree five is at most seven. Discrete Math. **162**, 199–214 (1996) https://doi.org/10.1016/0012-365X(95)00286-6

7. Sánchez-Arroyo, A.: Determining the total coloring number is NP-hard. Discrete Math. **78**, 315–319 (1989). https://doi.org/10.1016/0012-365X(89)90187-8

8. McDiarmid, J.H., Sánchez-Arroyo, A.: Total colorings regular bipartite graphs is NP-hard. Discrete Math. **124**, 155–162 (1994) https://doi.org/10.1016/0012-365X(92)00058-Y

9. Borodin, O.V., Kostochka, A.V., Woodall, D.R.: Total colorings of planar graphs with large maximum degree. J. Graph Theory **26**, 53–59 (1997) https://doi.org/10.1002/(SICI)1097-0118(199709)26:1⟨53::AID-JGT6⟩3.0.CO;2-G

10. Wang, W.F.: Total chromatic number of planar graphs with maximum degree ten. J. Graph Theory **54**, 91–102 (2007) https://doi.org/10.1002/jgt.20195

11. Kowalik, Ł., Sereni, J.-S., Škrekovski, R.: Total-colorings of plane graphs with maximum degree nine. SIAM J. Discrete Math. **22**, 1462–1479 (2008) https://doi.org/10.1137/070688389

12. Du, D.Z., Shen, L., Wang, Y.: Planar graphs with maximum degree 8 and without adjacent triangles are 9-totally-colorable, Discrete Appl. Math. **157**, 2778–2784 (2009) https://doi.org/10.1016/j.dam.2009.02.011

13. Hou, J.F., Zhu, Y., Liu, Z.G., Wu, J.L.: Total colorings of planar graphs without small cycles., Graphs Comb. **24**, 91–100 (2008) https://doi.org/10.1007/s00373-008-0778-8

14. Tan, X., Chen, H.Y., Wu, J.L.: Total colorings of planar graphs without adjacent 4-cycles., Lecture Notes on Operations Research, vol. 10, pp. 167–173 (2009)

15. Wang, H.J., Wu, L.D., Wu, J.L.: Total coloring of planar graphs with maximum degree 8., Theoret. Comput. Sci. **522**, 54–61 (2014) https://doi.org/10.1016/j.tcs.2013.12.006

16. Chang, J., Wang, H.J., Wu, J.L.: Total coloring of planar graphs with maximum degree 8 and without 5-cycles with two chords. Theoret. Comput. Sci. **476**, 16–23 (2013). https://doi.org/10.1016/j.tcs.2013.01.015

17. Shen, L., Wang, Y.Q.: Total colorings of planar graphs with maximum degree at least 8. Sci. China Ser A: Math. **52**, 1733–1742 (2009). https://doi.org/10.1007/s11425-008-0155-3

18. Sanders, D.P., Zhao, Y.: On total 9-coloring planar graphs of maximum degree seven. J. Graph Theory **31**, 67–73 (1999) https://doi.org/10.1002/(SICI)1097-0118(199905)31:1⟨67::AID-JGT6⟩3.0.CO;2-C

19. Xu, R.Y., Wu, J.L. Wang, H.J.: Total coloring of planar graphs without some chordal 6-cycles. Bull. Malays. Math. Sci. Soc. **520**, 124–129 (2014) https://doi.org/10.1007/s40840-014-0036-6

20. Wang, H.J., Gu, Y., Liu, B.: Total coloring of planar graphs without adjacent short cycles. J. Com. Optim. **33**(1), 265–274 (2017). https://doi.org/10.1007/s10878-015-9954-y

Logic and Machine Learning

Security on Ethereum: Ponzi Scheme Detection in Smart Contract

Hongliang Zhang[1], Jiguo Yu[2,3(✉)], Biwei Yan[1,2], Ming Jing[2,3], and Jianli Zhao[1,2,3]

[1] School of Computer Science and Technology, Qilu University of Technology (Shandong Academy of Sciences), Jinan 250353, Shandong, People's Republic of China
[2] Big Data Institute, Qilu University of Technology, Jinan 250353, Shandong, People's Republic of China
`jiguoyu@sina.com`
[3] Shandong Fundamental Research Center for Computer Science, Jinan 250300, People's Republic of China

Abstract. Ethereum has many transaction security issues such as Ponzi schemes, which are hidden in a large number of smart contracts. And they are difficult to be detected. Therefore, we propose a novel multi-granularity multi-scale convolutional neural network model (MM-CNN) to detect Ponzi schemes in smart contracts. A multi-granularity method is used to compress the smart contract opcodes with similar function to obtain multi-granularity frequency data of opcodes in MM-CNN. Then, we use a multi-scale convolution kernel to extract features of frequency data. The experiments show that the frequency features are the best measurements to represent the attributes of the Ponzi scheme. In the multi-granularity method, fine-grained opcode has a stronger ability to express Ponzi attributes. The recall rate of MM-CNN on the verification set is 98.07%, which shows the effectiveness of the scheme.

Keywords: Smart contract · Ponzi scheme · Multi-granularity · Multi-scale

1 Introduction

With the rapid development of Ethereum, users can use Ethereum to run distributed applications (Dapps) with various functions based on their wishes. However, it lacks security supervision and standardization. The most common method of fraud is to write a smart contract with a Ponzi scheme (Ponzi contracts). The "Ponzi scheme" created by Charles Ponzi is one of the most classic

This work was supported in part by the NSF of China under Grants 61832012 and 61771289, and the Key Research and Development Program of Shandong Province under Grant 2019JZZY020124, and the Pilot Project for Integrated Innovation of Science, Education and Industry of Qilu University of Technology (Shandong Academy of Sciences) under Grant 2020KJC-ZD02.

Q. Ni and W. Wu (Eds.): AAIM 2022, LNCS 13513, pp. 435–443, 2022.
https://doi.org/10.1007/978-3-031-16081-3_38

436 H. Zhang et al.

scams of the 20th century, which has now appeared on Ethereum. Ethereum currently has tens of thousands of smart contracts deployed and controls billions of dollars worth of Ethereum's cryptocurrency. Smart contract security incidents are also emerging in an endless stream, and the loss of funds is particularly serious. Therefore, detecting the Ponzi contract is currently an urgent task.

At present, many researchers use artificial intelligence to detect the security of smart contracts. The same is true for Ponzi contract detection. However, there are still many deficiencies in current research on Ponzi contract detection. Firstly, Peng et al. [1] and Chen et al. [2] convert the bytecode of the smart contract into an opcode and put frequency features of a single opcode into algorithm for detection. Although single opcode frequency can reflect part of the function of a smart contract, the single frequency construction cannot fully reflect the function of a smart contract. Therefore, more detailed frequency feature construction methods are needed. Secondly, the opcodes of smart contracts not only have frequency feature, but also have a sequence feature. Moreover, Wang et al. [3] found that the sequence information of the opcodes can also reflect the function of the smart contract. Therefore, other features of the smart contract need to be constructed to reflect the attributes of ponzi scheme.

To deal with these above deficiencies, we propose a novel multi-granularity multi-scale convolutional neural network model. The main contributions can be summarized as follows:

- We construct multiple models from the frequency feature, sequence feature and account feature to detect the Ponzi scheme. Among models, the proposed MM-CNN has the strongest detection ability for Ponzi contracts and the recall rate reached 98.07% in verification set.
- We construct multi-granularity data in opcodes, including fine-grained opcode, coarse-grained opcode, and coarse-fine-grained opcode. The experiments show that fine-grained opcode is the best expressive ability for detecting Ponzi contracts.
- We compare the expressivity of different modality of smart contract features for Ponzi schemes. Models that apply frequency features are more accurate than models that apply account or sequence features.

2   Related Work

Automated auditing is the most common method in code auditing methods, which is based on symbolic execution and abstract methods. Torres et al. developed a system called Honeybadger which used symbolic execution to expose smart contract [4]. The well-known automated auditing systems are Mythril [5], Oyente [6], Maian [7], and NeuCheck [8]. Compared with manual auditing methods, the efficiency of automated auditing is greatly improved.

AI auditing is a popular solution today. Chen et al. used the Random Forest (RF) algorithm for detection in a similar way a year later, which had a high accuracy rate. However, there was an overfitting phenomenon [9]. Jung et al. added account features on the basis of Chen's method, to provide new ideas for detecting Ponzi contracts, only machine learning algorithms were used to explore [10]. M. Bartolett et al. collected a large amount of data through multi-input heuristic address clustering, and extracted features related to the Ponzi scheme (for example, focusing on daily transaction volume, money flow, lifecycle, fraudulent operation, external money flow, and geopolitical information) [11], which did not use the features of source code for analysis and only extracted features from the transaction information. The information was greatly affected by people's subjective factors and cannot be a key factor in detecting Ponzi contracts.

We construct multi-granularity data on the Ponzi dataset and multiple models to detect the Ponzi scheme. Through experimental comparison with SVM and K-Nearest Neighbor (KNN) algorithms, the experimental results show that the obtained model has high accuracy.

3   The Experimental Scheme

In Fig. 1, the experimental scheme is divided into 6 steps. The first step is to obtain the dataset, and then analyze the dataset. The second step is to crawl features of smart contracts from the obtained dataset. The third step is to construct multi-granularity data of opcodes. The fourth step is to construct frequency features and sequence features. The fifth step is to balance the dataset with the SMOTETomek algorithm. The sixth step is to construct models.

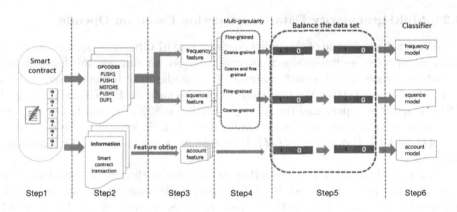

Fig. 1. Experimental process.

3.1   Feature Data Collection

The task of collecting the feature data can't depend on the source code of the smart contracts, because only 48.62% of those codes are open-source. So we need to obtain feature data that does not depend on the source code. We use etherscan to crawl opcodes and transactions on smart contracts. Opcodes and transactions are called opcode features and account features.

Opcode Features. We collecte 120 different opcodes with operands from crawling opcodes. We simplify the operand and only retain the opcode. For example, *PUSH* opcodes will have many opcodes with operands such as *PUSH*1, *PUSH*2, *PUSH*3, *PUSH*4, *PUSH*5, *PUSH*6, and we simplify them as *PUSH*. Therefore, there are 67 different opcodes.

Account Features. These transactions have three patterns: (*1*) When a Ponzi contract receives a fund, the fund will be sent to some other accounts immediately. (*2*) Very few investors' accounts will receive more funds than their investments. The creator of those Ponzi contracts usually receives funds through smart contracts. (*3*) The rapid return of investment funds will result in a lower balance of Ponzi contracts. So the transaction attributes of Ponzi contracts can be used to detect Ponzi schemes, and it is reasonable to construct smart contract account features..

Based on the transaction attributes, we gathere 7 types of features from transaction data and name them account features, which include the following 7 features: 1. The balance of the contract (Bal); 2. The number of investments ($InvNum$); 3 The number of transactions that have paid return ($PayNum$); 4. The number of investment funds ($InvFun$); 5. The number of paying return funds ($PayFun$); 6. The difference between the number of investment accounts and the accounts with paid return ($DifAcc$); 7. The difference between the investment funds and the returned funds ($DifFun$).

3.2 Multi-granularity Data Construction Based on Opcode

We convert the bytecode of the smart contract into 67 different opcodes and treat different opcodes as fine-grained construction. So we can get the fine-grained opcode sequence of each smart contract. We calculate the frequency of opcodes in the fine-grained data. We find that 3 opcodes such as *PUSH*, *SWAP*, and *DUP* frequently appear and the total proportion is 82%, frequently opcode will have a negative impact on Ponzi contract detection. So we will exclude *PUSH*, *SWAP*, and *DUP* opcodes. *OTHER* represents is not shown in the remaining opcodes.

Coarse-grained data is a collection of opcodes with similar functions. For example, grouping *ADD*, *SUB*, *MUL*, and *DIV* into one type, we call it *arithmetic* opcodes. *Arithmetic* represents coarse-grained data. We split 67 opcodes into 11 types of coarse-grained data, which are called *Arithmetic*, *Comparison*, *Bitwise*, *Cryptography*, *Jump*, *Storage*, *Memory*, *Stack*, *Block*, *Contract*, and *Others*. The *Others* include *PC*, *STOP*, and *UNKWON* (unrecognized opcode).

3.3 Feature Construction of Frequency and Sequence

We use different types of opcodes to construct frequency features. Frequency is the number of times that different types of opcodes appear in the smart contract.

The frequency features of opcodes can represent part of the code attributes by $l_m^i = \frac{t_m^i}{T^i}$. Among them, t_m^i is the number of the m-th opcode in the i-th smart contract, T^i is the sum of the number of opcodes in the i-th smart contract, and l_m^i is the frequency attributes of the m-th opcode in the i-th smart contract. Fine-grained frequency data use the number of 64 opcode types (exclude $PUSH$, $SWAP$, DUP) as the feature.

The logical sequence of codes can reflect the transaction properties of Ponzi schemes. Therefore, we propose the sequence feature of opcodes. Sequence data construction is to add sequences based on frequency features. We record the location of the first occurrence of different types of opcodes in the smart contract. Then, frequency features of the sorted opcode used as the sequence feature. Finally, we input sequenced frequency feature into the sequence model.

4   Model Construction

4.1   Frequency Model

We build a multi-scale convolution model based on frequency features. The core of which is to use multiple convolution kernels with different scales for frequency feature convolution. CNN can effectively learn corresponding features from a large number of samples, avoiding the complex feature extraction process. We employ the multi-granularity data to build a multi-scale convolutional neural network (MM-CNN).

Before inputting the frequency data into the model, the dimensionality of the frequency feature needs to be reduced. We use Principal Component Analysis (PCA) for dimensionality reduction and calculate the correlation coefficient between each frequency feature through the correlation coefficient method. The Eq. (2) for calculating the correlation coefficient of two random variables is as follows:

$$\rho_{XY} = \frac{\text{cov}(X,Y)}{\sigma_X \sigma_Y} = \frac{E\left[(X - \mu_X)(Y - \mu_Y)\right]}{\sigma_X \sigma_Y} = E\left[\left(\frac{X - \mu_X}{\sigma_X}\right)\left(\frac{Y - \mu_Y}{\sigma_Y}\right)\right]$$
(1)

Corresponding to our sample, we think that the value of a certain dimension feature x and y on each sample is a sampling of X and Y. Therefore, ρ_{XY} can be estimated with $\frac{1}{n}\sum_{i=1}^{n}\left(\frac{x_i - \mu_X}{\sigma_X}\right)\left(\frac{y_i - \mu_Y}{\sigma_Y}\right)$.

Then, the data is sent to the MM-CNN model. The data first enters three parallel convolutional layers in MM-CNN model, and each convolutional layer is composed of convolution and pooling to extract frequency features of different scales, the definition of multiple scale convolution is as follows:

$$y^{k,s} = Conv_{k \in K, s \in S}(X, W_k) + b$$
(2)

The three convolution kernels in $K = [k_1, k_2, k_3]$ are corresponding to the size of the three convolution kernels, and the three convolution strides are respectively $S = [s_1, s_2, s_3]$. k_1, k_2 and k_3 are increasingly larger convolution kernels. s_1, s_2 and

s_3 are larger and larger strides. X is the frequency feature data of the input, and y is the output value after a series of convolution kernel k and stride s.

We use pool layers of different scales to process data of different dimensions. After processing the same dimensions, the data of the three scales are added with the same weight, and the result is processed by the activation function Relu before entering the fully connected layer.

The Softmax layer performs a softmax operation on the output of the fully connected layer and converts the value to the probability of its corresponding classification. The softmax Eq. (4) is as follows:

$$\text{Softmax}\,(x_m) = \frac{e^{x_m}}{\sum_{n=1}^{N} e^{x_n}} \tag{3}$$

x_m is the m-th element of Softmax input x, N is the total number of elements. $Softmax(x_m)$ is the Softmax result of each classification corresponding to x_m element, the probability that the predicted result is the classification. To optimize the identification process of $Softmax$, the loss function is adopted in Eq. (5):

$$L_{loss} = -\sum_{x_m} P(x_m) \cdot \log Softmax(x_m) \tag{4}$$

$P(x_m)$ represents the sign of the x_m class (ture is 1, false is 0), and the sum of m softmax multiplied by $P(x_m)$ is loss.

4.2   Sequence Model and Account Model

Faced with problems and tasks that are sensitive to time series, LSTM is usually more suitable, so it is used to extract the features of opcode sequences. Therefore, we build a multi-granularity LSTM model (M-LSTM) to learn the patterns of the multi-granularity sequence features. The purpose is to compare with the frequency model and check the detection ability of the sequence features on the Ponzi scheme. The M-LSTM model consists of an input layer, a hidden layer, and an output layer. The hidden layer is the core part of the model. The hidden layer is the core part of the model.

We use a simple convolutional neural network for Ponzi scheme detection in account features (A-CNN). Because the account has fewer features, the core is a convolutional layer. The significance of the model is to determine whether the account features are reasonable for Ponzi contracts and contrast with the features of the other two models. Our purpose is to form a comparative experiment, put the account features into the traditional SVM and KNN algorithm for detection, which can judge whether the account features have reference value for identifying Ponzi contracts.

5   Experimental Analysis

To measure the detection performance of our models, extensive experiments are taken on Ponzi datasets. Next, we will introduce the experiments in detail.

We downloaded the Ponzi contract dataset from the website[1] Next we conducted experiments on frequency features, sequence features, and account features. We use multi-granularity data (fine-grained data, coarse-grained data, and coarse-fine-grained data) for comparison on the frequency model to select the best feature construction method. In the sequence model, fine-grained sequence data and coarse-grained sequence data are used for comparison. Account features are used in the account model.

After using the SMOTETomek algorithm to balance the dataset, we extract 70% of the dataset as the training dataset and the remaining 30% as the validation set. The running result is an average of ten runs. we use four evaluation indicators: accuracy, precision, recall, and F-score to evaluate the performance of the model.

5.1 Analysis of Experimental Results

We put data of different granularities into three models for experiments and analyze the experimental results of each model. The experimental result is the average of five measurements.

Table 1. All evaluation.

Feature	Model	Feature-grained	Accuracy	Precision	Recall	F-score
Frequency	MM-CNN	Fine-grained	94.81	92.87	97.14	94.95
		Coarse-grained	67.78	68.59	65.83	67.17
		Coarse-Fine-grained	95.76	93.81	98.07	95.89
	KNN	Fine-grained	81.52	87.64	73.64	79.96
		Coarse-grained	78.71	77.17	81.45	79.25
		Coarse-Fine-grained	79.87	86.29	72.92	84.30
	SVM	Fine-grained	84.39	95.00	72.73	82.39
		Coarse-grained	87.18	86.92	87.73	87.32
		Coarse-Fine-grained	83.50	88.37	78.45	83.11
Sequence	M-LSTM	Fine-grained	94.48	95.20	93.80	94.45
		Coarse-grained	83.73	82.11	86.68	84.15
Account	A-CNN		60.80	58.13	79.01	66.98
	KNN		72.43	87.74	51.92	65.23
	SVM		61.71	58.49	82.54	68.47

Frequency Model Analysis. Based on our MM-CNN model, machine learning algorithms such as KNN and SVM are added for comparison. The experimental results are shown in Table 1.

---

[1] http://xblock.pro.

The accuracy of the SVM algorithm and the KNN algorithm is basically the same in the fine-grained data, but MM-CNN has the highest accuracy. Explaining that fine-grained data can express the attributes of a Ponzi scheme. In the coarse-grained data, only SVM has a slight increase in accuracy and the KNN algorithm is lower than SVM, and the MM-CNN model performs worst. It shows that the MM-CNN model is difficult to extract key features from the coarse-grained data. The recall rates of the SVM and KNN in the three frequency data are relatively low, but the recall rate in MM-CNN is the highest, the verification set is 98.07%. In fraud detection, the recall rate should be increased as much as possible under the premise of ensuring the precision rate.

The coarse-grained frequency feature has the worst detection effect in models or algorithms, while the fine-grained frequency feature has a good detection effect. Among the coarse-fine-grained frequency feature, only the MM-CNN model has improved detection results compared to fine-grained data. Therefore, the fine-grained frequency feature highlights the universality.

Sequence and Account Model Analysis. In the sequence feature, only the M-LSTM model was used for experiments. In Table 1, it can be seen from the M-LSTM model that the performance of the model is ideal, and it can identify Ponzi contracts well. It can also reflect the sequence of Ponzi schemes to a certain extent. In the fine-grained sequence feature, the accuracy rate and precision rate are high, all four indicators have reached 95%. Sequence features can be regarded as a reference. In the coarse-grained sequence feature, indicators are not as good as the fine-grained sequence feature. The ability of multi-granularity sequence features to express the Ponzi contract is as follows: The expressive power of fine-grained sequence features is greater than that of coarse-grained sequence features.

In Table 1, the four reference indicators are not high in these algorithms or models. The performance of A-CNN is not outstanding, the account features are less, and it is difficult to accurately reflect Ponzi attributes. The main reason is that the different amounts of investment funds and the different propaganda intensity of smart contracts will lead to large differences in account features. Subjective factors can only reflect the attributes of Ponzi to a small extent. Explaining that account features can only be used as reference features for detecting Ponzi contracts.

6   Summary

We construct features from three modalities, namely the frequency features, sequence features, and account features. When building opcode features, we adopted a multi-granularity construction method to explore how to construct the best feature. When creating the model, the frequency feature uses the MM-CNN model, the sequence model uses the M-LSTM, and the account feature uses the A-CNN. Experiments prove that sequence features and account features are efficient for Ponzi attributes, but the frequency feature detection effect is the best. In the multi-granularity opcode, fine-grained opcode has a stronger ability to express Ponzi attributes.

References

1. Peng, J., Xiao, G.: Detection of smart Ponzi schemes using Opcode. In: Zheng, Z., Dai, H.-N., Fu, X., Chen, B. (eds.) BlockSys 2020. CCIS, vol. 1267, pp. 192–204. Springer, Singapore (2020). https://doi.org/10.1007/978-981-15-9213-3_15
2. Chen, W., Zheng, Z., Cui, J., Ngai, E., Zheng, P., Zhou, Y.: Detecting ponzi schemes on ethereum: Towards healthier blockchain technology. In: Proceedings of the 2018 World Wide Web Conference, pp. 1409–1418 (2018)
3. Wang, L., Cheng, H., Zheng, Z., Yang, A., Zhu, X.: Ponzi scheme detection via oversampling-based long short-term memory for smart contracts. Knowl. Based Syst. **228**, 107312 (2021)
4. Torres, C.F., Steichen, M., et al.: The art of the scam: Demystifying honeypots in ethereum smart contracts. In: 28th {USENIX} Security Symposium ({USENIX} Security 2019), pp. 1591–1607 (2019)
5. Bartoletti, M., Carta, S., Cimoli, T., Saia, R.: Dissecting ponzi schemes on ethereum: identification, analysis, and impact. Futur. Gener. Comput. Syst. **102**, 259–277 (2020)
6. Luu, L., Chu, D.-H., Olickel, H., Saxena, P., Hobor, A.: Making smart contracts smarter. In: Proceedings of the 2016 ACM SIGSAC Conference on Computer and Communications Security, pp. 254–269 (2016)
7. Nikolić, I., Kolluri, A., Sergey, I., Saxena, P., Hobor, A.: Finding the greedy, prodigal, and suicidal contracts at scale. In: Proceedings of the 34th Annual Computer Security Applications Conference, pp. 653–663 (2018)
8. Lu, N., Wang, B., Zhang, Y., Shi, W., Esposito, C.: Neucheck: a more practical ethereum smart contract security analysis tool. Soft. Practice Exp. **51**(10), 2065–2084 (2021)
9. Chen, W., Xu, Y., Zheng, Z., Zhou, Y., Yang, J.Y., Bian, J.: Detecting "pump & dump schemes" on cryptocurrency market using an improved apriori algorithm. In: 2019 IEEE In ternational Conference on Service-Oriented System Engineering (SOSE), pp. 293–2935. IEEE (2019)
10. Jung, E., Tilly, M.L., Gehani, A., Ge, Y.: Data mining-based ethereum fraud detection. In: 2019 IEEE International Conference on Blockchain (Blockchain), pp. 266–273. IEEE (2019)
11. Bartoletti, M., Pes, B., Serusi, S.: Data mining for detecting bitcoin ponzi schemes. In: 2018 Crypto Valley Conference on Blockchain Technology (CVCBT), pp. 75–84. IEEE (2018)

Learning Signed Network Embedding
via Muti-attention Mechanism

Zekun Lu, Qiancheng Yu$^{(\boxtimes)}$, Xiaofeng Wang, and Xiaoning Li

North Minzu University, Yinchuan 750021, China
1999019@num.edu.cn

Abstract. In consideration of most signed network embeddings only focusing on the low-order neighbors of the target node, they fail to make effective use of the high-order neighbors of the target node, and the link direction and sign of the node neighbors will affect the target node to varying degrees. Therefore, a SNEMA model using structure balance theory a multi head attention mechanism to aggregate high-order neighbors is proposed. The model gathers the information of high-order neighbors based on structural balance theory, captures node neighbors of different structure types through a multi head attention mechanism, obtains the low-dimensional feature vector representation of nodes through processing and learning, and applies the obtained network representation to the downstream task of link prediction. The experimental results on four real social network data sets show that the network representation obtained by the SNEMA model helps to improve the accuracy of link prediction, which shows that the SNEMA model has achieved better results in signed network representation learning.

Keywords: Signed network · Network embedding · Balance theory · Attention mechanism · Neighbor aggregation

1 Introduction

There are a large number of network structures in the real world, such as social networks, biological protein networks, citation networks, transportation networks, chemical molecular networks, and so on [1]. These network structures are huge and complex, which implies a lot of rich knowledge. Using complex network analysis methods to study these network structures will help people better mine the laws hidden in network data. Common complex network analysis tasks include: node classification, link prediction, community discovery, influence analysis, network propagation mechanism analysis, network evolution prediction, etc. [2]. In a large number of real networks, there are positive signs and negative signs on the edge of the network. The positive sign indicates a positive relationship, including friendship, support, trust, and so on; Negative signs indicate negative relationships, including: enemy, opposition, distrust, and other relationships; Such networks are defined as signed networks [3].

© The Author(s), under exclusive license to Springer Nature Switzerland AG 2022
Q. Ni and W. Wu (Eds.): AAIM 2022, LNCS 13513, pp. 444–455, 2022.
https://doi.org/10.1007/978-3-031-16081-3_39

Deep learning has made a breakthrough in the fields of natural language processing, machine vision, and so on. Therefore, people extend the deep learning method to graph data learning. The network representation learning based on graph deep learning has become a research hotspot in academic circles. Network representation learning [4] includes node embedding, edge embedding and graph embedding. Node embedding is to gather the characteristics of each node and its neighbor nodes, map the topological proximity to the low-dimensional representation space, and obtain the low-dimensional vector representation of the node. This representation retains the node characteristics and network topology as much as possible.

However, the existing network representation learning still can not mine the signed relationship of nodes and deal with the high-order neighbor information of nodes. Methods such as DeepWalk [5], GAT [6], and GraphSage [7] cannot handle negative links in the network; SNE [8], SINE [9], SIGAT [10] ignore node high-order neighbor information and are not suitable for processing sparse data sets; SGCN [11], SNEA [12] and others ignore the different structure types between node pairs and are not suitable for complex and dense data sets.

2 Preliminaries

In the following, we introduce some necessary definitions and Extended theory facilitate a better understanding of the problem and our proposed solution.

2.1 Structural Balance Theory

Authors should discuss the results and how they can be interpreted from the perspective of previous studies and of the working hypotheses. The findings and their implications should be discussed in the broadest context possible. Future research directions may also be highlighted.

Structural balance theory originated from the balance model of people's attitude towards things proposed by heidery [13]. Cartwright and Harary [14] further extended the theory proposed by hider to signed networks. The "+" and "-" on the edge represent positive and negative relations. At present, the structure balance theory is widely used in the fields of signed network embedding and link prediction algorithms.

(1) Structure balanced triangle: it can be judged by the product of the symbols of the three sides of the triangle: if it is positive, the triangle structure is balanced; Otherwise, the structure is unbalanced. From the perspective of sociology and psychology, the judgment of the structural balance of the above triangle is simply summarized as the following four intuitive understandings: a friend's friend is my friend; The enemy of a friend is my enemy; The enemy's friend is my enemy; The enemy of the enemy is my friend. The research shows that [15] in real signed networks, the number of structurally balanced triangles is much larger than that of structurally unbalanced ones, and the unbalanced network gradually evolves to a balanced network over time.

(2) Structural balance ring: if an L-ring (L ≥ 3) contains an even number of negative edges, the structure is balanced, otherwise the structure is unbalanced (Fig. 1).

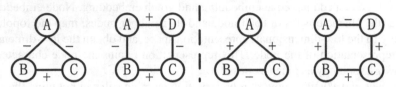

Fig. 1. Schematic diagram of structural balance (left) and structural imbalance (right)

2.2 Social Status Theory

Structural balance theory provides a theoretical basis for the analysis of unsigned networks, but there is a large deviation in directed signed networks. Subsequently, leskovec and Kleinberg [16] proposed a social status theory suitable for signed networks, which holds that if a positive edge points from a to B, a has a higher social status than B. If there is a negative side of a pointing to B, B has a higher social status than a, which is transitive (Fig. 2).

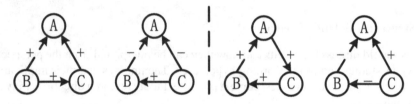

Fig. 2. Status balance (left) and status imbalance (right)

In the signed network composed of three nodes, the method to determine whether a triangle conforms to the social status theory is as follows: first, reverse the direction of all negative links in the triangle, and convert the symbols on the links into positive signs. If the final triangle can not form a loop, then the triangle conforms to the social status theory, otherwise it does not conform to the social status theory. If every member in a system follows the ranking method of the same status without status conflict, the symbol of the edge can be inferred as long as the direction of the edge is known [24–26].

3   Model Introduction

Based on the previous discussion on sociological theory in signed directed networks, we will introduce how to design our new SNEMA model in this section. Our model is divided into three parts: learning first-order neighbor information, aggregating high-order neighbor information and learning to get the final node vector representation, Finally, the quality of the model is tested by the experiment of link prediction.

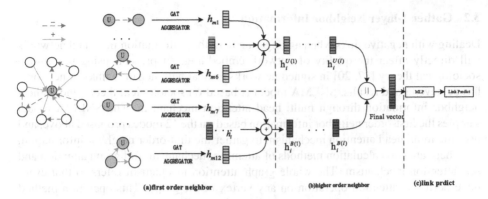

Fig. 3. Schematic diagram of SNEMA framework

3.1 Select Appropriate Node Neighbors

In social networks, the link behavior of user nodes will be affected by neighbor nodes, and different neighbors have different effects on target nodes. Therefore, when aggregating the neighbors of the target node, it is necessary to distinguish the impact of different neighbor types on the target node. According to the status balance theory [20], the proportion of triangular relationship models in line with the status balance theory reaches 80%–85%, which is significantly higher than the model of random combination of relationships. Moreover, the combination that does not comply with the triangular balance relationship is gradually transformed into the combination that complies with the triangular balance relationship with the passage of time, and finally the relationship of triangular combination tends to be stable. Therefore, taking eight patterns in line with the social status theory (the right of Fig. 4) as the sampling pattern of node neighbors can capture the different influence of neighbors on the target node to the greatest extent. In order to preserve the integrity of the target node, the neighbors of the target node that do not conform to the social status theory are classified according to the pattern on the left of Fig. 3. Finally, the neighbors of the target node sample according to the 12 modes proposed in Fig. 3. This sampling mode can capture the contribution of different types of neighbors to the target node.

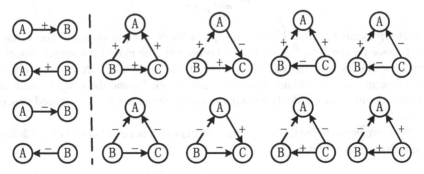

Fig. 4. Neighbor sampling mode

3.2 Gather 1-layer Neighbor Information

Dealing with negative links when aggregating neighbor information in signed networks will directly affect the quality of network embedding. At present, using the unique sociological theory [17, 20] in signed networks to deal with negative links in networks has achieved good results. SNEMA model is based on sociological theory and gathers neighbor information through multi head attention mechanism. The SNEMA model samples the first-order neighbor information based on the 12 modes proposed above, and uses the multi head attention mechanism to gather the first-order neighbor information.

There are two calculation methods of attention mechanism: full graph attention and self attention mechanism. The whole graph attention mechanism refers to that every node performs attention operation on any vertex of the graph. This operation method discards the feature of graph structure, and the effect will be very poor; The self attention mechanism is only carried out in the neighbor nodes, which retains the characteristics of the neighbor nodes to the greatest extent. This paper adopts self attention mechanism to aggregate node neighbors. Because different neighbor types have different effects on the target node, the node neighbors under different structures can be divided into different modes, and multi head attention is used to capture the neighbor types of different target nodes. Fully considering the characteristics of signed networks, the first-order neighbor sampling can be divided into 12 different structures. As shown in Fig. 4 (3), this red dashed arrow and the green solid arrow represent negative links and positive links respectively. Node u represents the target node and gray node represents the neighbors of the target node. For the target node neighbors in different modes, they use different attention to learn to obtain the node vector representation. The neighbors of nodes can be broadly divided into friends and enemies, which have different attribute information. We assign friend aggregators and enemy aggregators to learn the information of friends and enemies of nodes respectively. The first-order neighbor aggregation expression for the target node is as follows:

$$h_i^{B(1)} = \tanh(\frac{1}{M} \sum_{m=1}^{M} \sum_{j \in N_i^+} a_{ij}^m h_j^{(0)} W_m^{B(1)}) \tag{1}$$

$$h_i^{U(1)} = \tanh(\frac{1}{M} \sum_{m=1}^{M} \sum_{j \in N_i^-} a_{ij}^m h_j^{(0)} W_m^{U(1)}) \tag{2}$$

where M represents the total 12 modes, represents the weight attention coefficient of node j to node i under different modes, and represents the matrix parameters under the attention mechanics of balanced structure and unbalanced structure respectively. The values of parameters am and are trained under each mode, and the weight coefficients under different modes are calculated under balanced structure and unbalanced structure respectively.

In the aggregation of the first layer, the importance of node j to node i under different modes is defined as follows:

$$e_{ij}^{B^m} = a^m(h_i^{(0)} W_m^{B(1)}, h_j^{(0)} W_m^{B(1)}, B^m) \tag{3}$$

$$e_{ij}^{U^m} = a^m(h_i^{(0)} W_m^{U(1)}, h_j^{(0)} W_m^{U(1)}, U^m) \qquad (4)$$

The standardized attention coefficient can be calculated by the following formula:

$$a_{ij}^{B^m} = \frac{\exp(e_{ij}^{B^m})}{\sum_{t \in N_i^+} \exp(e_{it}^{B^m})} \qquad (5)$$

$$a_{ij}^{U^m} = \frac{\exp(e_{ij}^{U^m})}{\sum_{t \in N_i^-} \exp(e_{it}^{U^m})} \qquad (6)$$

3.3 Gather High Layer Neighbor Information

The structural balance theory is used to sample the higher-order neighbors of the target (the neighbors of layer L). Similarly, the balanced path (or unbalanced path) is expressed as containing an even (or odd) number of negative links. Therefore, a path from node to is an even (odd) number of negative links. Then the structural balance theory [17] believes that there is a positive (negative) link between node and. The neighbor nodes are aggregated into the balanced structure and the unbalanced structure in turn, and the nodes on each layer are classified into the structure balanced Bi and the unbalanced structure along the balanced (and unbalanced) path until all the nodes on the L layer of the target node are migrated. Figure 5 illustrates the schematic diagram of node i aggregating high-order neighbor information in a symbolic network.

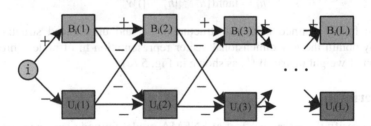

Fig. 5. Aggregate higher-order neighbors

Different from the convergence of the first layer The aggregation of high-level neighbors of the target node is more complex. When gathering high-order neighbors, the balanced structure gathers the ideas of neighbors: Friends of friends are friends and enemies of enemies are friends; The unbalanced structure gathers the thoughts of neighbors: the enemy of friends is the enemy, and the friend of the enemy is the enemy. Using this idea, all neighbor nodes of each layer are sampled recursively, as shown in the following formula:

$$h_i^{B(l)} = \tanh(\frac{1}{M} \sum_{m=1}^{M} \sum_{\substack{j \in N_i^+ \\ k \in N_i^-}} a_{ij}^m h_j^{B(l-1)} W_m^{B(l)} + a_{ik}^m h_k^{(l-1)} W_m^{B(l)}) \qquad (7)$$

$$\mathrm{h}_i^{U(l)} = \tanh(\frac{1}{M}\sum_{m=1}^{M}\sum_{\substack{j \in N_i^+ \\ k \in N_i^-}} a_{ij}^m h_j^{U(l-1)} W_m^{B(l)} + a_{ik}^m h_k^{B(l-1)} W_m^{B(l)}) \tag{8}$$

where $h^{B(l-1)}$ and $h^{U(l-1)}$ represents the vector representation of the balanced and unbalanced structure of the upper layer structure. In order to simplify the formula, ϕ represented $\{B^m, U^m\}$. a_{ij}^σ represents the importance of node i to node j in different modes, and calculates the weight coefficient between different nodes in different modes with formula (13):

$$e_{ij}^{\phi(l)} = a^\phi(h_i^{\phi(l-1)}W^\phi, h_j^{\phi(l-1)}W^\phi, \phi) \tag{9}$$

Calculate the attention coefficient under different modes with formula (14) and normalize it:

$$a_{ij}^\phi = \frac{\exp(e_{ij}^\phi)}{\sum_{\substack{k \in N_i^+ \\ \cup k \in N_i^-}} \exp(e_{ik}^\phi)} \tag{10}$$

For high-order neighbor sampling, based on the principle of structural balance, the target nodes of different modes are studied by attention mechanism with balanced structure and unbalanced structure to learn the shared weight matrix W^ϕ and attention coefficient in each mode. This can better deal with neighbors under different structures. When aggregating the neighbor information of layer L, the node information of the former (L-1) layer is aggregated to the balanced aggregation structure B^m and unbalanced aggregation structure U^m. Finally, the target node vector is expressed as:

$$h_i = \tanh([h_i^{B(l)}||h_i^{U(l)}])W^M \tag{11}$$

Splice the finally learned balanced structure $h_i^{B(l)}$ and unbalanced structure $h_i^{U(l)}$, and finally obtain the low dimensional vector representation hi of node i through the conversion of weight matrix W^M, as shown in Fig. 5 (c).

4   Experiment

In order to verify the learning effect of SNEMA model on node vector representation in signed network, the link prediction experiment is carried out and compared with the typical benchmark method in four real social network data sets to verify the learning effect of SNEMA model on node vector representation. In addition, the experiment is further extended to analyze the influence of super parameter selection on the performance of the model. SNEMA model is based on Python 3 6. The experimental environment is CPU i7-6700, six cores and twelve threads, the memory is 12 gb, and the graphics card is and R7 2 GB.

4.1   Datasets

Experiments in four real signed networks are carried out to verify the authenticity of the proposed framework. The statistical information of the four data sets is shown in Table 1:

Table 1. Dataset statistics

	Node	Positive links	Negative links	Proportion
Bit.Alphs	3775	12721	1399	0.90
Wikipedia	5875	18230	3529	0.83
Slashdot	37626	313543	513851	0.74
Epinions	45003	513851	102180	0.83

4.2 Baselines

- DeepWalk [4]: select the node sequence at the network nodes based on random walk, and then use skip gram model to model the probability of the nodes in the sequence.
- SINE [12]: use the characteristics of signed network to sample and model nodes with random walk method.
- SiGAT [13]: divide the node neighbors into 38 modes and use the multi head attention mechanism to gather the neighbor information.
- SGCN [14]: combine the balance theory to sample high-order neighbors, and use the idea of graph convolution to aggregate node neighbors.
- SNEA[15]: use the attention mechanism to learn different attention weight coefficients between node pairs and aggregate neighbor nodes with different weight coefficients.

4.3 Baseline Model Comparison Experiment

The results of SNEMA method and benchmark method in four real signed network data sets are shown in Table 2 and Table 3. The results show whether the link prediction between predicted users is correct. In this experiment, 80% of the data set is used as the training set and 20% as the test set. The evaluation indexes are AUC and F1 values. In the epinions experiment, the experiment trains 492824 edges and predicts 123207 edges. The AUC value is calculated according to the calculation results. The AUC value of the SNE-MA method proposed in this paper is 0.868, which is about 14%, 9%, 3%, 6% and 2% higher than that of deepwalk, sine, sigat, sgcn and snea respectively. This is because deepwalk is designed for unsigned networks. When the number of negative links is more, the effect is worse, and sigat fails to pay attention to high-order neighbors, Sgcn does not pay attention to the different influence of neighbors on the target node, and snea ignores the different types of neighbors of the node. It can be seen from the comparison that the SNEMA method proposed in this paper has a higher accuracy in the prediction of signed network links, especially in the data set with more nodes and edges. This is because the method proposed in this paper makes good use of the structure balance theory to collect samples of high-order neighbors, uses the multi head attention mechanism to capture neighbor nodes with different structures, and enhances the learning ability of vector representation between node pairs.

Table 2. AUC experimental results

Approach	BitAlphs	Wikipedia	Slashdot	Epinions
DeepWalk	0.780	0.761	0.723	0.720
SiNE	0.782	0.779	0.764	0.789
SiGAT	0.776	0.792	0.781	0.832
SGCN	**0.816**	0.801	0.768	0.802
SNEA	0.806	0.798	0.782	0.840
SNEMA	0.812	**0.813**	**0.791**	**0.868**

Table 3. F1 experimental results

Approach	BitAlphs	Wikipedia	Slashdot	Epinions
DeepWalk	0.813	0.820	0.782	0.780
SiNE	0.866	0.853	0.831	0.898
SiGAT	0.886	0.896	0.851	0.901
SGCN	0.912	0.899	0.849	0.908
SNEA	0.908	0.901	0.859	0.914
SNEMA	**0.918**	**0.911**	**0.862**	**0.928**

4.4 Superparametric Analysis

SNEMA parameters include aggregation layers and vector dimensions. This section mainly verifies the influence of different parameters on the performance of SNEMA through experiments. In the experiment, except for the verification parameters, other parameters are set to the default value.

Figures 6 (a) and 6(b) show that with the increase of training rounds, the loss value gradually decreases, the auc value increases, then gradually converges, and finally tends to be stable. Figure 6 (c) shows the effect of vector dimension on experimental performance. The SNEMA model has a preliminary effect when it is expressed in the 15 dimensional vector. Then, with the increase of the model dimension, the effect gradually increases, and the best effect is achieved in about 20 dimensions, and then tends to be stable. With the increase of dimension, the effect begins to decline. This situation can be understood as the introduction of relevant noise while maintaining more relationship information, so as to reduce the generalization ability.

Figure 6 (d) shows the effect of the number of aggregation neighbor layers L on the experimental performance. It is found that when layer $= 1$, the performance of the algorithm is the lowest, because the target node only aggregates the information of the first-order neighbors. Obviously, the insufficient number of aggregated neighbors will lead to poor results. Using the structure balance theory to aggregate the neighbors of the second layer, it can be seen from the fig that the effect is improved when the nodes

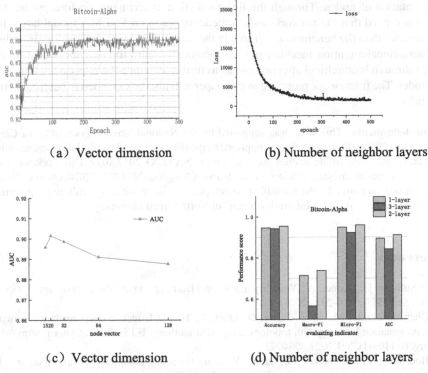

(a) Vector dimension

(b) Number of neighbor layers

(c) Vector dimension

(d) Number of neighbor layers

Fig. 6. Hyperparameter analysis

aggregate the neighbors of the second layer than that of only one layer. This is because the nodes aggregate more neighbor information, which makes the vector representation of the nodes richer, When the number of neighbor layers reaches 2, the effect is the best. When aggregating higher-level neighbors, the experimental performance is reduced, because the higher-level neighbors have less influence on the target node. With the increase of the number of high-level neighbors, the vector features of the nodes are scattered, but the data noise is increased, resulting in the reduction of the experimental effect.

5 Conclusion

This paper proposes a signed network embedding model based on multi head attention mechanism – SNEMA. The framework finds out the positive and negative neighbors of each order of nodes according to the structural balance theory, and comprehensively considers 12 neighbor sampling modes derived from the social status theory. The multi head attention mechanism is used to capture the contribution of neighbors in different modes to node embedding, and the aggregation of high-order neighbor characteristics of nodes is realized by stacking multiple convolution layers, so as to fuse the node's own attributes and each order neighbor attributes to obtain the low-dimensional vector representation of nodes. Through the link prediction experiments on four public data sets, it is proved that the network node embedding based on SNEMA model has better performance than the benchmark method. In the future work: we will consider introducing hierarchical attention mechanism to automatically learn the neighbors of different modes through hierarchical attention, so as to further enhance the expression ability of the model; The framework proposed in this paper will also be extended to heterogeneous networks.

Acknowledgements. This work was supported by the National Science Foundation of China (Grant Nos. 62062001); This work was supported by the Ningxia first-class discipline and scientific research projects (electronic science and technology, NXYLXK2017A07); This work was supported by the Provincial Natural Science Foundation of NingXia (NZ17111, 2020AAC03219) and this work was supported by the scientific research platform of "Digital Agriculture Empowering Ningxia Rural Revitalization Innovation Team" of North Minzu University.

References

1. O'Sullivan, D., Turner, A.: Visibility graphs and landscape visibility analysis. Int. J. Geogr. Inf. Sci. **15**(3), 221–237 (2001)
2. Chen, J., Zhong, M., Li, J., Wang, D., Qian, T., Tu, H.: Effective deep attributed network representation learning with topology adapted smoothing. IEEE Trans. Cybern. https://doi.org/10.1109/TCYB.2021.3064092
3. Boccaletti, S., Ivanchenko, M., Latora, V., et al.: Detecting complex network modularity by dynamical clustering. Phys. Rev. E **75**(4), 045102 (2007)
4. Tang, J., Chang, Y., Aggarwal, C., Liu, H.: A survey of signed network mining in social media. ACM Comput. Surv. (CSUR) **49**(3), 42 (2016)
5. Zhang, Z., Cui, P., Zhu, W.: Deep learning on graphs: a survey. IEEE Trans. Knowl. Data Eng. (2020)
6. Zhang, D., Yin, J., Zhu, X., et al.: Network representation learning: a survey. IEEE Trans. Big Data **6**(1), 3–28 (2018)
7. Zhou, J., Liu, L., Wei, W., et al.: Network representation learning: from preprocessing, feature extraction to node embedding. ACM Comput. Surv. (CSUR) **55**(2), 1–35 (2022)
8. Perozzi, B., Al-Rfou, R., Skiena, S.: DeepWalk: online learning of social representations, pp. 701–710. ACM Press (2014). ACMSI GKDD, 855–864 (2016)
9. Veličković, P., Cucurull, G., Casanova, A., et al.: Graph attention networks. arXiv preprint arXiv:1710.10903 (2017)
10. Hamilton, W., Ying, Z., Leskovec, J.: Inductive representation learning on large graphs. In: Advances in Neural Information Processing Systems, pp. 1025–1035 (2017)
11. Yuan, S., Wu, X., Xiang, Y.: SNE: signed network embedding. In: Kim, J., Shim, K., Cao, L., Lee, JG., Lin, X., Moon, YS. (eds.) PAKDD 2017. LNCS, vol. 10235, pp. 183–195. Springer, Cham (2017). https://doi.org/10.1007/978-3-319-57529-2_15
12. Wang, S., Tang, J., Aggarwal, C., et al.: Signed network embedding in social media. In: Proceedings of SIAM International Conference on Data Mining, pp. 327–335. SIAM Press, Houston (2017)
13. Heider, F.: Attitudes and cognitive organization. J. Psychol. **21**(1), 107–112 (1946)
14. Leskovec, J., Huttenlocher, D., Kleinberg, J.: Signed networks in social media. In: Proceedings of SIGCHI Conference on Human Factors in Computing Systems, pp. 1361–1370. ACM Press, New York (2010)

15. Girdhar, N., Bharadwaj, K.K.: Signed social networks: a survey. In: Singh, M., Gupta, P., Tyagi, V., Sharma, A., Ören, T., Grosky, W. (eds.) ICACDS 2016. CCIS, vol. 721, pp. 326–335. Springer, Singapore (2017). https://doi.org/10.1007/978-981-10-5427-3_35

16. Leskovec, J., Huttenlocher, D., Kleinberg, J.: Predicting positive and negative links in online social networks. In: Proceedings of the 19th International Conference on World Wide Web, pp. 641–650. ACM Press, New York (2010)

17. Chiang, K.Y., Natarajan, N., Tewari, A., et al.: Exploiting longer cycles for link prediction in signed networks. In: Proceedings of the 20th ACM International Conference on Information and Knowledge Management, pp. 1157–1162 (2011)

18. Vaswani, A., Shazeer, N., Parmar, N., et al.: Attention is all you need. In: Advances in Neural Information Processing Systems, vol. 30 (2017)

Three Algorithms for Converting Control Flow Statements from Python to XD-M

Jiarui Wang, Nan Zhang$^{(\boxtimes)}$, and Zhenhua Duan$^{(\boxtimes)}$

Institute of Computing Theory and Technology, and ISN Laboratory,
Xidian University, Xi'an 710071, China
`wangjiarui@stu.xidian.edu.cn`, `nanzhang@xidian.edu.cn`,
`zhhduan@mail.xidian.edu.cn`

Abstract. This paper presents an approach to show how to implement three complex statements: *continue, break, return* of Python to XD-M language. To this end, three algorithms for implementing three complex statements are given in detail. Further, three complex statements are implemented in XD-M. Finally, a program example is presented to illustrate how to use the proposed approach to build an XD-M program equivalent to the original Python program.

Keywords: Programming language · Complex control flow · Continue · Break · Return

1 Introduction

Modeling, Simulation and Verification Language (MSVL) is developed from tempol logic [9,10,14]. It is a logic programming language that can model, simulate and verify software and hardware systems. The underlying logic is Projection Tempol Logic(PTL) [18,20]. MSVL can be used for three purposes: (1)Modeling: Usually MSVL can model complex software and hardware systems, as well as multi-core parallel systems [19]; (2)Simulation: MSVL programs can also be executed like imperative programming languages such as C, C++, and Java; (3)Verification: More importantly, MSVL programs can be used to verify large-scale software and hardware systems [6,11]. It provides a new way of program verification, allowing developers to find and correct software and hardware bugs efficiently.

XD-M is a Python-like language. It is based on MSVL, and therefore, inherits advantages and features of MSVL. Basically it is designed to be an easy-to-use language and facilitated mathematical analysis and reasoning as well as

The research is supported by the National Key Research and Development Program of China (2018AAA0103202); National Natural Science Foundation of China (62172322, 61751207, 61732013); Shannxi Key Science and Technology Innovation Team Project (2019TD-001); Natural Science Basic Research Program of Shaanxi Province (2022JM-367).

Q. Ni and W. Wu (Eds.): AAIM 2022, LNCS 13513, pp. 456–465, 2022.
https://doi.org/10.1007/978-3-031-16081-3_40

implemented by more functionalities [8]. XD-M contains commonly used data types and language structures in Python, and XD-M is also an interpreted language [17].

Python is an interpreted, high-level and general-purpose programming language [3]. It is widely used in artificial intelligence and other scientific computing fields. The ever-increasing demand for Python programs means that their scales and structures become more and more complex. To ensure the correctness of Python programs, it is necessary to find out an effective and efficient way to verify Python programs. One available approach is to convert Python programs to XD-M programs so that Python programs can be formally verified using skills with XD-M.

There are three main complex control flows in Python: *continue* and *break* statements in loops, and *return* statement in a function definition [4,12]. The statements in XD-M have timing characteristics, so they are executed according to the state and interval of each statement. It is not able to support the jump execution of programs, i.e. *continue*, *break* and *return* statements cannot be implemented directly.

Therefore, we are required to use existing statements in XD-M language, such as sequential, conditional and loop statements, to implement the equivalent logic of *continue*, *break*, and *return* statements.

This paper presents an approach to show how to implement three complex statements: *continue*, *break*, *return* of Python in XD-M language. To this end, three algorithms for implementing three complex statements are given in detail. Further, three complex statements are implemented in XD-M. Finally, a program example is presented to illustrate how to use the proposed approach to build an XD-M program equivalent to the original Python program.

This paper is organized as follows. In the next section, we introduce the equivalent logic implementation of *continue*, *break* and *return* statements in XD-M. In Sect. 3, We present the validation results of the proposed approach and show one of the test programs. In Sect. 4, conclusions are drawn.

2   Converting Algorithms of Complex Control Flow

2.1   The Algorithm of Converting *continue* Statement

The *continue* statement is used to tell the program to skip the remaining statements in the current loop block and continue with the next loop [7]. The left-hand side of Fig. 1 shows the execution flow of a Python program that includes a *continue* statement. First of all, when the Python program executes the *continue* statement, it skips the remaining statements in the loop body, backs to the loop start, and performs the next loop. In the case where the loop has an *else* block, if the *while* loop is not terminated by *break* statement, after running all the loops, the program executes corresponding *else* block, and finally ends the execution of the entire *while* block.

To implement the function of *continue* statement equivalently in XD-M, we made some modifications to the original program structure. First of all, we

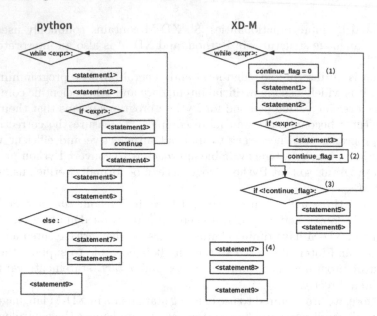

Fig. 1. An example of *continue* statement in Python and its equivalent implementation of XD-M

noticed that the Python program will return to the beginning of the loop after *continue* statement, and it will always be executed in the program order when it does not encounter *continue* statement. It can be seen that the program has two different execution states to indicate whether it encounters *continue* statement or not. To distinguish between these two states, we added a state machine called **continue_flag** to record the state of the program. We define that the state of **continue_flag** is 1 after the program encounters the *continue* statement, and 0 otherwise. The initial state of **continue_flag** is 0, and it will be reinitialized each time it goes back to the beginning of the loop. We divide the program into several parts and sort out the differences in the behavior of the program in different states, as shown in Table 1.

The right-hand side of Fig. 1 shows the equivalent implementation of the *continue* logic and its execution flow in the XD-M program. We deal with the codes of the six parts separately:

1. The first part of the codes will execute normally, but we need to initialize the **continue_flag** in time, so we added a statement that assigns the **continue_flag** to 0 at the beginning of the loop. As shown (1) in Fig. 1.
2. We do not make any changes to the part of the *if* statement that contains the *continue* statement before the *continue* statement.
3. We convert the *continue* statement to an assign statement that assigns **continue_flag** to 1. As shown (2) in Fig. 1.

Table 1. Behavior of the program in different states.

Program location	Program running behavior in continue_flag=0	Program running behavior in Continue_flag=1
Between the beginning of the loop and the *if* statement containing the *continue* statement	continue_flag must be 0, execution	continue_flag cannot be 1
The part before *continue* statement in the *if* statement that contains *continue* statement	continue_flag must be 0, and the code executes according to the if-conditions	continue_flag cannot be 1
Continue statement	State transition from 0 to 1	continue_flag is 1
The part after *continue* statement in the *if* statement that contains *continue* statement	continue_flag cannot be 0	continue_flag must be 1, no execution
From the next statement of the *if* statement containing *continue* statement to the end of the loop	Execution	No execution
Else block	Execution	Execution

4. The part after *continue* statement in the *if* statement containing *continue* statement, which is **statement4** in Fig. 1, will never be executed, so they are deleted directly.
5. For the part of the codes from the next statement of the *if* statement containing *continue* statement to the end of the loop, which is **statement5** and **statement6** in Fig. 1, we put this part of the statements into an *if* condition statement to judge the state of **continue_flag** at this time, **continue_flag**= 0 means normal execution, otherwise, no execution. As shown (3) in Fig. 1.
6. The *else* block, which is **statement7** and **statement8** in Fig. 1, will always execute regardless of the state, so we add all statements contained in the *else* block after the modified *while* statement. As shown (4) in Fig. 1.

2.2 The Algorithm of Converting *break* Statement

The *break* statement is used to break out of the loop body [5,15]. If the loop is terminated by a *break* statement, then any code in that loop block will not be executed. The left-hand side of Fig. 2 shows the execution flow of a Python program that includes a *break* statement. When the Python program executes the *break* statement, it jumps out of the loop body and jumps out of the *else* block corresponding to the loop. When the loop is not terminated by *break* statement, the *else* block corresponding to the loop executes normally.

Similar to the *continue* statement, there a state machine is also needed to check whether the *break* statement has been performed. We added a state

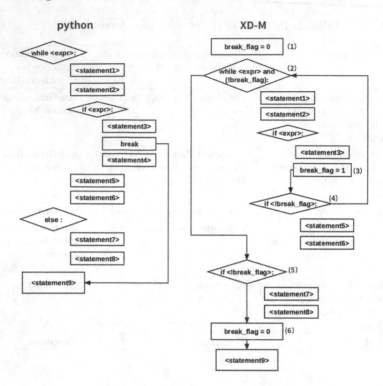

Fig. 2. An example of *break* statement in Python and its equivalent implementation of XD-M

machine called **break_flag** to record the state of the program. We define that the state of **break_flag** is 1 after the program encounters the *break* statement, and 0 otherwise. The initial state of **break_flag** is 0, and it needs to be initialized at both starting and ending of the loop, as shown (1) and (6) in Fig. 2, which will keep the program be correct in nested loops. We sort out the differences in the behavior of the program in different states, as listed in Table 2.

The right-hand side of Fig. 2 shows the equivalent way to implement the *break* logic and its execution flow in the XD-M program. Each segment of the code is discussed separately:

1. We found that as long as the program is in the state of **break_flag**=1, the entire loop body (including the corresponding *else* block) will not be executed. So we added a condition that negates the **break_flag** at the loop condition judgment so that when the **break_flag**=1, the loop will not be executed. We added a statement that assigns **break_flag** to 0 before entering the loop. We made this change for two reasons: 1) If it is not initialized before the loop body, the loop condition cannot be checked; 2) The implementation logic does not need to repeatedly initialize the **break_flag**. As shown (1) and (2) in Fig. 2.

Table 2. Behavior of the program in different states.

Program location	Program running behavior in **break_flag=0**	Program running behavior in **break_flag=1**
Between the beginning of the loop and the *if* statement containing the *break* statement	Execution	No execution
The part before *break* statement in the *if* statement that contains *break* statement	Execute according to the if-condition	No execution
break statement	State transition from 0 to 1	**break_flag** is 1
The part after *break* statement in the *if* statement that contains *break* statement	**break_flag** cannot be 0	**break_flag** must be 1, no execution
From the next statement of the *if* statement containing *break* statement to the end of the loop	Execution	No execution
Else block	Execution	No execution

2. We do not make any changes to the part of the *if* statement that contains the *break* statement before the *break* statement.
3. We convert the *break* statement into an assign statement that assigns **break_flag** to 1. As shown (3) in Fig. 2.
4. We deleted the code after the *break* statement in the *if* statement containing the *break* statement i.e. **statement4** in Fig. 2.
5. For the part of the codes from the next statement of the *if* statement containing *break* statement to the end of the loop, which is **statement5** and **statement6** in Fig. 2, we put this part of the statements into an *if* block. **break_flag** = 0 means normal execution, otherwise, no execution. As shown (4) in Fig. 2.
6. When the loop is terminated due to the execution of the *break* statement, the *else* block is not executed, whereas when the *break* statement is not executed, the loop exits normally, and the *else* statement block is executed normally. So we put *else* block, which is **statement7** and **statement8** in Fig. 2, into an *if* condition statement. If **break_flag**=0, it will be executed normally, otherwise it will not be executed. As shown (5) in Fig. 2.

2.3   The Algorithm of Converting *return* Statement

The *return* statement is used to exit the function [13,16]. The left hand side of Fig. 3 shows the execution flow of a Python program that includes a *return* statement. The Python program will quit the function when the *return* statement is executed. Similar to *continue* and *break* statements, the program's behavior is

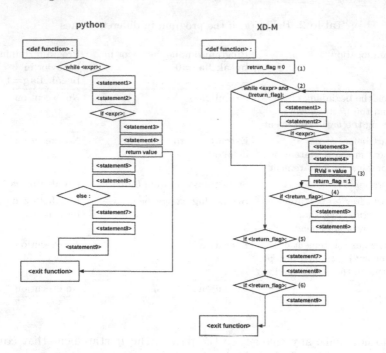

Fig. 3. An example of *return* statement in Python and its equivalent implementation of XD-M

related to whether the *return* statement has executed. We add a state machine named **return_flag**, and initialize it to 0 when entering the function body, as shown (1) in Fig. 3. We divide the program into several parts, as listed in Table 3.

The right hand side of Fig. 3 shows the equivalent implementation of the *return* logic and its execution flow in XD-M program. We deal with the codes of the six parts separately:

1. We add a condition th—at negates the **return_flag** at the loop condition checking, so that when the **return_flag**=1, the loop will not be executed. As shown (2) in Fig. 3.
2. We do not make any changes to the part of the *if* statement containing *return* statement before the *return* statement.
3. We record the return value of the function, and convert the *return* statement into an assign statement that assigns **return_flag** to 1. As shown (3) in Fig. 3.
4. For the part of the codes from the next statement of the *if* statement containing *return* statement to the end of the loop, which is **statement5** and **statement6** in Fig. 3, we put this part of the program into the condition statement, so that the operation of jump out of the loop is realized. As shown (4) in Fig. 3.

Table 3. Behavior of the program in different states.

Program location	Program running behavior in return_flag=0	Program running behavior in return_flag=1
From the beginning of the function to the *if* statement containing the *return* statement	Execution	No execution
The part before *return* statement in an *if* statement containing *return* statement	Execute according to the if-condition	No execution
return statement	Record the return value of the function, and state transition from 0 to 1	**return_flag** is 1
From the next statement of the *if* statement containing *return* statement to the end of the loop	Execution	No execution
Else bolck	Execution	No execution
From next statement of the *else* block to the end of the function	Execution	No execution

5. we put the *else* block corresponding to the loop, which is **statement7** and **statement8** in Fig. 3, into the condition statement, so that the operation of jump out of the *else* block is realized. As shown (5) in Fig. 3.
6. For the part of the program from the next statement of the *else* block to the end of the function, which is **statement9** in Fig. 3, we also put it into the condition statement, so that the operation of exiting the function is realized. As shown (6) in Fig. 3.

2.4 Put Them All Together

When three statements of *continue*, *break* and *return*, or two of them appear in an XD-M program at the same time, the conversion approach is the same as the one presented before. The state machine of the statement is initialized, the position of each state machine remains unchanged, and the changes of the respective state machines are set at the positions where the *continue*, *break* and *return* statements appear. Finally, we add the state detection of their respective state machines to the statements that might be skipped according to the top-down sequence of the program. There are two points we need to address:

– If the same segment of the program may be skipped by two or three complex control flow statements, then the state checking conditions of the respective state machines should be added to the segment at the same time, for example, the program certain piece of codes may be skipped by *break* statement or *return* statement, so it is necessary to add (**!break_flag and !return_flag**) judgment to this code. Accordingly, it can skip the code when the program executes *break* statement without executing *return* statement, or executes *return* statement without executing *break* statement.

– The state machine **break_flag** needs to be reset to 0 when exiting the loop. This is because in the nested loops, when the state machine of the inner loop is used and set as 1, and it is not reset to 0 in time, it will affect the jump logic of the outer loop body. This may cause the program to skip some code, which should be executed, resulting in a program error.

3   Verification

3.1   Test Environment

We developed a tool called P2M, which can be used to convert a Python program to a corresponding XD-M program in an automatical way. With P2M, we can check whether an XD-M program converted from a Python program containing three complex control flows is equivalent to the corresponding Python program.

To do so, we can print out the running results of the two programs at the same time. In this way, we can perform a large number of tests. The experimental results show that our approach is sound.

We use the AST library to finish syntax analysis of Python program [2]. The AST module helps Python applications process trees of the Python abstract syntax grammar. We use the *ast.parse()* of the library to generate the abstract syntax tree of the Python program, and use *ast.NodeTransformer* class to reconstruct the abstract syntax tree by defining the visitor method in this class. Finally, we use *ast.NodeVisitor* class to generate the corresponding XD-M code, by redefining the visitor method in this class and implementing the conversion scheme proposed in this paper.

3.2   Test Results

We use the aforementioned platform to test our approach. We constructed 60 Python programs with complex control flows, they have nested loops, nested *if* statements, and various combinations of *continue*, *break* and *return* statements. We list one of the more complex test programs in our repository [1]. This program includes loop nesting, *if* nesting, *if* and loop nesting, and three statements of *continue*, *break* and *return* appearing in one program at the same time. The test results of these programs show that the method proposed in this paper can construct programs with the same function as complex control flows and get the same running results. The proposed approach in this paper can be used not only with XD-M language but also with other programming languages without *continue*, *break* and *return* statements.

4   Conclusion

This paper proposes three algorithms which can be used to implement three complex control flow statements, *continue*, *break* and *return* in XD-M programs. The proposed approach has been realized in a tool P2M. This enables us to convert a Python program into an XD-M program in an automatical manner. However, the current version of P2M is only a draft version. It needs a lot of tests in the future. Furthermore, we need also to build a library of functions for XD-M.

References

1. P2M: A Python to XD-M translator. https://github.com/FairyJiar/P2M
2. Python AST Document. https://docs.python.org/3/library/ast.html
3. Python Homepage. https://www.python.org/
4. Allen, F.E.: Control flow analysis. ACM SIGPLAN Notices 5(7), 1–19 (1970)
5. Baker, B.S., Kosaraju, S.R.: A comparison of multilevel break and next statements. J. ACM (JACM) **26**(3), 555–566 (1979)
6. Bowman, H., Thompson, S.: A decision procedure and complete axiomatization of finite interval temporal logic with projection. J. Logic Comput. **13**(2), 195–239 (2003)
7. Chen, W.: Loop invariance with break and continue. Sci. Comput. Program. **209**, 102679 (2021)
8. Duan, Z.: Temporal Logic and Temporal Logic Programming. Science Press, Beijing (2005)
9. Duan, Z., Tian, C.: A practical decision procedure for propositional projection temporal logic with infinite models. Theoret. Comput. Sci. **554**, 169–190 (2014)
10. Duan, Z., Tian, C., Zhang, N.: A canonical form based decision procedure and model checking approach for propositional projection temporal logic. Theoret. Comput. Sci. **609**, 544–560 (2016)
11. Duan, Z., Yang, X., Koutny, M.: Framed temporal logic programming. Sci. Comput. Program. **70**(1), 31–61 (2008)
12. Hammond, M., Robinson, A.: Python Programming on Win32: Help for Windows Programmers. O'Reilly Media, Inc., Sebastapol (2000)
13. Lomet, D.B.: Control structures and the RETURN statement. In: Watson T.J., (ed.) IBM Research Division (1973)
14. Rosner, R., Pnueli, A.: A choppy logic. Weizmann Institute of Science, Department of Applied Mathematics (1986)
15. Sorva, J., Vihavainen, A.: Break statement considered. ACM Inroads **7**(3), 36–41 (2016)
16. Taft, S.T.: Implementing the extended return statement for ADA 2005. In: Proceedings of the 2008 ACM Annual International Conference on SIGAda Annual International Conference, pp. 97–104 (2008)
17. Tian, C., Duan, Z., Duan, Z.: Making CEGAR more efficient in software model checking. IEEE Trans. Softw. Eng. **40**(12), 1206–1223 (2014)
18. Wang, X., Tian, C., Duan, Z., Zhao, L.: MSVL: a typed language for temporal logic programming. Front. Comput. Sci. **11**(5), 762–785 (2017)
19. Zhang, N., Duan, Z., Tian, C.: A mechanism of function calls in MSVL. Theoret. Comput. Sci. **654**, 11–25 (2016)
20. Zhang, N., Duan, Z., Tian, C.: Model checking concurrent systems with MSVL. Sci. China Inf. Sci. **59**(11), 1–3 (2016)

Hyperspectral Image Reconstruction for SD-CASSI Systems Based on Residual Attention Network

Haobin Luo[✉], Guowei Su, Yi Wang, Jiajia Zhang, and Luobing Dong

Xidian University, No. 2 South Taibai Road, Xi'an 710071, Shaanxi, China
luo_haobin@stu.xidian.edu.cn

Abstract. Hyperspectral images contain both spatial and spectral information, which can be utilized to material identification. Therefore, they find significant advantages in object detection. Hyperspectral images are also believed to play an important part in geological survey and material classification. As the resolution of hyperspectral images increases, compressed sensing (CS) is proposed to reduce the data size, resulting in lower system latency. However, images after CS require reconstruction for further applications such as object detection. The idea of numerical optimization is adopted by conventional reconstruction algorithms. However, these algorithms are time-consuming in iteration. The efficiency and resulting image quality are also not satisfying. Therefore, deep neural networks (DNN) are expected to make better reconstruction algorithms. This paper proposes a novel reconstruction algorithm for hyperspectral images based on deep learning. The core idea is to apply a residual attention network. Firstly, convolution layers of different reception fields are applied to extract different features in hyperspectral images. Then the residual attention blocks satisfying the channel attention mechanism explore the inter-spectral correlation of hyperspectral images. Our proposed reconstruction model is tested to be effective and efficiency in experiments. Compared to three conventional algorithms, OMP, TwIST and GPSR, the proposed algorithm improves PSNR by over 8 db and reconstruction speed by 7 times. Moreover, the model achieves better reconstruction performance compared to a DNN-based model DNNnet.

Keywords: Deep neural networks · Hyperspectral image reconstruction · Attention

1 Introduction

Hyperspectral images refer to spectral images with resolution ranging within 10 nm, providing spatial and spectral information [1]. For a pixel in the hyperspectral space, the spectrum curve can be obtained by combining hyperspectral images of different wavelengths, which is useful in analyzing the components of a specific material. As a result, hyperspectral images are widely used in object detection, material identification, as well as surface identification, atmosphere components and so on. With the development of hyperspectral technologies, the resolution of hyperspectral images increases. However, the resulting explosion in data size not only challenges the storage system, but also requires high data throughput. Image compression at the terminals does not solve the challenge thoroughly. Compression during data collection is a better idea.

According to the Nyquist theorem [4], the minimum sample rate should be at least twice the highest frequency component. However, compressed sensing, proposed by Donoho [2], Candes, Romberg and Tao [3], proves the data can be sampled at a rate far less than the sample rate required in the Nyquist Theorem. Then reconstruction algorithms are applied to restore the original signals. By doing so, there is a significant drop in data size. CS provides new solutions for the hyperspectral compression and reconstruction.

Based on CS, coded aperture hyperspectral imaging is put forward to realize data compression at the front-end, which is significant for developing hyperspectral technologies. In practice, there are three approaches, single exposure, multiple exposure and push-broom [5], in coded aperture imaging. For single exposure, only one compressive sensed image is required, but the low data size makes it impractical for reliable reconstruction. Multiple exposure samples several times by the coded aperture mobile imaging system, requiring no relative motion between the imaging system and the object scenario. This approach is only applicable in limited scenarios. Push-broom approach iterates the coded aperture on mobile imaging system line-by-line or column-by-column, which increases the sampled data size and satisfies the requirements of aerospace imaging. Importantly, the quality of resulting image depends heavily on the CS reconstruction algorithm. In recent years, algorithms based on convex optimization have been proposed to reconstruct images through iterations. However, these algorithms are time-consuming and unable to deliver satisfying reconstructed images in all scenarios. This paper also studies how to reconstruct hyperspectral image effectively and efficiently.

With the increasing capability of parallel computing, DNN develops rapidly and plays an important role in computer vision (CV). The underlying features can be learned in the training stage. Both timeliness and effectiveness are guaranteed by using pretrained model for inference. In fields like object detection and image classification, DNN outperforms other algorithms. In recent years, hyperspectral images have been applied in CV to implement tasks including denoising [6], image classification [7] and object detection [8]. Nevertheless, as for hyperspectral image reconstruction based on CS, existing work does not take noise into consideration. And the inter-spectral correlation lacks attention.

In this paper, a novel reconstruction model based on residual attention network is proposed for hyperspectral images. Section 2 introduces existing work regarding reconstruction algorithms for hyperspectral images based on conventional approaches and DNN. Section 3 analyzes the design ideas and proposes a novel reconstruction model based on channel attention mechanisms. Experiment setup and results are specified in Sects. 4 and 5 respectively.

2 Related Work

The compressed sensing theory makes it possible for hyperspectral imaging systems to operate at a much lower sample rate, which helps hyperspectral imaging system in saving storage effectively. However, the image requires reconstruction and the reconstructed image quality heavily relies on reconstruction algorithm. Classical reconstruction approaches mainly adopt iterations to solve convex optimization problems. OMP [9], TwIST [10] and GPSR [11] are three representative solutions. In recent years, Wang et al. combined characteristics in space dimension and inter-spectral dimension to reconstruct hyperspectral images.

In the past few years, DNN is widely used in reconstruction algorithms for hyperspectral images. Wang et al. [12] proposed HyperReconNet for dual-dispatchers coded aperture snapshot spectral imaging (DD-CASSI) systems instead of single-dispatchers coded aperture snapshot spectral imaging (SD-CASSI) systems. To improve the reconstruction precision, coded apertures are optimized. They design the repetitive pattern based on forward modelling. The inner characteristics and correlation in the spectral dimension are also considered.

Jiang et al. [13] put forward DNNnet to reconstruct hyperspectral images based on DNN for DD-CASSI systems. The model includes an initial convolution layer and multiple residual blocks. The initial prediction is firstly obtained by preprocessing. Then the feature map is learned by the neural network during training. After training, the model is applied to reconstruct hyperspectral images.

Combining neural networks and classical approaches, Choi et al. [14] contributed a reconstruction model based on the prior knowledge on spectrum. The algorithm constructs a convolutional self-encoder for learning nonlinear spectral representations in real hyperspectral data that allows reconstructing its own input through its encoder and decoder network. When reconstructing, the method combines the fidelity of the regularized learned nonlinear spectral representation and the gradient in the spatial domain through a new fidelity prior sparsity of the gradient in the spatial domain.

Miao et al. [15] proposed a two-stage reconstruction model Lambda-Net. The coded aperture and compressive sampled values are required as inputs to the neural network. The first stage consists of a generative adversarial network (GAN) based on a self-attentive mechanism as well as a hierarchical channel reconstruction strategy. Hyperspectral 3D data are reconstructed in this stage. In the second stage, each channel of the hyperspectral image is refined.

Although DNN reaches state-of-the-art reconstruction performance for hyperspectral images, existing researches mainly focused on single exposure mode on DD-CASSI systems. The influence of noise has not been taken into consideration. Because SD-CASSI differs much from DD-CASSI, most reconstruction algorithms for DD-CASSI systems do not work on SD-CASSI systems [18]. Push-broom mode is widely used in aerospace and SD-CASSI systems are easier to implement. Therefore, it is of significance to develop reconstruction algorithms on such platforms.

3 Proposed Method

3.1 Network Design

The overall structure of the proposed residual attention network is shown in Fig. 1. The input of the network is the initial prediction obtained by the inverse transform. The network mainly consists of an initial convolution block and residual attention blocks. The initial convolution block is responsible for extracting the spatial and spectral information of the image, while the residual attention blocks are responsible for reconstructing the original hyperspectral image.

Figure 2 shows the initial convolution block. To extract different features from the input initial predictions, convolution kernels of sizes 3 * 3, 5 * 5 and 7 * 7 are applied. The convolution kernels provide various perceptual fields.

The structure of residual attention blocks is illustrated in Fig. 3. To prevent performance degradation resulted from gradient explosion or decay, a jump connection is introduced to fuse the initial feature map with the output of sigmoid function. Moreover, the real-time ability is improved and the training time is reduced with the presence of the jump connection.

In hyperspectral images, spectra contain different features. A channel attention module is added in the residual attention blocks to emphasize on such features, which assigns attention weights in the channel dimension to the feature maps. High weights will be assigned to key regions, therefore the feature information is amplified. In contrast, feature information in regions with low weights will be suppressed. In the channel attention module, the maximum pooling and the average pooling are performed respectively. Their results are passed through both upsampling and downsampling layer with the downsampling ratio t, and then served as inputs to the sigmoid activation function. The result of the sigmoid function acts as the adaptive weights, as described in Eq. (1), where Y is the input feature map. W_U and W_D denote the weights of upsampling layer and downsampling layer respectively. $\sigma(\cdot)$ and $\delta(\cdot)$ represent sigmoid and ReLU activation function respectively. AP stands for average pooling while MP stands for max pooling.

$$Q = \sigma(W_U \times \delta(W_D \times AP(Y)) + W_U \times \delta(W_D \times MP(Y))) \quad (1)$$

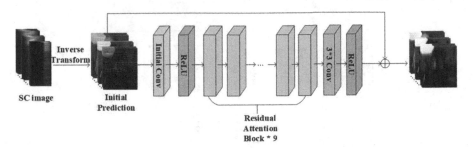

Fig. 1. Structure of the proposed neural network

In m^{th} residual attention block, the feature map is calculated according to Eq. (2), where Q is obtained by equation. W_1, W_2 and W_3 denote the weights of three convolution layers. F_{m-1} is the input feature map and F_m is the output feature map.

$$F_m = Q \times W_3 \times \delta(W_2 \times \delta(W_1 \times F_{m-1})) + F_{m-1} \tag{2}$$

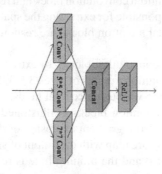

Fig. 2. Structure of initial convolution block

3.2 Loss Function

In regression tasks, the most common loss functions are L1 Loss, L2 Loss and SmoothL1 Loss. SmoothL1 Loss converges faster than L1 Loss and it is not sensitive to outliers or abnormal values compared to L2 Loss. Consequently, SmoothL1 Loss is adopted in this paper.

$$Loss(\theta) = \frac{1}{N} \sum_{i=1}^{N} \begin{cases} 0.5 \times (Y_i - Net(X_i|\theta))^2, & if\ |Y_i - Net(X_i|\theta)| < 1 \\ |Y_i - Net(X_i|\theta)| - 0.5, & otherwise \end{cases} \tag{3}$$

As shown in Eq. (3). Y_i denotes the original hyperspectral image and X_i denotes the image input to the network. $Net(\cdot)$ represents the network model and θ stands for the model parameters learned by the neural network. When the prediction is close to the ground truth, the loss function is calculated based on squared error, while absolute error is applied in other cases.

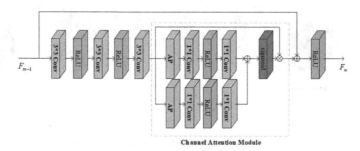

Fig. 3. Structure of the residual attention block

4 Experiment Setup

The hyperspectral imaging system based on compression sensing used in this paper is SD-CASSI as shown in Fig. 4. The system mainly consists of coded aperture, dispersive element and sensor imaging. When imaging, the object image is firstly modulated by the coded aperture. The modulated image is then scattered by the dispersive element. The spectral aliased picture is finally formed on the sensor. With given object image size M by N and the number of spectra L, and the imaging interval between neighboring spectra 1 pixel unit on the sensor, the size of aliased spectrum image is M*(N + L−1).

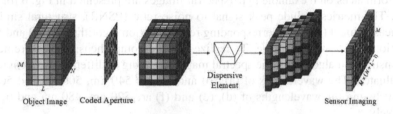

Fig. 4. The imaging process of SD-CASSI

The experiment is conducted on images in mat format from the open ICVL [16] hyperspectral image dataset. The size of these images is 1392-by-1300. The spectrum ranges from 400 nm to 700 nm, with an interval of 10 nm, making up 31 spectra, as shown in Fig. 5. Redundant images from duplicate scenarios are removed. The rest images are downsampled with padding 128, resulting in hyperspectral pictures in 256 * 256 * 31. The images are further normalized and sampled by push-broom simulation, introducing Poisson noises to the images. The initial predictions are then calculated using proposed theory. 80% of initial predictions are randomly selected as the train set and the rest ones serve as test set.

Fig. 5. Example images of ICVL dataset

The training of the model is realized on PyTorch [17] running on RTX3090. The downsampling ratio is set to 8. The initial learning rate is 0.0001. The model experiences 100 epoches with batch size 4, resulting in the total iteration time 400000. Adam optimizer is applied, the hyperparameters are $\beta_1 = 0.9$, $\beta_2 = 0.999$, $\epsilon = 1e - 5$ and *weight_decay* $= 0$ respectively. The performances of popular reconstruction algorithms OMP [9], GPSR [11], TwIST [10] and DNNnet [12] are also evaluated as comparison.

5 Result and Discussion

The performances on 6 example hyperspectral images are presented in Fig. 6 for comparison. The metrics include peak signal to noise ratio (PSNR), structural similarity index measurement (SSIM). Corresponding reconstruction algorithm, PSNR and SSIM are captioned below each image. To realize a more comprehensive measurement of the reconstruction algorithms, the spectral maps belonging to different wavelengths are also evaluated. The wavelengths of (a), (b) and (c) are 540 nm, 500 nm and 560 nm respectively and the wavelengths of (d), (e) and (f) are 520 nm, 480 nm and 620 nm respectively.

Figure 6 indicate that our proposed algorithm delivers reconstructed images of better quality, achieving higher PSNR and SSIM than the other 4 algorithms. These widely adopted algorithms introduce noise during reconstruction to different extents. Particularly, there are considerable noises produced by TwIST as well as noticeable loss in local features. By observing the boxed areas in Fig. 6, the proposed method performs clearer details and less distortion.

Generally, among conventional algorithms, OMP outperforms TwIST and GPSR. And our proposed model further increases PSNR by more than 8 db, improves SSIM by over 0.12. Our proposed method also reaches slightly better performance than the DNN-based model DNNnet.

The average reconstruction times are also illustrated in Table 1. The proposed method achieves over 7 times faster reconstruction than the conventional methods.

In summary, the proposed reconstruction algorithm not only outperforms OMP, TwIST and GPSR in both subjective and objective perspectives, but also reaches faster calculation speed, which proves the effectiveness and rationality of our solution with high accuracy and low delay.

Table 1. Average reconstruction time of different algorithms

Algorithm	OMP	TwIST	GPSR	DNNnet	Proposed
Average reconstruction time (ms)	38.91	26.39	15.18	2.11	2.14

Fig. 6. Comparison on 6 example hyperspectral images reconstructed by different algorithms

6 Conclusion

This paper proposed a reconstruction algorithm for hyperspectral images collected from push-broom SD-CASSI systems based on DNN. Residual attention blocks and channel attention mechanisms are utilized in model design to explore spatial and inter-spectral features. Comparing to conventional algorithms such as OMP, TwIST and GPSR, the proposed model achieves much higher performance as well as high computation efficiency in our experiment. It also outperforms the DNN-based model DNNnet in hyperspectral image reconstruction, indicating high effectiveness and efficiency. Compared to existing algorithms for DD-CASSI systems, it is of greater practical importance due to the broad use of SD-CASSI systems in aerospace fields.

References

1. Hyperspectral Image – an overview I ScienceDirect Topics. https://www.sciencedirect.com/topics/computer-science/hyperspectral-image. Accessed 7 June 2022
2. Donoho, D.L.: Compressed sensing. IEEE Trans. Inf. Theory **52**(4), 1289–1306 (2006)
3. Candès, E.J., Romberg, J., Tao, T.: Robust uncertainty principles: exact signal reconstruction from highly incomplete frequency information. IEEE Trans. Inf. Theory **52**(2), 489–509 (2006)
4. Nyquist rate – Wikipedia. https://en.wikipedia.org/wiki/Nyquist_rate. Accessed 7 June 2022
5. Liu, Y., et al.: A novel method of coded-aperture push-broom Compton scatter imaging: principles, simulations and experiments. Nucl. Instrum. Methods Phys. Res., Sect. A **1**(940), 30–39 (2019)
6. Wei, K., Fu, Y., Huang, H.: 3-D quasi-recurrent neural network for hyperspectral image denoising. IEEE Trans. Neural Netw. Learn. Syst. **32**(1), 363–375 (2020)
7. Qing, Y., Liu, W.: Hyperspectral image classification based on multi-scale residual network with attention mechanism. Remote Sens. **13**(3), 335 (2021)
8. Zhao, M., Yue, L., Hu, J., Du, S., Li, P., Wang, L.: Salient target detection in hyperspectral image based on visual attention. IET Image Proc. **15**(10), 2301–2308 (2021)
9. Davenport, M.A., Wakin, M.B.: Analysis of orthogonal matching pursuit using the restricted isometry property. IEEE Trans. Inf. Theory **56**(9), 4395–4401 (2010)
10. Bioucas-Dias, J.M., Figueiredo, M.A.: A new TwIST: two-step iterative shrinkage/thresholding algorithms for image restoration. IEEE Trans. Image Process. **16**(12), 2992–3004 (2007)
11. Chen, G., Li, D., Zhang, J.: Iterative gradient projection algorithm for two-dimensional compressive sensing sparse image reconstruction. Signal Process. **1**(104), 15–26 (2014)
12. Wang, L., Zhang, T., Fu, Y., Huang, H.: Hyperreconnet: Joint coded aperture optimization and image reconstruction for compressive hyperspectral imaging. IEEE Trans. Image Process. **28**(5), 2257–2270 (2018)
13. Yilin, J., Ran, S., Sanqiang, T.: Generative adversarial networks for hyperspectral image spatial super-resolution. J. China Univ. Posts Telecommun. **27**(4), 8 (2020)
14. Choi, I., Kim, M.H., Gutierrez, D., Jeon, D.S., Nam, G.: High-quality hyperspectral reconstruction using a spectral prior (2017)
15. Miao, X., Yuan, X., Pu, Y., Athitsos, V.: lambda-net: reconstruct hyperspectral images from a snapshot measurement. In 2019 IEEE InCVF International Conference on Computer Vision (ICCV), pp. 4058–4068 (2019)

16. Arad, B., Ben-Shahar, O.: Sparse recovery of hyperspectral signal from natural RGB images. In: Leibe, B., Matas, J., Sebe, N., Welling, M. (eds.) Computer Vision – ECCV 2016. LNCS, vol. 9911, pp. 19–34. Springer, Cham (2016). https://doi.org/10.1007/978-3-319-46478-7_2
17. Paszke, A., et al.: Pytorch: an imperative style, high-performance deep learning library. Adv. Neural Inf. Process. Syst. **32**, 1–14 (2019)
18. Multiframe Coded Aperture Spectral Imaging (CASSI) | Duke Information Spaces Project. https://disp.duke.edu/research/multiframe-coded-aperture-spectral-imaging-cassi. Accessed 7 June 2022

Author Index

Printed in the United States
by Baker & Taylor Publisher Services

Printed in the United States
by Baker & Taylor Publisher Services